Third Edition

INTRODUCTION TO GEOGRAPHIC INFORMATION SYSTEMS

Kang-tsung Chang

University of Idaho

Mc Graw Hill **Higher Education**

Boston   Burr Ridge, IL   Dubuque, IA   Madison, WI   New York   San Francisco   St. Louis
Bangkok   Bogotá   Caracas   Kuala Lumpur   Lisbon   London   Madrid   Mexico City
Milan   Montreal   New Delhi   Santiago   Seoul   Singapore   Sydney   Taipei   Toronto

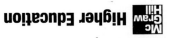
INTRODUCTION TO GEOGRAPHIC INFORMATION SYSTEMS, THIRD EDITION

Published by McGraw-Hill, a business unit of The McGraw-Hill Companies, Inc., 1221 Avenue of the Americas, New York, NY 10020. Copyright © 2006, 2004, 2002 by The McGraw-Hill Companies, Inc. All rights reserved. No part of this publication may be reproduced or distributed in any form or by any means, or stored in a database or retrieval system, without the prior written consent of The McGraw-Hill Companies, Inc., including, but not limited to, in any network or other electronic storage or transmission, or broadcast for distance learning.

Some ancillaries, including electronic and print components, may not be available to customers outside the United States.

This book is printed on acid-free paper.

1 2 3 4 5 6 7 8 9 0 QPD/QPD 0 9 8 7 6 5

ISBN 0–07–282682–7

Publisher: Margaret J. Kemp
Senior Sponsoring Editor: Daryl Bruflodt
Senior Developmental Editor: Lisa A. Bruflodt
Associate Marketing Manager: Todd L. Turner
Project Manager: April R. Southwood
Lead Production Supervisor: Sandy Ludovissy
Lead Media Project Manager: Judi David
Senior Media Technology Producer: Jeffry Schmitt
Senior Designer: David W. Hash
Cover Designer: Crystal Kadlec
Compositor: The GTS Companies
Typeface: 10/12 Times Roman
Printer: Quebecor World Dubuque, IA

Library of Congress Cataloging-in-Publication Data

Chang, Kang-tsung.
  Introduction to geographic information systems / Kang-tsung Chang. — 3rd ed.
      p. cm.
  Includes index.
  ISBN 0–07–282682–7 (hard copy : alk. paper)
  1. Geographic information systems. I. Title.

G70.212.C4735   2006
910'.285—dc22
                                        2005001288
                            CIP

www.mhhe.com

BRIEF CONTENTS

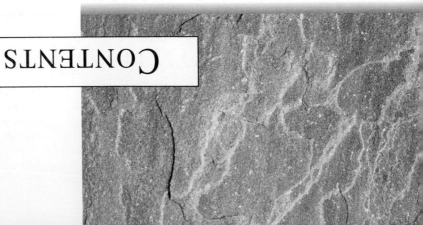

CONTENTS

CHAPTER 3

GEORELATIONAL VECTOR DATA MODEL 42

CHAPTER 4

OBJECT-BASED VECTOR DATA MODEL 61

CHAPTER 5

RASTER DATA MODEL 75

CHAPTER 6

DATA INPUT 96

CHAPTER 7

GEOMETRIC TRANSFORMATION 117

SPATIAL DATA EDITING   *133*

ATTRIBUTE DATA INPUT AND MANAGEMENT   *156*

CHAPTER 10

DATA DISPLAY AND CARTOGRAPHY   *175*

CHAPTER 11

DATA EXPLORATION   *204*

CHAPTER 12

VECTOR DATA ANALYSIS   229

CHAPTER 13

RASTER DATA ANALYSIS   252

CHAPTER 14

TERRAIN MAPPING AND ANALYSIS   271

CHAPTER 15

VIEWSHEDS AND WATERSHEDS 295

CHAPTER 16

SPATIAL INTERPOLATION 318

CHAPTER 17

GEOCODING AND DYNAMIC SEGMENTATION   *348*

CHAPTER 18

PATH ANALYSIS AND NETWORK APPLICATIONS   *367*

CHAPTER 19

GIS MODELS AND MODELING   *390*

PREFACE

THE IMPORTANCE OF GIS

A geographic information system (GIS) is a computer system for capturing, storing, querying, analyzing, and displaying geospatial data. Geospatial data describe both the locations and characteristics of spatial features such as roads, land parcels, and forests. The ability of GIS to handle and process both location and attribute data distinguishes GIS from other information systems and establishes GIS as a technology necessary for a wide variety of applications.

Since the 1970s GIS has been important in natural resource management including land-use planning, natural hazard assessment, wildlife habitat analysis, and timber management. More recently, GIS has been used for crime mapping and analysis, emergency planning, land records management, market analysis, and transportation planning. And integration of GIS with other geospatial technologies has found applications in location-based services, online interactive mapping, in-vehicle navigation systems, and precision farming. As of August 2004, geospatial technology is listed by the U.S. Department of Labor as one of the three most important emerging fields; the other two are nanotechnology and biotechnology.

GIS is not for specialists. Powerful and affordable computer hardware and software, the graphical user interface, and public digital data have brought GIS to mainstream use. It is not unusual to find students from more than 20 academic departments in an introductory GIS class.

The design of a GIS has involved fundamental concepts from fields such as geography, cartography, spatial analysis, surveying, mathematics, and database management. To capture GIS data, for example, we must consider the spatial referencing system, a concept long established in cartography and surveying. A spatial referencing system, either geographic or projected, determines how the locations of spatial features are measured. The design of a GIS has also kept pace with advances in computer technology. GIS vendors have incorporated object-oriented technology into the user interface design since the 1990s. The same technology has now been adopted for the GIS data model. To be proficient in GIS, we must be familiar with the basic concepts as well as new technologies.

ENHANCEMENTS TO THE THIRD EDITION

The third edition has 19 chapters. Chapters 1 to 5 explain GIS concepts and data models. Chapters 6 to 9 cover data input, editing, and management. Chapters 10 and 11 include data display and exploration. Chapters 12 and 13 provide an overview of core data analysis. Chapters 14 to 16 focus on surface mapping and analysis. Chapters 17 and 18 cover analysis of movement and linear features. Chapter 19 presents GIS models and modeling. Depending on course design and students needs, this third edition can be used as a text for a first GIS course, or a second GIS course, or both. The overall aim of the book is to provide students with a solid foundation in both concepts and practice.

For this edition I have added or expanded discussions on the following topics:

- Object-based vector data model (e.g., geodatabase) (Chapter 4)
- Wavelet transform (Chapter 5)
- Geometric transformation (Chapter 7)
- Topological rules and editing (Chapter 8)
- Spatial statistics (Chapter 12)
- Viewshed analysis and watershed analysis (Chapter 15)
- Geocoding and dynamic segmentation (Chapter 17)
- Path analysis (Chapter 18)
- Use of command line, ModelBuilder, and Python script in ArcToolbox (Chapters 15 and 19)

This edition also contains several new features. First, each chapter includes a set of review questions. Second, the applications section of each chapter now has a challenge task. These challenge tasks are designed to further develop problem-solving skills. Third, the applications section of each chapter includes a number of task-related questions to reinforce the learning process. We do not learn well if we just follow the instructions to complete a task without pausing and thinking about what we have done.

This edition retains those features that have proven to be useful in the first two editions. References drawn from many disciplines are up-to-date and resourceful. Boxes provide software-specific materials, worked examples, and additional relevant information. Figures are designed to complement the text. And new figures are added wherever appropriate.

COMBINING CONCEPTS WITH APPLICATIONS

This third edition continues to emphasize the practice of GIS. Each chapter again has problem-solving tasks in the applications section, complete with data sets and instructions. The number of tasks totals 77, with 3 to 6 tasks in each chapter. The instructions for performing the tasks correlate to ArcGIS 9.0.

All but two tasks in this edition use the 9.0 version of ArcGIS Desktop and its extensions of Spatial Analyst, 3D Analyst, Geostatistical Analyst, and ArcScan. The only two tasks that require use of ArcInfo Workstation deal with shortest path analysis and location–allocation in Chapter 18. They can be easily run in ArcGIS Desktop as soon as the Network Analyst extension is made available by ESRI, Inc.

The hands-on experience to be gained from the applications sections is intended to complement the discussions in the text and to enhance the working knowledge of GIS functions. We can learn GIS concepts from readings, but we can most fully grasp concepts behind the menus and icons through practice. Moreover, a regular requirement in the current GIS job market is to be skilled in using commercial GIS packages.

The website for the third edition, located at www.mhhe.com/changgis3e, contains a password-protected instructor's manual. Contact your McGraw-Hill sales representative for access information.

CREDITS

Data sets downloaded from the following websites are used for some tasks in this book:

Clearwater National Forest
http://www.fs.fed.us/r1/clearwater/gis/library/library_w.htm
Montana GIS data clearinghouse
http://www.nris.state.mt.us/
Northern California Earthquake Data Center
http://quake.geo.berkeley.edu/
University of Idaho Library
http://inside.uidaho.edu
Washington State Department of Transportation
GIS Data
http://www.wsdot.wa.gov/mapsdata/geodatacatalog/default.htm
Wyoming Geographic Information Advisory Council
http://wgiac2.state.wy.us/html/

ACKNOWLEDGMENTS

I would like to thank the following reviewers who provided many helpful comments on the third edition:

William Bajjali
University of Wisconsin–Superior

Gary W. Coutu
Temple College

Laurie A. B. Garo
University of North Carolina at Charlotte

Greg Gaston
University of North Alabama

Michael Harrison
University of Richmond

Miriam Helen Hill
Jacksonville State University

Michael E. Hodgson
University of South Carolina

Paul Franklin Hudson
University of Texas at Austin

Wei (Wayne) Jei
University of Missouri–Kansas City

Pete Kennedy
Greenville Technical College

James Knotwell
Wayne State College

Paul R. Larson
Southern Utah University

L. J. Layne
Alburquerque Technical Vocational Institute Community College

Peter Li
Tennessee Tech University

Ge Lin
West Virginia University

Max Lu
Kansas State University

Timothy Nyerges
University of Washington

Patrick J. Pellicane
Colorado State University

Jennifer Rahn
Baylor University

Binita Sinha
Diablo Valley College

Peter Siska
Stephen F. Austin University

Tim Strauss
University of Northern Iowa

I-Shian Suen
Iowa State University

Daniel Z. Sui
Texas A&M University

Selima Sultana
Auburn University

Ming-Hsiang Tsou
San Diego State University

Eugene Turner
California State University, Northridge

Fahui Wang
Northern Illinois University

Yu Zhou
Bowling Green State University

I have read their reviews carefully and incorporated many comments into the revision. Of course, I take full responsibility for the book.

I wish to thank Lisa Bruflodt, Daryl Bruflodt, Marge Kemp, David Hash, and April Southwood at McGraw-Hill for their guidance and assistance during various stages of this project.

Finally, this book is dedicated to the memory of Lillian R. Chang.

Kang-tsung Chang

INTRODUCTION

What technology can be useful for managing a forest, routing 911 vehicles, designing a cellular phone network, managing a city, designing a road, and providing famine relief? According to the U.S. Department of Labor website on the emerging fields, it is geospatial technology **(http://www. careervoyages.gov/).** As of August 2004, geospatial technology is listed by the U.S. Department of Labor as one of the three most important emerging fields, along with nanotechnology and biotechnology. Geospatial technology covers a number of fields

including remote sensing, cartography, surveying, and photogrammetry. But to integrate data from these different fields in geospatial technology, we rely on geographic information systems.

1.1 WHAT IS A GIS?

A **geographic information system (GIS)** is a computer system for capturing, storing, querying, analyzing, and displaying geographically referenced data. Also called geospatial data, **geographically referenced data** are data that describe both the locations and characteristics of spatial features such as roads, land parcels, and vegetation stands on the Earth's surface. The ability of a GIS to handle and process geographically referenced data distinguishes GIS from other information systems. It also establishes GIS as a technology important to a wide variety of applications, as shown at the U.S. Department of Labor's website.

1.1.1 GIS Applications

Since the beginning, GIS has been important in natural resource management including land-use planning, natural hazard assessment, wildlife habitat analysis, riparian zone monitoring, and timber management. Here are some examples on the Internet:

- The U.S. Geological Survey has the National Map program that provides nationwide geospatial data for applications in natural hazards, risk assessment, homeland security, and many other areas (http://nationalmap.usgs.gov/).

- The U.S. Census Bureau maintains an On-Line Mapping Resources website, where Internet users can map public geographic data of anywhere in the United States (http://www.census.gov/geo/www/maps/CP_OnLineMapping.htm).

- The U.S. Forest Service uses GIS and other computer technologies to map forest fires and to model fire behavior (http://www.fs.fed.us/fire/tools_tech/).

- The U.S. Department of Housing and Urban Development has a mapping program that combines housing development information with environmental data (http://www.hud.gov/offices/cio/emaps/index.cfm).

In more recent years GIS has been used for crime analysis, emergency planning, land records management, market analysis, and transportation applications. Here are some examples on the Internet:

- The National Institute of Justice uses GIS to map crime records and to analyze their spatial patterns by location and time (http://www.ojp.usdoj.gov/nij/maps/).

- The Federal Emergency Management Agency links a flood insurance rate map database to physical features in a GIS database (http://www.fema.gov/fhm/mm_main.shtm).

- Larimer County, Colorado allows public access to the county's land records in a GIS database (http://www.larimer.org/).

Integration of GIS with global positioning system (GPS), wireless technology, and the Internet has also introduced new and exciting applications. Here are some examples:

- Location-based services (LBS) technology allows mobile phone users to be located and to receive location information, such as nearby ATMs and restaurants.

- Interactive-mapping websites let users select map layers for display and make their own maps.

- In-car navigation systems find the shortest route between an origin and destination and provide turn-by-turn directions to drivers.

- Precision farming promotes site-specific farming activities such as herbicide or fertilizer application.

Box 1.1 includes additional information on the above GIS applications. As GIS becomes better known through the public media, there will be new applications in the future, both conventional and imaginative.

1.1.2 Components of a GIS

Like any other information technology, GIS requires the following four components to work with geographically referenced data:

- Computer System. The computer system includes the computer and the operating system to run GIS. Typically the choices are PCs that use the Windows operating system (e.g., Windows 2000, Windows XP) or workstations that use the UNIX or Linux operating system. Additional equipment may include monitors for display, digitizers and scanners for spatial data input, GPS receivers and mobile devices for fieldwork, and printers and plotters for hard-copy data display.

- GIS Software. The GIS software includes the program and the user interface for driving the hardware. Common user interfaces in GIS are menus, graphical icons, command lines, and scripts.

Box 1.1 | **More Interesting Examples of GIS Applications**

- The National Association of Realtors has an interactive mapping service that can assist potential homebuyers in finding listings that meet their criteria and viewing these listings on a street map (**http://www.realtor.com/**).
- Keyhole has an "internet 3-D earth visualization" program that allows users to access high-resolution satellite images from the company's server for many parts of the Earth's surface (**http://www.keyhole.com/**).

- Dodgeball provides services to mobile phone users that will locate their friends and friends-of-friends within 10 blocks (**http://www. dodgeball.com/**).
- Eyebeam, a nonprofit arts-based technology firm, has a website called Fundrace that shows the geocoded locations of campaign contributions in major cities of the United States (**http://www.fundrace.org**).

- Brainware. Equally important as the computer hardware and software, the brainware refers to the purpose and objectives, and provides the reason and justification for using GIS.
- Infrastructure. The infrastructure refers to the necessary physical, organizational, administrative, and cultural environments that support GIS operations. The infrastructure includes requisite skills, data standards, data clearinghouses, and general organizational patterns.

1.2 A BRIEF HISTORY OF GIS

GIS is not new. Since the late 1960s computers have been used to store and process geographically referenced data. Early examples of GIS-related work from the late 1960s and 1970s include the following:

- Computer mapping at the University of Edinburgh, the Harvard Laboratory for Computer Graphics, and the Experimental Cartography Unit (Coppock 1988; Chrisman 1988; Rhind 1988).
- Canada Land Inventory and the subsequent development of the Canada Geographic Information System (Tomlinson 1984).
- Publication of Ian McHarg's *Design with Nature* and its inclusion of the map overlay method for suitability analysis (McHarg 1969).

- Introduction of an urban street network with topology in the U.S. Census Bureau's DIME (Dual Independent Map Encoding) system (Broome and Meixler 1990).

For many years, though, GIS has been considered to be too difficult, expensive, and proprietary. The advent of the graphical user interface (GUI), powerful and affordable hardware and software, and public digital data has broadened the range of GIS applications and brought GIS to mainstream use in the 1990s. Box 1.2 shows a list of GIS software producers and their main products.

Various trade reports suggest that ESRI, Inc. and Intergraph Corp. lead the GIS industry in terms of the software market and software revenues. The main software product from ESRI, Inc. is ArcGIS, a scalable system with ArcView, ArcEditor, and ArcInfo. All three versions of the system operate on the Windows platforms and share the same applications and extensions, but they differ in their capabilities. ArcView has data integration, query, display, and some analysis capabilities. ArcEditor has additional functionalities for data editing. And ArcInfo has more data conversion and analysis capabilities than ArcView and ArcEditor. Intergraph Corp. has two main products: GeoMedia and MGE. GeoMedia is a desktop GIS package for data integration and visualization and is compatible with standard Windows development tools. MGE, which operates on the Windows operating systems or UNIX, consists of a series of products for data production and analysis. Data can be migrated from GeoMedia to MGE, and vice versa.

GRASS is unique among GIS software packages because it is free. Originally developed by the U.S. Army Construction Engineering Research Laboratories, GRASS is currently maintained in both the United States (Baylor University) and Germany (University of Hannover). Some GIS packages are targeted at certain user groups. TransCAD, for example, is a package designed for use by transportation professionals. Microsoft, Oracle, and IBM have also entered the GIS industry. Microsoft MapPoint targets business analysts who need to analyze and map geographic data (http://www.microsoft.com/). Oracle Spatial can store, access, and manage spatial data in Oracle's relational database management system (http://www.oracle.com/). IBM offers Spatial DataBlade, an extension that allows location-based data to be stored in IBM's relational database (http://www-306.ibm.com/software/data/informix/blades/spatial/).

1.3 GEOGRAPHICALLY REFERENCED DATA

Geographically referenced data separate GIS from other information systems. Therefore, before discussing GIS operations, we must understand the nature of geographically referenced data. Take the example of roads. To describe a road, we refer to its location (i.e., where it is) and its characteristics (e.g., length, name, speed limit, and direction), as shown in Figure 1.1. The location, also called geometry or shape, represents **spatial data**, whereas the characteristics are **attribute data**. Thus a road, like any geographically referenced data, has the two components of spatial data and attribute data.

Along with the proliferation of GIS activities, numerous GIS textbooks have been published, and several journals and trade magazines are now devoted to GIS and GIS applications.

Box 1.2 **A List of GIS Software Producers and Their Main Products**

The following is a list of GIS software producers and their main products as of August 2004:

- Environmental Systems Research Institute (ESRI), Inc. (http://www.esri.com/): **ArcGIS, ArcView 3.x**
- Autodesk Inc. (http://www3.autodesk.com/): **Autodesk Map**
- Land Management Information Center at Minnesota Planning (http://www.lmic.state.mn.us/EPPL7/): **EPPL7**
- Baylor University, Texas (http://grass.baylor.edu/): **GRASS**
- Clark Labs (http://www.clarklabs.org/): **IDRISI**
- International Institute for Aerospace Survey and Earth Sciences, the Netherlands (http://www.itc.nl/ilwis/): **ILWIS**

- Manifold.net (http://www.manifold.net/): **Manifold System**
- MapInfo Corporation (http://www.mapinfo.com/): **MapInfo**
- Keigan Systems (http://www.keigansystems.com/): **MFworks, Keigan Grid**
- Intergraph Corporation (http://www.intergraph.com/): **MGE, GeoMedia**
- Bentley Systems, Inc. (http://www2.bentley.com/): **Microstation**
- PCI Geomatics (http://www.pcigeomatics.com/): **Geomatica**
- Caliper Corporation (http://www.caliper.com/): **TransCAD, Maptitude**

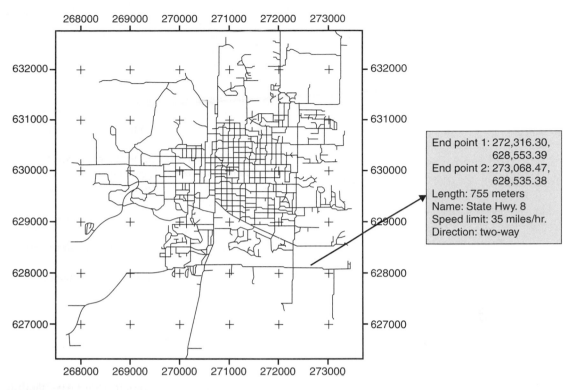

Figure 1.1
An example of geographically referenced data. The street network is based on a plane coordinate system. The box on the right lists the *x*- and *y*-coordinates of the end points and other attributes of a street segment.

1.3.1 Spatial Data

Spatial data describe the locations of spatial features, which may be discrete or continuous. **Discrete features** are individually distinguishable features that do not exist between observations. Discrete features include points (e.g., wells), lines (e.g., roads), and areas (e.g., land-use types). **Continuous features** are features that exist spatially between observations. Examples of continuous feature are elevation and precipitation. A GIS represents these spatial features on the Earth's surface as map features on a plane surface. This transformation involves two main issues: the spatial reference system and the data model.

The locations of spatial features on the Earth's surface are based on a geographic coordinate system with longitude and latitude values, whereas the locations of map features are based on a plane coordinate system with *x*-, *y*-coordinates. **Projection** is the process that can transform the Earth's spherical surface to a plane surface and bridge the two spatial reference systems. But because the transformation always involves some distortion, hundreds of plane coordinate systems that have been developed to preserve certain spatial properties are in use. To align with one another spatially for GIS operations, map layers must be based on the same coordinate system. A basic understanding of projection and coordinate systems is therefore crucial to users of spatial data.

The data model defines how spatial features are represented in a GIS (Figure 1.2). The **vector** data model uses points and their x-, y-coordinates to construct spatial features of points, lines, and areas. The **raster data model** uses a grid and grid cells to represent the spatial variation of a feature. The two data models differ in concept: vector data are ideal for representing discrete features; raster data are better suited for representing continuous features. They also differ in data structure. The raster data model uses a simple data structure with rows and columns and fixed cell locations. The vector data model may be geo-relational or object-based, may or may not involve topology, and may include simple or composite features.

The **georelational data model** uses a split system to store spatial and attribute data. The **object-based data model,** on the other hand, stores spatial data and attribute data in a single system. Recent trends suggest that GIS vendors have adopted the object-based data model in their software development. For example, the **geodatabase data model,** introduced by ESRI, Inc. in ArcGIS, is object-based. The shift of the data model mainly reflects advances in computer technology. But to GIS users, it means new concepts, new data formats, and new interfaces. The object-based data model uses objects to organize spatial data and methods to define the characteristics and behaviors of spatial objects. The geodatabase data model is built on ArcObjects, a collection of thousands of objects, properties, and

methods. Therefore, when we work in ArcGIS, we interact with these objects through the user interface (Figure 1.3).

Topology expresses explicitly the spatial relationships between features, such as two lines meeting perfectly at a point and a directed line having an explicit left and right side. Topological or topology-based data are useful for detecting and correcting digitizing errors in geographic data sets and are necessary for some GIS analyses. But nontopological data can display faster. To distinguish between them, the Ordnance Survey in Great Britain offers topological and nontopological data separately to accommodate end users with different needs. Likewise, users of ESRI software recognize **coverages** as topological data, **shapefiles** as nontopological data, and geodatabases as with or without topology (Figure 1.4).

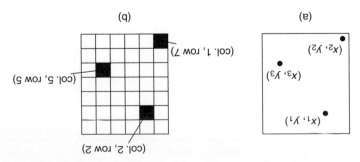

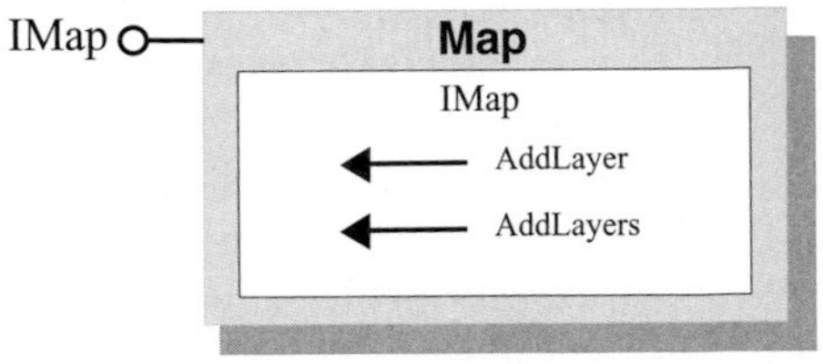

Figure 1.3
When we use the add data button to add a layer to a map, we are actually using the *AddLayer* method on the *IMap* interface that the *Map* class supports.

Figure 1.2
The vector data model uses x-, y-coordinates to represent point features (a), and the raster data model uses cells in a grid to represent point features (b).

	Geo-relational	Object-based
Topological	Coverage	Geodatabase
Non-topological	Shapefile	Geodatabase

Figure 1.4
A classification of data formats used in ESRI software by topology and data model.

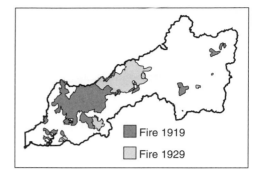

Fire 1919
Fire 1929

Figure 1.6
An example of the regions model. Two regions, one for burned areas in 1919 and the other in 1929, overlap each other in some areas. Each region is also comprised of spatially disjoint parts.

Composite features are built on simple features of points, lines, and polygons. The **triangulated irregular network (TIN),** which approximates the terrain with a set of nonoverlapping triangles, is made of nodes (points) and edges (lines) (Figure 1.5). The **regions** data model, which allows regions to overlap and to have spatially disjoint components, is built on polygons (Figure 1.6). The **dynamic segmentation model,** which has a linear measure system, is built on top of linear features. Composite features are useful in GIS because they can handle more complex spatial relationships. For example, dynamic segmentation allows highway rest areas, which are typically recorded in linear measures, to be plotted on a highway map, which is based on a coordinate system (Figure 1.7).

1.3.2 Attribute Data

Attribute data describe the characteristics of spatial features. For raster data, each cell has a value that corresponds to the attribute of the spatial feature at that location. A cell is tightly bound to

Figure 1.5
An example of the TIN model.

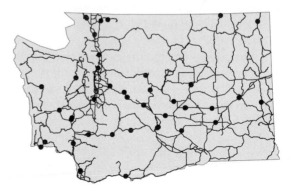

Figure 1.7
An example of the dynamic segmentation model. Rest areas are plotted as point features on highway routes in Washington State.

its cell value. For vector data, the amount of attribute data to be associated with a spatial feature can vary significantly. A road segment may only have the attributes of length and speed limit, whereas a soil polygon may have dozens of physical and chemical properties, interpretations, and performance data. How to join spatial and attribute data is therefore important in the case of vector data.

1.3.3 Joining Spatial and Attribute Data

The georelational data model stores attribute data separately from spatial data in a split system. The two data components are linked through the feature IDs. The object-based data model stores spatial data as an attribute along with other attributes in a single system. Thus the object-based data model eliminates the complexity of coordinating and synchronizing two sets of data files as required in a single system. It also brings GIS closer to other nonspatial information systems because spatial data files are no longer needed.

Whether spatial and attribute data are stored in a split or single system, the relational database model is the norm for data management in GIS. A **relational database** is a collection of tables (relations). The connection between tables is made through a key, a common field whose values can uniquely identify a record in a table. For example, the feature ID serves as the key in the georelational data model to link spatial data and attribute data.

A relational database is efficient and flexible for data search, data retrieval, data editing, and creation of tabular reports. Each table in the database can be prepared, maintained, and edited separately from other tables. And the tables can remain separate until a query or an analysis requires that attribute data from different tables be linked or joined together.

With GIS increasingly becoming part of an organization's much larger information system, attribute data needed for a GIS project are likely to come from an enterprisewide database. If spatial data distinguish GIS as a special type of informa-

Spatial data input	1. Data entry: use existing data, create new data 2. Data editing 3. Geometric transformation 4. Projection and reprojection
Attribute data management	1. Data entry and verification 2. Database management 3. Attribute data manipulation
Data display	1. Cartographic symbolization 2. Map design
Data exploration	1. Attribute data query 2. Spatial data query 3. Geographic visualization
Data analysis	1. Vector data analysis: buffering, overlay, distance measurement, spatial statistics, map manipulation 2. Raster data analysis: local, neighborhood, zonal, global, raster data manipulation 3. Terrain mapping and analysis 4. Viewshed and watershed 5. Spatial interpolation 6. Geocoding and dynamic segmentation 7. Path analysis and network applications
GIS modeling	1. Binary models 2. Index models 3. Regression models 4. Process models

Figure 1.8
A classification of GIS operations.

tion system, attribute data tend to connect GIS with other information management systems. This unique combination enhances the role of GIS in an organization.

1.4 GIS OPERATIONS

Although GIS activities no longer follow a set sequence, to explain what we do in GIS, we can group GIS activities into spatial data input, attribute data management, data display, data exploration, data analysis, and GIS modeling (Figure 1.8). This section provides an overview of GIS operations. It also serves as a preview of the chapters to follow by highlighting each group of GIS activities.

1.4.1 Spatial Data Input

The most expensive part of a GIS project is data acquisition. Two basic options for data acquisition are (1) use existing data and (2) create new data. Digital data clearinghouses have become commonplace on the Internet in recent years. The strategy for a GIS user is to look at what exists in the public domain before deciding to either buy data from private companies or create new data.

New digital spatial data can be created from satellite images, GPS data, field surveys, street addresses, and text files with x-, y-coordinates. But paper maps remain the dominant data source. Manual digitizing or scanning can convert paper maps into digital format. A newly digitized map typically requires editing and geometric transformation. Editing removes digitizing errors, which may relate to the location of spatial data such as missing polygons and distorted lines, or the topology such as dangling arcs and unclosed polygons. Geometric transformation converts a newly digitized map, which has the same physical dimension as its source map, into a real-world coordinate system. Geometric transformation can also convert satellite images, which are recorded in rows and columns, into projected coordinates. Because a geometric transformation operates on a set of control points, we often have to adjust the control points to reduce the amount of transformation error to an acceptable level.

1.4.2 Attribute Data Management

To complete a GIS database, we must enter and verify attribute data through digitizing and editing. Attribute data reside as tables in a relational database. An attribute table is organized by row and column. Each row represents a spatial feature, and each column or field describes a characteristic. Attribute tables in a database must be designed to facilitate data input, search, retrieval, manipulation, and output. Two basic elements in the design of a relational database are the key and the type of data relationship: the key establishes a connection between corresponding records in two tables, and the type of data relationship dictates how the tables are actually joined or linked. In practice, attribute data management also includes such tasks as adding or deleting fields and creating new fields from existing fields.

1.4.3 Data Display

Because maps are most effective in communicating spatial information, mapmaking is a routine GIS operation. We derive maps from data query and analysis. And we prepare maps for data visualization and presentation. A map for presentation usually has a number of elements: title, subtitle, body, legend, north arrow, scale bar, acknowledgment, neatline, and border. These elements work together to bring spatial information to the map reader.

The first step in mapmaking is to assemble map elements. Windows-based GIS packages have simplified this process by providing choices for each map element through menus and palettes. But we must be aware of "default options." Without a basic understanding of map symbols, colors, and typology, we can easily produce a bad map according to cartographic standards. The second step is map design. Map design is a creative process that cannot be easily replaced by default templates and computer code. The mapmaker must experiment with the layout and visual hierarchy. A poorly designed map can confuse the map reader and even distort the information intended by the mapmaker.

1.4.4 Data Exploration

Usually a precursor to data analysis, **data exploration** involves the activities of exploring the general trends in the data, taking a close look at data subsets, and focusing on possible relationships between data sets. Effective data exploration requires interactive and dynamically linked visual tools. A Windows-based GIS package is ideal for data exploration. We can display maps, graphs, and tables in multiple but dynamically linked windows so that, when we select a data subset from a table, it automatically highlights the corresponding features in a graph and a map. This kind of interactivity increases our capacity for information processing and synthesis.

Because geographically referenced data consist of spatial data and attribute data, data exploration can be approached from spatial data, or attribute data, or both. The importance of maps in GIS operations has added geographic visualization to the list of data exploration activities. **Geographic visualization** sets up a context for visual information processing by relying on maps and map-based tools such as data classification, data aggregation, and map comparison.

1.4.5 Data Analysis

Figure 1.8 classifies data analysis into seven groups. The two first groups include basic analytical tools. For vector data, these tools include buffering, overlay, distance measurement, spatial statistics, and map manipulation. Buffering creates buffer zones by measuring straight-line distances from select features. Overlay, recognized by many as the most important GIS tool, combines geometries and attributes from different layers to create the output (Figure 1.9). Distance measurement calculates distances between spatial features. Spatial statistics detect spatial dependence and patterns of concentration among features. And map manipulation tools manage and alter layers in a database.

Common tools for analyzing raster data are traditionally grouped into local, neighborhood, zonal, and global operations. A local operation operates on individual cells; a neighborhood operation, a specified neighborhood such as a 3-by-3 window; a zonal operation, a group of cells with same values or like features; and a global operation, the entire raster. Often a raster data operation relates the input to the output with a mathematical function. For example, a local operation can compute the average of the input rasters on a cell-by-cell basis (Figure 1.10). Other tools that are not in the above groups can perform such tasks as data extraction and data generalization.

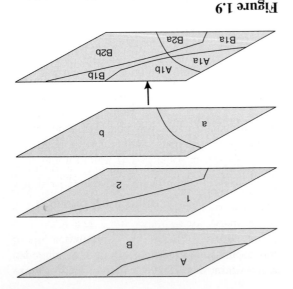

Figure 1.9

A vector-based overlay operation combines spatial data and attribute data from different layers to create the output.

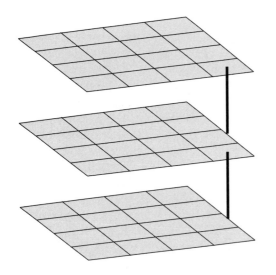

Figure 1.10

A raster data operation with multiple rasters can take advantage of the fixed cell locations.

The terrain has been the object for mapping and analysis for hundreds of years. Mapping techniques such as contouring, profiling, hill shading, hypsometric tinting, and 3-D views are useful for visualizing the land surface. Topographic measures including slope, aspect, and surface curvature are important for studies of timber management, soil erosion, hydrologic modeling, wildlife habitat suitability, and many other fields. Analyses of viewshed and watershed are extensions of terrain analysis. A viewshed analysis determines areas of the land surface that are visible from one or more observation points. A watershed analysis can derive topographical features such as flow direction, stream networks, and watershed boundaries for hydrologic applications.

Precipitation, snow accumulation, water table, and many other spatial phenomena are similar to the terrain in terms of visualization and data analysis. But unlike the terrain, data for these phenomena are only available at a small number of sample points. To construct surfaces from sample points, we need **spatial interpolation,** the process of using points with known values to estimate values at other points. Spatial interpolation covers a variety of methods including trend surface models, Thiessen polygons, kernel density estimation, inverse distance weighted, splines, and kriging. Kriging is also known as a geostatistical method because it can not only predict unknown values but also estimate prediction errors.

Geocoding converts street addresses or street intersections into point features. Dynamic segmentation plots linearly referenced data on a coordinate system. The two techniques are similar in that both can locate data from a data source that lacks x-, y-coordinates and both use linear features (e.g., streets, highways) as reference data. Data generated from these two techniques are necessary inputs for certain data analyses. Geocoded data are crucial for crime mapping and analysis, and dynamic segmentation layers are useful for managing and analyzing highway-related data.

Path analysis finds the least cost path between cells by using a cost raster that defines the cost to move through each cell. A network is a system of topologically connected linear features that also has the appropriate attributes for the flow of objects. Among various network applications is shortest path analysis, which finds the shortest path between stops on a network. Path analysis and shortest path analysis share the same objective and problem-solving algorithm but differ in applications. Path analysis is raster-based and works with "virtual" paths whereas shortest path analysis is vector-based and works with an existing network.

1.4.6 GIS Models and Modeling

A model is a simplified representation of a phenomenon or a system, and **GIS modeling** refers to the use of a GIS and its functionalities in building a model with geographically referenced data (i.e., a spatially explicit model). GIS models can be grouped into four general types: binary, index, regression, and process models.

Binary and index models are similar in that both rely on overlay, either vector- or raster-based, to combine geographically referenced data from different layers for multicriteria evaluation. But they differ in the evaluation procedure and the output. A binary model queries the overlay output to separate areas that satisfy the criteria from those that do not. An index model calculates the index value from the overlay output and ranks areas based on the index value. A regression model also relies on overlay to combine data for the analysis of the statistical relationship between a dependent variable and independent variables. If statistically significant, a regression model can be used for predicting the dependent variable. A process model integrates existing knowledge about the environmental processes in the real world and quantifies the processes with a set of relationships and equations. Examples of process models include those for predicting soil erosion, nonpoint source pollution, water quality and quantity, and sediment. A GIS can assist environmental modelers in the areas of data visualization, database management, and data exploration.

1.5 ORGANIZATION OF THIS BOOK

The book is organized into seven sections: GIS data and data models (Chapters 2–5), data input and management (Chapters 6–9), data display and exploration (Chapters 10–11), core data analysis (Chapters 12–13), surface analysis (Chapters 14–16), line analysis (Chapters 17–18), and GIS models and modeling (Chapter 19).

Chapter 2 discusses coordinate systems, a topic essential to geographically referenced data. Common coordinate systems including the transverse Mercator projection, the Lambert conformal conic projection, the Universal Transverse Mercator (UTM) grid system, and the State Plane Coordinate (SPC) system are covered in Chapter 2 along with their parameters. Chapters 3 and 4 examine the georelational data model and the object-based data model respectively—both chapters also deal with topology and composite features. Chapter 5 focuses on the raster data model and discusses different types of raster data, raster data structure, data compression, and conversion between raster and vector data.

Chapter 6 describes online digital data, metadata (information about data), data conversion, and creation of new data from satellite images, field data, and paper maps. Chapter 7 concentrates on the geometric transformation of newly digitized maps and satellite images and interpretation of the transformation results. Chapter 8 covers spatial data editing related to location and topological errors. Chapter 8 also discusses spatial data accuracy standards, edgematching, line simplification, and line smoothing. Chapter 9 examines attribute data input and management including the relational database model and its design, relational database examples, and attribute data entry and manipulation.

Chapter 10 deals with data display and cartography. Starting with the elements of cartographic symbolization, Chapter 10 proceeds to discuss type of map, text and text placement, map design, and map production. Chapter 11 covers data exploration. The topics include exploratory data analysis, attribute data query, spatial data query, and map-based geographic visualization. Chapters 12 and 13 provide the basic tools for GIS analysis and their applications. Chapter 12 includes the vector-based tools of buffering, overlay, distance measurement, pattern analysis statistics, and map manipulation. Chapter 13 covers local, neighborhood, zonal, and global (physical distance measure) operations for analyzing raster data

and additional data extraction and generalization operations. Chapter 13 also has a section that uses overlay and buffering as examples to compare vector- and raster-based data analysis.

Chapter 14 examines terrain mapping and analysis. Terrain mapping includes contouring, profiling, hill shading, hypsometric tinting, and perspective view. Terrain analysis includes slope, aspect, and surface curvature. Worked examples are provided for the computing algorithms whenever appropriate, and a comparison is made between elevation rasters and TINs for terrain analysis. Chapter 15 focuses on viewshed analysis, watershed analysis, and their applications. Chapter 15 also discusses the parameters and options for both analyses. Chapter 16 covers spatial interpolation including the basic elements of spatial interpolation, global methods, local methods, kriging, and cross-validation statistics. Throughout Chapter 16, a small data set and worked examples are used to illustrate how the interpolation algorithms work.

Chapter 17 discusses geocoding and dynamic segmentation. Both techniques are popular in GIS because they can convert data that lack x-, y-coordinates into point and line features. Chapter 18 examines path analysis and network applications. Worked examples are provided to illustrate the same shortest path algorithm, first in raster format and then in vector format. Chapters 17 and 18 are linked by their focus on linear features, which serve as either the reference data or as inputs in the operations.

Chapter 19 gives an overview of GIS models and modeling. Following an introduction to the classification of models, the modeling process, and the role of GIS in modeling, Chapter 19 describes the basics of building binary, index, regression, and process models. Chapter 19 also offers numerous examples from a variety of disciplines for each type of model.

1.6 CONCEPTS AND PRACTICE

Each chapter in this book has two main sections. The first section covers a set of topics and concepts, and the second section covers applications with three to six problem-solving tasks. Additional materials to the first section include information boxes, websites, key concepts and terms, and review questions. The applications section provides step-by-step instructions as well as questions to reinforce the learning process. We do not learn well if we just follow the instructions to complete a task without pausing and thinking about the process. A challenge task is also included at the end of each applications section to further develop the necessary skills for problem solving. Each chapter is complete with an extensive, updated bibliography.

This book stresses both concept and practice. GIS concepts explain the purpose and objectives of GIS operations and the interrelationship among GIS operations. A basic understanding of map projection, for example, explains why we must project map layers into a common coordinate system before using them together and why we must input numerous projection parameters. Our knowledge of map projection is long lasting because the knowledge will neither change with the technology nor become outdated with new versions of a GIS package.

GIS is a problem-solving tool (Wright et al. 1997). To apply the tool correctly and efficiently, we must become proficient in using the tool. Although a Windows-based GIS package has im-proved the human-computer interaction over a command-driven package, it still requires us to sort out a multitude of menus, buttons, and tools and to know how and when to use them. Practice, which is a regular feature in mathematics and statistics textbooks, is really the only way for us to become proficient in using GIS. Practice can also help us grasp GIS concepts. For instance, the root mean square (RMS) error, a measure of the goodness of geometric transformation, may be difficult to com-prehend mathematically. But after a couple of geometric transformations, the RMS error starts to make more sense because we can see how the error measure changes each time with a different set of control points.

Practice sections in a GIS textbook require data sets and GIS software. Many data sets used in this book are from GIS classes taught at the University of Idaho over the past 18 years. Instructions accompanying the exercises correlate to ArcGIS 9.0. All tasks except for two in Chapter 18 use ArcGIS Desktop and the extensions of Spatial Analyst, 3D Analyst, Geostatistical Analyst, and ArcScan. The only two tasks that require use of ArcInfo Workstation deal with shortest path analysis and location-allocation. They can be easily run in ArcGIS Desktop as soon as the Network Analyst extension is made available by ESRI, Inc.

KEY CONCEPTS AND TERMS

Attribute data: Data that describe the characteristics of spatial features.

Continuous features: Spatial features that exist between observations.

Coverage: An ESRI data format for topological vector data.

Data exploration: Data-centered query and analysis.

Discrete features: Spatial features that do not exist between observations, form separate entities, and are individually distinguishable.

Dynamic segmentation model: A data model that allows the use of linearly measured data on a coordinate system.

Geodatabase data model: An ESRI data model that is object-based.

Geographic information system (GIS): A computer system for capturing, storing, querying, analyzing, and displaying geographically referenced data.

Geographic visualization: Use of maps for data exploration and visual information processing.

Geographically referenced data: Data that describe both the locations and characteristics of spatial features on the Earth's surface.

GIS modeling: The process of using GIS in building models with spatial data.

Georelational data model: A vector data model that uses a split system to store spatial data and attribute data.

Object-based data model: A data model that uses objects to organize spatial data and stores spatial data and attribute data in a single system.

Projection: The process of transforming from a geographic grid to a plane coordinate system.

Raster data model: A spatial data model that uses a grid and cells to represent the spatial variation of a feature.

Regions: Higher-level vector data that can have spatially disjoint components and can overlap one another.

Relational database: A collection of tables, which can be connected to each other by attributes whose values can uniquely identify a record in a table.

Shapefile: An ESRI data format for nontopological vector data.

Spatial data: Data that describe the geometry of spatial features.

Spatial interpolation: A process of using points with known values to estimate values at other points.

Topology: A subfield of mathematics that is applied in GIS to ensure that the spatial relationships between features are expressed explicitly.

Triangulated irregular network (TIN): A data model that approximates the terrain with a set of nonoverlapping triangles.

Vector data model: A spatial data model that uses points and their x-, y-coordinates to construct spatial features of points, lines, and areas.

REVIEW QUESTIONS

1. Define geographically referenced data.
2. Describe an example of GIS application from your discipline.
3. Go to the U.S. Department of Housing and Urban Development website (http://www.hud.gov/offices/cio/emaps/index.cfm) and plot an emap around your university. Are there any hazardous sites?
4. Go to the National Institute of Justice website (http://www.ojp.usdoj.gov/nij/maps/). What does it mean by "hot spot" analysis?
5. Location-based services are probably the most commercialized GIS-related field. What do location-based services provide?
6. What kinds of software and hardware are you currently using for GIS classes and projects?
7. Search for MapPoint at Microsoft's website (http://www.microsoft.com/). What are the main functions of MapPoint?
8. Define spatial data and attribute data.
9. Explain the difference between vector data and raster data.
10. Explain the difference between the georelational data model and the object-based data model.
11. How does data exploration differ from data analysis?
12. Suppose you are required to do a GIS project for a class. What kinds of activities or operations do you have to perform to complete the project?
13. Name two tools or techniques for vector data analysis.
14. Name two types of operations for raster data analysis.
15. Describe an example from your discipline in which a GIS can provide useful tools for building a model.

APPLICATIONS: INTRODUCTION

ArcGIS uses a single, scalable architecture and user interface. ArcGIS has three versions: ArcView, ArcEditor, and ArcInfo. ArcView is the simplest version and has fewer capabilities than the other two versions. All three versions use the same applications of ArcCatalog and ArcMap and share the same extensions such as Spatial Analyst, 3D Analyst, and Geostatistical Analyst. ArcInfo has the additional ArcInfo Workstation, a command-driven application similar to ARC/INFO version 7.x.

You can tell which version of ArcGIS you are using by looking at the title of an application. For example, the title of ArcCatalog may appear as ArcCatalog–ArcView or ArcCatalog–ArcInfo, depending on if you are using ArcView or ArcInfo. To check the availability of the extensions, you can do the following: close all ArcGIS applications, click the Start menu, point to Programs, point to ArcGIS, and select Desktop Administrator. Double-click Availability in the next menu to display the available licenses.

This applications section covers three tasks. Task 1 introduces ArcCatalog and ArcToolbox, Task 2 ArcMap and the Spatial Analyst extension, and Task 3 ArcInfo Workstation. As mentioned in the text, only two tasks in Chapter 18 use ArcInfo Workstation. All other tasks use ArcCatalog and ArcMap. Typographic conventions used in the instructions include italic typeface for data sets (e.g., *emidalat*) and boldface for questions (e.g., **Q1**).

Task 1: Introduction to ArcCatalog

What you need: *emidalat,* an elevation raster; and *emidastrm.shp,* a stream shapefile.

Task 1 introduces ArcCatalog, an application for managing data sets.

1. Start ArcCatalog. ArcCatalog lets you set up connections to your data sources, which may reside in a folder on a local disk or a database on the network. For Task 1, you will make connection to the folder containing the Chapter 1 database (e.g., chap1). Click the Connect to Folder button.

Navigate to the chap1 folder and click OK. The chap1 folder now appears in the Catalog tree. Expand the folder to view the data sets.

2. Click *emidalat* in the Catalog tree. Click the Preview tab to view the elevation raster. Click the Metadata tab, and select FGDC for the stylesheet. The narrative shows that *emidalat* is a raster dataset in ESRI GRID format and that the raster is projected onto the Universal Transverse Mercator (UTM) coordinate system.

Q1. What does FGDC stand for?

3. Click *emidastrm.shp* in the Catalog tree. On the Preview tab, you can preview the geography and table of *emidastrm.shp*.

4. ArcCatalog has tools for various data management tasks. You can access these tools by right-clicking a data set to open its context menu. Right-click *emidastrm.shp*, and the menu shows Copy, Delete, Rename, Create Layer, Export, and Properties. Using the context menu, you can copy *emidastrm.shp* and paste it to a different folder or delete it. A layer, or a layer file, is a visual representation of a data set. The export tool can export a shapefile to a geodatabase or a coverage. The properties dialog shows the data set information.

5. This step lets you create a personal geodatabase and then import *emidalat* and *emidastrm.shp* to the geodatabase. Right-click the Chapter 1 database in the Catalog tree, point to New, and select Personal Geodatabase. Click the new geodatabase and rename it *Task1.mdb*. If the extension .mdb does not appear, select Options from the Tools menu and on the General tab uncheck the box to hide file extensions.

6. There are two options for importing *emidalat* and *emidastrm.shp* to *Task1.mdb*. You will try both options. Right-click *Task1.mdb*, point to Import, and select Raster Datasets. In the next dialog, navigate to *emidalat*, add it for the input raster, and click OK to import.

7. Now you will use the second option, ArcToolbox, to import *emidastrm.shp* to *Task1.mdb*. ArcCatalog's standard toolbar has a button called Show/Hide ArcToolbox Window. Click the button to open ArcToolbox. Dock the ArcToolbox window so that you can see both the window and the Catalog tree. Right-click ArcToolbox and select Environments. The Environment Settings dialog has five sets of settings. One setting that is used in most operations is the working directory. Click the dropdown arrow for General Settings. Navigate to the Chapter 1 database and set it as the current workspace. Tools in ArcToolbox are organized into a hierarchy. The tool you need for importing *emidastrm.shp* resides in the Conversion Tools/To Geodatabase toolset. Double-click Feature Class to Feature Class to open the tool. Select *emidastrm.shp* for the input features, select *Task1.mdb* for the output location, specify *emidastrm* for the output feature class name, and click OK. Expand *Task1.mdb* to make sure that the import operations are complete.

Q2. The number of usable tools in ArcToolbox varies depending on which version of ArcGIS you are using. Go to ArcGIS Desktop Help / Geoprocessing (including ArcToolbox) / Geoprocessing tool reference / Geoprocessing Commands Quick Reference Guide. Open the document. Is the Feature Class to Feature Class tool for Task 1 available to all three versions of ArcGIS?

Task 2: Introduction to ArcMap

What you need: *emidalat* and *emidastrm.shp,* same as Task 1.

In Task 2, you will learn the basics of working with ArcMap.

1. You can start ArcMap by clicking the Launch ArcMap button in ArcCatalog or from the Programs menu. ArcMap is the main application for data display, data query, data analysis, and data output. ArcMap organizes data sets into data frames (also called maps). You open a new data frame called Layers when you launch ArcMap. Right-click Layers and select Properties. On the General tab, change the name Layers to Task 2 and click OK.

2. Next add *emidalat* and *emidastrm.shp* to Task 2. Click the Add Data button in ArcMap, navigate to the Chapter 1 database, and select *emidalat* and *emidastrm.shp.* To select more than one data set to add, click the first data set and then click other data sets while holding down the Ctrl key. An alternative to the Add Data button is the drag-and-drop method. You can add a data set in ArcMap by dragging it from the Catalog tree and dropping it in ArcMap's view window.

3. A warning message states that one or more layers are missing spatial reference information. Click OK to dismiss the dialog. *emidastrm.shp* does not have the projection information, although it is based on the UTM coordinate system, same as *emidalat.* You will learn in Chapter 2 how to define a coordinate system.

4. Both *emidastrm* and *emidalat* are highlighted in the table of contents, meaning that they are both active. You can deactivate by clicking on the empty space. The table of contents has three tabs: Display, Source, and Selection. On the Display tab, you can change the drawing order of the layers by dragging and dropping a layer up or down. The Source tab shows the data source of each layer. (The Source tab also lists tables that have been added.) The Selection tab lets you choose the selectable layer.

Q3. Does ArcMap draw the top layer in the table of contents first?

5. The standard toolbar in ArcMap has such tools as Zoom In, Zoom Out, Pan, Select Elements, and Identify. When you hold the mouse point over a tool, a ToolTip appears in a floating box to tell you the name of the tool and a short message about the use of the tool appears at the bottom of the ArcMap window.

6. ArcMap has two views: Data View and Layout View. (The buttons for the two views are located at the bottom of the view window.) Data View is for viewing data, whereas Layout View is for viewing the map product for printing and plotting. For this task, you will stay with Data View.

7. This step is to change the symbol for *emidastrm*. Click the symbol for *emidastrm* in the table of contents to open the Symbol Selector dialog. You can either select a preset symbol (e.g., river) or make up your own symbol for *emidastrm* by specifying the color, width, and properties of the symbol. Choose the preset symbol for river.

8. Next classify *emidalat* into the elevation zones of <900, 900–1000, 1000–1100, 1100–1200, 1200–1300, and >1300 meters. Right-click *emidalat*, and select Properties. Click the Symbology tab. Click Classified in the Show frame. Change the number of classes to 6, and click the Classify button. The Method dropdown list shows six methods. Select Manual. There are two ways to set the break values for the elevation zones manually. To use the first method, you will check the box to snap breaks to data values and then click the first break line and drag it to the intended value of 900. Then set the other break lines at 1000, 1100, 1200, 1300, and 1337. To use the second method, you will click the first cell in the Break Values frame and enter 900. Then enter 1000, 1100, 1200, and 1300 for the next four cells. (If the break value you entered is changed to a different value, reenter it.) Click OK to dismiss the Classification dialog.

Q4. List the other classification methods besides Manual that are available in ArcMap.

9. You can change the color scheme for *emidalat* by using the Color Ramp dropdown list in the Layer Properties dialog. Sometimes it is easier to select a color scheme using words instead of graphic views. In that case, you can right-click inside the Color Ramp box and uncheck Graphic View. The Color Ramp dropdown list now shows White to Black, Yellow to Red, etc. Select Elevation #1. Click OK to dismiss the dialog.

10. ArcMap has access to several extensions including Spatial Analyst. Select Extensions from the Tools menu and check Spatial Analyst. Then select Toolbars from the View menu and check Spatial Analyst. The Spatial Analyst toolbar should now appear in ArcMap. Click the Spatial Analyst dropdown arrow, point to Surface Analysis, and select Slope. In the Slope dialog, select *emidalat* for the input surface and click OK to run the command. *Slope of emidalat* is the slope layer of *emidalat*. An alternative for ArcInfo users is to use ArcToolbox in ArcMap. The Slope tool in the Spatial Analyst Tools/Surface toolset can perform the same operation except that you have to save the output on disk. Spatial Analyst Tools are not available for ArcView users.

Q5. To use an extension such as Spatial Analyst, you must work with the Tools menu and the View menu. Why?

11. You can save Task 2 as a map document before exiting ArcMap. Select Save As from the File menu in ArcMap. Navigate to the Chapter 1 database, enter *chap1* for the file name, and click Save. ArcMap automatically adds the extension .mxd to chap1. Data sets displayed in Task 2 are now saved with *chap1.mxd*. To re-open *chap1.mxd*, *chap1.mxd* must reside in the same folder as the data sets it references. You can save a map document with the relative path name option (e.g., without the drive name). Select Map Properties from ArcMap's File menu. In the next dialog, click on Data Source Options. The Data Source Options dialog has the options of full path names and relative path names.

12. To make sure that *chap1.mxd* is saved correctly, first select Exit from ArcMap's File menu. Then launch ArcMap again. *chap1.mxd* should appear in the ArcMap dialog. If not, select an existing map, navigate to the Chapter 1 database, and double-click on *chap1.mxd*.

Task 3: Introduction to ArcInfo Workstation

What you need: *emidalat*, same as Task 1; and *breakstrm*, a stream coverage.

ArcInfo Workstation organizes commands into modules such as Arc, ArcPlot, ArcEdit, and Tables. Arc is the start-up module; ArcPlot displays coverages, grids, and TINs; ArcEdit includes editing functions; and Tables works with tables or INFO files. While working in ArcInfo Workstation, you need to move between modules, depending on the task you want to perform. Task 3 uses ArcInfo Workstation to display *emidalat* and *breakstrm*. *breakstrm* shows the same streams as *emidastrm.shp* in Task 1 except that *breakstrm* is a coverage. In the following instructions, the explanation of a command comes after /*.

1. Use the Programs menu to start ArcInfo Workstation. After the Arc prompt appears, use the Workspace (or W) command to change the directory to the Chapter 1 database, where *emidalat* and *breakstrm* reside.

2. First, you need to go to Arcplot and set up the environment to display *emidalat* and *breakstrm*:

   Arc: arcplot
   Arcplot: display 9999 /*define the computer monitor as the display device
   Arcplot: mapextent emidalat /*define the map extent of *emidalat*

3. The next command displays *emidalat* by using a lookup table (emidalat.lut):

   Arcplot: gridpaint emidalat value emidalat.lut /*display *emidalat* by elevation zone
   Arcplot: arclines breakstrm 3 /*display *breakstrm* in green

4. Use Quit to exit Arcplot and Arc:

   Arcplot: quit
   Arc: quit

Challenge Task

What you need: *menan-buttes*, an elevation raster.

This challenge task asks you to display *menan-buttes* in 10 elevation zones and save the map along with Task 2 in *chap1.mxd*.

1. Open *chap1.mxd*. Select Data Frame from ArcMap's Insert menu. Rename the new data frame Challenge, and add *menan-buttes* to Challenge.

2. Display *menan-buttes* in 10 elevation zones by using the elevation #2 color ramp and the following break values: 4800, 4900, 5000, 5100, 5200, 5300, 5400, 5500, 5600, and 5619 (feet).

3. Save Challenge with Task 2 in *chap1.mxd*.

REFERENCES

Broome, F. R., and D. B. Meixler. 1990. The TIGER Data Base Structure. *Cartography and Geographic Information Systems* 17: 39–47.

Chrisman, N. 1988. The Risks of Software Innovation: A Case Study of the Harvard Lab. *The American Cartographer* 15: 291–300.

Coppock, J. T. 1988. The Analogue to Digital Revolution: A View from an Unreconstructed Geographer. *The American Cartographer* 15: 263–75.

McHarg, I. L. 1969. *Design with Nature*. New York: Natural History Press.

Rhind, D. 1988. Personality as a Factor in the Development of a Discipline: The Example of Computer-Assisted Cartography. *The American Cartographer* 15: 277–89.

Tomlinson, R. F. 1984. Geographic Information Systems: The New Frontier. *The Operational Geographer* 5: 31–35.

Wright, D. J., M. F. Goodchild, and J. D. Proctor. 1997. Demystifying the Persistent Ambiguity of GIS as "Tool" versus "Science." *Annals of the Association of American Geographers* 87: 346–62.

2

COORDINATE SYSTEMS

CHAPTER OUTLINE

A basic principle in GIS is that map layers to be used together must align spatially. Obvious mistakes can occur if they do not. For example, Figure 2.1 shows the road maps of Idaho and Montana downloaded separately from the Internet. Obviously, the two maps do not register spatially. To connect the road networks across the state border, we must convert them to a common spatial reference system. Chapter 2 deals with coordinate systems, which are the basis for the spatial reference.

GIS users typically work with map features on a plane such as Figure 2.1. These map features represent spatial features on the Earth's surface. The locations of map features are based on a plane co-ordinate system expressed in *x*- and *y*-coordinates, whereas the locations of spatial features on the Earth's surface are based on a geographic coordinate system expressed in longitude and latitude values. A map projection bridges the two types of coordinate systems. The process of projection transforms the Earth's surface to a plane, and the outcome is a map projection, ready to be used for a plane or projected coordinate system.

We regularly download data sets from the Internet or get them from government agencies for GIS projects. Some digital data sets are measured in longitude and latitude values, whereas others are in different projected coordinate systems. Invariably, these data sets must be processed before they can be used together. Processing in this case means projection and reprojection. **Projection** converts data sets from geographic coordinates to projected coordinates, and **reprojection** converts from one type of projected coordinates to another type. Typically, projection and reprojection are among the initial tasks performed in a GIS project.

Chapter 2 is divided into the following five sections. Section 2.1 describes the geographic

19

coordinate system. Section 2.2 discusses projection, types of map projections, and map projection parameters. Sections 2.3 and 2.4 cover commonly used map projections and coordinate systems respectively. Section 2.5 discusses how to work with coordinate systems in a GIS package.

2.1 GEOGRAPHIC COORDINATE SYSTEM

The **geographic coordinate system** is the location reference system for spatial features on the Earth's surface (Figure 2.2). The geographic coordinate system is defined by **longitude** and **latitude**. Both longitude and latitude are angular measures: longitude measures the angle east or west from the prime meridian, and latitude measures the angle north or south of the equatorial plane (Figure 2.3). **Meridians** are lines of equal longitude. The prime meridian passes through Greenwich, England and has the reading of 0°. Using the prime meridian as a reference, we can measure the longitude value of a point on the Earth's surface as 0° to 180° east or west of the prime meridian. Meridians are therefore used for measuring location in the E–W direction. **Parallels** are lines of equal latitude. Using the equator as 0° latitude, we can measure the latitude value of a point as 0° to 90° north or south of the equator. Parallels are therefore used for measuring location in the N–S direction. A point location denoted by (120°W, 60°N) means that it is 120° west of the prime meridian and 60° north of the equator. The prime meridian and the equator serve as the baselines of the geographic coordinate system. The notation of geographic coordinates is therefore

Figure 2.1
The top map shows the road networks in Idaho and Montana based on different coordinate systems. The bottom map shows the road networks based on the same coordinate system.

Interstate
U.S.
State

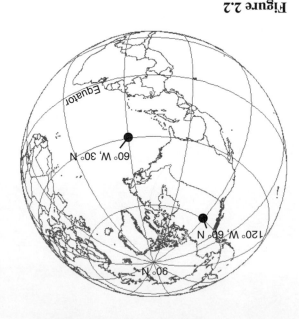

Figure 2.2
The geographic coordinate system.

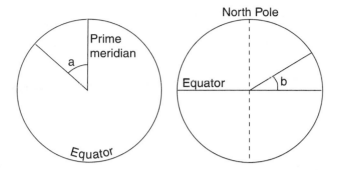

Figure 2.3
A longitude reading is represented by *a* on the left, and a latitude reading is represented by *b* on the right. Both longitude and latitude readings are angular measures.

like plane coordinates: longitude values are equivalent of *x* values and latitude values are equivalent of *y* values. And it is conventional in GIS to enter longitude and latitude values with positive or negative signs. Longitude values are positive in the eastern hemisphere and negative in the western hemisphere. Latitude values are positive if north of the equator, and negative if south of the equator.

The angular measures of longitude and latitude may be expressed in **degrees-minutes-seconds (DMS), decimal degrees (DD),** or radians (rad). Given that 1 degree equals 60 minutes and 1 minute equals 60 seconds, we can easily convert between DMS and DD. For example, a latitude value of $45°52'30''$ would be equal to $45.875°$ $(45 + 52/60 + 30/3600)$. Radians are typically used in computer programs. One radian equals $57.2958°$, and one degree equals 0.01745 rad.

2.1.1 Approximation of the Earth

The first step to map spatial features on the Earth's surface is to select a model that approximates the shape and size of the Earth. The simplest model is a sphere, which is typically used in discussing map projections (Section 2.3). But the Earth is not a perfect sphere: the Earth is wider along the equator than between the poles. Therefore a better approximation of the shape of the Earth is a **spheroid,** also called **ellipsoid,** an ellipse rotated about its minor axis.

A spheroid has its major axis (*a*) along the equator and its minor axis (*b*) connecting the poles (Figure 2.4). A parameter called the flattening (*f*), defined by $(a - b)/a$, measures the difference between the two axes of a spheroid. Geographic coordinates based on a spheroid are known as **geodetic coordinates,** which are the basis for all mapping systems (Iliffe 2000). In this book, we will use the generic term geographic coordinates.

The geoid, an even closer approximation of the Earth, has an irregular surface, which is affected by irregularities in the density of the Earth's crust and mantle. The geoid surface is treated as the surface of mean sea level, which is important

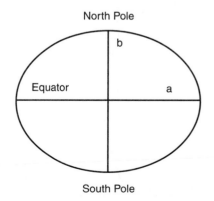

Figure 2.4
The flattening is based on the difference between the semimajor axis *a* and the semiminor axis *b*.

for measuring elevations or heights. For example, elevation readings from a GPS (global positioning system) receiver are measured from the surface of the geoid (Chapter 6). This chapter is mainly concerned with spheroids.

2.1.2 Datum

A **datum** is a mathematical model of the Earth, which serves as the reference or base for calculating the geographic coordinates of a location (Burkard 1984; Moffitt and Bossler 1998). The definition of a datum consists of an origin, the parameters of the spheroid selected for the computations, and the separation of the spheroid and the Earth at the origin. Many countries have developed their own datums for local surveys. Among these local datums are the European Datum, the Australian Geodetic Datum, the Tokyo Datum, the Indian Datum (for India and several adjacent countries), and the Hu Tzu Shan Datum (for Taiwan).

Until the late 1980s, **Clarke 1866**, a ground-measured spheroid, was the standard spheroid for mapping in the United States. Clarke 1866's semimajor axis (equatorial radius) and semiminor axis (polar radius) measure 6,378,206.4 meters (3962.96 miles) and 6,356,583.8 meters (3949.21 miles) respectively, with the flattening of 1/294.979. **NAD27** (North American Datum of 1927) is a local datum based on the Clarke 1866 spheroid, with its origin at Meades Ranch in Kansas. In 1986 the National Geodetic Survey (NGS) introduced **NAD83** (North American Datum of 1983), an Earth-centered (also called geocentered) datum based on the **GRS80** (Geodetic Reference System 1980) spheroid. GRS80's semimajor axis and semiminor axis measure 6,378,137.0 meters (3962.94 miles) and 6,356,752.3 meters (3949.65 miles) respectively, with the flattening of 1/298.257. In the case of GRS80, the shape and size of the Earth was determined through measurements made by Doppler satellite observations.

The horizontal shift from NAD27 to NAD83 can be substantial (Figure 2.5). Positions of points can change between 10 and 100 meters in the conterminous United States, more than 200 meters in Alaska, and in excess of 400 meters in Hawaii. For example, for the Ozette quadrangle map from the Olympic

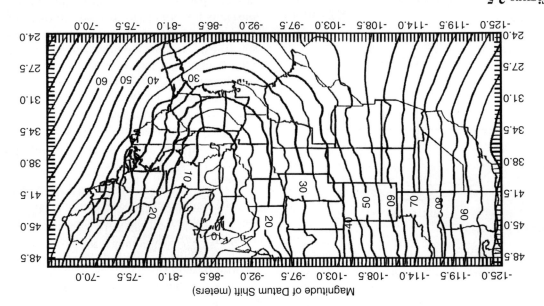

Magnitude of Datum Shift (meters)

Figure 2.5
The isolines show the magnitudes of the horizontal shift from NAD27 to NAD83 in meters. See Section 2.1.2 for the definition of the horizontal shift. (By permission of the National Geodetic Survey.)

Box 2.1 | Conversion Between Datums

Conversion between datums involves transformation and computation of geographic coordinates. Free software packages can be downloaded from the Internet for datum conversion. For example, Nadcon is a software package developed by the National Geodetic Survey (NGS) for conversion between NAD27 and NAD83. Nadcon can be downloaded at the NGS website (**http://www.ngs.noaa.gov/TOOLS/**

Nadcon/Nadcon.html) or at the Topographic Engineering Center of the U.S. Army Corps of Engineers website (**http://crunch.tec.army.mil/software/ corpscon/corpscon.html**). GPS receivers usually have the options to read coordinates based on different datums (other than the default of WGS84). Many GIS packages offer a large number of datums and spheroids to accommodate users from different countries.

Peninsula in Washington, the shift is 98 meters to the east and 26 meters to the north. The horizontal shift is therefore 101.4 meters ($\sqrt{98^2 + 26^2}$).

Many GIS users in the United States have migrated from NAD27 to NAD83, while others are still in the process of adopting NAD83. The same is true with data sets downloadable from GIS data clearinghouses: some are based on NAD83, and others NAD27. Until the switch from NAD27 to NAD83 is complete, we must keep watchful eyes on the datum because digital layers based on the same coordinate system but different datums will not register correctly.

WGS84 (World Geodetic System 1984) is a reference system or datum established by the National Imagery and Mapping Agency (NIMA, now the National Geospatial-Intelligence Agency or NGA) of the U.S. Department of Defense (Kumar 1993). WGS84 agrees with GRS80 in terms of measures of the semimajor and semiminor axes. But WGS84 has a set of primary and secondary parameters. The primary parameters define the shape and size of the Earth, whereas the secondary parameters refer to local datums used in different countries (National Geospatial-Intelligence Agency 2000; Iliffe 2000). WGS84 is the datum for GPS readings. The satellites used by GPS send their positions in WGS84 coordinates and all calculations internal to GPS receivers are based on WGS84.

Migrating from NAD27 to NAD83 or from NAD27 to WGS84 requires a datum transformation, which recomputes longitude and latitude values

from one geographic coordinate system to another. A commercial GIS package may offer several transformation methods such as three-parameter, seven-parameter, Molodensky, and abridged Molodensky. A good reference on datum transformation and its mathematical methods is an online report from the National Geospatial-Intelligence Agency (2000). Free software packages for data conversion are also available online (Box 2.1).

Although the migration from NAD27 to NAD83 is still underway, other reference systems that are more accurate than NAD83 have already been developed for local surveys (Kavanagh 2003). In the late 1980s, the NGS began a program of using GPS technology to establish the High Accuracy Reference Network (HARN) on a state-by-state basis. In 1994, the NGS started the Continuously Operating Reference Stations (CORS) network, a network of over 200 stations that provide measurements for the postprocessing of GPS data. The positional difference of a control point may be up to a meter between NAD83 and HARN but less than 10 centimeters between HARN and CORS (Snay and Soler 2000).

This section has focused on the use of datums and reference systems for measuring horizontal positions (i.e., geographic coordinates). Before leaving the topic, it should be noted that the concept of datum also applies to measurements of elevations or heights. The National Geodetic Vertical Datum (NGVD) of 1929 was based on observations at tidal stations on the Atlantic, Pacific, and Gulf of Mexico shorelines. Refinement of the 1929 datum including gravimetric and other anomalies has resulted in the North American Vertical Datum of 1988 (NAVD88). NAVD88 is now the reference vertical datum for elevation readings in North America.

2.2 MAP PROJECTIONS

The process of projection transforms the spherical Earth's surface to a plane (Robinson et al. 1995; Dent 1999). The outcome of this transformation process is a **map projection**: a systematic arrangement of parallels and meridians on a plane surface representing the geographic coordinate system.

We can use data sets based on geographic coordinates directly in a GIS and, in fact, we are seeing more maps made with such data sets. But a map projection provides a couple of distinctive advantages. First, a map projection allows us to use two-dimensional maps, either paper or digital, instead of a globe. Second, a map projection allows us to work with plane or projected coordinates rather than longitude and latitude values. Computations with geographic coordinates are more complex and yield less accurate distance measurements (Box 2.2).

2.2.1 Types of Map Projections

Map projections can be grouped by either the preserved property or the projection surface. Cartographers group map projections by the preserved property into the following four classes: conformal, equal area or equivalent, equidistant, and azimuthal. A **conformal projection** preserves local angles and shapes. An **equivalent projection** represents areas in correct relative size. An **equidistant projection** maintains consistency of scale along certain lines. And an **azimuthal projection** retains certain accurate directions. The preserved property of a map projection is often included in its name such as the Lambert conformal conic projection or the Albers equal-area conic projection.

But the transformation from the Earth's surface to a flat surface always involves distortion and no map projection is perfect. This is why hundreds of map projections have been developed for map-making (Maling 1992; Snyder 1993). Every map projection preserves certain spatial properties while sacrificing other properties.

 Box 2.2 How to Measure Distances on the Earth's Surface

The equation for measuring distances on a plane coordinate system is:

$$D = \sqrt{(x_1 - x_2)^2 + (y_1 - y_2)^2}$$

where x_i and y_i are the coordinates of point i.

This equation, however, cannot be used for measuring distances on the Earth's surface. Because meridians converge at the poles, the length of 1-degree latitude does not remain constant but gradually decreases from the equator to 0 at the pole. The standard and simplest method for calculating the shortest distance between two points on the Earth's surface uses the equation:

$$\cos (d) = \sin (a) \sin (b) + \cos (a) \cos (b) \cos (c)$$

where d is the angular distance between points A and B in degrees, a is the latitude of A, b is the latitude of B, and c is the difference in longitude between A and B. To convert d to a linear distance measure, one can multiply d by the length of 1 degree at the equator, which is 111.32 kilometers or 69.17 miles. This method is accurate unless d is very close to zero (Snyder 1987).

Most commercial data producers deliver spatial data in geographic coordinates so that they can be used with any projected coordinate system the end user needs to work with. But more GIS users are using spatial data in geographic coordinates directly for data display and even simple analysis. Distance measurements from such spatial data are usually derived from the shortest spherical distance between points.

The conformal and equivalent properties are mutually exclusive. Otherwise a map projection can have more than one preserved property, such as conformal and azimuthal. The conformal and equivalent properties are global properties, meaning that they apply to the entire map projection. The equidistant and azimuthal properties are local properties and may be true only from or to the center of the map projection.

The preserved property is important for selecting an appropriate map projection for thematic mapping. For example, a population map of the world should be based on an equivalent projection. By representing areas in correct size, the population map can create a correct impression of population densities. In contrast, an equidistant projection would be better for mapping the distance ranges from a missile site.

Cartographers often use a geometric object and a globe (i.e., a sphere) to illustrate how to construct a map projection. For example, by placing a cylinder tangent to a lighted globe, one can draw a projection by tracing the lines of longitude and latitude onto the cylinder. The cylinder in this example is the projection surface, also called the developable surface, and the globe is called the **reference globe.** Other common projection surfaces include a cone and a plane. Therefore, map projections can be grouped by their projection surfaces into cylindrical, conic, and azimuthal. A map projection is called a **cylindrical projection** if it can be constructed using a cylinder, a **conic projection** if using a cone, and an **azimuthal projection** if using a plane.

The use of a geometric object helps explain two other projection concepts: case and aspect. For a conic projection, the cone can be placed so that it is tangent to the globe or intersects the globe (Figure 2.6). The first is the simple case, which results in one line of tangency, and the second is the secant case, which results in two lines of tangency. A cylindrical projection behaves the same way as a conic projection in terms of case. An azimuthal projection, on the other hand, has a point of tangency in the simple case and a line of tangency in the secant case. Aspect describes the

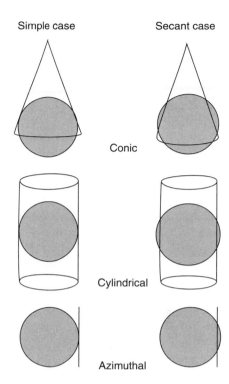

Figure 2.6
Case and projection.

placement of a geometric object relative to a globe. A plane, for example, may be tangent at any point on a globe. A polar aspect refers to tangency at the pole, an equatorial aspect at the equator, and an oblique aspect anywhere between the equator and the pole (Figure 2.7).

2.2.2 Map Projection Parameters

The concept of case relates directly to the standard line, a common parameter in defining a map projection. A **standard line** refers to the line of tangency between the projection surface and the reference globe. For cylindrical and conic projections the simple case has one standard line whereas the secant case has two standard lines. The standard line is called the **standard parallel** if it follows a parallel, and the **standard meridian** if it follows a meridian.

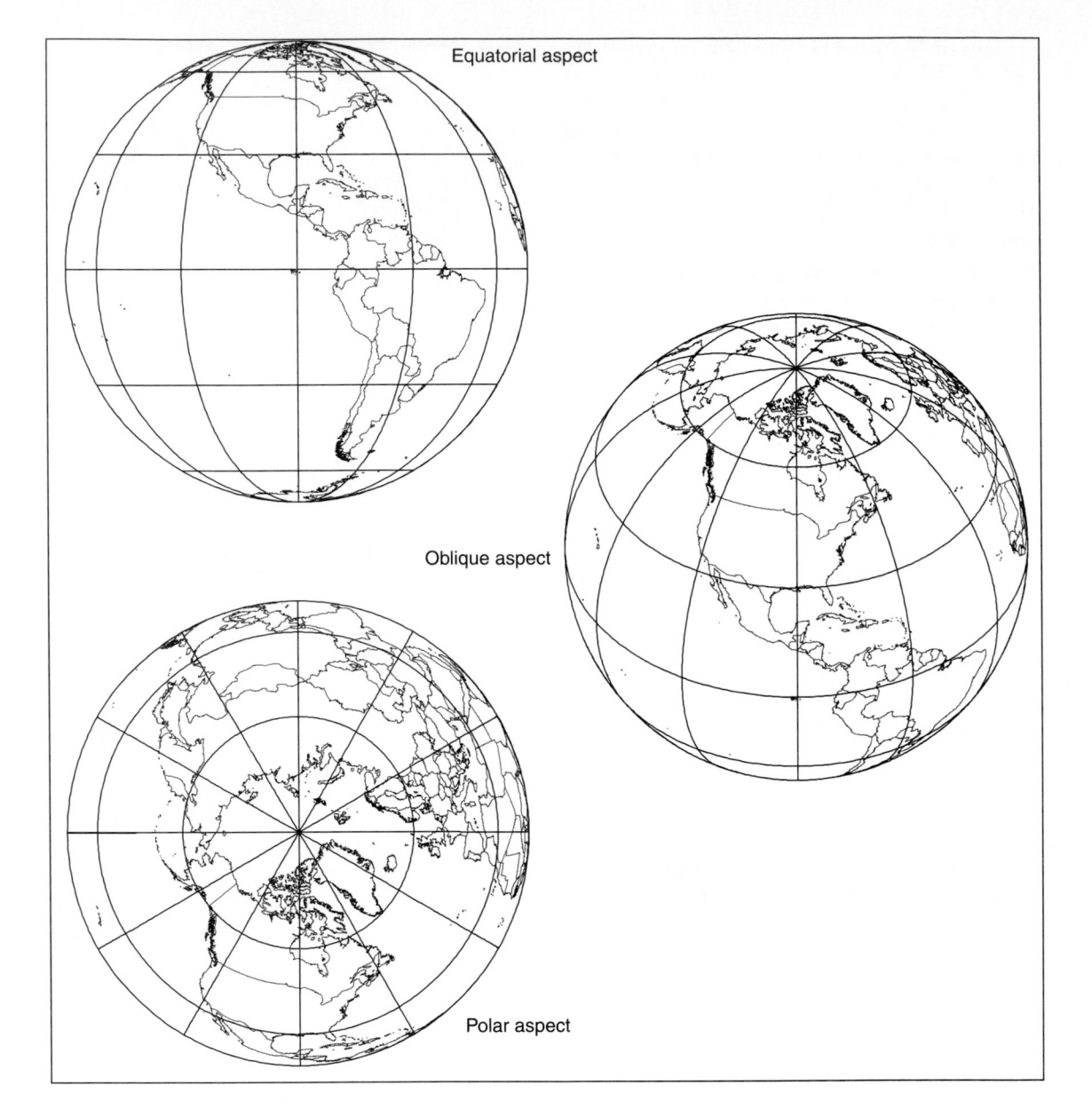

Equatorial aspect

Oblique aspect

Polar aspect

Figure 2.7
Aspect and projection.

Because the standard line is the same as on the reference globe, it has no distortion from the projection process. Away from the standard line, projection distortion can result from tearing, shearing, or compression of the spherical surface to meet the projection surface. A common measure of projection distortion is scale, which is defined as the ratio of a distance on a map (or globe) to its corresponding ground distance. The **principal scale,** or the scale of the reference globe, can

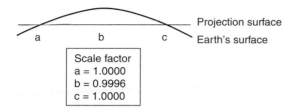

Figure 2.8
The central meridian in this secant case transverse Mercator projection has a scale factor of 0.9996. The two standard lines on either side of the central meridian have a scale factor of 1.0.

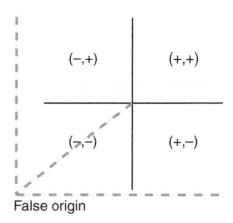

Figure 2.9
The central parallel and the central meridian divide a map projection into four quadrants. Points within the NE quadrant have positive x- and y-coordinates, points within the NW quadrant have negative x-coordinates and positive y-coordinates, points within the SE quadrant have positive x-coordinates and negative y-coordinates, and points within the SW quadrant have negative x- and y-coordinates. The purpose of having a false origin is to place all points within the NE quadrant.

therefore be derived from the ratio of the globe's radius to the Earth's radius (3963 miles or 6378 kilometers). For example, if a globe's radius is 12 inches, then the principal scale is 1:20,924,640 (1:3963 × 5280).

The principal scale applies only to the standard line in a map projection. This is why the standard parallel is sometimes called the latitude of true scale. The local scale applies to other parts of the map projection. Depending on the degree of distortion, the local scale can vary across a map projection. The **scale factor** is the normalized local scale, which is defined as the ratio of the local scale to the principal scale. The scale factor is 1 along the standard line and becomes either less than 1 or greater than 1 away from the standard line.

The standard line should not be confused with the central line. Whereas the standard line dictates the distribution pattern of projection distortion, the **central lines** (the central parallel and meridian) define the center of a map projection. The central parallel, sometimes called the latitude of origin, often differs from the standard parallel. Likewise, the central meridian often differs from the standard meridian. A good example showing the difference between the central meridian and the standard line is the transverse Mercator projection. Normally a secant projection, a transverse Mercator projection is defined by its central meridian and two standard lines on either side. The standard line has a scale factor of 1, and the central meridian has a scale factor of less than 1 (Figure 2.8).

When a map projection is used as the basis of a coordinate system, the center of the map projec-

tion, as defined by the central parallel and the central meridian, becomes the origin of the coordinate system and divides the coordinate system into four quadrants. The x-, y-coordinates of a point are either positive or negative, depending on where the point is located (Figure 2.9). To avoid having negative coordinates, we can assign x-, y-coordinate values to the origin of the coordinate system. The **false easting** is the assigned x-coordinate value and the **false northing** is the assigned y-coordinate value. Essentially, the false easting and false northing create a false origin so that all points fall within the NE quadrant and have positive coordinates (Figure 2.9).

2.3 COMMONLY USED MAP PROJECTIONS

Hundreds of map projections are in use. Commonly used map projections in GIS are not necessarily the same as those we see in classrooms or in magazines. For example, the Robinson projection is a popular projection for general mapping at the global scale because it is aesthetically pleasing (Dent 1999). But the Robinson projection is not suitable for GIS applications. A map projection for GIS applications usually has one of the preserved properties mentioned earlier, especially the conformal property. Because it preserves local shape and angles, a conformal projection allows adjacent maps to join correctly at the corners. This is important in developing a map series such as the U.S. Geological Survey (USGS) quadrangle maps.

2.3.1 Transverse Mercator

The **transverse Mercator projection,** also known as the Gauss-Kruger, is a variation of the Mercator projection, probably the best-known projection for mapping the world (Figure 2.10). The Mercator

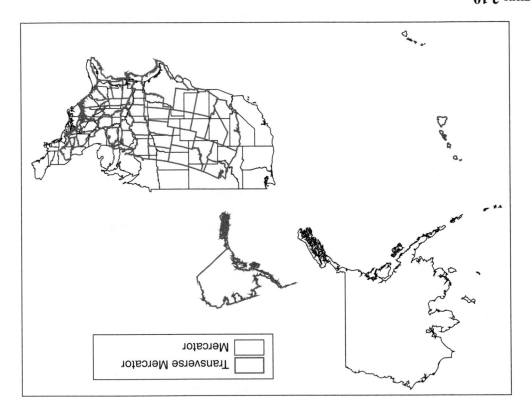

Figure 2.10
The Mercator and the transverse Mercator projections of the United States. For both projections, the central meridian is 90° W and the latitude of true scale is the equator.

projection uses the standard parallel, whereas the transverse Mercator projection uses the standard meridian. Both projections are conformal.

The transverse Mercator is the basis for two common coordinate systems to be discussed later in Section 2.4. The definition of the projection requires the following parameters: scale factor at central meridian, longitude of central meridian, latitude of origin (or central parallel), false easting, and false northing.

2.3.2 Lambert Conformal Conic

The **Lambert conformal conic projection** is a standard choice for mapping a midlatitude area of greater east–west than north–south extent, such as the state of Montana or the conterminous United States

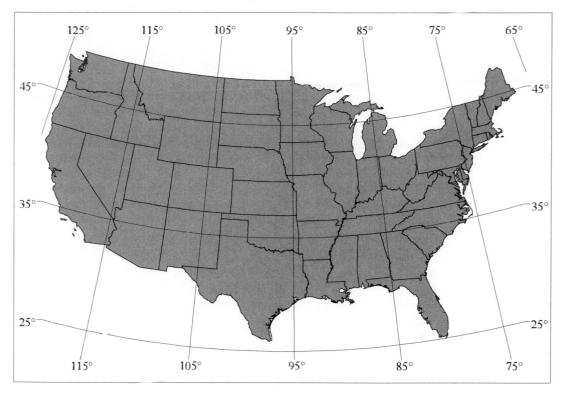

Figure 2.11
The Lambert conformal conic projection of the conterminous United States. The central meridian is 96° W, the two standard parallels are 33° N and 45° N, and the latitude of projection's origin is 39° N.

(Figure 2.11). The USGS has used the Lambert conformal conic for many topographic maps since 1957.

Typically used as a secant projection, the Lambert conformal conic is defined by the parameters of the first and second standard parallels, central meridian, latitude of projection's origin, false easting, and false northing.

2.3.3 Albers Equal-Area Conic

The Albers equal-area conic projection has the same parameters as the Lambert conformal conic projection. In fact, the two projections are quite similar except that one is equal area and the other is conformal. The Albers equal-area conic is the projection for the National Land Cover Data 1992 for the conterminous United States.

2.3.4 Equidistant Conic

The equidistant conic projection is also called the simple conic projection. The projection preserves the distance property along all meridians and one or two standard parallels. It uses the same parameters as the Lambert conformal conic.

2.4 PROJECTED COORDINATE SYSTEMS

A **projected coordinate system,** also called a plane coordinate system, is built on a map projection. Projected coordinate systems and map projections are often used interchangeably. For example, the Lambert conformal conic is a map projection but it can also refer to a coordinate system. In practice,

Three coordinate systems are commonly used in the United States: the Universal Transverse Mercator (UTM) grid system, the Universal Polar Stereographic (UPS) grid system, and the State Plane Coordinate (SPC) system. As a group, coordinates of these common systems are sometimes called real-world coordinates. This section also includes the Public Land Survey System (PLSS). Although the PLSS is a land partitioning system and not a coordinate system, it is the basis for land parcel mapping. Additional readings on these systems can be found in Robinson et al. (1995) and Muehrcke et al. (2001).

However, projected coordinate systems are designed for detailed calculations and positioning, and are typically used in large-scale mapping such as at a scale of 1:24,000 or larger (Box 2.3). Accuracy in a feature's location and its relative position to other features is therefore a key consideration in the design of a projected coordinate system.

To maintain the level of accuracy desired for measurements, a projected coordinate system is often divided into different zones, with each zone defined by a different projection center. Moreover, a projected coordinate system is defined not only by the parameters of the map projection it is based on but also the parameters (e.g., datum) of the geographic coordinate system that the map projection is derived from. As mentioned earlier, all mapping systems are based on a spheroid rather than a sphere. The difference between a spheroid and a sphere may not be a concern for general mapping at small map scales but a matter of importance in the detailed mapping of land parcels, soil polygons, or vegetation stands.

2.4.1 The Universal Transverse Mercator (UTM) Grid System

Used worldwide, the UTM **grid system** divides the Earth's surface between 84° N and 80° S into 60 zones. Each zone covers 6° of longitude, and is numbered sequentially with zone 1 beginning at 180° W. Each zone is further divided into the northern and southern hemispheres. The designation of a UTM zone therefore carries a number and a letter. For example, UTM Zone 10N refers to the zone between 126° W and 120° W in the northern hemisphere. The inside of this book's front cover has a list of the UTM zone numbers and their longitude ranges. Figure 2.12 shows the UTM zones in the conterminous United States.

Because datum is part of the definition of a projected coordinate system, the UTM grid system may be based on NAD27, NAD83, or WGS84. To complete the above example, if UTM Zone 10N is based on NAD83, then its full designation reads NAD 1983 UTM Zone 10N.

Each UTM zone is mapped onto a secant case transverse Mercator projection, with a scale factor of 0.9996 at the central meridian and the equator as the latitude of origin. The standard meridians are

---

Box 2.3 Map Scale

Map scale is the ratio of the map distance to the corresponding ground distance. This definition applies to different measurement units. A 1:24,000 scale map can mean that a map distance of 1 centimeter represents 24,000 centimeters (240 meters) on the ground. A 1:24,000 scale map can also mean that a map distance of 1 inch represents 24,000 inches (2000 feet) on the ground. Regardless of its measurement unit, 1:24,000 is a larger map scale than 1:100,000 and a 1:24,000 scale map shows more details in a smaller area than a 1:100,000 scale map. Some cartographers consider maps with a scale of 1:24,000 or larger as large-scale maps.

Map scale should not be confused with spatial scale, a term commonly used in natural resource management. Spatial scale refers to the size of area or extent. Unlike map scale, spatial scale is not rigidly defined. A large map scale, spatial scale simply means that it covers a larger area than a small spatial scale. A large spatial scale to an ecologist is therefore a small map scale to a cartographer.

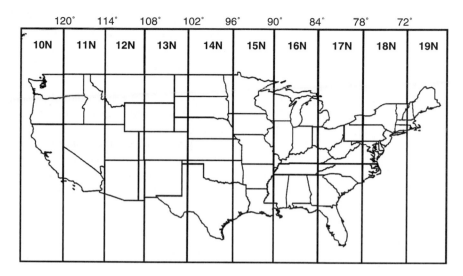

Figure 2.12
UTM zones range from zone 10N to 19N in the conterminous United States.

180 kilometers to the east and the west of the central meridian (Figure 2.13). The use of a projection per UTM zone is designed to maintain the accuracy of at least one part in 2500 (i.e., distance measured over a 2500-meter course on the UTM grid system would be accurate within a meter of the true measure) (Muehrcke et al. 2001).

In the northern hemisphere, UTM coordinates are measured from a false origin located at the equator and 500,000 meters west of the UTM zone's central meridian. In the southern hemisphere, UTM coordinates are measured from a false origin located at 10,000,000 meters south of the equator and 500,000 meters west of the UTM zone's central meridian.

The use of a false origin means that UTM coordinates are very large numbers. For example, the NW corner of the Moscow East, Idaho quadrangle map has the UTM coordinates of 500,000 and 5,177,164 meters. To preserve data precision for computations with coordinates, we can apply **x-shift** and **y-shift** values to all coordinate readings to reduce the number of digits. For example, if the x-shift value is set as −500,000 meters and the y-shift value as −5,170,000 meters for the previous quadrangle map, the coordinates for its NW corner become 0 and 7164 meters. Small numbers such as 0 and 7164

reduce the chance of having truncated computational results. The x-shift and y-shift are therefore important if coordinates are stored in single precision (i.e., up to seven significant digits). Like false easting and false northing, x-shift and y-shift change the values of x-, y-coordinates in a data set. They must be documented along with the projection parameters in the metadata (information about data, Chapter 6), especially if the map is to be shared with other users.

2.4.2 The Universal Polar Stereographic (UPS) Grid System

The **UPS grid system** covers the polar areas. The stereographic projection is centered on the pole and is used for dividing the polar area into a series of 100,000-meter squares, similar to the UTM grid system.

2.4.3 The State Plane Coordinate (SPC) System

The **SPC system** was developed in the 1930s to permanently record original land survey monument locations in the United States. To maintain the required accuracy of one part in 10,000 or less, a state may have two or more SPC zones. As examples,

Because of the switch from NAD27 to NAD83, there are SPC27 and SPC83. Besides the change of the datum, SPC83 has a few other changes. SPC83 coordinates are published in meters instead of feet. The states of Montana, Nebraska, and South Carolina have each replaced multiple zones by a single SPC zone. California has reduced SPC zones from seven to six. And Michigan has changed from transverse Mercator to Lambert conformal conic projections. A list of SPC83 is available on the inside of this book's back cover.

Some states in the United States have developed their own statewide coordinate system. Montana, Nebraska, and South Carolina all have a single SPC zone, which can serve as the statewide coordinate system. Idaho is another example. Idaho is divided into two UTM zones (11 and 12) and three SPC zones (West, Central, and East). These zones work well as long as the study area is within a single zone. When a study area covers two or more zones, the data sets must be converted to a single zone for spatial registration. But the conversion to a single zone also means that the data sets can no longer maintain the accuracy level designed for the UTM or the SPC coordinate system. The Idaho statewide coordinate system, adopted in 1994 and modified in 2003, is still based on a transverse Mercator projection but its central meridian passes through the center of the state (114° W). (A complete list of parameters of the Idaho statewide coordinate system is included in Task 1 of the applications section.) Changing the location of the central meridian means one zone for the entire state.

2.4.4 The Public Land Survey System (PLSS)

The PLSS is a land partitioning system (Figure 2.15). Using the intersecting township and range lines, the system divides the lands mainly in the central and western states into 6 × 6 mile squares or townships. Each township is further partitioned into 36 square-mile parcels of 640 acres, called sections. (In reality, many sections are not exactly 1 mile by 1 mile in size.)

Land parcel layers are typically based on the PLSS. The Bureau of Land Management (BLM) is

Oregon has the North and South SPC zones and Idaho has the West, Central, and East SPC zones (Figure 2.14). Each SPC zone is mapped onto a map projection. Zones that are elongated in the north-south direction (e.g., Idaho's SPC zones) use the transverse Mercator and zones that are elongated in the east-west direction (e.g., Oregon's SPC zones) use the Lambert conformal conic. (The only exception is zone 1 of Alaska, which uses the oblique Mercator to cover the panhandle of Alaska.) Point locations within each SPC zone are measured from a false origin located to the southwest of the zone.

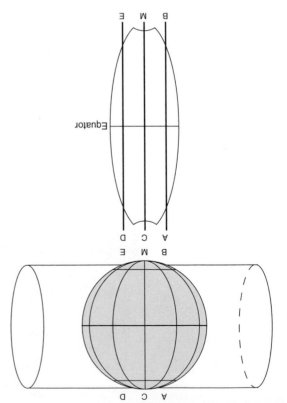

Figure 2.13
A UTM zone represents a secant case transverse Mercator projection. CM is the central meridian, and AB and DE are the standard meridians. The standard meridians are placed 180 kilometers west and east of the central meridian. Each UTM zone covers 6° of longitude and extends from 84° N to 80° S. The size and shape of the UTM zone are exaggerated for illustration purposes.

Figure 2.14
SPC83 zones in the conterminous United States. The thinner lines are county boundaries, and the bold lines are state boundaries. This map corresponds to the SPC83 table on the inside of this book's back cover.

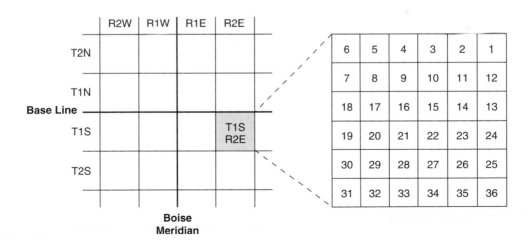

Figure 2.15
The shaded survey township has the designation of T1S, R2E. T1S means that the survey township is south of the base line by one unit. R2E means that the survey township is east of the Boise (principal) meridian by 2 units. Each survey township is divided into 36 sections. Each section measures 1 mile by 1 mile and has a numeric designation.

T he following projection file is used by ArcGIS to store information on the NAD 1983 UTM Zone 11N coordinate system:

PROJCS ["NAD_1983_UTM_Zone_11N", GEOGCS["GCS_North_American_1983",
DATUM["D_North_American_1983",SPHEROID["GRS_1980",6378137.0,298.257222101]],
PRIMEM["Greenwich",0.0], UNIT["Degree",0.0174532925199433]],
PROJECTION ["Transverse_Mercator"], PARAMETER["False_Easting",500000.0],
PARAMETER["False_Northing",0.0], PARAMETER["Central_Meridian",−117.0],
PARAMETER["Scale_Factor",0.9996], PARAMETER["Latitude_Of_Origin",0.0],
UNIT["Meter",1.0]]

The information comes in three parts. The first part defines the geographic coordinate system: NAD83 for the datum, GRS80 for the spheroid, the prime meridian of 0° at Greenwich, and units of degrees. The file also lists the major axis (6378137.0) and the denominator of the flattening (298.257222101) for the spheroid. The number of 0.0174532925199433 is the conversion factor from degree to radian (an angular unit typically used in computer programming). The second part defines the map projection parameters of name, false easting, false northing, central meridian, scale factor, and latitude of origin. And the third part defines the linear unit in meters.

2.5 WORKING WITH COORDINATE SYSTEMS IN GIS

Basic GIS tasks with coordinate systems involve defining a coordinate system, projecting geographic coordinates to projected coordinates, and reprojecting projected coordinates from one system to another.

A GIS package typically has many options of datums, spheroids, and coordinate systems. For example, Autodesk Map offers 3000 global systems, presumably 3000 combinations of coordinate system, datum, and spheroid. A constant challenge for us is how to work with this large number of coordinate

systems. Recent trends suggest that commercial GIS companies have tried to provide assistance in the following three areas: projection file, predefined coordinate systems, and on-the-fly projection.

2.5.1 Projection File

A projection file is a text file that stores information on the coordinate system that a data set is based on. Box 2.4, for example, shows a projection file for the NAD 1983 UTM Zone 11N coordinate system. The projection file contains information on the geographic coordinate system, the map projection parameters, and the linear unit.

Besides identifying a data set's coordinate system, a projection file serves at least two other purposes: it can be used for projecting or reprojecting the data set, and it can be exported to other data sets that are based on the same coordinate system.

2.5.2 Predefined Coordinate Systems

A GIS package typically groups coordinate systems into predefined and custom (Table 2.1). A predefined coordinate system, either geographic or projected, means that its parameter values are known

developing a **Geographic Coordinate Data Base (GCDB)** of the PLSS for the western United States **(http://www.blm.gov/gcdb/).** Generated from BLM survey records, the GCDB contains coordinates and other descriptive information for section corners and monuments recorded in the PLSS. Legal descriptions of a parcel layer can then be entered using, for example, bearing and distance readings originating from section corners.

| Box 2.5 | Coordinate Systems in ArcGIS |

ArcGIS divides coordinate systems into geographic and projected. The user can define a coordinate system by selecting a predefined coordinate system, importing a coordinate system from an existing data set, or creating a new (custom) coordinate system. The parameters that are used to define a coordinate system are stored in a projection file. A projection file is provided for a predefined coordinate system. For a new coordinate system, a projection file can be named and saved for future use or for projecting other data sets.

The predefined geographic coordinate systems in ArcGIS have the main options of world, continent, and spheroid-based. WGS84 is one of the world files. Local datums are used for the continental files. For example, the Indian Datum and Tokyo Datum are available for the Asian continent. The spheroid-based

options include Clarke 1866 and GRS80. The predefined projected coordinate systems have the main options of world, continent, polar, national grids, UTM, State Plane, and Gauss-Kruger (one type of the transverse Mercator projection mainly used in Russia and China). For example, the Mercator is one of the world projections; the Lambert conformal conic and Albers equal-area are among the continental projections; and the UPS is one of the polar projections.

A new coordinate system, either geographic or projected, is user-defined. The definition of a new geographic coordinate system requires a datum including a selected spheroid and its major and minor axes. The definition of a new projected coordinate system must include a datum and the parameters of the projection such as the standard parallels and the central meridian.

TABLE 2.1	A Classification of Coordinate Systems in GIS Packages

	Predefined	Custom
Geographic	NAD27, NAD83	Undefined local datum
Projected	UTM, State Plane	IDTM

and are already coded in the GIS package. The user can therefore select a predefined coordinate system without defining its parameters. Examples of predefined coordinate systems include NAD27 (based on Clarke 1866) and Minnesota SPC83, North (based on a Lambert conformal conic projection). In contrast, a custom coordinate system requires its parameter values to be specified by the user. The Idaho statewide coordinate system (IDTM) is an example of a custom coordinate system.

2.5.3 On-the-Fly Projection

On-the-fly projection is a feature that has been heavily advertised by GIS vendors. On-the-fly

projection is designed for displaying data sets that are based on different coordinate systems. The software package uses the projection files available and automatically converts the data sets to a common coordinate system. This common coordinate system is by default the coordinate system of the first data set in display. If a data set has an unknown coordinate system, the GIS package may use an assumed coordinate system. For example, ArcGIS uses NAD27 as the assumed geographic coordinate system.

On-the-fly projection does not actually change the coordinate system of a data set. Thus it cannot replace the task of projecting and reprojecting data sets in a GIS project. If a data set is to be used frequently in a different coordinate system, we should reproject the data set. And if the data sets to be used in spatial analysis have different coordinate systems, we should convert them to the same coordinate system to obtain the most accurate results.

Like other GIS packages, ArcGIS provides a suite of tools to work with coordinate systems. Box 2.5 is a summary of how to work with coordinate systems in ArcGIS.

KEY CONCEPTS AND TERMS

Azimuthal projection: One type of map projection that retains certain accurate directions. Azimuthal also refers to one type of map projection that uses a plane as the projection surface.

Central lines: The central parallel and the central meridian. Together, they define the center or the origin of a map projection.

Clarke 1866: A ground-measured spheroid, which is the basis for the North American Datum of 1927 (NAD27).

Conformal projection: One type of map projection that preserves local shapes.

Conic projection: One type of map projection that uses a cone as the projection surface.

Cylindrical projection: One type of map projection that uses a cylinder as the projection surface.

Datum: The basis for calculating the geographic coordinates of a location. A spheroid is a required input to the derivation of a datum.

Decimal degrees (DD) system: A measurement system for longitude and latitude values such as 42.5°.

Degrees-minutes-seconds (DMS) system: A measuring system for longitude and latitude values such as 42°30'00", in which 1 degree equals 60 minutes and 1 minute equals 60 seconds.

Ellipsoid: A model that approximates the Earth. Also called *spheroid*.

Equidistant projection: One type of map projection that maintains consistency of scale for certain distances.

Equivalent projection: One type of map projection that represents areas in correct relative size.

False easting: A value applied to the origin of a coordinate system to change the x-coordinate readings.

False northing: A value applied to the origin of a coordinate system to change the y-coordinate readings.

Geographic Coordinate Data Base (GCDB): A database developed by the U.S. Bureau of Land Management (BLM) to include longitude and latitude values and other descriptive information for section corners and monuments recorded in the PLSS.

Geodetic coordinates: Geographic coordinates that are based on a spheroid.

Geographic coordinate system: A location reference system for spatial features on the Earth's surface.

GRS80: A satellite-determined spheroid for the Geodetic Reference System 1980.

Lambert conformal conic projection: A common map projection, which is the basis for the SPC system for many states.

Latitude: The angle north or south of the equatorial plane.

Longitude: The angle east or west from the prime meridian.

Map projection: A systematic arrangement of parallels and meridians on a plane surface.

Meridians: Lines of longitude that measure locations in the E–W direction on the geographic coordinate system.

NAD27: North American Datum of 1927, which is based on the Clarke 1866 spheroid and has its center at Meades Ranch, Kansas.

NAD83: North American Datum of 1983, which is based on the GRS80 spheroid and is measured from the center of the spheroid.

Parallels: Lines of latitude that measure locations in the N–S direction on the geographic coordinate system.

Principal scale: Same as the scale of the reference globe.

Projected coordinate system: A plane coordinate system that is based on a map projection.

Projection: The process of transforming the spatial relationship of features on the Earth's surface to a flat map.

Public Land Survey System (PLSS):   A land partitioning system used in the United States.

Reference globe: A reduced model of the Earth, from which map projections are made. Also called a *nominal* or *generating globe.*

Reprojection: Projection of spatial data from one projected coordinate system to another.

Scale factor: Ratio of the local scale to the scale of the reference globe. The scale factor is 1.0 along a standard line.

Spheroid: A model that approximates the Earth. Also called *ellipsoid.*

Standard line: Line of tangency between the projection surface and the reference globe. A standard line has no projection distortion and has the same scale as that of the reference globe.

Standard meridian: A standard line that follows a meridian.

Standard parallel: A standard line that follows a parallel.

State Plane Coordinate (SPC) system: A coordinate system developed in the 1930s to permanently record original land survey monument locations in the United States. Most states have more than one zone based on the SPC27 or SPC83 system.

Transverse Mercator projection: A common map projection, which is the basis for the UTM grid system and the SPC system.

Universal Polar Stereographic (UPS) grid system: A grid system that divides the polar area into a series of 100,000-meter squares, similar to the UTM grid system.

Universal Transverse Mercator (UTM) grid system: A coordinate system that divides the Earth's surface between 84° N and 80° S into 60 zones, with each zone further divided into the northern hemisphere and the southern hemisphere.

WGS84:   A satellite-determined spheroid for the World Geodetic System 1984.

***x*-shift:**   A value applied to *x*-coordinate readings to reduce the number of digits.

***y*-shift:**   A value applied to *y*-coordinate readings to reduce the number of digits.

REVIEW QUESTIONS

1. Describe the three levels of approximation of the shape and size of the Earth for GIS applications.

2. Why is the datum important in GIS?

3. Describe two common datums used in the United States.

4. Pick up a USGS quadrangle map of your area. Examine the information on the map margin. If the datum is changed from NAD27 to NAD83, what is the expected horizontal shift?

5. Go to the NGS-CORS website **(http://www. ngs.noaa.gov/CORS/cors-data.html).** How many continuously operating reference stations do you have in your state? Use the links at the website to learn more about CORS.

6. Explain the importance of map projection.

7. Describe the four types of map projections by the preserved property.

8. Describe the three types of map projections by the projection or developable surface.

9. Explain the difference between the standard line and the central line.

10. How is the scale factor related to the principal scale?

11. Name two commonly used projected coordinate systems that are based on the transverse Mercator projection.

12. Find the GIS data clearinghouse for your state at the Geospatial One-Stop website (**http://www.geo-one-stop.gov/**). Go to the clearinghouse website. Does the website use a common coordinate system for the statewide data sets? If so, what is the coordinate system? What are the parameters values for the coordinate system? And, is the coordinate system based on NAD27 or NAD83?

13. Explain how a UTM zone is defined in terms of its central meridian, standard meridian, and scale factor.

14. Which UTM zone are you in? Where is the central meridian of the UTM zone?

15. How many SPC zones does your state have? What map projections are the SPC zones based on?

16. Describe how on-the-fly projection works.

APPLICATIONS: COORDINATE SYSTEMS

This applications section has four tasks. Task 1 shows you how to project a shapefile from a geographic coordinate system to a custom projected coordinate system. In Task 2, you will also project a shapefile from a geographic to a projected coordinate system but use the coordinate systems already defined in Task 1. In Task 3, you will create a shapefile from a text file containing point locations in geographic coordinates and project the shapefile onto a predefined projected coordinate system. In Task 4, you will see how on-the-fly projection works and then reproject a shapefile from one projected coordinate system to another.

All four tasks use the Define Projection and Project tools in ArcToolbox, which are available in ArcCatalog as well as ArcMap. The Define Projection tool defines a coordinate system. The Project tool projects a geographic or projected coordinate system. ArcToolbox has three options for defining a coordinate system: selecting a predefined coordinate system, importing a coordinate system from an existing data set, or creating a new (custom) coordinate system. A predefined coordinate system has a projection file. A new coordinate system can be saved into a projection file, which can then be used to define or project other data sets.

This applications section uses shapefiles (i.e., feature classes) for all four tasks. ArcToolbox has separate projection tools in the Coverage Tools/ Data Management/Projections toolset to work with coverages. These tools use projection files to define coordinate systems. ArcToolbox also has a separate tool in the Data Management Tools/ Projections and Transformations/Raster toolset for projecting the coordinate system of a raster.

Task 1: Project a Feature Class from a Geographic to a Projected Coordinate System

What you need: *idll.shp*, a shapefile measured in geographic coordinates and in decimal degrees. *idll.shp* is an outline layer of Idaho.

For Task 1, you will first define *idll.shp* by selecting a predefined geographic coordinate system and then project the shapefile onto the Idaho transverse Mercator coordinate system (IDTM). IDTM is not a predefined system. IDTM has the following parameter values:

Projection Transverse Mercator
Datum NAD83
Units meters
Parameters
scale factor: 0.9996
central meridian: −114.0
reference latitude: 42.0
false easting: 2,500,000
false northing: 1,200,000

1. Start ArcCatalog, and make connection to the Chapter 2 database. Highlight *idll.shp* in the Catalog tree. On the Metadata tab, the summary information lists the coordinate system as geographic. Click the link to Spatial Reference Information. The information shows that the coordinate system is GCS_Assumed_Geographic_1, an assumed coordinate system.

2. First define the coordinate system for *idll.shp*. Click Show/Hide ArcToolbox Window to open the ArcToolbox window in ArcCatalog. Right-click ArcToolbox and select Environments. Click the General Setting dropdown arrow and select the Chapter 2 database for the current workspace. Double-click the Define Projection tool in the Data Management Tools/Projections and Transformations toolset. Select *idll.shp* for the input feature class. The dialog shows that *idll.shp* already has a coordinate system. But it is an assumed coordinate system. Click the button for the coordinate system to open the Spatial Reference Properties dialog. Click Select. Double-click Geographic Coordinate Systems, North America, and North American Datum 1927.prj. Click OK to dismiss the dialogs. Check the spatial reference information of *idll.shp* again. The Metadata tab should show GCS_North_American_1927.

3. Next project *idll.shp* to the IDTM coordinate system. Double-click the Project tool in the Data Management Tools/Projections and Transformations/Feature toolset. In the Project dialog, select *idll.shp* for the input feature class, specify *idtm.shp* for the output feature class, and click the button for the output coordinate system to open the Spatial Reference Properties dialog. Click the New dropdown arrow and select Projected. In the New Projected Coordinate System dialog, first enter idtm for the Name. Then you need to provide projection information in the Projection frame and for the Geographic Coordinate System. In the Projection frame,

select Transverse_Mercator from the Name dropdown list. Enter the following parameter values: 2500000 for False_Easting, 1200000 for False_Northing, −114 for Central_Meridian, 0.9996 for Scale_Factor, and 42 for Latitude_Of_Origin. Make sure that the Linear Unit is Meter. Click Select for the Geographic Coordinate System. Double-click North America, and North American Datum 1983.prj. Click OK to dismiss the New Projected Coordinate System dialog. Click Save As in the Spatial Reference Properties dialog, and enter *idtm83.prj* as the file name. Dismiss the Spatial Reference Properties dialog.

4. A green dot appears next to Geographic Transformation in the Project dialog. This is because *idll.shp* is based on NAD27 and IDTM is based on NAD83. The green dot indicates that the projection requires a geographic transformation. Click Geographic Transformation's dropdown arrow and select NAD_1927_To_NAD_1983_NADCON. Click OK to run the command.

5. On the Metadata tab, you can verify if *idll.shp* has been successfully projected to *idtm.shp*.

Q1. Summarize in your own words the steps you have followed to complete Task 1.

Task 2: Import a Coordinate System

What you need: *stationsll.shp*, a shapefile measured in longitude and latitude values and in decimal degrees. *stationsll.shp* contains snow courses in Idaho.

In Task 2, you will complete the projection of *stationsll.shp* by importing the projection information on *idll.shp* and *idtm.shp* from Task 1.

1. On the Metadata tab, verify that *stationsll.shp* has an assumed geographic coordinate system. Double-click the Define Projection tool. Select *stationsll.shp* for the input feature class. Click the button for the coordinate system. Click Import in the Spatial Reference Properties dialog. Double-click *idll.shp* to add. Dismiss the dialogs.

Q2. Describe in your own words what you have done in Step 1.

2. Double-click the Project tool. Select *stationsl.shp* for the input feature class, specify *stationsm.shp* for the output feature class, and click the button for the output coordinate system. Click Import in the Spatial Reference Properties dialog. Double-click *idtm.shp* to add. Dismiss the Spatial Reference Properties dialog. Click the Geographic Transformation's dropdown arrow and select NAD_1927_To_NAD_1983_NADCON. Click OK to complete the operation. *stationsm.shp* is now projected onto the same projected (IDTM) coordinate system as *idtm.shp*.

Task 3: Project a Shapefile by Using a Predefined Coordinate System

What you need: *snow.txt*, a text file containing the geographic coordinates of 40 snow courses in Idaho.

In Task 3, you will first create an event layer from *snow.txt*. Then you will project the event layer, which is still measured in longitude and latitude values, to a predefined projected (UTM) coordinate system and save the output into a shapefile.

1. Launch ArcMap. Rename the new data frame Tasks 3&4 and add *snow.txt* to Tasks 3&4. (Notice that the table of contents is on the Source tab.) Click the Tools menu and select Add XY Data. In the next dialog, make sure that *snow.txt* is the input table, longitude is the X field, and latitude is the Y field. The dialog shows that the spatial reference of the input coordinates is an unknown coordinate system. Click the Edit button to open the Spatial Reference Properties dialog. Click Select. Double-click Geographic Coordinate Systems, North America, and North American Datum 1983.prj. Dismiss the dialogs.

2. *snow.txt Events* is added to ArcMap. You can now project *snow.txt Events* and save the output to a shapefile. Click Show/Hide ArcToolbox Window to open the ArcToolbox window in ArcMap. Double-click the Project tool in the Data Management Tools/Projections and Transformations/Feature toolset. Select *snow.txt Events* for the input dataset, and specify *snowutm83.shp* for the output feature class. Click the button for the output coordinate system. Click Select in the Spatial Reference Properties dialog. Double-click Projected Coordinate Systems, Utm, Nad 1983, and NAD 1983 UTM Zone 11N.prj. Click OK to project the dataset.

Q3. You did not have to ask for a geographic transformation in Step 2. Why?

Task 4: Convert from One Coordinate System to Another

What you need: *idtm.shp* from Task 1 and *snowutm83.shp* from Task 3.

Task 4 first shows how on-the-fly projection works in ArcMap and then asks you to convert *idtm.shp* from the IDTM coordinate system to the UTM coordinate system.

1. Right-click Tasks 3&4, and select Properties. The Coordinate System tab shows GCS_North_American_1983 to be the current coordinate system. ArcMap assigns the coordinate system of the first layer (i.e., *snow.txt Events*) to be the data frame's coordinate system. You can change it by clicking Import in the Data Frame Properties dialog. In the next dialog, double-click *snowutm83.shp*. Dismiss the dialogs. Now Tasks 3&4 is based on the NAD 1983 UTM Zone 11N coordinate system.

2. Add *idtm.shp* to Tasks 3&4. Although *idtm.shp* is based on the IDTM coordinate system, it registers spatially with *snowutm83* in ArcMap. (A couple of snow courses are supposed to be outside the Idaho border.) ArcGIS can reproject a data set on-the-fly (Section 2.5.3). It uses the spatial reference information available to project *idtm* to the coordinate system of the data frame.

3. The rest of Task 4 is to project *idtm.shp* to the UTM coordinate system and to create a new shapefile. Double-click the Project tool. Select *idtm* for the input feature class, specify *idutm83.shp* for the output feature class, and click the button for the output coordinate system. Click Select in the Spatial Reference Properties dialog. Double-click Projected Coordinate Systems, Utm, Nad 1983, and NAD 1983 UTM Zone 11N.prj. Click OK to dismiss the dialogs.

Q4. Can you use Import instead of Select in Step 3? If yes, how?

4. Although *idutm83* looks exactly the same as *idtm* in ArcMap, it has been projected to the UTM grid system.

Challenge Task

What you need: *idroads.shp* and *mtroads.shp*.

The Chapter 2 database includes *idroads.shp* and *mtroads.shp*, the road shapefiles for Idaho and

Montana respectively. *idroads.shp* is projected onto the IDTM, but it has the wrong false easting (500,000) and false northing (100,000) values. *mtroads.shp* is projected onto the NAD 1983 State Plane Montana FIPS 2500 coordinate system in meters, but it does not have a projection file.

1. Use the Project tool and the IDTM information from Task 1 to reproject *idroads.shp* with the correct false easting (2,500,000) and false northing (1,200,000) values, while keeping the other parameters the same. Name the output *idroads2.shp*.

2. Use the Define Projection tool to first define the coordinate system of *mtroads.shp*. Then use the Project tool to reproject *mtroads.shp* to the IDTM and name the output *mtroads_idtm.shp*.

3. Use the Metadata tab in ArcCatalog to verify that *idroads2.shp* and *mtroads_idtm.shp* have the same spatial reference information.

REFERENCES

Burkard, R. K. 1984. *Geodesy for the Layman.* Washington, DC: Defense Mapping Agency. Available at **http://www.ngs.noaa.gov/PUBS_LIB/Geodesy4Layman/ TR80003A.HTM#ZZ0/.**

Dent, B. D. 1999. *Cartography: Thematic Map Design,* 5th ed. Dubuque, IA: Wm. C. Brown.

Iliffe, J. 2000. *Datums and Map Projections for Remote Sensing, GIS, and Surveying.* Boca Raton, FL: CRC Press.

Kavanagh, B. F. 2003. *Geomatics.* Upper Saddle River, NJ: Prentice Hall.

Kumar, M. 1993. World Geodetic System 1984: A Reference Frame for Global Mapping, Charting and Geodetic

Applications. *Surveying and Land Information Systems* 53: 53–56.

Maling, D. H. 1992. *Coordinate Systems and Map Projections,* 2d ed. Oxford, England: Pergamon Press.

Moffitt, F. H., and J. D. Bossler. 1998. *Surveying,* 10th ed. Menlo Park, CA: Addison-Wesley.

Muehrcke, P. C., J. O. Muehrcke, and A. J. Kimerling. 2001. *Map Use: Reading, Analysis, and Interpretation.* Madison, WI: JP Publishers.

National Geospatial-Intelligence Agency. 2000. *Department of Defense World Geodetic System 1984: Its Definition and Relationships with Local Geodetic Systems,* 3rd ed. NIMA TR8350.2.

Amendment 1, January 3, 2000. Available at **http://earth-info.nga.mil/GandG/tr8350_2.html/.**

Robinson, A. H., J. L. Morrison, P. C. Muehrcke, A. J. Kimerling, and S. C. Guptill. 1995. *Elements of Cartography,* 6th ed. New York: Wiley.

Snay, R. A., and T. Soler. 2000. Modern Terrestrial Reference Systems. Part 2: The Evolution of NAD 83. *Professional Surveyor,* February 2000.

Snyder, J. P. 1987. *Map Projections—A Working Manual.* Washington, DC: U.S. Geological Survey Professional Paper 1395.

Snyder, J. P. 1993. *Flattening the Earth: Two Thousand Years of Map Projections.* Chicago: University of Chicago Press.

CHAPTER 3

GEORELATIONAL VECTOR DATA MODEL

Looking at a paper map, we can tell what features are like and how they are spatially related to one another. For example, we can easily see in Figure 3.1 that Idaho borders Montana, Wyoming, Utah, Nevada, Oregon, Washington, and Canada, and contains several Indian reservations. How can the computer "see" the same features and their spatial relationships? Chapter 3 and Chapter 4 attempt to answer the question from the perspective of vector data.

The vector data model prepares data in two basic steps so that the computer can process the data. First, it uses points and their x-, y-coordinates to represent spatial features as points, lines, and areas.

Second, it organizes geometric objects and their spatial relationships into digital data files that the computer can access, interpret, and process.

The vector data model has undergone more changes over the past two decades than any other topics in GIS. As an example, ESRI, Inc. has introduced a new vector data model with each new software package: coverage with Arc/Info, shapefile with ArcView, and geodatabase with ArcGIS. The coverage and shapefile are examples of the georelational data model, whereas the geodatabase is an example of the object-based data model. The evolution of the vector data model reflects primarily advances in computer technology and the competitive nature of the GIS market. But to GIS users, a new data model means accepting a whole new set of concepts, terms, and data file structures. This is why the vector data model is divided in two chapters: Chapter 3 covers the georelational data model, and Chapter 4 covers the object-based data model.

Chapter 3 has the following five sections. Section 3.1 introduces the georelational data model. Section 3.2 covers the representation of simple

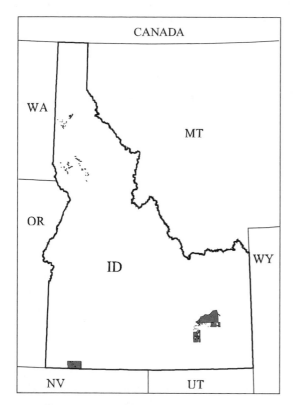

Figure 3.1
A reference map showing Idaho and lands held in trust by the United States for Native Americans.

features as points, lines, and areas. Section 3.3 introduces topology and topological data structures. Section 3.4 discusses nontopological vector data. Section 3.5 covers spatial features that are better represented as composites of points, lines, and areas.

3.1 GEORELATIONAL DATA MODEL

Geographically referenced data comprise the spatial and attribute components. Spatial data describe the locations of spatial features, whereas attribute data describe the characteristics of spatial features. The **georelational data model** stores spatial and attribute data separately in a split system: spatial data ("geo") in graphic files and attribute data ("relational") in a relational database. Typically, a

Graphic Files	INFO File

Graphic Files:
Polygon/arc list
Arc-coordinate list
Left/right list
⋮

Polygon-ID	Field 1	· · ·
1		
2		
3		

Figure 3.2
Based on the georelational data model, an ArcInfo coverage has two components: graphic files for spatial data and INFO files for attribute data. The feature ID connects the two components.

georelational data model uses the feature label or ID to link the two components (Figure 3.2). The two components must be synchronized so that they can be queried, analyzed, and displayed in unison.

The georelational data model has been used in GIS for over two decades. In recent years, however, it has faced challenges from the object-based data model. Advances in computer technology have made it possible to store both spatial data and attribute data in a single system, thus eliminating the problem of data synchronization in a split system. Although its future is uncertain, the georelational data model is still a dominant vector data model at present.

Chapter 3 emphasizes the spatial component of the georelational data model. Specifically, it covers the graphic representation of spatial features and use of data files to describe the location and the geometric relationship of spatial features. Chapter 9 covers the attribute component.

3.2 REPRESENTATION OF SIMPLE FEATURES

The **vector data model** uses the geometric objects of point, line, and area to represent simple spatial features (Figure 3.3). Dimensionality and property distinguish the three types of geometric objects as well as the features they represent.

A **point** has 0 dimension and has only the property of location. A point may also be called a node, vertex, or 0-cell. A point feature is made of a point or a set of separate points. Wells, benchmarks, and gravel pits are examples of point features.

A **line** is one-dimensional and has the property of length. A line has two end points and points in between to mark the shape of the line. The shape of a line may be a smooth curve or a connection of straight-line segments. Smooth curves are typically fitted by mathematical equations such as splines. Straight-line segments may represent human-made features such as canals and streets, or they may simply be approximations of curves. A line is also called an edge, link, chain, or 1-cell. A line feature is made of lines. Roads, streams, and contour lines are examples of line features.

An **area** is two-dimensional and has the properties of area (size) and perimeter. Made of connected lines, an area may be alone or share boundaries with other areas. An area may contain holes, such as a national forest containing private land parcels (holes). The existence of holes means that the area has both the external and internal boundaries. An area is also called a polygon, face, zone, or 2-cell. An area feature is made of polygons. Examples of area features include timber stands, land parcels, and water bodies.

The representation of simple features using points, lines, and areas is not always straightforward because it can depend on map scale. For example, a city on a 1:1,000,000 scale map may appear as a point, but the same city may appear as an area on a 1:24,000 scale map. Occasionally, the representation of vector data can also depend on the criteria established by government mapping agencies (Robinson et al. 1995). A stream may appear as a single line near its headwaters but as an area along its lower reaches. In this case, the width of the stream determines how it should be represented on a map. The U.S. Geological Survey (USGS) uses single lines to represent streams less than 40 feet wide on 1:24,000 scale topographic maps and double lines for larger streams. Therefore, a stream may appear as a line or an area depending on its width and the criterion used by the government agency.

3.3 TOPOLOGY

For some GIS applications, the conceptual representation of spatial features as points, lines, and areas is just the first step in building the vector data model. The next step is to turn to topology to express explicitly the spatial relationships between features.

Topology is the study of those properties of geometric objects that remain invariant under certain transformations such as bending or stretching (Massey 1967). For example, a rubber band can be stretched and bent without losing its intrinsic property of being a closed circuit, as long as the transformation is within its elastic limits.

Point Feature

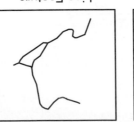
Line Feature

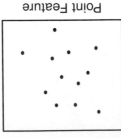
Area Feature

Figure 3.3
Point, line, and area features.

Box 3.1   Adjacency and Incidence

If a line joins two points, the points are said to be adjacent and incident with the line, and the adjacency and incidence relationships can be expressed explicitly in matrices. Figure 3.4 shows an adjacency matrix and an incidence matrix for a digraph. The row and column numbers of the adjacency matrix correspond to the node numbers, and the numbers within the matrix refer to the number of arcs joining the corresponding nodes in the digraph. For example, 1 in (11,12) means one arc joint from node 11 to node 12, and 0 in (12,11) means no arc joint from node 12 to node 11. The direction of the arc determines if 1 or 0 should be assigned.

The row numbers of the incidence matrix correspond to the node numbers in Figure 3.4, and the column numbers correspond to the arc numbers. The number 1 in the matrix means an arc is incident from a node, -1 means an arc is incident to a node, and 0 means an arc is not incident from or to a node. Take the example of arc 1. It is incident from node 13, incident to node 11, and not incident to all the other nodes. Thus the matrices express the adjacency and incidence relationships mathematically.

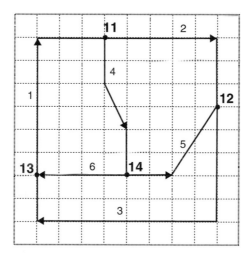

Adjacency matrix

	11	12	13	14
11	0	1	0	1
12	0	0	1	0
13	1	0	0	0
14	0	1	1	0

Incidence matrix

	1	2	3	4	5	6
11	-1	1	0	1	0	0
12	0	-1	1	0	-1	0
13	1	0	-1	0	0	-1
14	0	0	0	-1	1	1

Figure 3.4
The adjacency matrix and incidence matrix for a digraph.

Topology is often explained through graph **theory,** a subfield of mathematics that uses diagrams or graphs to study the arrangements of geometric objects and the relationships between geometric objects (Wilson and Watkins 1990). Important to the vector data model are digraphs (directed graphs), which include points and directed lines. The directed lines are called **arcs,** and the points where arcs meet or intersect are called **nodes.** Adjacency and incidence are two relationships that can be established between nodes and arcs in digraphs (Box 3.1). If an arc joins two nodes, the nodes are said to be adjacent and incident with the arc.

3.3.1 TIGER

An early application of topology in preparing geographically referenced data is the TIGER (Topologically Integrated Geographic Encoding and Referencing) database from the U.S. Census Bureau (Broome and Meixler 1990). The TIGER database contain legal and statistical area boundaries such as counties, census tracts, and block groups, which can be linked to the census data, as well as roads, railroads, streams, water bodies, power lines, and pipelines. The database also includes the address range on each side of a street segment.

In the TIGER database, points are called 0-cells, lines 1-cells, and areas 2-cells (Figure 3.5). Each 1-cell in a TIGER file is a directed line, meaning that the line is directed from a starting point toward an end point with an explicit left and right side. Each 2-cell and 0-cell has knowledge of the 1-cells associated with it. In other words, the TIGER database includes the spatial relationships between points, lines, and areas. Using the built-in spatial relationships, we can associate a block group with the streets or roads that make up its boundary. Likewise, we can identify an address on either the right side or the left side of a street (Figure 3.6).

The TIGER database, with its Census 2000 version, is available for download at the Census Bureau's website (http://www.census.gov/). Topol-ogy used in the TIGER database is also included in the USGS digital line graph (DLG) products. DLGs are digital representations of point, line, and area features from the USGS quadrangle maps including roads, streams, boundaries, and contours.

3.3.2 ESRI's Coverage Model

Besides work by federal agencies, commercial GIS vendors such as ESRI, Inc. have also used topology to develop proprietary vector data formats.

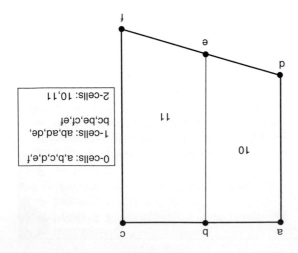

Figure 3.5
Topology in the TIGER database involves 0-cells or points, 1-cells or lines, and 2-cells or areas.

0-cells: a,b,c,d,e,f
1-cells: ab,ad,de, bc,be,cf,ef
2-cells: 10,11

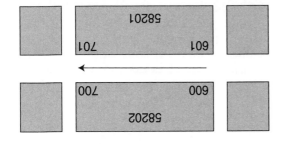

Figure 3.6
Address ranges and ZIP codes in the TIGER database have the right- or left-side designation based on the direction of the street.

ESRI, Inc. introduced the coverage model in the 1980s to separate GIS from CAD (computer-aided design) at the time. AutoCAD by Autodesk was, and still is, the leading CAD package. A data format used by AutoCAD for transfer of data files is called DXF (drawing exchange format). DXF maintains data in separate layers and allows the user to draw each layer using different line symbols, colors, and text. But DXF files do not support topology.

Coverage is a topology-based vector data format. A coverage can be a point coverage, line coverage, or polygon coverage. The coverage model supports three basic topological relationships (Environmental Systems Research Institute, Inc. 1998):

- **Connectivity:** Arcs connect to each other at nodes.
- **Area definition:** An area is defined by a series of connected arcs.
- **Contiguity:** Arcs have directions and left and right polygons.

Other than the use of terms, the above three topological relationships are similar to the topological relationships in the TIGER database.

For example, a road network for traffic volume analysis is typically a topology-based line coverage. The connectivity relationship ensures that roads (arcs) meet perfectly at road junctions (nodes). And the contiguity relationship makes it possible to distinguish northbound from southbound roads and to associate traffic analysis zones on each side of the road.

3.3.3 Coverage Data Structure

The coverage model incorporates the topological relationships into the structure of feature data. The data structure of a point coverage is simple: It contains feature identification numbers (IDs) and pairs of x- and y-coordinates (Figure 3.7).

Figure 3.8 shows the data structure of a line coverage. The starting point of an arc is the from-node and the end point the to-node. The arc-node list sorts out the arc–node relationship. For example, arc 2 has 12 as the from-node and 13 as the to-node.

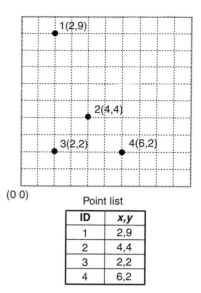

Point list

ID	x,y
1	2,9
2	4,4
3	2,2
4	6,2

Figure 3.7
The data structure of a point coverage.

The arc-coordinate list shows the x-, y-coordinates of the from-node, the to-node, and other points (vertices) that make up each arc. For example, arc 3 consists of the from-node at (2, 9), the to-node at (4, 2), and two vertices at (2, 6) and (4, 4). Arc 3 therefore has three line segments.

Figure 3.9 shows the data structure of a polygon coverage. The polygon/arc list shows the relationship between polygons and arcs. For example, arcs 1, 4, and 6 connect to define polygon 101. Polygon 104 differs from the other polygons because it is surrounded by polygon 102. To show that polygon 104 is a hole within polygon 102, the arc list for polygon 102 contains a zero to separate the external and internal boundaries. Polygon 104 is also an isolated polygon consisting of only one arc (7). A node (15) is placed along the arc to be the beginning and end node. Polygon 100, which is outside the map area, is the external or universe polygon.

The left/right list in Figure 3.9 shows the relationship between arcs and their left and right polygons. For example, arc 1 is a directed line from node 13 to node 11 and has polygon 100 as the polygon on the left and polygon 101 as the polygon

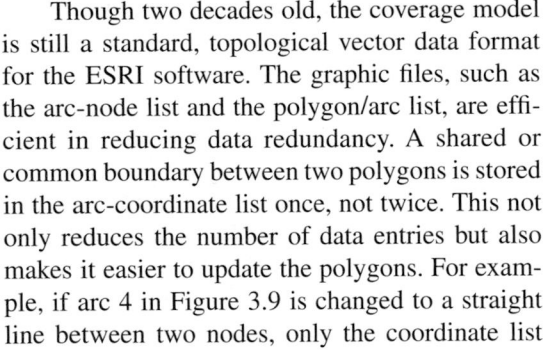

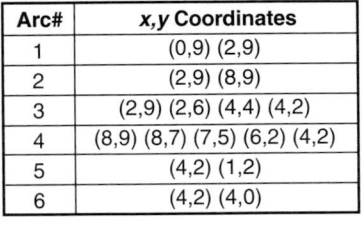

Arc-node list

Arc#	F-node	T-node
1	11	12
2	12	13
3	12	15
4	13	15
5	15	14
6	15	16

Arc-coordinate list

Arc#	x,y Coordinates
1	(0,9) (2,9)
2	(2,9) (8,9)
3	(2,9) (2,6) (4,4) (4,2)
4	(8,9) (8,7) (7,5) (6,2) (4,2)
5	(4,2) (1,2)
6	(4,2) (4,0)

Figure 3.8
The data structure of a line coverage.

on the right. The arc-coordinate list in Figure 3.9 shows the nodes and vertices that make up each arc.

Though two decades old, the coverage model is still a standard, topological vector data format for the ESRI software. The graphic files, such as the arc-node list and the polygon/arc list, are efficient in reducing data redundancy. A shared or common boundary between two polygons is stored in the arc-coordinate list once, not twice. This not only reduces the number of data entries but also makes it easier to update the polygons. For example, if arc 4 in Figure 3.9 is changed to a straight line between two nodes, only the coordinate list for arc 4 needs to be changed.

3.3.4 Importance of Topology

Topology-based data sets require additional data files to store the spatial relationships between features. This naturally raises the question: What are the advantages of having topology built into a data set? Increasingly, GIS users, who need to put together a database, are asking the question of whether to have topology or not (Box 3.2).

Topology has at least two main advantages. The first advantage is the assurance of data quality. In fact, data quality was the main reason that the TIGER database turned to topology in the first place. The topological relationships enable us to detect errors such as lines that do not meet correctly or polygons that are not closed properly. These kinds of errors must be corrected to avoid incomplete features and to ensure data integrity. For example, a shortest path analysis requires roads to meet correctly. If a gap exists on a supposedly continuous road, the analysis will take a circuitous route to avoid the gap.

Second, topology can enhance GIS analysis. A current popular GIS application is address geocoding, a process of plotting street addresses as point features on a map. Address geocoding providers typically use the TIGER database as a reference because the database not only has address ranges but also separates them according to the left or right side of the street. The built-in topology in the TIGER database makes it possible to plot street addresses correctly.

There are other types of analyses that can benefit from using topological data sets. Traffic volume

Box 3.2 | **Topology or No Topology**

T he decision on topology depends on the GIS project. For some projects, topological functions are not necessary; for others they are a must. For example, a producer of GIS data will find it absolutely necessary to use topology for error checking and for ensuring that lines meet correctly and polygons are closed properly. Likewise, a GIS analyst working with transportation and utility networks will want to use topological data for data analysis.

Ordnance Survey (OS) is perhaps the first major GIS data producer to offer both topological and nontopological data to end users. OS MasterMap is a new framework for the referencing of geographic information in Great Britain (**http://www.ordnancesurvey. co.uk/oswebsite/**). MasterMap has two types of polygon data: independent and topological polygon data. Independent polygon data duplicate the coordinate geometry shared between polygons. In contrast, topological polygon data include the coordinate geometry shared between polygons only once and reference each polygon by a set of line features. The reference of a polygon by line features is similar to the polygon/arc list discussed in Section 3.3.3.

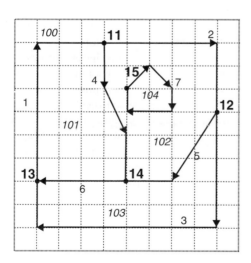

Left/right list

Arc#	L-poly	R-poly
1	100	101
2	100	102
3	100	103
4	102	101
5	103	102
6	103	101
7	102	104

Polygon/arc list

Polygon #	Arc#
101	1,4,6
102	4,2,5,0,7
103	6,5,3
104	7

Arc-coordinate list

Arc#	x,y Coordinates
1	(1,3) (1,9) (4,9)
2	(4,9) (9,9) (9,6)
3	(9,6) (9,1) (1,1) (1,3)
4	(4,9) (4,7) (5,5) (5,3)
5	(9,6) (7,3) (5,3)
6	(5,3) (1,3)
7	(5,7) (6,8) (7,7) (7,6) (5,6) (5,7)

Figure 3.9
The data structure of a polygon coverage.

analysis is similar to address geocoding because traffic volumes are also directional. Another example is deer habitat analysis, which often involves edges between habitat types, especially edges between old growth and clear-cuts (Chang et al. 1995). If edges are coded with left and right polygons in a topology-based data set, specific habitat types along edges can be easily tabulated and analyzed.

Recent developments in GIS suggest a renewed interest in topology. As discussed in Chapter 4, the geodatabase data model has introduced a set of 25 topology rules. Some of these rules apply to two or more data sets. Therefore, one can build topology between a county map and a census tract map to ensure that counties and census tracts share common (coincident) boundaries. Although the data model is different, topology continues to play the role of ensuring the quality of GIS data.

3.4 NONTOPOLOGICAL VECTOR DATA

GIS companies introduced topology two decades ago to separate GIS from CAD. But in less than a decade, the same companies adopted nontopological data format as one of the standard, nonproprietary data formats. As in many areas of GIS, the choice of data format is driven by the competitive nature of the GIS market. When one company adopts a new data format, other companies follow. As a result, commercial GIS packages such as ArcView, ArcGIS, MapInfo, and GeoMedia all have adopted nontopological data formats since the 1990s.

Shapefile is a standard nontopological data format used in ESRI products. Although the shapefile treats a point as a pair of x-, y-coordinates, a line as a series of points, and a polygon as a series of lines, no files describe the spatial relationships between these geometric objects. Shapefile polygons actually have duplicate arcs for the shared boundaries and can overlap one another. Rather than having multiple files as for an ArcInfo coverage, the geometry of a shapefile is stored in two basic files: the *.shp* file stores the feature geometry,

and the *.shx* file maintains the spatial index of the feature geometry.

Nontopological data such as shapefiles have two main advantages. First, they can display more rapidly on the computer monitor than topology-based data (Theobald 2001). This advantage is particularly important for people who use, rather than produce, GIS data. Second, they are nonproprietary and interoperable, meaning that they can be used across different software packages (e.g., MapInfo can use shapefiles and ArcView can use MapInfo Interchange Format files). GIS users have pushed for interoperability since the early 1990s. The push has resulted in the establishment of Open GIS Consortium, Inc., a nonprofit, international, voluntary consensus standards organization in 1994 (**http://www.opengis.org/**). Interoperability has been a primary mission of Open GIS Consortium, Inc. from the start. The introduction of nontopological data in the early 1990s was perhaps a response to the call for interoperability.

Shapefiles can be converted to coverages, and vice versa. The conversion from a shapefile to a coverage requires the building of topological relationships and the removal of duplicate arcs. The conversion from a coverage to a shapefile is simpler. But if a coverage has topological errors—such as lines not joined perfectly—the errors can lead to problems of missing features in the shapefile. Figure 3.10a shows a coverage that has errors in line joining. After the coverage is converted to a shapefile, all lines that have errors disappear in the shapefile (Figure 3.10b). Figure 3.10 serves as an example of the importance of topology in maintaining data integrity.

3.5 DATA MODELS FOR COMPOSITE FEATURES

Composite features refer to those spatial features that are better represented as composites of points, lines, and polygons. ESRI's coverage model, for example, includes such composite features as TINs (triangulated irregular networks), regions, and routes.

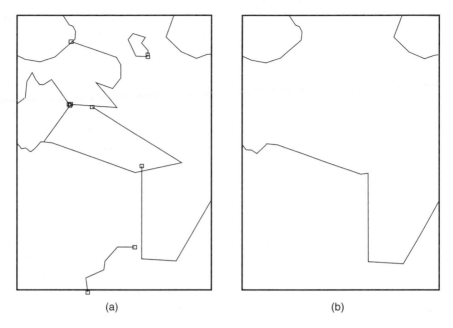

(a) (b)

Figure 3.10

A polygon coverage, shown in *(a),* has topological errors. Each small square symbol represents an error caused by lines that do not meet correctly. The shapefile, shown in *(b),* is converted from the polygon coverage.

3.5.1 TIN

The **TIN** data model approximates the terrain with a set of nonoverlapping triangles (Figure 3.11). Each triangle in a TIN assumes a constant gradient. Flat areas of the land surface have fewer but larger triangles. Areas with high variability in elevation have denser but smaller triangles.

The inputs to a TIN include point, line, and area features. Point features are elevation points with x, y, and z values. The x, y values represent the location of a point, and the z value represents the elevation at the point. Elevation points can come from a variety of sources including stereo air photos, survey data, and digital elevation models (DEMs). From these irregularly sampled elevation points, an initial TIN can be constructed using **Delaunay triangulation,** an iterative process of connecting points with their two nearest neighbors to form triangles that are as equiangular as possible (Watson and Philip 1984; Tsai 1993). This initial TIN can then be improved in terms of its

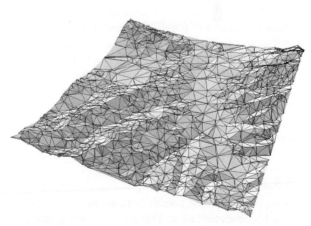

Figure 3.11

A TIN uses a series of nonoverlapping triangles to approximate the terrain.

approximation of the surface by incorporating line features such as streams, ridge lines, and roads and by area features such as lakes and reservoirs.

A TIN prepared from the above procedure comprises three types of geometric objects: triangles (faces), points (nodes), and lines (edges). The TIN data structure therefore includes the triangle number, the number of each adjacent triangle, and data files showing the lists of points, edges, as well as the x, y, and z values of each elevation point (Figure 3.12).

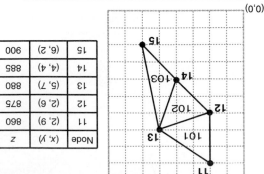

(0,0)

Node	(x, y)	z
11	(2, 9)	860
12	(2, 6)	875
13	(5, 7)	880
14	(4, 4)	885
15	(6, 2)	900

Triangle	Node list	Neighbors
101	11, 13, 12	--, 102, --
102	13, 14, 12	103, --, 101
103	13, 15, 14	--, --, 102

Figure 3.12
The data structure of a TIN.

The TIN is a data model designed for terrain mapping and analysis. It is an alternative to DEMs and contour lines for representing the land surface. Chapter 14 on terrain mapping and analysis has more detailed discussions on the making and use of TINs.

3.5.2 Regions

A region is defined here as a geographic area with similar characteristics. The concept of regions is well established in such disciplines as geography, landscape ecology, and forestry (Berry 1968; Bailey 1983; Forman and Godron 1986). Uniform regions can apply to a variety of physical and cultural phenomena including landforms, natural vegetation, forest fires, ethnic groups, and religion. The concept of regions has also been incorporated into the structure of hierarchical regions, a structure that divides the Earth's surface into progressively smaller regions of increasingly uniform characteristics. Well-known examples of hierarchical regions are census units (Figure 3.13), hydrologic units, and ecological units (Box 3.3).

A data model for geographic regions must be able to handle two spatial characteristics: a region may have spatially joint or disjoint areas, and regions can overlap or cover the same area (Figure 3.14). The simple polygon coverage cannot handle either characteristic. Therefore, regions are organized as subclasses in polygon coverages and, through additional data files, regions are related to the underlying polygons and arcs.

Figure 3.15 shows the file structure for an example with four polygons, five arcs, and two regions. The region–polygon list relates the regions to the polygons. Region 101 consists of polygons 11 and 12. Region 102 has two components: one includes polygons 12 and 13, which are spatially joint, and the other consists of polygon 14, which is spatially disjoint from the other polygons. Region 101 overlaps region 102 in polygon 12. The region–arc list links the regions to the arcs. Region 101 has only one ring, which connects arcs 1 and 2. Region 102 has two rings: one connects arcs 3 and 4, and the other consists of arc 5.

Because region subclasses can be built on existing polygons and arcs, many government agencies have used the regions data model to create and store additional data layers for distribution. For example, the Clearwater National Forest uses region subclass historical fire perimeters, one region subclass per year (http://www.fs.fed.us/r1/clearwater/gis/library/library_w.htm). Another example is the National Hydrography Dataset (NHD), which uses region subclasses to store lakes, stream reaches, and inundation areas (http://nhd.usgs.gov/).

The regions data model has lost its importance in recent years. Both the shapefile and the geodatabase (Chapter 4) can have polygons with multiple spatially disjoint components. Therefore the newer data models have taken away the reason for having

Box 3.3 Hierarchical Regions

The U.S. Census Bureau compiles and distributes census data by state, county, census tract, block group, and block. These area units form a nested hierarchy: blocks are within a block group; block groups are within a census tract; and so on. The U.S. Geological Survey also organizes hydrologic units into a nested hierarchy. The system divides the United States at four spatial levels: regions, subregions, accounting units, and cataloging units. Each hydrologic unit has a unique code of two to eight digits, depending on the unit's classification within the hierarchy. An effort is now under way to increase the levels of hydrologic units from four to six.

Ecological units are more complex examples of hierarchical regions (Bailey 1983; Cleland et al. 1997). The National Hierarchical Framework of Ecological Units, adopted by the U.S. Forest Service in 1993, systematically maps areas of the Earth's surface at the scales of ecoregion, subregion, landscape, and land unit. The delineation of ecological units is based on the association of ecological factors. Climate and landform define ecoregion and subregion boundaries. Relief, geologic parent materials, and potential natural communities define landscape boundaries. And topography, soil characteristics, and plant associations define land unit boundaries.

Hierarchical regions are useful for incorporating spatial scale into GIS analysis. The modifiable areal unit problem (MAUP) suggests that the "modifiable" nature of area units used in spatial analysis can influence the analysis and modeling results (Openshaw and Taylor 1979). Therefore, users of census data must be aware of the scale effect while choosing area units for analysis. MAUP effects have been reported in a variety of fields including demographic analysis (Fotheringham and Wong 1991), wildlife management (Svancara et al. 2002), and transportation planning (Chang et al. 2002).

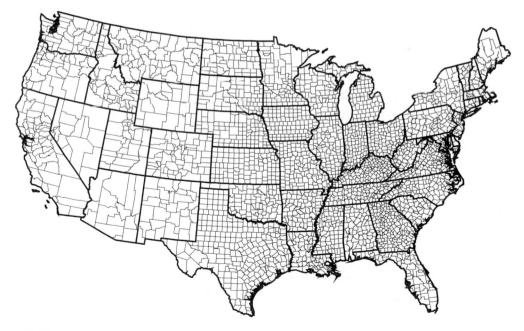

Figure 3.13
A hierarchy of counties and states in the conterminous United States.

along roads. Natural resource agencies also use linear measures to record water quality and fishery conditions along streams. These linear attributes, called **events**, must be associated with routes so that they can be displayed and analyzed with other spatial features.

The coverage model stores routes as subclasses in a line coverage in the same way as region subclasses in a polygon coverage. A route subclass is a collection of sections. A **section** refers directly to lines (i.e., arcs) in a line coverage and positions along lines. Because lines are made of a series of x-, y-coordinates based on a coordinate system, this means that a section is also measured in coordinates and its length can be derived from its reference lines.

Figure 3.16 shows a route (Route-ID = 1) in a thick shaded line, which is built on a line coverage. The route has three sections, and the section table relates them to the arcs in the line coverage. Section 1 (Section-ID = 1) covers the entire length of arc 7; therefore, the from-position (F-POS) is 0 percent and the to-position (T-POS) is 100 percent. The from-measure (F-MEAS) of section 1 is 0 because it is the beginning point of the route. The to-measure (T-MEAS) of section 1 is 40 units (meters or feet, depending on the coverage's measurement

region subclasses in the coverage model. The concept of regions, however, is still important in GIS.

3.5.3 Routes

A **route** is a linear feature such as a highway, a bike path, or a stream but, unlike other linear features, a route has a measurement system that allows linear measures to be used on a projected coordinate system. Transportation agencies normally use linear measures from known points such as the beginning of a highway, a milepost, or a road intersection to locate accidents, bridges, and pavement conditions

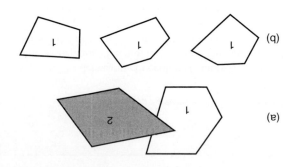

Figure 3.14
The regions data model allows overlapped regions (a) and spatially disjoint components (b).

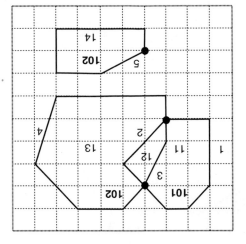

Region-polygon list

Region #	Polygon #
101	11
101	12
102	12
102	13
102	14

Region-arc list

Region #	Ring #	Arc #
101	1	1
101	1	2
102	1	3
102	1	4
102	2	5

Figure 3.15
The data structure of a region subclass.

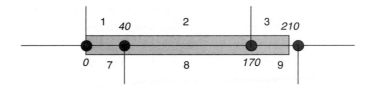

Route-ID	Section-ID	Arc-ID	F-MEAS	T-MEAS	F-POS	T-POS
1	1	7	0	40	0	100
1	2	8	40	170	0	100
1	3	9	170	210	0	80

Figure 3.16
The data structure of a route subclass.

unit), which is measured from the line coverage. Section 2 also covers the entire length of arc 8 over a distance of 130 units. Its from-measure and to-measure continue from section 1. Section 3 covers 80 percent of arc 9, thus its to-position is coded 80. The to-measure of section 3 is computed by adding 80 percent of the length of arc 9 (i.e., 80% of 50 units) to its from-measure. Combining the three sections that are directly linked to the arcs in the line coverage, the route has a total length of 210 units (40 + 130 + 40).

The shapefile data format allows measured polylines to be used as routes. A measured polyline is a set of lines that store *x*- and *y*-coordinates as well as a measure (*m*) value. The *m* values provide the linear measures for plotting events along lines. But because the *m* values are usually entered manually, it is difficult to use polyline shapefiles as routes.

Routes are useful for a variety of applications. A public transit agency can build different bus routes on the same street network. A state transportation department can use different routes for different classes of highways such as interstate, U.S., and state. And, although the term *route* may not be most appropriate, a mapping agency can include different types of political boundaries (e.g., state and county boundaries) in different route subclasses, mainly for the purpose of data distribution. Chapter 17 has more detailed discussions on the creation and use of routes and events.

KEY CONCEPTS AND TERMS

Arc: A line connected to two end points.

Area: A spatial feature that is represented by a series of lines and has the geometric properties of size and perimeter. Also called *polygon, face,* or *zone.*

Area definition: A topological relationship used in ESRI's coverage model, which stipulates that an area is defined by a series of connected arcs.

Connectivity: A topological relationship used in ESRI's coverage model, which stipulates that arcs connect to each other at nodes.

Contiguity: A topological relationship used in ESRI's coverage model, which stipulates that arcs have directions and left and right polygons.

Coverage: A topological vector data format used in ESRI products.

Delaunay triangulation: An iterative process of connecting points with their two nearest neighbors to form triangles as equiangular as possible in a triangulated irregular network (TIN).

Event: An attribute that can be associated and displayed with a route.

Georelational data model: A GIS data model that stores spatial data and attribute data in two separate but related file systems.

Graph theory: A subfield of mathematics that uses diagrams or graphs to study the arrangements of objects and the relationships between objects.

Line: A spatial feature that is represented by a series of points and has the geometric properties of location and length. Also called *arc, edge, link,* or *chain.*

Node: The beginning or end point of a line.

Point: A spatial feature that is represented by a pair of coordinates and has only the geometric property of location. Also called *node* or *vertex.*

Route: A linear feature that allows linear measures to be used on a projected coordinate system.

Section: A part of a route that refers directly to the underlying arcs and positions along arcs.

Shapefile: A nontopological vector data format used in ESRI products.

Topology: A subfield of mathematics that studies properties of geometric objects, which remain invariant under certain transformations such as bending or stretching.

Triangulated irregular network (TIN): A vector data model that approximates the terrain with a set of nonoverlapping triangles.

Vector data model: A data model that uses points and their *x*-, *y*-coordinates to construct spatial features.

REVIEW QUESTIONS

1. Find the GIS data clearinghouse for your state at the Geospatial One-Stop website (**http://www.geo-one-stop.gov/**). Go to the clearinghouse website. What data format(s) does the website use for delivering vector data?

2. Continue with question 1. Name two data sets available at your statewide website for each feature type of point, line, and area.

3. Name the three data formats that ESRI, Inc. has developed for vector data over the past 20 years.

4. The georelational data model uses a split system to store vector data. What does "split system" mean?

5. Name the three types of simple features used in GIS and their geometric properties.

6. Draw a stream network, and show how the topological relationships of connectivity and contiguity can be applied to the network.

7. How many arcs connect at node 12 in Figure 3.8?

8. How many line segments does arc 4 have in Figure 3.8?

9. Suppose an arc (arc 8) is added to Figure 3.9 from node 13 to node 11. Write the polygon/arc list for the new polygons and the left/right list for arc 8.

10. Describe how the data file structure of a polygon coverage enforces the coverage model's topological relationships.

11. Explain the importance of topology in GIS.

12. What are the main advantages of using shapefiles?

13. Draw a small TIN to illustrate that it is a composite of simple features.

14. In what ways does a region of the regions data model differ from a polygon of the coverage model.

15. Explain how a route is related to arcs through sections in Figure 3.16.

APPLICATIONS: GEORELATIONAL VECTOR DATA MODEL

This applications section consists of three tasks. In Task 1, you will import an ArcInfo interchange file to a coverage, convert the coverage to a geodatabase feature class, and convert the geodatabase feature class to a shapefile. Task 1 also lets you examine the data structure of a coverage, a geodatabase feature class, and a shapefile. In Task 2, you will view regions and routes that are part of the national hydrography data sets in the United States. In Task 3, you will view a TIN in ArcCatalog and ArcMap.

Task 1: Examine the Data File Structure of Coverage and Shapefile

What you need: *land.e00*, an ArcInfo interchange file.

In Task 1, you will use the interchange file *land.e00* to create a coverage and a shapefile. You will then view the data sets associated with these two data models in ArcCatalog and examine their data file structure using the Windows Explorer.

1. Start ArcCatalog, and make connection to the Chapter 3 database. Click the Show/Hide ArcToolbox Window button to open the ArcToolbox window. Double-click the Import From Interchange File tool in the Coverage Tools/Conversion/To Coverage toolset. Navigate to the Chapter 3 database and select *land.e00* for the input interchange file. Specify *land* for the output dataset. Click OK to execute the conversion. An alternative is to use the Import from Interchange File tool of ArcView 8x Tools, available in ArcCatalog's View menu.

2. The coverage *land* now appears in the Catalog tree. (If not, select Refresh from the View menu in ArcCatalog.) Click the plus sign to expand *land*. The coverage contains four feature classes: *arc, label, polygon,* and *tic*. On the Preview tab, you can preview each class by first highlighting it in the Catalog

tree. *arc* shows lines (arcs); *label,* the label point for each polygon; *polygon,* polygons; and *tic,* the tics or control points in *land*. Notice that the symbols for the four classes correspond to the feature type.

3. Right-click *land* in the Catalog tree and select Properties. The Coverage Properties dialog has four tabs: General, Projection, Tics and Extent, and Tolerances. The General tab shows that the topology exists for the polygon feature class. The Projection tab shows an unknown coordinate system. The Tics and Extent tab shows the tics and area extent of the coverage. And the Tolerances tab shows various tolerance values for building topology and editing.

4. Right-click *polygon* and select Properties. The Coverage Feature Class Properties dialog has the General, Items, and Relationships tabs. The Items tab shows the fields in the attribute table.

5. Data files associated with *land* reside in two folders in the Chapter 3 database: land and INFO. You can use the Windows Explorer to view these files. The land folder contains arc data files (.adf). Some of these graphic files are recognizable by name, such as arc.adf for the arc list and pal.adf for the polygon/arc list. The INFO folder, which is shared by other coverages in the same database, contains attribute data files such as arc0000.dat, arc0000.nit, and so on. All files in both folders are binary files and cannot be read.

6. This step is to convert *land* to a polygon shapefile. There are at least two options for the conversion. First, you can use the Feature Class to Shapefile (multiple) tool in the Conversion Tools/To Shapefile toolset. The tool can convert a coverage's feature classes (but not those with preliminary topology) to shapefiles. Second, you can use the Export function in a data set's context menu. Here

you will use the second option. Right-click *land_polygon* (the polygon feature class of *land*), point to Export, and select To Shapefile (single). In the next dialog, select the Chapter 3 database for the output location, and enter *land_polygon* for the output feature class name. Click OK. The operation creates *land_polygon.shp* and adds the shapefile to the Catalog tree.

7. Right-click *land_polygon.shp* in the Catalog tree and select Properties. The Shapefile Properties dialog has the General, Fields, and Indexes tabs. The Fields tab shows the fields in the shapefile. The Indexes tab shows that the shapefile has a spatial index, which can increase the speed for drawing and data query.

8. The *land_polygon* shapefile is associated with a number of data files. You can use the Windows Explorer to view these files in the Chapter 3 database. Among these files, *land_polygon.shp* is the shape (geometry) file, *land_polygon.dbf* is an attribute data file in dBASE format, and *land_polygon.shx* is the spatial index file.

Q1. Describe in your own words the difference between a coverage and a shapefile in terms of data structure?

Q2. The coverage data model uses a split system to store the spatial and attribute data. Use *land* as an example and name the two systems.

Task 2: View Regions and Routes

What you need: *nhd*, a hydrography data set for the 8-digit watershed (18070105) in Los Angeles, California.

Hydrography data sets are available for download from the National Hydrography Dataset website maintained by the U.S. Geological Survey (USGS) and the U.S. Environmental Protection Agency (EPA) (**http://nhd.usgs.gov/**). *nhd* is a coverage with built-in regions and routes. Task 2 lets you view these composite features as well as the simple features of arcs and polygons.

1. Expand *nhd* in the Catalog tree. The *nhd* coverage contains 11 layers: *arc, label, node, polygon, region.lm, region.rch, region.wb, route.drain, route.lm, route.rch,* and *tic.*

2. Launch ArcMap. Rename the data frame *nhd1,* and add *polygon, region.lm, region.rch,* and *region.wb* to *nhd1.* The *polygon* layer consists of all polygons, on which the three region subclasses are built. Right-click *nhd region.lm* and select Open Attribute Table. The field FTYPE shows that *nhd region.lm* consists of inundation areas.

Q3. Regions from different region subclasses may overlap. Do you see any overlaps between the three region subclasses of the *nhd* coverage?

3. Insert a new data frame and rename it *nhd2.* Add *arc, route.drain, route.lm,* and *route.rch* to *nhd2.* The *arc* layer consists of all arcs, on which the three route subclasses are built. Right-click *nhd route.rch* and select Open Attribute Table. Each record in the table represents a reach, a segment of surface water having a unique identifier. These reaches provide the link between water-related data available from the EPA and the drainage network.

Q4. Different route subclasses can be built on the arcs. Do you see any arcs used by different route subclasses of the *nhd* coverage?

4. Each layer in *nhd* can be exported to a shapefile or a geodatabase feature class. For example, you can right-click *nhd route.rch,* point to Data, and select Export Data. The Export Data dialog lets you save the data set as either a shapefile or a geodatabase feature class.

Task 3: View TIN

What you need: *emidalat,* a TIN prepared from a digital elevation model.

1. Click *emidalat* in the Catalog tree. The Contents tab shows that *emidalat* is a TIN.

2. Insert a new data frame in ArcMap. Rename the data frame Task 3, and add *emidatin* to Task 3. Right-click *emidatin,* and select Properties. On the Source tab, the Data Source frame shows the number of nodes and triangles as well as the Z (elevation) range.

Q5. How many triangles does *emidatin* have?

3. On the Symbology tab, uncheck Elevation and click the Add button in the Show frame. In the next dialog, highlight Edges with the same symbol, click Add, and then click Dismiss. Click OK to dismiss the Layer Properties. The ArcMap window now shows the triangles that make up *emidatin*. Using the same procedure above, you can view nodes that make up *emidatin*.

Challenge Task

What you need: *fire* and *highway*.

The Chapter 3 database includes *fire,* a polygon coverage with region subclasses, and *highway,* a line coverage with a route subclass.

1. Launch ArcMap if necessary. Insert a data frame and rename it Fire. Add *polygon, regions.fire1*, *regions.fire12*, and *regions. fire123* of the *fire* coverage to Fire. The three region subclasses contain regions that were burned once, twice, and thrice respectively.

Q1. Are regions in the *fire* coverage made of spatially disjoint components?

Q2. Do regions in the *fire* coverage overlap?

2. Insert a new data frame in ArcMap. Rename the data frame Highway and add *arc* and *route.fastroute* of *highway* to the data frame. Use a thicker line symbol for *highway route.fastroute*.

Q3. How many line segments is *highway route.fastroute* made of? (Tip: Zoom in around *highway route.fastroute*, and use the Identify tool to identify *highway route.fastroute*.)

REFERENCES

Bailey, R. G. 1983. Delineation of Ecosystem Regions. *Environmental Management* 7: 365–73.

Berry, B. J. L. 1968. Approaches to Regional Analysis: A Synthesis. In B. J. L. Berry and D. F. Marble, eds., *Spatial Analysis: A Reader in Statistical Geography,* pp. 24–34. Englewood Cliffs, NJ: Prentice-Hall.

Broome, F. R., and D. B. Meixler. 1990. The TIGER Data Base Structure. *Cartography and Geographic Information Systems* 17: 39–47.

Chang, K., D. L. Verbyla, and J. J. Yeo. 1995. Spatial Analysis of Habitat Selection by Sitka Black-Tailed Deer in Southeast

Alaska. *Environmental Management* 19: 579–89.

Chang, K., Z. Khatib, and Y. Ou. 2002. Effects of Zoning Structure and Network Detail on Traffic Demand Modeling. *Environment and Planning B* 29: 37–52.

Cleland, D. T., R. E. Avers, W. H. McNab, M. E. Jensen, R. G. Bailey, T. King, and W. E. Russell. 1997. National Hierarchical Framework of Ecological Units. In M. S. Boyce and A. Haney, eds., *Ecosystem Management Applications for Sustainable Forest and Wildlife Resources,* pp. 181–200. New Haven, CT: Yale University Press.

Environmental Systems Research Institute, Inc. 1998. *Understanding GIS: The ARC/INFO Method.* Redlands, CA: ESRI Press.

Forman, R. T. T., and M. Godron. 1986. *Landscape Ecology.* New York: Wiley.

Fotheringham, A. S., and D. W. S. Wong. 1991. The Modifiable Areal Unit Problem in Multivariate Statistical Analysis. *Environment and Planning A* 23: 1025–44.

Massey, W. S. 1967. *Algebraic Topology: An Introduction.* New York: Harcourt, Brace & World.

Openshaw, S., and P. J. Taylor. 1979. A Million or So Correlation

Coefficients: Three Experiments on the Modifiable Areal Unit Problem. In N. Wrigley, ed., *Statistical Applications in the Spatial Sciences,* pp. 127–44. London: Pion.

Robinson, A. H., J. L. Morrison, P. C. Muehrcke, A. J. Kimerling, and S. C. Guptill. 1995. *Elements of Cartography,* 6th ed. New York: Wiley.

Svancara, L. K., E. O. Garton, K. Chang, J. M. Scott, P. Zager, and M. Gratson. 2002. The Inherent Aggravation of Aggregation: An Example with Elk Aerial Survey Data. *Journal of Wildlife Management* 66: 776–87.

Theobald, D. M. 2001. Topology Revisited: Representing Spatial Relations. *International Journal of Geographical Information Science* 15: 689–705.

Tsai, V. J. D. 1993. Delaunay Triangulations in TIN Creation: An Overview and a Linear Time Algorithm. *International Journal of Geographical Information Systems* 7: 501–24.

Watson, D. F., and G. M. Philip. 1984. Systematic Triangulations. *Computer Vision, Graphics, and Image Processing* 26: 217–23.

Wilson, R. J., and J. J. Watkins. 1990. *Graphs: An Introductory Approach.* New York: Wiley.

OBJECT-BASED VECTOR DATA MODEL

The georelational data model covered in Chapter 3 represents one type of the vector data model. Still a dominant model in GIS, the georelational data model uses a split system to store spatial data and attribute data. Recent advances in computer technology, however, have made it possible to store spatial data and attribute data together in a single system. This new data model is called the object-based data model.

The object-based data model uses objects to represent and organize spatial features. For example, an object may represent a timber stand. The geometry (spatial data) of the stand is stored as an attribute along with other attributes (e.g., age class, crown diameter) in a record. This eliminates the use of a split system and the need for data synchronization. More importantly, an object can have properties and methods. For example, the stand object can have the properties of shape and extent. These properties and methods can not only capture the characteristics of the real-world objects but also enhance the way we interact with these objects in a GIS.

A new data model means a new set of concepts, terms, and data formats to the user. The new object-based data model is no exception. But the migration from the georelational to the object-based data model should be relatively easy because it is intuitive to think of spatial features as objects.

Chapter 4 has five sections. Section 4.1 introduces the object-based data model, classes, and relationships between classes. Section 4.2 focuses on the geodatabase data model, a new data model from ESRI, Inc. Section 4.3 explains interfaces. Section 4.4 describes topology rules. Section 4.5 covers the advantages of using the geodatabase over the traditional coverage model.

4.1 OBJECT-BASED DATA MODEL

Object-oriented technology has impacted many fields for more than two decades. User interface such as the Windows environment is perhaps the most obvious application example of object-oriented technology. We use menus, icons, and dialog boxes, instead of command lines, to interact with a software package such as Microsoft Word. When we click on a menu, the menu shows available operations for the object (e.g., copy or delete a Word document). This kind of user interface is efficient for both the user and the software developer: The user can simply point and click to complete a task, and the software developer can restrict operation choices to only those that are specifically defined for the object, thus avoiding unnecessary operational errors.

Companies specializing in utilities and communications have adopted object-oriented technology in their software development since the early 1990s. They include SmallWorld (now GESmallWorld, **http://www.gepower.com/prod_serv/subst_ntwk. htm**) and Laser-Scan (**http://www.laserscan.com/**). Popular GIS software packages such as MapInfo and ArcView have also incorporated object-oriented technology into the user interface design and the macro language (e.g., Avenue for ArcView and MapX for MapInfo). The use of the graphical user interface in these software packages has in fact helped bring GIS to mainstream use in the 1990s.

The introduction of the geodatabase data model by ESRI, Inc. in 1999 is different from previous applications of object-oriented technology in GIS. It represents a new beginning, at least for ESRI products, of embracing an object-based, rather than a georelational, data model (Zeiler 1999). The geodatabase data model may have also promoted increased research activities in developing complex object-based models for 3-D, transportation, and other applications (Konoz and Adams 2002; Huang 2003; Shi et al. 2003).

The object-based data model treats spatial data as objects. An **object** can represent a spatial feature such as a road, a timber stand, or a hydro-logic unit. An object can also represent a road layer or the coordinate system that the road layer is based on. In fact, almost everything we work with in GIS can be represented as an object. Chapter 4 focuses on objects that are related to the vector data model.

The object-based data model differs from the georelational data model in two important aspects. First, the object-based data model stores both the spatial and attribute data of spatial features in a single system rather than a split system. The spatial data are stored in a special field along with other fields for attribute data. Figure 4.1, for example, shows a land-use data set. The field called *shape* stores the geometry, or the spatial data, of each land-use polygon. Using a single system to store both spatial and attribute data is considered a major breakthrough by GIS software developers. This breakthrough is important because GIS software developers must regularly deal with the issues of data storage and data file structure.

Second, the object-based data model allows a spatial feature (object) to be associated with a set of properties and methods. A **property** describes an attribute or characteristic of an object. A **method** performs a specific action. As described earlier, a timber stand (a feature object) can have

Objectid	Shape	Landuse_ID	Category	Shape_Length	Shape_Area
1	Polygon	1	5	14,607.7	5,959,800
2	Polygon	2	8	16,979.3	5,421,216
3	Polygon	3	5	42,654.2	21,021,728

Figure 4.1

The object-based data model stores each land-use polygon in a record. The Shape field stores the spatial data of land-use polygons. Other fields store attribute data such as Landuse_ID and Category.

the properties of shape and extent. It can also have the method of delete. Properties and methods directly impact how GIS operations are performed. Our work in an object-based GIS is in fact dictated by the properties and methods that have been defined for the objects in the GIS.

4.1.1 Classes

A **class** is a set of objects with similar attributes. Unless objects are organized into some hierarchical structure, the management of properties and methods in a GIS can become difficult, if not impossible. Operationally, a class defines the properties and methods of objects that are members of the class. For example, a class called *feature* can cover point, line, and polygon feature objects. The *feature* class defines the same properties (e.g., shape and extent) and same methods (e.g., delete) for all three types of objects.

Since the early stage of adopting object-oriented technology in GIS, the literature has proposed the principles of generalization and specialization for the grouping and differentiation of objects (Worboys et al. 1990; Egenhofer and Frank 1992; Worboys 1995). Generalization uses the commonality among objects to group objects into a higher-order class. In the previous example, the three feature types of point, line, and polygon are grouped into the same class through generalization. Specialization, on the other hand, differentiates objects of a given class by an attribute or attributes. For example, geometry can differentiate the three feature types (Figure 4.2).

The use of generalization and specialization therefore creates a hierarchical structure, in which objects are grouped into classes and classes are grouped into superclasses and subclasses.

4.1.2 Relationships Between Classes

Following the grouping of objects into classes, the next step is to sort out the relationships between classes. The object-oriented software design literature has suggested the relationships of association, composition, aggregation, type inheritance, and instantiation that can be established between classes (Larman 1997; Zeiler 1999).

Association defines how many instances of one class can be associated with the other class through multiplicity expressions at both ends of the relationship. Common multiplicity expressions are one (default), and one or more (1..*). For example, a street can be associated with one or more signal lights but it can only be associated with one coordinate system (Figure 4.3).

Aggregation describes the whole-part relationship between classes. Aggregation is a type of association except that the multiplicity at the composite ("whole") end is typically 1 and the multiplicity at the other ("part") end is zero or any positive integer. For example, a census tract is an aggregate of a number of census blocks.

Composition is similar to aggregation in that the multiplicity at the composite end is 1 and the multiplicity at the other end is zero or any positive integer. But the composite in a composition relationship solely owns the part. For example, a highway can have zero or any number of roadside rest areas. The highway controls the lifetime of the roadside rest area.

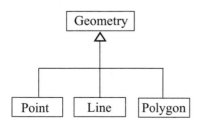

Figure 4.2
The *Geometry* property of the *Feature* class can differentiate the object types of *point, line,* and *polygon.*

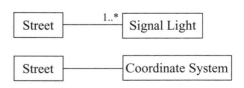

Figure 4.3
Two examples of class associations.

4.2 THE GEODATABASE DATA MODEL

The **geodatabase** data model is the third major data model offered by ESRI, Inc., following the coverage model in the 1980s and the shapefile in the 1990s. The geodatabase is part of **ArcObjects**, a collection of thousands of objects, properties, and methods. Based on the Component Object Model (COM) technology, ArcObjects provides the foundation for ArcGIS Desktop (Zeiler 2001; Ungerer and Goodchild 2002). ArcObjects is organized into over 50 groups of objects. Chapter 4 deals primarily with objects in the geodatabase group.

Instantiation means that an object of a class can be created from an object of another class. For example, a high-density residential area object can be created from a residential area object (Figure 4.5).

Type inheritance defines the relationship between a superclass and a subclass. A subclass is a member of a superclass and inherits the properties and methods of the superclass. But a subclass can have additional properties and methods to separate itself from other members of the superclass. For example, residential area is a member of built-up area but it can have properties such as lot size that separate residential area from commercial or industrial area (Figure 4.4).

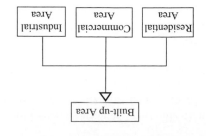

Figure 4.4
An example of type inheritance.

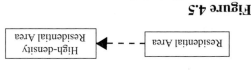

Figure 4.5
An example of instantiation.

4.2.1 Geometric Representation of Spatial Feature

The geodatabase data model uses the geometries of point, polyline, and polygon to represent vector-based spatial features (Zeiler 1999). A point feature may be a simple feature with a point or a multipoint feature with a set of points. A polyline feature is a set of line segments, which may or may not be connected. A polygon feature may be made of one or many rings. A ring is a set of connected, closed, nonintersecting line segments. The geodatabase is similar to the shapefile in terms of feature geometries (Box 4.1).

The geodatabase is also similar to the coverage model in terms of simple features. The difference between the two models lies mainly in the composite features of regions and routes (Box 4.1). The region subclass in the coverage model is no longer supported in the geodatabase. But the geometric characteristics of the region subclass are retained by the geodatabase because multipart polygons in a geodatabase can have spatially joint or disjoint components and can overlap each other. The route subclass in the coverage model is replaced by polylines with m (measure) values in the geodatabase data model. Instead of working through sections and arcs, the geodatabase uses m values for linear measures along a route. This is similar to the shapefile's measured polyline. But unlike the shapefile, the geodatabase stores the m values directly with the x- and y-coordinates in the geometry field (Figure 4.6).

Figure 4.7 shows an example of a three-part route in a geodatabase. The measure field directly records 0, 40, 170, and 210 along the route. These measures are based on a predetermined starting point, which, in this case, is the end point on the left.

Box 4.1 | **Geodatabase, Shapefile, and Coverage for ESRI Software**

The following table compares the terms used to describe geometries of the geodatabase, shapefile, and coverage.

Geodatabase	Shapefile	Coverage
Point	Point	Point
Multipoints	Multipoints	NA
Polyline	Polyline	Arc
Polygon	Polygon	Polygon
Polygon	Polygon	Region
Polyline with measure	Polyline with measure	Route
Feature class	Shapefile	Point, arc, polygon, tic
Feature dataset	NA	Coverage

	x	y	m
0	1,135,149	1,148,350	47.840
1	1,135,304	1,148,310	47.870
2	1,135,522	1,148,218	47.915

Figure 4.6
The linear measures (m) of a route are stored with x- and y-coordinates in a geodatabase. In this example, the m values are in miles, whereas the x- and y-coordinates are in feet.

4.2.2 Data Structure

The geodatabase data model distinguishes between "feature classes" and "feature datasets" in data structure (Figure 4.8). A **feature class** stores spatial data of the same geometry type. A **feature dataset** stores feature classes that share the same coordinate system and area extent. A feature class does not have to be included in a feature dataset. In that case, the feature class is called a standalone feature class. Feature classes that are included in a feature dataset typically participate in topological relationships with each other. (Chapter 8 introduces use of feature datasets for topological editing.)

A feature class is like a shapefile in having simple features, and a feature dataset is similar to a

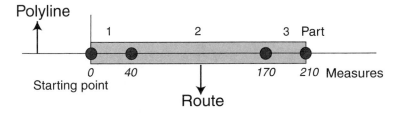

Figure 4.7
A route, shown here as a thicker, gray line, is built on a polyline with linear measures in a geodatabase.

In a geodatabase, feature classes can be standalone feature classes or members of a feature dataset.

Figure 4.8

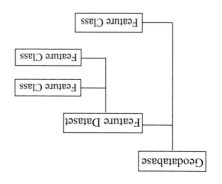

coverage in having multiple data sets that are based on the same coordinate system and area extent. But a feature dataset can contain different theme layers, whereas a coverage contains only different parts of a single layer such as arcs, nodes, and tics.

The hierarchical structure of feature dataset and feature class can be exploited for the purpose of data organization and management. For example, if a project involves two study areas, we can use two feature datasets, one for each study area. This simplifies data management operations such as copy and delete. Moreover, any intermediate data created through data query and analysis in the project will automatically be defined with the same coordinate system as the feature dataset, thus saving the time for defining the coordinate system of each new feature class.

4.3 INTERFACE

An **interface** represents a set of externally visible operations of an object. As mentioned earlier, an object can have properties and methods. But these properties and methods are hidden in the object. Therefore, to use the properties and methods of an object, we do not work with the object directly but work with an interface that has been implemented on the object (Figure 4.9).

Software developers can use thousands of objects, properties, and methods in ArcObjects to develop customized applications (Chang 2005). Most

Recent trends suggest that mapping agencies are also taking advantage of the hierarchical structure of feature dataset and feature class for data delivery. The National Hydrography Dataset (NHD) program, for example, distributes data in two feature datasets (Box 4.2) (**http://nhd.usgs.gov/data.html**). One feature dataset includes feature classes for stream reach applications, and the other includes feature classes of hydrologic units.

A geodatabase can be designed for single or multiple users. A personal geodatabase is for a single user. It stores data in a Microsoft Access database. A multiuser geodatabase stores data in a database management system such as Oracle, Microsoft SQL Server, IBM DB2, and Informix. Chapter 4 deals mainly with personal geodatabases.

Box 4.2 NHDinGEO

The National Hydrography Dataset program uses the acronym NHDinGEO for their new data in geo-databases. As of early 2004, the program's website offers a geodatabase called *NHD_Geo_July3.mdb* as sample data that can be downloaded. The geodatabase includes two feature datasets and a number of attribute tables. The Hydrography feature dataset has feature classes such as NHDFlowline, NHDWaterbody, and NHDPoint for stream reach applications. The

HydrologicUnits feature dataset has feature classes of basin, region, subbasin, subregion, subwatershed, and watershed. This geodatabase model is replacing the program's coverage model (called NHDinARC). The coverage model uses the subclasses of regions and routes to store some of the same feature classes in NHDinGEO. NHD may be setting a precedent for data delivery in the future.

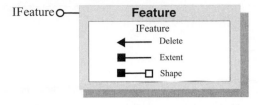

Figure 4.9

A *Feature* object implements the *IFeature* interface. *IFeature* has access to the properties of *Extent* and *Shape* and the method of *Delete*. Object-oriented technology uses symbols to represent interface, property, and method. The symbols for the two properties are different in this case because *Extent* is a read-only property whereas *Shape* is a read and write (by reference) property.

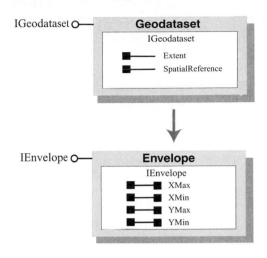

Figure 4.10

A *Geodataset* object supports *IGeodataset* and an *Envelope* object supports *IEnvelope*. Box 4.4 has a code fragment that shows how to use the interfaces to derive the area extent of a feature layer.

ArcGIS users, however, do not have to deal with ArcObjects directly. (Box 4.3 describes situations in which one may encounter ArcObjects while working with routine operations in ArcGIS.) This is because graphical user interfaces such as menus, icons, and dialogs have already been developed by ESRI, Inc. to access ArcObjects. Nevertheless, a basic knowledge of the interface is still useful for understanding the working of ArcGIS Desktop. The following covers three basic principles—encapsulation, inheritance, and polymorphism—underlying the concept of interface.

Encapsulation refers to the mechanism to hide the properties and methods of an object so that the object can only be accessed through the

predefined interfaces. Figure 4.10 shows how we can use two interfaces to derive the area extent of a feature layer (a type of *Geodataset*). We first access the *Extent* property via the *IGeodataset* interface that a *Geodataset* object supports. The *Extent* property returns an *Envelope* object, which implements *IEnvelope*. We then use the properties of *XMin*, *XMax*, *YMin*, and *YMax* on *IEnvelope* to derive the area extent. Box 4.4 shows a code fragment of how this is done using ArcObjects.

 Box **4.3** | **ArcObjects and ArcGIS**

ArcGIS Desktop is built on ArcObjects, a collection of objects. In ArcGIS, we typically access ArcObjects through the graphical user interface rather than the objects directly. Both ArcCatalog and ArcMap, however, have embedded Visual Basic Editor that can run macros using ArcObjects. Using macros, one can create customized commands, menus, and tools. The Object Browser in the Developer Tools and the ArcGIS Desktop Help offer information on ArcObjects and its use.

One dialog that we use frequently in ArcMap is to calculate field values such as the area and perimeter values of polygons. The dialog has an advanced option, which allows the user to enter VBA (Visual Basic for Applications) code. Task 1 of the applications section of Chapter 12 uses this advanced option.

Inheritance means that an object inherits properties and methods from the class to which the object belongs. Likewise, a subclass inherits properties and methods from the superclass to which the subclass belongs. Suppose that residential area is a superclass and low-density area and high-density area are its subclasses. All properties and methods that have been defined for the residential area class are inherited by both subclasses. Inheritance is therefore a principle that directly links the structural aspects and the behavioral aspects of objects. It saves software developers from having to define the same properties and methods twice, once for low-density areas and the other for high-density areas.

Polymorphism means that the same method can produce different effects, depending on the kind of object to which the method is being applied. Suppose *AddLayer* is a method for adding a data set to view. The outcome from the method depends on the data source. If it is a coverage data source, the method adds a coverage. If it is a shapefile data source, the method adds a shapefile. Like inheritance, polymorphism saves software developers from having to repeat the same method for different classes.

4.4 TOPOLOGY RULES

Topology, introduced in the coverage model but disappeared with the shapefile, is back with a new look in the geodatabase data model. Instead of enforcing topology in geographically referenced data such as the coverage model, the geodatabase defines topology as relationship rules and lets the user choose the rules, if any, to be implemented in a feature dataset.

Table 4.1 shows 25 topology rules by feature type. Some rules apply to features within a feature

Box 4.4 How to Use an Interface

The following is a code fragment that shows how to use the interface to derive the spatial reference (i.e., coordinate system) and extent information from a feature layer. The code fragment is written in VBA using ArcObjects. All interfaces start with the letter *I*, and their respective variables start with the letter *p*. An apostrophe means that the line is a comment line. An underscore means line continuation.

```
'Declare variables by pointing them to interfaces.
Dim pGeoDataset As IGeoDataset
Dim pEnvelope As IEnvelope
Dim MinX, MaxX, MinY, MaxY As Double
'Derive the extent of the geodataset.
Set pEnvelope = pGeoDataset.Extent
'Get minx, miny, maxx, and maxy from pEnvelope.
MinX = pEnvelope.XMin
MinY = pEnvelope.YMin
MaxX = pEnvelope.XMax
MaxY = pEnvelope.YMax
'Report the geodataset's extent.
MsgBox "Minimum X is: " & MinX & " Minimum Y is: " & MinY & Chr$(10) & _
    "Maximum X is: " & MaxX & " Maximum Y is: " & MaxY
```

TABLE 4.1	Topology Rules in the Geodatabase Data Model

Feature Type	Rule
Polygon	Must not overlap, must not have gaps, must not overlap with, must be covered by feature class of, must cover each other, must be covered by, boundary must be covered by, area boundary must be covered by boundary of, and contains point
Line	Must not overlap, must not intersect, must not have dangles, must not have pseudo-nodes, must not intersect or touch interior, must not overlap with, must be covered by feature class of, must be covered by boundary of, endpoint must be covered by, must not self overlap, must not self intersect, and must be single part
Point	Must be covered by boundary of, must be properly inside polygons, must be covered by endpoint of, and must be covered by line

class, while others apply to two or more participating feature classes. Rules that apply to feature geometries within a feature class are functionally similar to the built-in topology for the coverage model. Rules that apply to two or more feature classes are new with the geodatabase.

The following lists some real-world applications of topology rules:

- Counties must not overlap.
- County must not have gaps.
- County boundary must not have dangles (i.e., must be closed).
- Census tracts and counties must cover each other.
- Voting district must be covered by county.
- Contour lines must not intersect.
- Interstate route must be covered by feature class of reference line.
- Milepost markers must be covered by reference line.
- Label points must be properly inside polygons.

Some rules such as no gap, no overlap, and no dangles are general in nature and can probably apply to many polygon feature classes. Some, such as the relationship between milepost markers and reference line, are specific to transportation applications. Examples of topology rules that have been proposed for data models from different disciplines are available at the ESRI website (**http://support.esri.com/datamodels**). Chapter 8

has a more detailed discussion of topological rules and how to use them for correcting digitizing errors.

4.5 ADVANTAGES OF THE GEODATABASE DATA MODEL

Using the geodatabase as an example, Section 4.5 discusses the advantages of the object-based data model. We can use coverages, shapefiles, and geodatabases in ArcGIS. We can also export or import from one data format into another. The question is then: What are the advantages of migrating from the traditional coverage model to the geodatabase data model? Unless the advantages outweigh the cost of data conversion, we may not want to adopt the new data model.

The geodatabase data model is built on Arc-Objects. A geodatabase can therefore take advantage of new functionalities from object-oriented technology. For example, ArcGIS provides four general validation rules for the grouping of objects: attribute domains, relationship rules, connectivity rules, and custom rules (Zeiler 1999). Attribute domains group objects into subtypes by a valid range of values or a valid set of values for an attribute. (Task 1 in Chapter 9 uses an attribute domain for attribute data entry.) Relationship rules such as topology rules connect objects that are associated. Connectivity rules let users build geometric networks such as streams, roads, and water and electric utilities. (Chapter 18 deals with connectivity in the case of a street network.) And custom rules allow users to create custom features for advanced applications. These validation rules, which are useful for some applications, are not available for the coverage model.

A geodatabase provides a convenient framework for storing and managing different types of GIS data. Besides vector data (the topic of Chapter 4), a geodatabase can also store raster data, triangulated irregular networks (TINs), location data, and attribute tables. As mentioned in Section 4.2, at least one federal program (i.e., the National Hydrography Dataset program) is switching from the coverage model to the geodatabase data model for data delivery. Their justification for the switch is that the geodatabase data model is better for web-based data access and query and for data downloading. Some state agencies such as the New York State Department of Transportation are also offering transportation data in geodatabase format (**http://www.nysgis.state.ny.us/inventories/thruway.htm**).

The geodatabase data model eliminates the complexity of coordinating between the spatial and attribute components, thus reducing the processing overhead. Additionally, because a geodatabase stores both spatial and attribute data in tables, it can operate fully within a relational database environment and bring GIS together with other information technology (IT) applications (Shekhar and Chawla 2003). Thousands of objects, properties, and methods in ArcObjects are available for GIS users to develop customized applications (Burke 2003; Chang 2005). Customized applications can reduce the amount of repetitive work (e.g., define the project the coordinate system of each data set in a project), streamline the workflow (e.g., combine defining and projecting coordinate systems into one step), and even produce functionalities that are not easily available in ArcGIS.

Finally, the geodatabase data model allows custom objects to be developed for different industries and applications. Real-world objects all have different properties and behaviors. It is therefore impossible to apply, for example, the properties and methods of transportation-related objects to forestry-related objects. The geodatabase data model provides the foundation for the development of different geodatabases for different disciplines. As of July 2004, the ESRI website on data models (**http://support.esri.com/datamodels**) has posted over 25 data models by industry.

KEY CONCEPTS AND TERMS

Aggregation: A class relationship that defines an object from one class is an aggregate of objects from another class.

ArcObjects: A collection of objects used by ArcGIS Desktop.

Association: A class relationship that defines how many instances of one class can be associated with the other class.

Class: A set of objects with similar attributes.

Composition: A class relationship that defines an object from one class as a "whole" and objects from another class as "parts" and that the whole controls the lifetime of the parts.

Encapsulation: A principle used in object-oriented technology to hide the properties and methods of an object so that the object can only be accessed through the predefined interfaces.

Feature class: A data set that stores features of the same geometry type in a geodatabase.

Feature dataset: A collection of feature classes in a geodatabase that share the same coordinate system and area extent.

Geodatabase: An object-based data model developed by ESRI, Inc.

Inheritance: A principle used in object-oriented technology that stipulates that an object can inherit properties and methods from the class that the object belongs to.

Instantiation: A class relationship stating that an object of a class can be created from an object of another class.

Interface: A set of externally visible operations of an object.

Method: A specific action that an object can perform.

Object: An entity such as a land parcel that has a set of properties and can perform operations upon requests.

Object-based data model: A data model that uses objects to organize spatial data.

Polymorphism: A principle used in object-oriented technology that stipulates that the same method, if applied to different kinds of objects, can produce different effects.

Property: An attribute or characteristic of an object.

Type inheritance: A class relationship that stipulates that a class inherits the properties and methods of another class.

REVIEW QUESTIONS

1. Explain the difference between the georelational data model and the object-based data model.

2. Provide an example of an object from your discipline and suggest the kinds of properties and methods that the object can have.

3. Provide an example of a hierarchical structure of class and object from your discipline.

4. Explain the class relationship of type inheritance.

5. Explain the class relationship of instantiation.

6. What is ArcObjects?

7. Describe the difference between the geodatabase data model and the coverage (traditional) model in terms of the geometric representation of spatial features.

8. Compare Figure 4.7 with Figure 3.16 in Chapter 3, and explain the difference between the geodatabase data model and the coverage model in handling the route data structure.

9. Explain the relationship between geodatabase, feature dataset, and feature class.

10. How does a standalone feature class differ from a feature class in a feature dataset?

11. Feature dataset is useful for data management. Can you think of an example in which you want to organize data by feature dataset?

12. What is an interface?

13. Define the encapsulation principle in object-oriented technology.

14. Define the polymorphism principle in object-oriented technology.

15. Table 4.1 shows "must not overlap" as a topology rule for polygon features. Provide an example from your discipline that can benefit from enforcing this topology rule.

16. "Must not intersect" is a topology rule for line features. Provide an example from your discipline that can benefit from enforcing this topology rule.

17. The text covers several advantages of adopting the geodatabase data model. Can you think of an example in which you would prefer the geodatabase to the coverage model for a GIS project?

This applications section consists of three tasks. In Task 1, you will work with the basic elements of the geodatabase data model. Task 2 shows how you can update the area and perimeter values of a polygon shapefile by converting it to a geodatabase feature class. In Task 3, you will view routes in the form of polylines with *m* values.

Task 1: Create Geodatabase, Feature Dataset, and Feature Class

What you need: *elevzone.shp* and *stream.shp*, two shapefiles that have the same coordinate system and extent.

In Task 1, you will first create a personal geodatabase and a feature dataset. You will then import the shapefiles into the feature dataset as feature classes and examine their data file structure. The name of a feature class in a geodatabase must be unique. In other words, you cannot use the same name for both a standalone feature class and a feature class in a feature dataset.

1. Start ArcCatalog, and make connection to the Chapter 4 database. This step creates a personal geodatabase. Right-click the Chapter 4 database in the Catalog tree, point to New, and select Personal Geodatabase. Rename the new personal geodatabase *Task1.mdb*.

2. Next create a new feature dataset. Right-click *Task1.mdb*, point to New, and select Feature Dataset. In the next dialog, enter *Area 1* for the name and click on Edit to edit the feature dataset's spatial reference. Import the coordinate system of *stream.shp* to be the feature dataset's coordinate system.

3. *Area 1* should now appear in *Task1.mdb*. Right-click *Area 1*, point to Import, and select Feature Class (multiple). Use the browse button or the drag-and-drop method to select *elevzone.shp* and *stream.shp* for the input features. Make sure that the output

geodatabase points to *Area 1*. Click OK to run the import operation.

4. Right-click *Task1.mdb* in the Catalog tree and select Properties. The Database Properties dialog has the General and Domains tabs. A domain can be used to establish valid values or a valid range of values for an attribute to minimize data entry errors.

5. Right-click *elevzone* and select Properties. The Feature Class Properties dialog has the General, Fields, Indexes, Subtypes, and Relationships tabs. The Fields tab shows fields in the feature class. A feature class can have subtypes, each with one or more attributes for differentiation. A feature class can also have relationships that relate objects in a subtype to another.

6. You can use the Windows Explorer to find *Task1.mdb* in the Chapter 4 database. *Task1.mdb* is a Microsoft Access database. Double-click *Task1.mdb* to see the contents of the database. Among the tables is *elevzone*. You can view the table by double-clicking on it. The table is the same as *elevzone*'s preview table in ArcCatalog. There are many other tables with the GDB prefix. These are tables reserved for use by the geodatabase.

Q1. *elevzone* in *Task1.mdb* contains the same polygon features as *elevzone.shp*. But the contents (tabs) of their property dialog differ. Why?

Task 2: Convert a Shapefile to a Geodatabase Feature Class

What you need: *landsoil.shp*, a polygon shapefile that does not have the correct area and perimeter values.

When shapefiles are used as inputs in an overlay operation, ArcGIS Desktop does not automatically update the area and perimeter values of the output shapefile. *landsoil.shp* represents such an output shapefile. In this task, you will update the area and perimeter values of *landsoil.shp* by converting it into a feature class in a geodatabase.

1. Click *landsoil.shp* in the Catalog tree. On the Preview tab, change the preview type to Table. The table shows two sets of area and perimeter values. Moreover, each field contains duplicate values. Obviously *landsoil.shp* does not have the updated area and perimeter values.

2. Use the same procedure as in Task 1 to create a new personal geodatabase and name it *Task2.mdb*. Right-click *Task2.mdb,* point to Import, and select Feature Class (single). In the next dialog, select *landsoil.shp* for the input features. Make sure that *Task2.mdb* is the output location. Enter *landsoil* for the output feature class name. Click OK to create *landsoil* as a standalone feature class in *Task2.mdb*.

Q2. Besides shapefiles (feature classes), what other types of data can be imported to a geodatabase?

3. Now preview the table of *landsoil* in *Task2.mdb*. On the far right of the table, the fields of Shape_Length and Shape_Area show the correct perimeter and area values respectively.

Task 3: Examine Polylines with Measures

What you need: *decrease24k.shp*, a shapefile showing Washington state highways.

decrease24k.shp is a shapefile downloaded from the Washington State Department of Transportation (WDOT) website. The shapefile contains polylines with measure (*m*) values. In other words, the shapefile contains highway routes. Originally in geographic coordinates, *decrease24k.shp* has been projected onto the Washington State Plane, South Zone, NAD83, and Units feet.

1. Launch ArcMap. Rename the data frame Task 3, and add *decrease24k.shp* to Task 3.

Open the attribute table of *decrease24k*. The Shape field in the table suggests that *decrease24k* is a polyline shapefile with measures. The SR field stores the state route identifiers. Close the table.

2. This step is to add the Identify Route Locations tool. The tool does not appear on any toolbar by default. You need to add it. Select Customize from the Tools menu. On the Commands tab, select the category of Linear Referencing. The Commands frame shows five commands. Drag and drop the Identify Route Locations command to a toolbar (e.g., the Tools toolbar). Close the Customize dialog.

3. Use the Select Features tool to select a highway from *decrease24k.shp*. Click the Identify Route Locations tool, and then click a point along the selected highway. This opens the Identify Route Location Results dialog and shows the measure value of the point you clicked as well as minimum measure, maximum measure, and other information.

Q3. Can you tell the direction in which the route mileage is accumulated?

Challenge Task

NHD_Geo_July3 is a geodatabase downloaded from the National Hydrography Dataset program (**http://nhd.usgs.gov/data.html**).

Q1. Name the feature datasets included in the geodatabase.

Q2. Name the feature classes contained in each of the feature datasets.

Q3. *NHD_Geo_July3* contains the same types of hydrologic data as *nhd* in the Chapter 3 database. *NHD_Geo_July3* is based on the geodatabase data model, whereas *nhd* is based on the coverage model. Compare the two data sets and describe in your own words the difference between them.

REFERENCES

Burke, R. 2003. *Getting to Know ArcObjects: Programming ArcGIS with VBA*. Redlands, CA: ESRI Press.

Chang, K. 2005. *Programming ArcObjects with VBA: A Task-Oriented Approach*. Boca Raton, FL: CRC Press.

Egenhofer, M. F., and A. Frank. 1992. Object-Oriented Modeling for GIS. *Journal of the Urban and Regional Information System Association* 4: 3–19.

Huang, B. 2003. An Object Model with Parametric Polymorphism for Dynamic Segmentation. *International Journal of Geographic Information Science* 17: 343–60.

Koncz, N. A., and T. M. Adams. 2002. A Data Model for Multi-Dimensional Transportation Applications. *International Journal of Geographic Information Science* 16: 551–70.

Larman, C. 1997. *Applying UML and Patterns: An Introduction to Object-Oriented Analysis and Design*. Upper Saddle River, NJ: Prentice Hall PTR.

Shekhar, S., and S. Chawla. 2003. *Spatial Databases: A Tour*. Upper Saddle River, NJ: Prentice Hall.

Shi, W., B. Yang, and Q. Li. 2003. An Object-Oriented Data Model for Complex Objects in Three-Dimensional Geographic Information Systems. *International Journal of Geographical Information Science* 17: 411–30.

Ungerer, M. J., and M. F. Goodchild. 2002. Integrating Spatial Data Analysis and GIS: A New Implementation Using the Component Object Model (COM). *International Journal of Geographical Information Science* 16: 41–53.

Worboys, M. F., H. M. Hearnshaw, and D. J. Maguire. 1990. Object-Oriented Data Modelling for Spatial Databases. *International Journal of Geographical Information Systems* 4: 369–83.

Worboys, M. F. 1995. *GIS: A Computing Perspective*. London: Taylor & Francis.

Zeiler, M. 1999. *Modeling Our World: The ESRI Guide to Geodatabase Design*. Redlands, CA: ESRI Press.

Zeiler, M., ed. 2001. *Exploring ArcObjects*. Redlands, CA: ESRI Press.

RASTER DATA MODEL

The vector data model uses the geometric objects of point, line, and area to represent spatial features. Although ideal for discrete features with well-defined locations and shapes, the vector data model does not work well with spatial phenomena that vary continuously over the space such as precipitation, elevation, and soil erosion (Figure 5.1). A better option for representing continuous phenomena is the raster data model. The raster data model uses a regular grid to cover the space. The value in each grid cell corresponds to the characteristic of a spatial phenomenon at the cell location.

And the changes in the cell value reflect the spatial variation of the phenomenon.

Unlike the vector data model, the raster data model has not changed in terms of its concept or data format for the past three decades. Research on the raster data model has instead concentrated on data structure and data compression. A wide variety of data used in GIS are encoded in raster format. They include digital elevation data, satellite images, digital orthophotos, scanned maps, and graphic files. Raster data tend to require large amounts of computer memory. Therefore, issues dealing with data storage and retrieval are important to raster data users.

Commercial GIS packages can display raster and vector data simultaneously, and can easily convert between these two types of data. In many ways, raster and vector data are complementary of each other. Integration of these two types of data has therefore become a common and desirable feature in a GIS project.

Chapter 5 is divided into the following six sections. Section 5.1 covers the basic elements of raster data including cell value, cell size, bands, and spatial reference. Section 5.2 presents differ- ent types of raster data. Section 5.3 describes three different raster data structures. Section 5.4 focuses on data compression methods. Section 5.5 dis- cusses data conversion between raster and vector data. Section 5.6 offers examples of data integra- tion in GIS.

Figure 5.1
A continuous elevation raster with darker shades for higher elevations.

5.1 ELEMENTS OF THE RASTER DATA MODEL

A raster data model is variously called a grid, a raster map, a surface cover, or an image in GIS. Raster is adopted in Chapter 5. A raster consists of rows, columns, and cells. Cells are also called pix- els with images. The origin of rows and columns is typically at the upper-left corner of the raster. Rows function as y-coordinates and columns as x-coordinates. Each cell in the raster is explicitly defined by its row and column position.

Raster data represent points by single cells, lines by sequences of neighboring cells, and areas by collections of contiguous cells (Figure 5.2). Although the raster data model has its weakness in representing the precise location of spatial fea- tures, it has the distinctive advantage in having fixed cell locations (Tomlin 1990). In computing algorithms, a raster can be treated as a matrix with rows and columns, and its cell values can be stored into a two-dimensional array. All commonly used programming languages can easily handle arrayed variables. Raster data are therefore much easier for data manipulation, aggregation, and analysis than vector data.

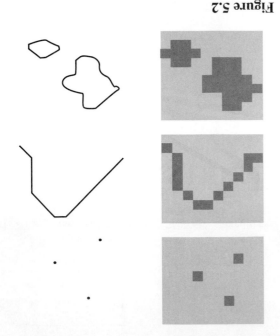

Figure 5.2
Representation of point, line, and area features: raster format on the left and vector format on the right.

5.1.1 Cell Value

Each cell in a raster carries a value, which repre- sents the characteristic of a spatial phenomenon at the location denoted by its row and column. Depending on the coding of its cell values, a raster can be either an **integer** or a **floating-point** raster. An

---

Box 5.1 | **Rules in Determining a Categorical Cell Value**

If a large cell covers forest, pasture, and water on the ground, which category should be entered for the cell? The most common method is to enter the category that occupies the largest area percentage of the cell. But in some situations the majority rule may not be the best option. For example, studies of endangered species are more inclined to use the presence/absence rule than the majority rule (Chrisman 2001). As long

as an endangered species has been recorded in a cell, no matter how much of the cell the species occupies, the cell will be coded with a value for presence. Similarly, the determination of cell values may be based on a ranking of the importance of spatial features to the study. If pasture is deemed to be more important than forest and water, a mixed cell of all three features will be coded as pasture.

---

integer value has no decimal digits, whereas a floating-point value does. Integer cell values usually represent categorical data, which may or may not be ordered. A land cover raster may use 1 for urban land use, 2 for forested land, 3 for water body, and so on. A wildlife habitat raster, on the other hand, may use the same integer numbers to represent ordered categorical data of optimal, marginal, and unsuitable habitats. Floating-point cell values represent continuous, numeric data. For example, a precipitation raster may have precipitation values of 20.15, 12.23, and so forth.

A floating-point raster requires more computer memory than an integer raster. This difference can become an important factor for a GIS project that covers a large area. There are a couple of other differences. We can access the cell values of an integer raster through a value attribute table. But a floating-point raster may not have a value attribute table because of its potentially large number of records. We can use individual cell values to query and display an integer raster. But the same operation on a floating-point raster should be based on value ranges, such as 12.0 to 19.9, because the chance of finding a specific value is small.

Where does the cell value apply within the cell? The answer depends on raster data operation. Typically the cell value applies to the center of the cell in operations that involve distance measurements. Examples include resampling pixel values (Chapter 7) and calculating physical distances

(Chapter 13). Many other raster data operations are cell-based, instead of point-based, and assume that the cell value applies to the entire cell.

5.1.2 Cell Size

The cell size determines the resolution of the raster data model. A cell size of 10 meters means that each cell measures 100 square meters (10×10 meters). A cell size of 30 meters, on the other hand, means that each cell measures 900 square meters (30×30 meters). Therefore a 10-meter raster has a finer (higher) resolution than a 30-meter raster.

A large cell size cannot represent the precise location of spatial features, thus increasing the chance of having mixed features such as forest, pasture, and water in a cell (Box 5.1). These problems lessen when a raster uses a smaller cell size. But a small cell size increases the data volume and the data processing time.

5.1.3 Raster Bands

A raster may have a single band or multiple bands. Each cell in a multiband raster is associated with more than one cell value. An example of a multiband raster is a satellite image, which may have five, seven, or more bands at each cell location. Each cell in a single-band raster has only one cell value. An example of a single-band raster is an elevation raster, which has one elevation value at each cell location.

5.1.4 Spatial Reference

Raster data must have the spatial reference information so that they can align spatially with other data sets in a GIS. For example, to superimpose an elevation raster on a vector-based soil layer, we must first make sure that both data sets are based on the same coordinate system. A raster that has been processed to match a projected coordinate system is often called a **georeferenced raster.**

Two adjustments are necessary in associating a projected coordinate system with a raster. First, the origin of a projected coordinate system is at the lower-left corner whereas the origin of a raster is typically at the upper-left corner. Second, projected coordinates must correspond to the rows and columns of the raster. The following example illustrates what these two adjustments mean.

Suppose an elevation raster has the following information on the number of rows, number of columns, cell size, and area extent expressed in UTM (Universal Transverse Mercator) coordinates:

- Rows: 463, columns: 318, cell size: 30 meters
- x-, y-coordinates at the lower-left corner: 499995, 5177715
- x-, y-coordinates at the upper-right corner: 509535, 5191065

We can verify that the numbers of rows and columns are correct by using the bounding UTM coordinates and the cell size:

- Number of rows = (5191065 − 5177715)/30 = 463
- Number of columns = (509535 − 499995)/30 = 318

We can also derive the UTM coordinates that define each cell. For example, the cell of (row 1, column 1) has the following UTM coordinates (Figure 5.3):

- 499995, 5191035 or (5191065 − 30) at the lower-left corner
- 500025 or (499995 + 30), 5191065 at the upper-right corner
- 500010 or (499995 + 15), 5190050 or (5191065 − 15) at the cell center

5.2 TYPES OF RASTER DATA

A large variety of data that we use in GIS are encoded in raster format. These data all share the same basic elements of the raster data model.

5.2.1 Satellite Imagery

Remotely sensed satellite data are familiar to GIS users. The spatial resolution of a satellite image relates to the ground pixel size. For example, a spatial resolution of 30 meters means that each pixel in the satellite image corresponds to a ground pixel of 900 square meters. The pixel value, also called the brightness value, represents light energy reflected or emitted from the Earth's surface (Jensen 1996; Lillesand and Kiefer 2000). The measurement of light energy is based on spectral bands from a continuum of wavelengths known as the electromagnetic spectrum. Panchromatic images are comprised of a single spectral band, whereas multispectral images are comprised of multiple bands.

The U.S. **Landsat** satellites, started by the National Aeronautics and Space Administration (NASA) with the cooperation of the U.S. Geological Survey (USGS) in 1972, have produced the most widely used imagery worldwide. Landsat 1, 2, and 3 acquired images by the Multispectral Scanner (MSS) with a spatial resolution of about 79 meters. When launched in 1982, Landsat 4 carried a new sensor, the Thematic Mapper (TM) scanner. TM images have a spatial resolution of

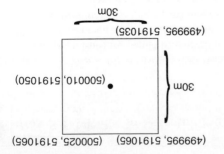

(499995, 5191035)

(499995, 5191065)   (500025, 5191065)

(500010, 5191050)

30m

Figure 5.3

UTM coordinates for the extent and the center of a 30-meter cell.

30 meters and seven spectral bands (blue, green, red, near infrared, midinfrared I, thermal infrared, and midinfrared II). A second TM was launched aboard Landsat 5 in 1984. Landsat 6 failed to reach its orbit after launch in 1993.

Landsat 7 was launched successfully in April 1999, carrying an Enhanced Thematic Mapper Plus (ETM+) sensor (**http://landsat.usgs.gov/**). The enhanced sensor is designed to seasonally monitor small-scale processes on a global scale, such as cycles of vegetation growth, deforestation, agricultural land use, erosion and other forms of land degradation, snow accumulation and melt, and urbanization. The spatial resolution of Landsat 7 imagery is 15 meters in the panchromatic band; 30 meters in the six visible, near infrared, and shortwave infrared bands; and 60 meters in the thermal infrared band. In June 2003, NASA discovered an instrument malfunction of the Scan Line Corrector (SLC) on board Landsat 7. Since then, Landsat 7 ETM+ has been collecting image data in the "SLC-off" mode. Reports from the Landsat program suggest that the center of an SLC-off data product should be similar in quality to previous Landsat 7 data.

In December 1999, NASA's Earth Observing System launched the Terra spacecraft to study the interactions among the Earth's atmosphere, lands, oceans, life, and radiant energy (heat and light) (**http://terra.nasa.gov/About/**). Terra carries five sensors, of which ASTER (Advanced Spaceborne Thermal Emission and Reflection Radiometer) is the only high-spatial-resolution instrument. ASTER's spatial resolution is 15 meters in the visible and near infrared range, 30 meters in the shortwave infrared band, and 90 meters in the thermal infrared band. A major application of ASTER data products is land cover classification and change detection.

The U.S. National Oceanic and Atmospheric Administration (NOAA) uses weather satellites as an aid to weather prediction and monitoring. NOAA's Polar Orbiting Environmental Satellites (POES) carry the AVHRR (Advanced Very High Resolution Radiometer) scanner, which provides data useful for large-area land cover and vegetation mapping (**http://edc.usgs.gov/products/satellite/**

avhrr.html). AVHRR data have a spatial resolution of 1.1 kilometers, which may be too coarse for some GIS projects. But the coarse spatial resolution is offset by the daily coverage and reduced volume of AVHRR data.

A number of countries have programs similar to Landsat in the United States. The French **SPOT** satellite series began in 1986 (**http://www.spot.com/**). Each SPOT satellite carries two types of sensors. SPOT 1 to 4 acquire single-band imagery with a 10-meter spatial resolution and multiband imagery with a 20-meter resolution. SPOT 5, launched successfully in May 2002, sends back higher-resolution imagery: 5 and 2.5 meters in single-band, and 10 meters in multiband. High spatial resolution makes SPOT images good spatial data sources for GIS projects. Other important satellite programs have also been established since the late 1980s in India (**http://www.isro.org/**) and Japan (**http://www.jaxa.jp/index_e.html**).

The privatization of the Landsat program in the United States in 1985 has opened the door for private companies to gather and market remotely sensed data using various platforms and sensors. (The French space agency handed over responsibilities of commercial operation of SPOT 5 to Spot Image in 2002.) Ikonos, a satellite designed by Space Imaging (**http://www.spaceimaging.com/**) to acquire panchromatic images with a 1-meter spatial resolution and multispectral images with a 4-meter resolution, was successfully launched in September 1999. A resolution of 1 meter is high enough to detect ground objects such as cars, small houses, fires, and troop deployments. Digital-Globe's QuickBird reached its orbit successfully in October 2001 (**http://www.digitalglobe.com/**). QuickBird collects panchromatic images with a 61-centimeter spatial resolution and multispectral images with a 2.44-meter resolution.

5.2.2 USGS Digital Elevation Models (DEMs)

A **digital elevation model (DEM)** consists of an array of uniformly spaced elevation data. A DEM is point-based, but it can easily be converted to raster data by placing each elevation point at the center of a cell. Most GIS users in the United States use DEMs from the USGS. USGS DEMs include the 7.5-minute DEM, 30-minute DEM, 1-degree DEM, and Alaska DEM.

The 7.5-minute DEMs provide elevation data at a spacing of 30 meters or 10 meters on a grid measured in UTM coordinates. Each DEM covers a 7.5-by-7.5-minute block that corresponds to a USGS 1:24,000 scale quadrangle (Box 5.2). The USGS groups the 7.5-minute DEMs into three levels by data accuracy and production method, with level 1 having the least accurate data. Only level-1 and level-2 DEMs are currently available for most states. Level-1 DEMs have the vertical accuracy of ±15 meters or better, whereas level-2 DEMs have the vertical accuracy of ±7 meters or better. Level-2 DEMs, which have been smoothed and processed to remove identifiable systematic errors, can have resolutions at 30 meters or 10 meters.

The 30-minute DEMs provide elevation data at a spacing of 2 arc-seconds on the geographic grid (about 60 meters in the midlatitudes). Each DEM covers a 30-by-30-minute block, corresponding to the east or west half of a USGS 30-by-60-minute 1:100,000 scale quadrangle. The vertical accuracy of the 30-minute DEMs is ±25 meters or better.

The 1-degree DEMs provide elevation data at a spacing of 3 arc-seconds on the geographic grid (about 100 meters in the midlatitudes). Each DEM covers a 1-by-1-degree block, corresponding to the east or west half of a USGS 1-by-2-degree 1:250,000 scale quadrangle. The Defense Mapping Agency (DMA, now the National Geospatial-Intelligence Agency or NGA) originally produced the 1-degree DEMs by interpolation from digitized contour lines. The vertical accuracy of elevation data is about ±30 meters.

The USGS provides the 7.5-minute and 15-minute Alaska DEMs, with the point spacing referenced to latitude and longitude. The spacing for the 7.5-minute Alaska DEMs is 1 arc-second of latitude by 2 arc-seconds of longitude, and the spacing for the 15-minute Alaska DEMs is 2 arc-seconds of latitude by 3 arc-seconds of longitude.

5.2.3 Non–USGS DEMs

A basic method for producing DEMs is to use a stereoplotter and aerial photographs with overlapped areas. The stereoplotter creates a 3-D model, which allows the operator to compile elevation data. Although this method can produce highly accurate DEM data at a finer resolution than USGS DEMs, it is expensive for coverage of large areas.

There are alternatives to the use of a stereoplotter. A popular alternative with GIS users is to generate DEMs from satellite imagery such as the SPOT stereo model. Software packages for extracting elevation data from SPOT images on the personal computer are commercially available. Besides imagery data, the data extraction process requires ground control points, which can be measured in the field by GPS (global positioning system). The quality of

---

  Box 5.2 No-Data Slivers in 7.5-Minute DEM

A 7.5-minute DEM is supposed to correspond to a USGS 1:24,000 scale quadrangle, but a sliver of no data often shows up along the border of the DEM. This is because the bounding coordinates of the 7.5-minute DEM do not form a rectangle. The 7.5-minute DEM data are stored as a series of profiles, with a constant spacing of 30 meters along and between each profile. When DEMs are cast on the UTM coordinate system, the bounding coordinates of a 7.5-minute DEM form a quadrilateral, rather than a rectangle. Thus it cannot match the quadrangle perfectly. The no-data value is used to fill all the uneven rows and columns.

such DEMs, however, depends on the software package and the quality of the inputs.

Stereo radar data can also be the source for producing DEMs. An active remote sensor, radar emits microwave pulses and measures returned energy from ground objects. Radar can penetrate cloud cover, which usually presents a major problem with aerial photography. As an example, Intermap Technologies produces elevation data from stereo radar data at a spatial resolution of 5 to 15 meters and with a vertical resolution of 2 to 5 meters **(http://www.intermaptechnologies.com/).**

LIDAR (light detection and ranging) is a new technology for producing DEMs (Turner 2000). The basic components of a LIDAR system include a laser scanner mounted in an aircraft, GPS, and an Inertial Measurement Unit (IMU). A LIDAR sensor emits rapid laser pulses over an area and uses the time lapse of the pulse to measure distance. At the same time, the location and orientation of the laser source are determined by GPS and IMU, respectively. One application of LIDAR technology is the creation of high-resolution DEMs, with a spatial resolution of 0.5 to 2 meters and a vertical accuracy of ±15 centimeters (Flood 2001) (Figure 5.4). These DEMs are already georeferenced and can have different height levels such as ground elevation and canopy elevation.

5.2.4 Global DEMs

DEMs at different resolutions are now available on the global scale. ETOPO5 (Earth Topography–5 Minute) data cover both the land surface and ocean floor of the Earth, with a grid spacing of 5 minutes of latitude by 5 minutes of longitude **(http://edcwww.cr.usgs.gov/glis/hyper/guide/ etopo5).** Both GTOPO30 **(http://lpdaac.usgs. gov/gtopo30/gtopo30.asp)** and GLOBE **(http:// www.ngdc.noaa.gov/seg/topo/globe.shtml/)** offer global DEMs with a horizontal grid spacing of 30 arc-seconds or approximately 1 kilometer.

Global DEMs such as GTOPO30 and GLOBE were derived from raster data such as satellite imagery and vector data such as contour lines. The vertical accuracy of GLOBE is estimated to be within 30 meters from raster sources and within 160 meters from vector sources.

5.2.5 Digital Orthophotos

A **digital orthophoto quad (DOQ)** is a digitized image prepared from an aerial photograph or other remotely sensed data, in which the displacement caused by camera tilt and terrain relief has been removed (Figure 5.5). The USGS began producing DOQs in 1991 from 1:40,000 scale aerial photographs of the National Aerial Photography Program.

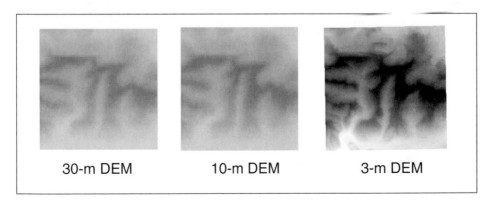

30-m DEM   10-m DEM   3-m DEM

Figure 5.4
DEMs at three resolutions: 30 meters, 10 meters, and 3 meters. The 30-meter and 10-meter DEMs are USGS DEMs. The 3-meter DEM, which contains more topographic details than the other two, is a derived product from LIDAR data.

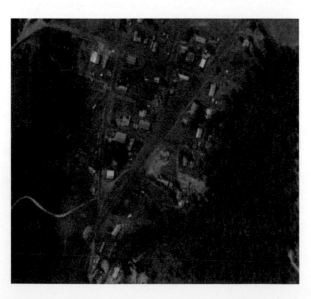

Figure 5.5
A rural settlement is shown in this USGS 1-meter black-and-white DOQ.

These USGS DOQs are georeferenced and can be registered with topographic and other maps.

The standard USGS DOQ format is either a 3.75-minute quarter quadrangle or 7.5-minute quadrangle in the form of a black-and-white, color infrared, or natural color image. Most GIS users use 3.75-minute quarter quadrangles in black and white. These quarter quadrangles have a 1-meter spatial resolution and have pixel values representing 256 gray levels, similar to a single-band satellite image. A color orthophoto is a multiband image, each band representing red, green, or blue light. Easily integrated in a GIS, DOQs are the ideal background for data display and data editing.

5.6 Bi-Level Scanned Files

A **bi-level scanned file** is a scanned image containing values of 1 or 0 (Figure 5.6). In GIS, bi-level scanned files are usually made for the purpose of digitizing (Chapter 6). They are scanned from paper or Mylar maps that contain boundaries of soils, parcels, and other features. A GIS package has tools for converting bi-level scanned files into

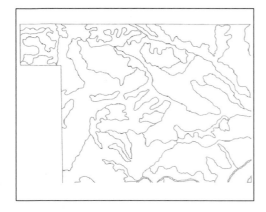

Figure 5.6
A bi-level scanned file showing soil lines.

vector-based features. Maps to be digitized are typically scanned at 300 or 400 dots per inch (dpi).

5.7 Digital Raster Graphics (DRGs)

A **digital raster graphic (DRG)** is a scanned image of a USGS topographic map (Figure 5.7). The USGS scans the 7.5-minute topographic map at 250 dpi, thus producing a DRG with a ground

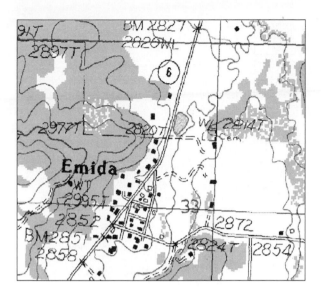

Figure 5.7
The same rural settlement shown in Figure 5.5 is depicted in this USGS DRG.

resolution of 2.4 meters. The USGS uses up to 13 colors on each 7.5-minute DRG. Because these 13 colors are based on an 8-bit (256) color palette, they may not look exactly the same as on the paper maps. USGS DRGs are georeferenced to the UTM coordinate system.

5.2.8 Graphic Files

Maps, photographs, and images can be stored as digital graphic files. Many popular graphic files are in raster format, such as TIFF (tagged image file format), GIF (graphics interchange format), and JPEG (Joint Photographic Experts Group). The USGS distributes DOQs in TIFF or GeoTIFF. Geo-TIFF is a georeferenced version of TIFF. By having the spatial reference information of the image, DOQs can be readily used with other GIS data.

5.2.9 GIS Software-Specific Raster Data

GIS packages use raster data that are imported from DEMs, satellite images, scanned images, graphic files, and ASCII files or are converted from vector data. These raster data are named differently. For example, ESRI, Inc. names raster data grids.

Based on a proprietary format, **ESRI grids** are either integer or floating-point. An integer grid has a value attribute table that stores cell values. A floating-point grid may not have a table, because of its potentially large number of records. ESRI software can convert a floating-point grid to an integer grid, and vice versa.

5.3 RASTER DATA STRUCTURE

Raster data structure refers to the storage of raster data so that they can be used and processed by the computer. Over the past three decades, several raster data structures have been proposed. This section examines three common structures—cell-by-cell encoding, run-length encoding, and quad tree.

5.3.1 Cell-by-Cell Encoding

The **cell-by-cell encoding** method provides the simplest raster data structure. A raster is stored as a matrix, and its cell values are written into a file by row and column (Figure 5.8). Functioning at the cell level, this method is an ideal choice if the cell values of a raster change continuously.

DEMs use the cell-by-cell data structure because the neighboring elevation values are rarely the same. Satellite images also use the cell-by-cell encoding method for data storage. With multiple spectral bands, however, each pixel in a satellite image has more than one value. Multiband imagery is typically stored in the following three formats (Jensen 1996). The band sequential (.bsq) method stores the values of an image band as one file. Therefore, if an image has seven bands, the data set has seven consecutive files, one file per band. The band interleaved by line (.bil) method stores, row by row, the values of all the bands in one file. Therefore the file consists of row 1, band 1; row 1, band 2 . . . row 2, band 1; row 2, band 2 . . . and so on. The band interleaved by pixel (.bip) method stores the values of all the bands by pixel in one file. The file is therefore comprised of pixel (1, 1), band 1; pixel (1, 1), band 2 . . . pixel (2, 1), band 1; pixel (2, 1), band 2 . . . and so on.

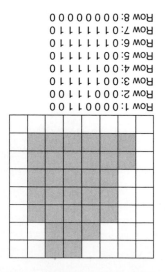

Figure 5.8
The cell-by-cell data structure records each cell value by row and column. The gray cells have the cell value of 1.

Row 1: 0 0 0 1 1 0 0
Row 2: 0 0 1 1 0 0
Row 3: 0 0 1 1 1 1 0
Row 4: 0 0 1 1 1 1 0
Row 5: 0 0 1 1 1 0
Row 6: 0 1 1 1 1 1 0
Row 7: 0 1 1 1 1 1 0
Row 8: 0 0 0 0 0 0 0

5.3.2 Run-Length Encoding

The cell-by-cell encoding method becomes inefficient if a raster contains many redundant cell values. For example, a bi-level scanned file from a soil map has many 0s representing non-inked areas and only occasional 1s representing the inked soil lines. Raster models with many repetitive cell values can be more efficiently stored using the **run-length encoding (RLE)** method, which records the cell values by row and by group. A group refers to adjacent cells with the same cell value. Figure 5.9 shows the run-length encoding of the polygon in gray. For each row, the starting cell and the end cell denote the length of the group ("run") that falls within the polygon.

A bi-level scanned file of a 7.5-minute soil quadrangle map, scanned at 300 dpi, can be over 8 megabytes (MB) if it is stored on a cell-by-cell basis. But using the RLE method, the file is reduced to about 0.8 MB at a 10:1 compression ratio. RLE is therefore a method for encoding as well as compressing raster data. Many GIS packages use RLE in addition to the cell-by-cell encoding method for storing raster data. They include GRASS, IDRISI, and ArcGIS.

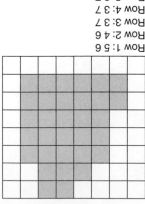

Figure 5.9
The run-length encoding method records the gray cells by row. Row 1 has two adjacent gray cells in columns 5 and 6. Row 1 is therefore encoded with one run, beginning in column 5 and ending in column 6. The same method is used to record other rows.

Row 1: 5 6
Row 2: 4 6
Row 3: 3 7
Row 4: 3 7
Row 5: 3 7
Row 6: 2 7
Row 7: 2 7

5.3.3 Quad Tree

Instead of working along one row at a time, **quad tree** uses recursive decomposition to divide a raster into a hierarchy of quadrants (Samet 1990). Recursive decomposition refers to a process of continuous subdivision until every quadrant in a quad tree contains only one cell value.

Figure 5.10 shows a raster with a polygon in gray, and a quad tree that stores the feature. The quad tree contains nodes and branches (subdivisions). A node represents a quadrant. Depending on the cell value(s) in the quadrant, a node can be a nonleaf node or a leaf node. A nonleaf node represents a quadrant that has different cell values. A nonleaf node is therefore a branch point, meaning that the quadrant is subject to subdivision. A leaf node, on the other hand, represents a quadrant that has the same cell value. A leaf node is therefore an

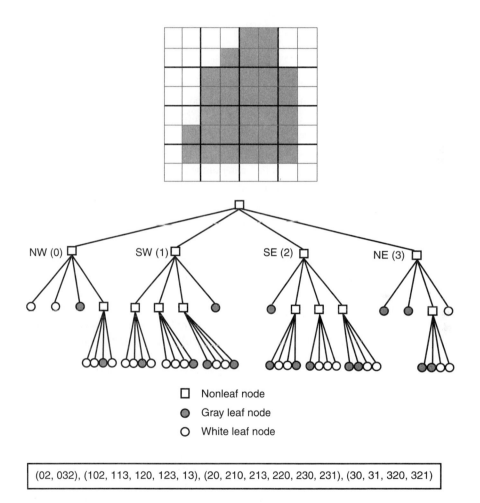

☐ Nonleaf node

● Gray leaf node

○ White leaf node

(02, 032), (102, 113, 120, 123, 13), (20, 210, 213, 220, 230, 231), (30, 31, 320, 321)

Figure 5.10

The regional quad tree method divides a raster into a hierarchy of quadrants. The division stops when a quadrant is made of cells of the same value (gray or white). A quadrant that cannot be subdivided is called a leaf node. In the diagram, the quadrants are indexed spatially: 0 for NW, 1 for SW, 2 for SE, and 3 for NE. Using the spatial indexing method and the hierarchical quad tree structure, the gray cells can be coded as 02, 032, and so on. See Section 5.3.3 for more explanation.

end point, which can be coded with the value of the homogeneous quadrant (gray or white). The depth of a quad tree, or the number of levels in the hierarchy, can vary depending on the complexity of the two-dimensional feature.

After the subdivision is complete, the next step is to code the two-dimensional feature using the quad tree and a spatial indexing method. For example, the level-1 NW quadrant (with the spatial index of 0) in Figure 5.10 has two gray leaf nodes. The first, 02, refers to the level-2 SE quadrant, and the second, 032, refers to the level-3 SE quadrant of the level-2 NE quadrant. The string of (02, 032) and others for the other three level-1 quadrants completes the coding of the two-dimensional feature.

Regional quad tree is an efficient method for storing area data, especially if the data contain few categories. This method is also efficient for data processing (Samet 1990). Quad tree has other uses in GIS as well. Researchers have proposed using a hierarchical quad tree structure for storing and indexing global data (Tobler and Chen 1986; Dutton 1999; Ottoson and Hauska 2002). One advantage of using quad tree for global data is that the quad tree can be directly applied to the spherical surface of the Earth. Quad tree is also useful as a spatial index method. Spatial indexing helps locate spatial data, both raster and vector, easily and quickly. Oracle, for example, uses quad tree as a method in indexing spatial data in Oracle Spatial.

5.3.4 Header File

To import raster data such as a DEM or a satellite image, a GIS package requires information about the raster, such as the data structure, area extent, cell size, number of bands, and value for no data. Often this information is contained in a header file. Functionally, a header file is similar to a projection file for a coordinate system (Box 5.3).

There may be other files besides the header file that accompany a raster data set. A satellite image, for example, may have two optional files. The statistics file describes file statistics such as minimum,

---

Box 5.3 A Header File Example

The following is a header file example for a GTOPO30 DEM. The explanation of each entry in the file is given after /*.

```
BYTEORDER M        /* byte order in which image pixel values are stored. M = Motorola byte order.
LAYOUT BIL         /* organization of the bands in the file. BIL = band interleaved by line.
NROWS 6000         /* number of rows in the image.
NCOLS 4800         /* number of columns in the image.
NBANDS 1           /* number of spectral bands in the image. 1 = single band.
NBITS 16           /* number of bits per pixel.
BANDROWBYTES 9600  /* number of bytes per band per row.
TOTALROWBYTES 9600 /* total number of bytes of data per row.
BANDGAPBYTES 0     /* number of bytes between bands in a BSQ format image.
NODATA −9999       /* value used for masking purpose.
ULXMAP −99.9958333333334  /* longitude of the center of the upper-left pixel (decimal degrees).
ULYMAP 39.9958333333333   /* latitude of the center of the upper-left pixel (decimal degrees).
XDIM 0.00833333333333     /* x dimension of a pixel in geographic units (decimal degrees).
YDIM 0.00833333333333     /* y dimension of a pixel in geographic units (decimal degrees).
```

maximum, mean, and standard deviation for each spectral band in an image. And the color file associates colors with different pixel values in an image. An ESRI grid also has additional files to store information on the descriptive statistics of the cell values and, in the case of an integer grid, the number of cells that have the same cell value.

5.4 DATA COMPRESSION

Raster data sets typically contain large amounts of data and require considerable memory space. The average file size is about 1.1 megabytes (MB) for a 30-meter DEM; 9.9 MB for a 10-meter DEM; 5 to 15 MB for a 7.5-minute DRG; and 45 MB for a 3.75-minute quarter DOQ in black and white. An uncompressed 7-band TM scene requires almost 200 MB for storage (Sanchez and Canton 1999). The memory requirement becomes even higher for high-resolution satellite images.

Data compression refers to the reduction of data volume, a topic particularly important for data delivery and Internet mapping. We are familiar with data compression programs, such as WinZip for PCs and gzip for the UNIX environment. These programs can work with any type of data files and maintain the original structure of files and folders. This section, however, focuses on image compression.

A variety of techniques are available for image compression. Compression techniques can be lossless or lossy. A **lossless compression** method allows the original image to be precisely reconstructed. The run-length encoding (RLE) method discussed in Section 5.3.2 is an example of lossless compression. PackBits compression is a more efficient variation of RLE. The GIF format uses a lossless compression method called LZW (Lempel-Ziv-Welch). Although the GIF format is technically lossless, it does not retain the quality of the original image because it uses a maximum of 256 (8-bit) colors. The TIFF data format offers both PackBits and LZW for image compression. The USGS uses TIFF for delivering DRGs and DOQs.

A **lossy compression** method cannot reconstruct fully the original image but can achieve high-compression ratios. The common JPEG format uses a lossy compression method. The method divides an image into blocks of 64 (8 × 8) and processes each block independently. The colors in each block are shifted and simplified to reduce the amount of data encoding. This block-based processing usually results in the "blocky" appearance. Image degradation through lossy compression can affect GIS-related tasks such as extracting ground control points from air photographs or satellite images for the purpose of georeferencing (Chapter 7).

Like GIS techniques, image compression techniques change constantly. One of the newer techniques is MrSID. MrSID (Multi-resolution Seamless Image Database) is a compression technique originally developed at the Los Alamos National Laboratory and awarded in the late 1990s to LizardTech Inc. (**http://www.lizardtech.com/**). Multiresolution means that MrSID has the capability of recalling the image data at different resolutions or scales. Seamless means that MrSID can compress a large image such as a DOQ or a satellite image with subblocks and eliminates the artificial block boundaries during the compression process. MrSID is supported by major GIS companies and is used by federal agencies such as USGS and NGA for delivering DOQs and satellite images.

Although MrSID is a proprietary format, it is well known that MrSID uses the wavelet transform for data compression. JPEG 2000, an updated version of the popular open format, also uses the wavelet transform (Christopoulos et al. 2000). The wavelet transform therefore appears to be the latest choice for image compression.

The **wavelet transform** treats an image as a wave and progressively decomposes the wave into simpler wavelets (Addison 2002). Using a wavelet (mathematical) function, the transform repetitively averages groups of adjacent pixels (e.g., 2, 4, 6, 8, or more) and, at the same time, records the differences between the original pixel values and the average. The differences, also called wavelet coefficients, can be 0, greater than 0, or less than 0. In parts of an image that have few significant variations, most pixels will have coefficients of 0 or very close to 0. To save storage, these parts of the image can be stored at

Both MrSID and JPEG 2000 can perform either lossless or lossy compression. A lossless compression saves the wavelet coefficients and uses them to reconstruct the original image. A lossy compression, on the other hand, stores only the averages and those coefficients that did not get rounded off to 0. Trade reports have shown that JPEG 2000 can achieve compression ratios from 20:1 to 300:1 for color images and from 10:1 to 50:1 for gray scale images without a perceptible difference in the quality of an image. A recent study suggests that if JPEG 2000 compression is at or under a 10:1 ratio, it should be

possible to extract ground control points from air photographs or satellite images for georeferencing (Li et al. 2002).

5.5 DATA CONVERSION

Data conversion is a standard functionality in a GIS package. We need to use this functionality to take advantage of having both vector and raster data for a GIS project. The conversion of vector data to raster data is called **rasterization**, and the conversion of raster to vector data is called **vectorization** (Figure 5.12). These two conversion methods use different computer algorithms (Piwowar et al. 1990).

The simpler of the two conversion methods, rasterization involves three basic steps (Clarke 1995). The first step sets up a raster with a specified

---

Box 5.4 A Simple Wavelet Example: The Haar Wavelet

A Haar wavelet consists of a short positive pulse followed by a short negative pulse (Figure 5.11a). Although the short pulses result in jagged lines rather than smooth curves, the Haar function is excellent for illustrating the wavelet transform because of its simplicity. Figure 5.11b shows an image with darker pixels near the center. The image is encoded as a series of numbers. Using the Haar function, we take the average of each pair of adjacent pixels. The averaging results in the string (2, 8, 8, 4) and retains the quality of the original image at a lower resolution. But if the process continues, the averaging results in the string (5, 6) and loses the darker center in the original image.

Suppose that the process stops at the string (2, 8, 8, 4). The wavelet coefficients will be −1 (1 − 2), −1 (7 − 8), 0 (8 − 8), and 2 (6 − 4). By rounding off these coefficients to 0, it would save the storage space by a factor of 2 and still retain the quality of the original image. If, however, a lossless compression

is needed, we can use the coefficients to reconstruct the original image. For example, 2 − 1 = 1 (the first pixel), 2 − (−1) = 3 (the second pixel), and so on.

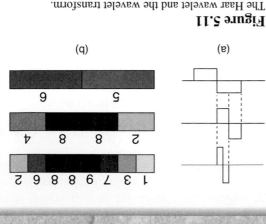

Figure 5.11
The Haar wavelet and the wavelet transform. (a) Three Haar wavelets at three scales (resolutions). (b) A simple example of the wavelet transform.

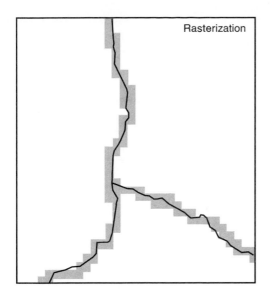 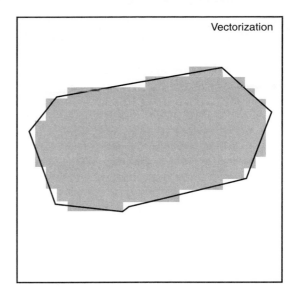

Figure 5.12
On the left is an example of conversion from vector to raster data, or rasterization. On the right is an example of conversion from raster to vector data, or vectorization.

cell size to cover the area extent of vector data and assigns initially all cell values as zeros. The second step changes the values of those cells that correspond to points, lines, or polygon boundaries. The cell value is set to 1 for a point, the line's value for a line, and the polygon's value for a polygon boundary. The third step fills the interior of the polygon outline with the polygon value. Errors from rasterization are usually related to the design of the computer algorithm and the size of raster cell (Bregt et al. 1991).

Vectorization involves three basic elements: line thinning, line extraction, and topological reconstruction (Clarke 1995). Lines in the vector data model have length but no width. Raster lines in a scanned file, however, usually occupy several pixels in width. Raster lines must be thinned, ideally to a 1-cell width, for vectorization. Line extraction is the process of determining where individual lines begin and end. Topological reconstruction is to connect extracted lines as well as to show where digitizing errors exist. Results of raster-to-vector conversion often show steplike

features along diagonal lines. A subsequent line smoothing operation can eliminate those artifacts from raster data.

5.6 INTEGRATION OF RASTER AND VECTOR DATA

Given a wide variety of raster data, integration of raster data and vector data can take place in different forms in a GIS project. DOQs, DRGs, and graphic files are useful as the background for data display, and as the source for digitizing or editing vector data (Box 5.5). Bi-level scanned files are inputs for digitizing line or polygon features (Chapter 6). DEMs are the most important data source for deriving such topographic features as contour, slope, aspect, drainage network, and watersheds (Chapters 14 and 15). These topographic features can be stored in either raster or vector format.

Satellite images play an important role in data integration. Georeferenced satellite images are like DOQs, useful for displaying with other spatial

Box 5.5   Linking Vector Data with Images

L inking a vector-based feature with an image is probably the most common and useful type of data integration. An example of this hyperlink application is to link a map showing houses for sale with pictures of the houses so that, when a house (point) is clicked, it

opens a picture of the house in a graphic file. The hyperlink tool in ArcMap can link a point, line, or polygon feature in vector format with an image. To open the image, the tool uses an attribute that specifies the path to the image.

---

features such as roads, stands, and parcels. But satellite images also contain quantitative spectral data that can be processed to create layers such as land cover, vegetation, urbanization, snow accumulation, and environmental degradation (Mesev et al. 1995; Hinton 1996). As an example, the National Land Cover Data Set for the contermitmous United States is based primarily on Landsat 5 TM data (Vogelmann et al. 2001).

Vector data are regularly used as ancillary information for processing satellite images (Ehlers et al. 1989; Ehlers et al. 1991; Hinton 1996; Jensen 1996; Wilkinson 1996). Image stratification is one example. It uses vector data to divide the landscape into major areas of different characteristics and then treats these areas separately in image processing and classification. Another example is the use of vector data in selecting control points for the georeferencing of remotely sensed data (Couloigner et al. 2002).

Recent developments suggest that a closer integration of GIS and remote sensing has taken place. A GIS package can read files created in an image-processing package, and vice versa. For example, ArcGIS supports files created in ERDAS (IMAGINE, GIS, and LAN files) **(http://gis.leica-geosystems.com/)** and ER Mapper **(http://www.ermapper.com/)**. There are also extensions to a GIS package that can process satellite images. As examples, the Image Analyst extension to ArcGIS has some of the same image analysis functions as in ERDAS IMAGINE and the Feature Analyst extension to ArcGIS can extract features in shapefile format directly from satellite images, especially those of high resolution **(http://www.featureanalyst.com/)**. As high-resolution satellite images gain more acceptances among GIS users, we can expect the tie the between GIS and remote sensing to grow even stronger.

Key Concepts and Terms

Bi-level scanned file: A scanned file containing values of 1 or 0.

Cell-by-cell encoding: A raster data structure that stores cell values in a matrix by row and column.

Data compression: Reduction of data volumes, especially for raster data.

Digital elevation model (DEM):   A digital model with an array of uniformly spaced elevation data in raster format.

Digital orthophoto quad (DOQ):   A digitized image prepared from an aerial photograph or other remotely sensed data, in which the displacement caused by camera tilt and terrain relief has been removed.

Digital raster graphic (DRG):   A scanned image of a USGS topographic map.

ESRI grid: A proprietary ESRI format for raster data.

Floating-point raster: A raster that contains cells of continuous values.

Georeferenced raster: A raster that has been processed to align with a projected coordinate system.

Integer raster: A raster that contains cell values of integers.

Landsat: An orbiting satellite that provides repeat images of the Earth's surface. Landsat 7 was launched in April 1999.

Lossless compression: One type of data compression that allows the original image to be precisely reconstructed.

Lossy compression: One type of data compression that can achieve high-compression ratios but cannot reconstruct fully the original image.

Quad tree: A raster data structure that divides a raster into a hierarchy of quadrants.

Raster data model: A data model that uses rows, columns, and cells to construct spatial features.

Rasterization: Conversion of vector data to raster data.

Run-length encoding (RLE):   A raster data structure that records the cell values by row and by group. A run-length encoded file is also called run-length compressed (RLC) file.

SPOT: A French satellite that provides repeat images of the Earth's surface. SPOT 5 was launched in May 2002.

Vectorization: Conversion of raster data to vector data.

Wavelet transform: A new image compression technique that treats an image as a wave and progressively decomposes the wave into simpler wavelets.

REVIEW QUESTIONS

1. What are the basic elements of the raster data model?

2. Explain the advantages and disadvantages of the raster data model vs. the vector data model.

3. Name two examples each for integer rasters and floating-point rasters.

4. Explain the relationship between cell size, raster data resolution, and raster representation of spatial features.

5. You are given the following information on a 30-meter DEM:
   • UTM coordinates in meters at the lower-left corner: 560635, 4816399
   • UTM coordinates in meters at the upper-right corner: 570595, 4830380
   How many rows does the DEM have? How many columns does the DEM have? And, what are the UTM coordinates at the center of the (row 1, column 1) cell?

6. Go to either the Space Imaging website (**http://www.spaceimaging.com/**) or the DigitalGlobe website (**http://www.digitalglobe.com/**), and take a look at their high-resolution sample imagery.

7. How many 7.5-minute DEMs are covered by one 30-minute DEM?

8. You can find the status graphics for USGS DEMs, DRGs, and DOQs at the following website: **http://statgraph.cr.usgs.gov/viewer.htm.** How much of your state is covered for each of these USGS digital data products?

9. Find the GIS data clearinghouse for your state at the Geospatial One-Stop website (**http://www.geo-one-stop.gov/**). Go to the clearinghouse website. Does the website offer USGS DEMs, DRGs, and DOQs online? Does the website offer both 30-meter and 10-meter USGS DEMs?

10. Use a diagram to explain how the run-length encoding method works.

11. Refer to the figure below, draw a quad tree, and code the spatial index of the shaded (spatial) feature.

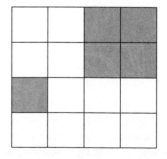

12. A number of government agencies have used MrSID to store digital images such as DOQs and DRGs. When these images are made available online, we can see them at different resolutions by zooming in or out. Go to the Massachusetts GIS website **(http://www.state.ma.us/mgis/mrsid.htm)** and view the DOQs and DRGs at different resolutions.

13. Explain the difference between lossless and lossy compression methods.

14. What is vectorization?

15. Use an example from your discipline and explain the usefulness of integrating vector and raster data.

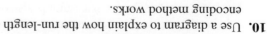

APPLICATIONS: RASTER DATA MODEL

This applications section has three tasks. The first two tasks let you view two types of raster data: DEM and Landsat TM imagery. Task 3 covers the conversion of two shapefiles, one line and one polygon, to raster data.

Task 1: View USGS DEM Data

What you need: *Task1*, a folder that contains data files for a USGS 7.5-minute DEM in SDTS (spatial data transfer standard) format.

In Task 1, you will import a USGS 7.5-minute DEM to a grid and use ArcCatalog to examine the grid's properties.

1. Start ArcCatalog and make connection to the Chapter 5 database. Click Show/Hide ArcToolbox Windows to open the ArcToolbox window in ArcCatalog. Double-click the Import From SDTS tool in the Coverage Tools/Conversion/To Coverage toolset. An alternative is to use the SDTS Raster to Grid tool of ArcView 8x Tools, available in ArcCatalog's View menu.

2. In the Import From SDTS dialog, use the browse button to navigate to the data files in the *Task1* folder. The data files all have the 8146 prefix. Double-click any of the files. The Input SDTS Transfer File Prefix in the dialog should list 8146. Change the output grid name to *Menan-Buttes* and save it in the Chapter 5 database. Click OK to execute the conversion. The conversion creates an elevation grid and ten tables associated with the grid.

3. This step examines *Menan-Buttes*, an elevation raster created in Step 2. Right-click *Menan-Buttes* in the Catalog tree and select Properties. The General tab shows information about *Menan-Buttes* in five categories: data source, raster information, extent, spatial reference, and statistics. An integer grid with a cell size of 30 (meters), *Menan-Buttes* is projected onto the NAD_1927_UTM_Zone_12N coordinate system. *Menan-Buttes* has a minimum elevation of 4771 (feet) and a maximum elevation of 5619.

Q1. How many rows and columns does *Menan-Buttes* have?

Q2. What are the *x*- and *y*-coordinates at the upper-left corner of *Menan-Buttes*?

4. Launch ArcMap. Rename the data frame Task 1, and add *Menan-Buttes* to Task 1. Right-click *Menan-Buttes* and select Properties. On the Symbology tab, right-click on the Color Ramp box and uncheck Graphic View. Then select Elevation #1 from the Color Ramp's dropdown menu. Dismiss the Properties dialog. ArcMap now displays the dramatic landscape of the twin buttes.

Task 2: View a Satellite Image in ArcMap

What you need: *tmrect.bil*, a Landsat TM image comprised of the first five bands.

Task 2 lets you view a Landsat TM image with five bands. By changing the color assignments of the bands, you can alter the view of the image.

1. Right-click *tmrect.bil* in ArcCatalog and select Properties. The General tab shows that *tmrect.bil* has 366 rows, 651 columns, and 5 bands.

Q3. Can you verify that *tmrect.bil* is stored in the band interleaved by line format?

Q4. What is the pixel size (in meters) of *tmrect.bil*?

2. Launch ArcMap. Insert a new data frame and rename it Task 2. Add *tmrect.bil* to Task 2. The table of contents shows *tmrect.bil* as a RGB Composite with Red for Band_1, Green for Band_2, and Blue for Band_3.

3. Select Properties from the context menu of *temrect.bil*. On the Symbology tab, use the dropdown menus to change the RGB composite: Red for Band_3, Green for Band_2, and Blue for Band_1. Click OK. You should see the image as a color photograph.

4. Next, use the following RGB composite: Red for Band_4, Green for Band_3, and Blue for Band_2. You should see the image as a color infrared photograph.

Task 3: Convert Vector Data to Raster Data

What you need: *nwroads.shp* and *nwcounties.shp*, shapefiles showing major highways and counties in the Pacific Northwest, respectively.

In Task 3, you will convert a line shapefile (*nwroads.shp*) and a polygon shapefile (*nwcounties.shp*) to grids. Covering Idaho, Washington, and Oregon, both shapefiles are projected onto a Lambert conformal conic projection and are measured in meters.

1. Insert a new data frame in ArcMap and rename it Task 3. Add *nwroads.shp* and *nwcounties.shp* to Task 3. Open ArcToolbox.

2. Double-click the Feature to Raster tool in the Conversion Tools/To Raster toolset. Select *nwroads* for the input features, select RTE_NUM1 for the field, save the output raster as *nwroads_gd*, enter 5000 for the output cell size, and click OK to run the conversion. *nwroads_gd* appears in the map in different colors. Each color represents a numbered highway. The highways look blocky because of the large cell size (5000 meters).

3. Double-click the Feature to Raster tool. Select *nwcounties* for the input features, select FIPS for the field, save the output raster as *nwcounties_gd*, enter 5000 for the output cell size, and click OK. *nwcounties_gd* appears in the map with symbols representing the classified values of 1 to 119 (119 is the number of counties). Double-click *nwcounties_gd* in the table of contents. On the Symbology tab, select Unique Values in the Show frame and click OK. Now the map shows *nwcounties_gd* with a unique symbol for each county.

Q5. *nwcounties_gd* has 157 rows and 223 columns. If you had used 2500 for the Output cell size, how many rows would the output grid have?

Challenge Task

What you need: *emidalat*, an elevation raster; and *idtm.shp*, a polygon shapefile.

A USGS DEM, *emidalat* is projected onto the UTM coordinate system. *idtm.shp*, on the other hand, is based on the Idaho Transverse Mercator (IDTM) coordinate system. This challenge task asks you to project *emidalat* onto the IDTM coordinate system. It also asks you for the layer information about *emidalat*.

1. Use the Metadata tab in ArcCatalog to read the spatial reference information about *emidalat* and *idtm.shp*, including the datum.

2. Use the Project Raster tool in the Data Management Tools/Projections and Transformations/Raster toolset to project *emidalat* onto the IDTM coordinate system. Use the default resampling technique and a cell size of 30 (meters). Rename the output raster *emidatm*.

3. Launch ArcMap. Rename the new data frame Challenge, and add *idtm.shp* and *emidatm* to Challenge. *emidatm* should appear as a very small rectangle in northern Idaho.

Q1. What is the maximum elevation in *emidatm*?

Q2. Is *emidatm* a floating-point grid or an integer grid?

Q3. How many rows and columns does *emidatm* have?

REFERENCES

Addison, P. S. 2002. *The Illustrated Wavelet Transform Handbook.* Bristol, UK: Institute of Physics Publishing.

Bregt, A. K., J. Denneboom, H. J. Gesink, and Y. Van Randen. 1991. Determination of Rasterizing Error: A Case Study with the Soil Map of the Netherlands. *International Journal of Geographical Information Systems* 5: 361–67.

Chrisman, N. 2001. *Exploring Geographic Information Systems,* 2d ed. New York: Wiley.

Christopoulos, C., A. Skodras, and T. Ebrahimi. 2000. The JPEG2000 Still Image Coding System: An Overview. *IEEE Transactions on Consumer Electronics* 46 (4): 1103–1127.

Clarke, K. C. 1995. *Analytical and Computer Cartography,* 2d ed. Englewood Cliffs, NJ: Prentice Hall.

Couloigner, I., K. P. B. Thomson, Y. Bedard, B. Moulin, E. LeBlanc, C. Djima, C. Latouche, and

N. Spicher. 2002. Towards Automating the Selection of Ground Control Points in Radarsat Images Using a Topographic Database and Vector-Based Data Matching. *Photogrammetric Engineering and Remote Sensing* 68: 433–40.

Dutton, G. H. 1999. *A Hierarchical Coordinate System for Geoprocessing and Cartography.* Berlin: Springer-Verlag.

Ehlers, M., G. Edwards, and Y. Bedard. 1989. Integration of Remote Sensing with Geographical Information Systems: A Necessary Evolution. *Photogrammetric Engineering and Remote Sensing* 55: 1619–27.

Ehlers, M., D. Greenlee, T. Smith, and T. Star. 1991. Integration of Remote Sensing and GIS: Data and Data Access. *Photogrammetric Engineering and Remote Sensing* 57: 669–75.

Flood, M. 2001. Laser Altimetry: From Science to Commercial LIDAR Mapping.

Photogrammetric Engineering and Remote Sensing 67: 1209–17.

Hinton, J. E. 1996. GIS and Remote Sensing Integration for Environmental Applications. *International Journal of Geographical Information Systems* 10: 877–90.

Jensen, J. R. 1996. *Introductory Digital Image Processing: A Remote Sensing Perspective,* 2d ed. Upper Saddle River, NJ: Prentice Hall.

Li, Z., X. Yuan, and K. W. K. Lam. 2002. Effects of JPEG Compression on the Accuracy of Photogrammetric Point Determination. *Photogrammetric Engineering and Remote Sensing* 68: 847–53.

Lillesand, T. M., and R. W. Kiefer. 2000. *Remote Sensing and Image Interpretation,* 4th ed. New York: Wiley.

Mesev, T. V., P. A. Longley, M. Batty, and Y. Xie. 1995. Morphology from Imagery—Detecting and

Measuring the Density of Urban Land Use. *Environment and Planning A* 27: 759–80.

Ottoson, P., and H. Hauska. 2002. Ellipsoidal Quadtrees for Indexing of Global Geographical Data. *International Journal of Geographical Information Science* 16: 213–26.

Piwowar, J. M., E. F. LeDraw, and D. J. Dudycha. 1990. Integration of Spatial Data in Vector and Raster Formats in a Geographic Information System Environment. *International Journal of Geographical Information Systems* 4: 429–44.

Samet, H. 1990. *The Design and Analysis of Spatial Data Structures*. Reading, MA: Addison-Wesley.

Sanchez, J., and M. P. Canton. 1999. *Space Image Processing*. Boca Raton, FL: CRC Press.

Tobler, W., and Z. Chen. 1986. A Quadtree for Global Information Storage. *Geographical Analysis* 18: 360–71.

Tomlin, C. D. 1990. *Geographic Information Systems and Cartographic Modeling*. Englewood Cliffs, NJ: Prentice Hall.

Turner, A. K. 2000. LIDAR Provides Better DEM Data. *GEOWorld* 13(11): 30–31.

Vogelmann, J. E., S. M. Howard, L. Yang, C. R. Larson, B. K. Wylie, and N. Van Driel. 2001. Completion of the 1990s National Land Cover Data Set for the Conterminous United States from Landsat Thematic Mapper Data and Ancillary Data Sources. *Photogrammetric Engineering and Remote Sensing* 67: 650–62.

Wilkinson, G. G. 1996. A Review of Current Issues in the Integration of GIS and Remote Sensing Data. *International Journal of Geographical Information Systems* 10: 85–101.

DATA INPUT

CHAPTER OUTLINE

The most expensive part of a GIS project is database construction. Converting from paper maps to digital maps used to be the first step in constructing a database. But in recent years this situation has changed as digital data clearinghouses have become commonplace on the Internet. Now we can look at what are available in the public domain before deciding to create new data. Many government agencies at the federal, state, regional, and local levels have set up clearinghouses for distributing GIS data. Private companies have also been active in marketing GIS data. Some of them produce new GIS data for their customers; others produce value-added GIS data from public data.

The proliferation of online GIS data, both vector- and raster-based, has made it easier to organize a GIS project. But the topic of data input is still important for two reasons. First, public data, as the term suggests, are intended for all GIS users rather than users of a particular software package. Therefore, we must be familiar with metadata and data conversion from one format to another. Second, the proliferation of online data suggests that government agencies and private companies are continually producing new data. New data must be produced to be put on the Internet or to be sold to customers. Even for GIS users, we may not find the digital data we need for a project and may have to create our own data.

New GIS data can be created from a variety of data sources. They include satellite images, field data, street addresses, text files with x and y coordinates, and paper maps. New data can also be created using manual digitizing, scanning, or on-screen digitizing. Our knowledge of data sources and production methods can better prepare us in putting together a GIS database.

Chapter 6 is presented in the following four sections. Section 6.1 discusses existing GIS data on the Internet, including examples from different levels of government and private companies. Sections 6.2 and 6.3 cover metadata and the data exchange methods respectively. Section 6.4 provides an overview of creating new GIS data from different data sources and using different production methods.

6.1 EXISTING GIS DATA

To find existing GIS data for a project is often a matter of knowledge, experience, and luck. Since the early 1990s, government agencies at different levels in the United States have set up websites for sharing public data and for directing users to the source of the desired information (Onsrud and Rushton 1995). The Internet is also a medium for finding existing data from nonprofit organizations and private companies. But searching for GIS data on the Internet can be difficult. A keyword search with GIS will probably result in thousands of matches, but most hits are irrelevant to the user. Internet addresses may be changed or discontinued. Data on the Internet may be in a format that is incompatible with the GIS package used for a project, or to be usable for a project, the data may need extensive processing such as clipping the study area from a large data set or merging several data sets.

Most GIS data on the Internet are data that many organizations regularly use for GIS activities. These are called **framework data,** which typically include seven basic layers: geodetic control (accurate positional framework for surveying and mapping), orthoimagery (rectified imagery such as orthophotos), elevation, transportation, hydrography, governmental units, and cadastral information **(http://www.fgdc.gov/framework/framework.html/).**

Public data are downloadable from the Internet. Most data are free or at the cost of processing fees. All levels of government let GIS users access their public data through clearinghouses in the United States. The following sections present public data available at the federal, state, regional, metropolitan, and county levels as of August 2004 as well as data from private companies.

6.1.1 Federal Geographic Data Committee

The **Federal Geographic Data Committee (FGDC)** is a 19-member interagency committee **(http://www.fgdc.gov/).** Two main activities of the FGDC are the development of metadata standards and coordination of the National Spatial Data Infrastructure (NSDI). The NSDI is aimed at the sharing of geospatial data throughout all levels of government, the private and nonprofit sectors, and the academic community. The FGDC website provides a link to the Geospatial Data Clearinghouse, a collection of 250 spatial data nodes in the United States and overseas. Also available is a listing of websites that provide mostly free U.S. geospatial and attribute data.

6.1.2 Geospatial One-Stop

The **Geospatial One-Stop** is a new geospatial data portal established by the Federal Office of Management and Budget in 2003 as an *e*-government initiative **(http://www.geo-one-stop.gov/).** The Geospatial One-Stop website provides links to the NSDI Clearinghouse as well as statewide, regional, and local clearinghouse websites. One of the main objectives of the Geospatial One-Stop is to expand collaborative partnerships at all levels of government to help leverage investments in geospatial data and to reduce duplication of data.

6.1.3 U.S. Geological Survey

Through its National Map program, the U.S. Geological Survey (USGS) is the major provider of GIS data in the United States. Its website **(http://geography.usgs.gov/)** offers pathways to USGS national mapping and remotely sensed data and to thematic data clearinghouses on biological, geologic, and water resources data. Public data available from the USGS include both vector and raster data.

Digital line graphs (DLGs) are digital representations of point, line, and area features from the USGS quadrangle maps at the scales of 1:24,000, 1:100,000, and 1:2,000,000. DLGs include such data categories as hypsography (i.e., contour lines and spot elevations), hydrography, boundaries, transportation, and the U.S. Public Land Survey System. DLGs contain attribute data and are topologically structured. It should be noted that the term DLG also refers to a data format (Section 6.3.2).

National Land Cover Data (NLCD) 1992 include 21 thematic classes for the conterminous United States. NLCD 1992 were compiled from the Thematic Mapper (TM) imagery of the early 1990s and other geospatial ancillary data sets. The 21 classes resemble the Anderson level II land use/land cover scheme used by the USGS in the 1970s and early 1980s (Anderson et al. 1976). A new project called National Land Cover Characterization 2001 (NLCD 2001) uses the Landsat 7 ETM+ imagery to compile land cover data for all 50 states and Puerto Rico. Information on both NLCD 1992 and 2001 is available at **http://landcover.usgs.gov/**.

USGS digital elevation models (DEMs) can be downloaded at three designated websites (**http://data.geocomm.com/, http://www.mapmart.com/, http://www.atdi-us.com/**). They include 7.5-minute, 15-minute, and 30-minute DEMs. The 7.5-minute DEMs have either a 30-meter or 10-meter resolution. The **National Elevation Dataset (NED)** is a recent effort made by the USGS to provide 1:24,000 scale DEMs nationwide (1:63,360 scale DEMs for Alaska) (**http://gisdata.usgs.gov/ned/**). NED uses a seamless data distribution system so that DEMs to be downloaded are based on user-defined areas.

Other GIS-related data available from the USGS include Landsat 7 ETM+ data, TM data, digital orthophoto quads (DOQs), digital raster graphics (DRGs), and air photos from the National Aerial Photography Program. The USGS has recently initiated AmericaView, a new program designed to make satellite data from the U.S. government more accessible to the public through a network of state consortia (**http://americaview. usgs.gov/**). The pilot consortium, OhioView, offers Landsat 7 and ASTER data for the State of Ohio and elsewhere (**http://www.ohioview.org/**). The USGS expects to expand the program to all 50 states.

6.1.4 U.S. Census Bureau

The U.S. Census Bureau offers the TIGER/Line files, which are extracts of geographic/cartographic information from its TIGER (**T**opologically **I**nte**g**rated **G**eographic **E**ncoding and **R**eferencing) database. The TIGER/Line files contain legal and statistical area boundaries such as counties, census tracts, and block groups, which can be linked to the census data, as well as roads, railroads, streams, water bodies, power lines, and pipelines (Sperling 1995). TIGER/Line attributes include the address range on each side of a street segment that can be used for address matching.

Several versions of the TIGER/Line files are available for download at the Census Bureau's website (**http://www.census.gov/**). The latest version corresponds to Census 2000 and covers all counties and county equivalents for each state.

6.1.5 Natural Resources Conservation Service

The Natural Resources Conservation Service (NRCS) of the U.S. Department of Agriculture distributes soils data nationwide through its website (**http://soils.usda.gov/**). There are two soil databases: STATSGO and SSURGO. Compiled at 1:250,000 scales, the STATSGO (**State Soil Geographic**) database is suitable for broad planning and management uses. Compiled from field mapping at scales ranging from 1:12,000 to 1:63,360, the SSURGO (**Soil Survey Geographic**) database is designed for uses at the farm, township, and county levels.

6.1.6 Statewide Public Data: An Example

The Geospatial One-Stop website provides a link to every state in the United States for statewide GIS data. For example, the Montana State Library maintains a GIS data clearinghouse in Montana (**http://www.nris.state.mt.us/**). This clearinghouse offers both statewide and regional data. Statewide data include such categories as administrative and political boundary, biological and ecologic, environmental, inland water resources, and transportation networks. These data are available in ArcInfo export files and shapefiles for downloading.

6.1.7 Regional Public Data: An Example

The Greater Yellowstone Area Data Clearinghouse (GYADC) **(http://www.sdvc.uwyo.edu/gya/)** is a FGDC data node sponsored by a group of federal agencies, state agencies, universities, and nonprofit organizations. This data clearinghouse focuses on basic framework data for Yellowstone and Grand Teton National Parks.

6.1.8 Metropolitan Public Data: An Example

Sponsored by 19 local governments in the San Diego region, the San Diego Association of Governments (SANDAG) **(http://www.sandag.cog. ca.us/)** is an example of a metropolitan data clearinghouse. Data that can be downloaded from SANDAG's website include administrative boundaries, base map features, district boundaries, land cover and activity centers, transportation, and sensitive lands/natural resources.

6.1.9 County-Level Public Data: An Example

Many counties in the United States offer GIS data for sale. Clackamas County in Oregon, for example, distributes data in ArcInfo export files, shapefiles, and DXF files through its GIS division **(http://www.co.clackamas.or.us/gis/)**. Examples of data sets include zoning boundaries, flood zones, tax lots, school districts, voting precincts, park districts, and fire districts.

6.1.10 GIS Data from Private Companies

Many GIS companies are engaged in software development, technical service, consulting, and data production. Some also provide free data or samples of commercial data on the Internet or can direct GIS users to suitable sources. ESRI, Inc., for example, offers the Geography Network **(http:// www.geographynetwork.com/)**, a clearinghouse

with data provided by organizations worldwide. (The Geography Network can be accessed directly from ArcMap.)

Some companies provide specialized GIS data for their customers. GDT (Geographic Data Technology) offers street, address, postal, and census databases at its website **(http://www.geographic. com/home/index.cfm)**. Tele Atlas North America **(http://www.teleatlas.com/landingpage.jsp)** offers road and address databases for urban centers and rural areas. (Tele Atlas acquired GDT in April 2004.) In contrast, online GIS data stores tend to carry a variety of geospatial data. Examples of GIS data stores include GIS Data Depot **(http://data. geocomm.com/)**, Map-Mart **(http://www. mapmart. com/)**, and LAND INFO International **(http://www. landinfo.com/)**.

6.2 METADATA

Metadata provide information about geospatial data (Guptill 1999). They are therefore an integral part of GIS data and are usually prepared and entered during the data production process. Metadata are important to anyone who plans to use public data for a GIS project. First, metadata let us know if the data meet our specific needs for area coverage, data quality, and data currency. Second, metadata show us how to transfer, process, and interpret geospatial data. Third, metadata include the contact for additional information.

The FGDC has developed the content standards for metadata and provide detailed information about the standards at their website **(http:// www.fgdc.gov/metadata/metadata.html/)**. These standards have been adopted by federal agencies in developing their public data. FGDC metadata standards describe a data set based on the following categories:

- Identification information—basic information about the data set, including title, geographic data covered, and currency.
- Data quality information—information about the quality of the data set, including positional and attribute accuracy, completeness,

consistency, sources of information, and methods used to produce the data.

- Spatial data organization information—information about the data representation in the data set, such as method for data representation (e.g., raster or vector) and number of spatial objects.
- Spatial reference information—description of the reference frame for, and means of encoding coordinates in the data set, such as the parameters for map projections or coordinate systems, horizontal and vertical datums, and the coordinate system resolution.
- Entity and attribute information—information about the content of the data set, such as the entity types and their attributes and the domains from which attribute values may be assigned.
- Distribution information—information about obtaining the data set.
- Metadata reference information—information on the currency of the metadata information and the responsible party.

FGDC standards do not specify how metadata should be organized in a computer system or in a data transfer, or how metadata should be transmitted to the user. To assist the entry of metadata, many metadata tools have been developed over the past several years (**http://www.fgdc. gov/metadata/metatool.html/**). These tools are typically designed for different operating systems. Some tools are free and some are designed for specific GIS packages. As examples, ESRI, Inc. offers a metadata creation tool in ArcCatalog and Intergraph offers the Spatial Metadata Management System (SMMS) for GeoMedia.

6.3 CONVERSION OF EXISTING DATA

Public data are delivered in a variety of data formats. Unless the data format is compatible with the GIS package in use, we must first convert the data. **Data conversion** is defined here as a mechanism for converting GIS data from one format to another. Data conversion can be easy or difficult; it depends upon the specificity of the data format. Proprietary data formats require special translators for data conversion, whereas neutral or public formats require a GIS package that has translators to work with the formats.

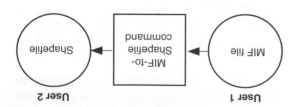

Figure 6.1
The MIF to Shapefile tool in ArcGIS converts a MapInfo file to a shapefile.

6.3.1 Direct Translation

Direct translation uses a translator in a GIS package to directly convert geospatial data from one format to another (Figure 6.1). Direct translation used to be the only method for data conversion before the development of data standards and open GIS. Many users still prefer direct translation because it is easier to use than other methods. ArcToolbox in ArcGIS, for example, can translate ArcInfo's interchange files, MGE and Microstation's DGN files, Auto-CAD's DXF and DWG files, and MapInfo files into shapefiles or geodatabases. Likewise, GeoMedia can access and integrate data from ArcGIS, AutoCAD, MapInfo, MGE, and Microstation.

6.3.2 Neutral Format

A **neutral format** is a public or de facto format for data exchange. DLG is a common neutral format. The USGS provides DLG files, and the Natural Resources Conservation Service (NRCS) uses DLG as one of the formats for distributing digital soil maps (Box 6.1). There are two DLG formats: DLG standard and DLG optional. Most commercial GIS packages have translators to work with the DLG formats (Figure 6.2).

The **Spatial Data Transfer Standard (SDTS)** is a neutral format approved by the Federal Information Processing Standards (FIPS) Program in 1992 (**http://mcmcweb.er.usgs.gov/sdts/**). Several

Box 6.1 | SSURGO Data Delivery Formats

As of July 2004, there are two options to download SSURGO (Soil Survey Geographic) data from the NRCS (Natural Resources Conservation Service) website. The first option is to download a SSURGO data set, which includes map data, attribute data, and metadata. Map data are available in modified digital line graph (DLG-3) optional and Arc interchange file formats. Attribute data are available in ASCII format with DLG-3 map data and in Arc interchange format with Arc interchange map files. And metadata are in ASCII format.

The second option, which supersedes the first option, is to download both spatial and attribute data through the Soil Data Mart. Spatial data are available in shapefile, coverage, and Arc interchange file formats. Attribute data are in ASCII format. The Soil Data Mart provides a template database so that attribute data can be easily imported as tables into an Access database.

federal agencies have converted some of their data to SDTS format. They include the USGS, U.S. Army, U.S. Army Corps of Engineers, Census Bureau, and U.S. National Oceanic and Atmospheric Administration. The USGS, for example, has converted many DLG files into SDTS format. These files are sometimes called SDTS/DLG files. GIS vendors such as ESRI, Inc., Intergraph, and MapInfo provide translators in their software packages for importing SDTS data.

In practice, SDTS uses "profiles" to transfer spatial data. Each profile is targeted at a particular type of spatial data. Currently there are five SDTS profiles:

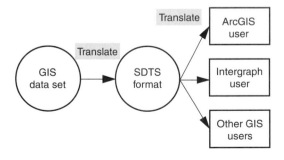

Figure 6.2
To accommodate users of different GIS packages, a government agency can translate public data into a neutral format such as SDTS format. Using the translator in the GIS package, the user can convert the public data into the format used in the GIS.

- The Topological Vector Profile (TVP) covers DLG, TIGER, and other topology-based vector data.
- The Raster Profile and Extensions (RPE) accommodate digital orthophotos, digital elevation models, and other raster data.
- The Transportation Network Profile (TNP) covers vector data with network topology.
- The Point Profile supports geodetic control point data.
- The Computer Aided Design and Drafting Profile (CADD) supports vector-based CADD data, with or without topology.

USGS 7.5-minute DEMs that can be downloaded online are typically in SDTS format. So are USGS DLG files. Creating an elevation raster from a SDTS raster profile transfer is relatively straightforward. But creating a topology-based vector data set from a SDTS topological vector profile transfer can be challenging because a topological vector profile transfer may contain composite features such as routes and regions in addition to topology.

The **vector product format (VPF)** is a standard format, structure, and organization for large geographic databases that are based on the georelational data model. The National Geospatial-Intelligence Agency (NGA) uses VPF for digital vector products developed at a variety of scales **(http://www.nga.mil/)**. NGA's vector products for drainage system, transportation, political boundaries, and populated places are also part of the global database that is being developed by the International Steering Committee for Global Mapping (ISCGM)(**http://www.iscgm.org/html4/index.html**). Similar to a SDTS topological vector profile, a VPF file may contain composite features of regions and routes. Box 6.2 summarizes tools in ArcGIS for importing DLG, SDTS, TIGER, and VPF files.

Although a neutral format is typically used for public data from government agencies, it can also be found with "industry standards" in the private sector. A good example is the DXF (drawing interchange file) format of AutoCAD. Another example is the ASCII format. Many GIS packages can import ASCII files, which have point data with x-, y-coordinates, into digital data sets.

6.4 CREATING NEW DATA

Section 6.4 covers a variety of data sources that we can use to create new geospatial data. Address geocoding, also called address matching, can create point features from street addresses. Street addresses are therefore an important data source for creating new data. Address geocoding is covered in Chapter 17.

6.4.1 Remotely Sensed Data

Satellite images can be digitally processed to produce a wide variety of thematic data for a GIS project. Land use/land cover data such as USGS National Land Cover Data are typically derived from satellite images. Other types of data include vegetation types, crop health, eroded soils, geologic features, the composition and depth of water bodies, and even snowpack. Satellite images provide timely data and, if collected at regular intervals, they can also provide temporal data that are valuable for recording and monitoring changes in the terrestrial and aquatic environments.

Some GIS users felt in the past that satellite images did not have sufficient resolution, or were not accurate enough, for their projects. This is no longer the case with high-resolution satellite images (Chapter 5). Ikonos and QuickBird images can now be used to extract detailed features such as roads, trails, buildings, trees, riparian zones, and impervious surfaces.

Digital orthophoto quads (DOQs) are digitized aerial photographs that have been differentially rectified or corrected to remove image displacements by camera tilt and terrain relief. DOQs therefore combine the image characteristics of a photograph with the geometric qualities of a map. Black-and-white DOQs have a 1-meter ground resolution (i.e., each pixel in the image measures 1-by-1 meter on the ground) and pixel values representing 256 gray levels (Figure 6.3). DOQs can

---

Box 6.2 Importing DLG, SDTS, TIGER, and VPF Files in ArcGIS

Tools for importing DLG, SDTS, TIGER, and VPF files are all available to ArcInfo users from the Coverage Tools/Conversion/To Coverage toolset. The Import From DLG tool can convert a DLG file, either standard or optional, into a coverage. The Import From SDTS tool can convert a Topological Vector Profile or a Point Profile into a coverage and a Raster Profile into an ESRI grid. Although all SDTS files carry the same DDF extension, the tool can automatically determine which type of profile is being converted through the header file information. The Advanced Tiger Conversion tool can convert a set of TIGER/Line files into a set of coverages. The tool also has an option for the user to choose which version of TIGER/Line files (e.g., 2000) to convert. The Import from VPF tool can convert an un-tiled VPF coverage or VPF tile into a coverage. If a VPF coverage has multiple linear feature classes, it is converted into a line coverage with route subclasses. And if a VPF coverage has multiple area feature classes, it is converted into a polygon coverage with region subclasses.

ArcView users can use the SDTS Point to Coverage and SDTS Raster to Grid tools to import SDTS files. These ArcView 8x tools are available in ArcCatalog's View menu.

Figure 6.3
A digital orthophoto quad (DOQ) can be used as the background for digitizing or updating geospatial data.

be effectively used as a background for digitizing or updating of new roads, new subdivisions, and timber harvested areas.

6.4.2 Field Data

Two important types of field data are survey data and global positioning system (GPS) data. Survey data consist primarily of distances, directions, and elevations. Distances can be measured in feet or meters using a tape or an electronic distance measurement instrument. The direction of a line can be measured in azimuth or bearing using a transit, theodolite, or total station. An azimuth is an angle measured clockwise from the north end of a meridian to the line. Azimuths range in magnitude from 0° to 360°. A bearing is an acute angle between the line and a meridian. The bearing angle always has the accompanied letters that locate the quadrant (i.e., NE, SE, SW, or NW) in which the line falls. In the United States, most legal plans use bearing directions. An elevation difference between two points can be measured in feet or meters using levels and rods.

In GIS, field survey typically provides data for determining parcel boundaries. An angle and a

distance can define a parcel boundary between two stations (points). For example, the description of N45°30′W 500 feet means that the course (line) connecting the two stations has a bearing angle of 45 degrees 30 minutes in the NW quadrant and a distance of 500 feet (Figure 6.4). A parcel represents a close traverse, that is, a series of

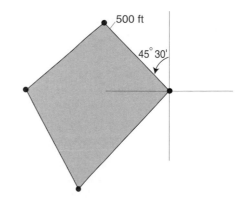

Figure 6.4
A bearing and a distance determine a course between two stations.

established stations tied together by angle and distance (Kavanagh 2003). A close traverse also begins and ends at the same point. **Coordinate geometry (COGO)**, a study of geometry and algebra, provides the methods for creating digital spatial data of points, lines, and polygons from survey data.

Using GPS satellites in space as reference points, a GPS receiver can determine its precise position on the Earth's surface (Moffitt and Bossler 1998). **GPS data** include the horizontal location based on a geographic or projected coordinate system and, if chosen, the elevation of the point location (Box 6.3). A collection of GPS positions along a line can determine a line feature, and a series of lines measured by GPS can determine an area feature. This is why GPS has become a useful tool for spatial data input (Kennedy 1996). The GPS receiver measures its distance (range) from a satellite using the travel time and speed of signals it receives from the satellite. With three satellites simultaneously available, the re-

ceiver can determine its position in space (x, y, z), relative to the center of mass of the Earth. But to correct timing errors, a fourth satellite is required to get precise positioning (Figure 6.5). The receiver's position in space can then be converted to latitude, longitude, and height based on a reference spheroid (i.e., WGS84).

The U.S. military maintains a constellation of 24 NAVSTAR (Navigation Satellite Timing and Ranging) satellites in space, and each satellite follows a precise orbit. This constellation gives GPS users between five and eight satellites visible from any point on the Earth's surface. Data transmitted by GPS satellites are modulated onto two carrier waves, referred to as L1 and L2. Two binary codes are in turn modulated onto the carrier waves. These two codes are the coarse acquisition (C/A) code, which is available to the public, and the precise (P) code, which is designed for military use exclusively. In addition to NAVSTAR satellites, there are also the Russian GLONASS system and the forthcoming European Galileo system.

Box 6.3 An Example of GPS Data

The following printout is an example of GPS data. The header information shows that the datum used is NAD27 (North American Datum of 1927) and the coordinate system is UTM (Universal Transverse Mercator). The GPS data include seven point locations. The record for each point location includes the UTM zone number (i.e., 11), Easting (x-coordinate), and Northing (y-coordinate). The GPS data do not include the height or Alt value.

H R DATUM
M G NAD27 CONUS
H Coordinate System
U UTM UPS

H IDNT	Zone	Easting	Northing	Alt	Description
W 001	11T	0498884	5174889	−9999	09-SEP-98
W 002	11T	0498093	5187334	−9999	09-SEP-98
W 003	11T	0509786	5209401	−9999	09-SEP-98
W 004	11T	0505955	5222740	−9999	09-SEP-98
W 005	11T	0504529	5228746	−9999	09-SEP-98
W 006	11T	0505287	5230364	−9999	09-SEP-98
W 007	11T	0501167	5252492	−9999	09-SEP-98

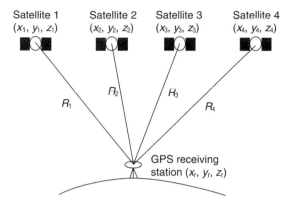

Figure 6.5
Use four GPS satellites to determine the coordinates of a receiving station. x_i, y_i, and z_i are coordinates relative to the center of mass of the Earth. R_i represents the distance (range) from a satellite to the receiving station.

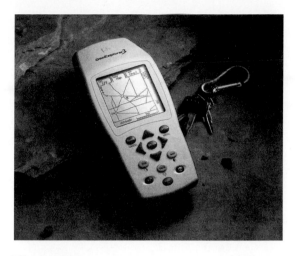

Figure 6.6
A portable GPS receiver. (Courtesy of Trimble.)

An important aspect of using GPS for spatial data entry is the need to correct errors in GPS data. The first type may be described as noise errors, which include ephemeris (positional) error, clock errors (orbital errors between monitoring times), atmospheric delay errors, and multipath errors (signals bouncing off obstructions before reaching the receiver). The second type of error is intentional. For example, to make sure that no hostile force could get precise GPS readings, the U.S. military used to degrade their accuracy under the policy called "Selective Availability" or "SA" by introducing noise into the satellite's clock and orbital data. SA was switched off in May 2000, and GPS accuracies in basic point positioning improved from 100 meters to about 10 to 20 meters.

With the aid of a reference or base station, **differential correction** can significantly reduce noise errors. Located at points that have been accurately surveyed, reference stations are operated by private companies and by public agencies such as those participating in the National Geodetic Survey (NGS) Continuously Operating Reference System (CORS). Using its known position, the reference receiver can calculate what the travel time of the GPS signals should be. The difference between the predicted and actual travel times thus

becomes an error correction factor. The reference receiver computes error correction factors for all visible satellites. These correction factors are then available to GPS receivers within about 300 miles (500 kilometers) of the reference station. GIS applications usually do not need real-time transmission of error correction factors. Differential correction can be made later as long as records are kept of measured positions and the time each position is measured.

Equally important as correcting errors in GPS data is the type of GPS receiver. Most GIS users use code-based receivers (Figure 6.6). With differential correction, code-based GPS readings can easily achieve the accuracy of 3 to 5 meters, and some newer receivers are even capable of submeter accuracy. Carrier phase receivers and dual-frequency receivers are mainly used in surveying and geodetic control. They are capable of subcentimeter differential accuracy (Lange and Gilbert 1999).

GPS data can include elevation readings (also called orthometric heights) at point locations. These readings are measured from the surface of the geoid, which is treated as the surface of mean sea level (Figure 6.7). The geoid is a closer approximation of the Earth than a spheroid (Chapter 2). But the geoid cannot be modeled mathematically. Wherever the

6.4.3 Text Files with x-, y-Coordinates

Digital spatial data can be generated from a text file that contains x-, y-coordinates. The x-, y-coordinates can be geographic (in decimal degrees) or projected. Each pair of x-, y-coordinates create a point. Therefore, we can create spatial data from a file that records the locations of weather stations, epicenters, or a hurricane track.

6.4.4 Digitizing Using a Digitizing Table

Digitizing is the process of converting data from analog to digital format. Manual digitizing uses a digitizing table (Figure 6.8). A **digitizing table** has a built-in electronic mesh, which can sense the position of the cursor. To transmit the x-, y-coordinates of a point to the connected computer, the operator simply clicks on a button on the cursor after lining up the cursor's cross hair with the point. Large-size digitizing tables typically have an absolute accuracy of 0.001 inch (0.003 centimeter).

Many GIS packages have a built-in digitizing module for manual digitizing. The module is likely to have commands that can help move or snap a feature (i.e., a point or line) to a precise location. Figure 6.9 shows a line can be snapped to an existing line within a user-specified tolerance or distance.

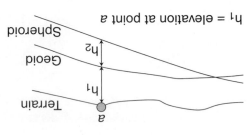

h_1 = elevation at point a

h_2 = geoid undulation at point a

$h_1 + h_2$ = spheroid height at point a

Figure 6.7
Elevation readings from a GPS receiver are measured from the surface of the geoid rather than the spheroid.

mass of the Earth's crust changes, the geoid's gravitational potential also changes, resulting in an unpredictable surface. The separation between the geoid surface and the spheroid surface, also called geoid undulation, must be measured locally at specific sites. In the United States, the NGS provides GEOID99, a geoid-elevation estimation model referenced to the GRS80 spheroid. The model can produce geoid-spheroid separation data as well as contours of these data.

(a)

(b)

Figure 6.8
A large digitizing table (a) and a cursor with a 16-button keypad (b). (Courtesy of GTCO Calcomp, Inc.)

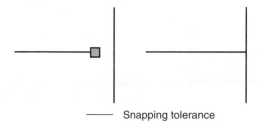

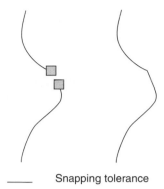

—— Snapping tolerance

Figure 6.9
The end of a new line can be automatically snapped to an existing line if the gap is smaller than the specified snapping tolerance.

——— Snapping tolerance

Figure 6.10
A point can be automatically snapped to another point if the gap is smaller than the specified snapping tolerance.

Likewise, Figure 6.10 shows a point (a node or vertex) can be snapped to another point, again within a specified distance.

Digitizing usually begins with a set of control points (also called tics), which are later used for converting the digitized map to real-world coordinates (Chapter 7). Digitizing point features is simple: each point is clicked once to record its location. Digitizing line features can follow either point mode or stream mode. The operator selects points (nodes and vertices) to digitize in point mode. In stream mode, lines are digitized at a preset time or distance interval. For example, lines can be automatically digitized at a 0.01-inch interval. Point mode is preferred if features to be digitized have many straight-line segments. Because the vector data model treats a polygon as a series of lines, digitizing polygon features is the same as digitizing line features. The coverage model requires each polygon to have a label, which is treated as a feature similar to the polygon's boundary.

Although digitizing itself is mostly manual, the quality of digitizing can be improved with planning and checking. An integrated approach is useful in digitizing different layers of a GIS database that share common boundaries. For example, soils, vegetation types, and land-use types may share some common boundaries in a study area. Digitizing these boundaries only once and using them on each layer not only saves time in digitizing but also ensures the matching of the layers.

A rule of thumb in digitizing line or polygon features is to digitize each line once and only once to avoid duplicate lines. Duplicate lines are seldom on top of one another because of the high accuracy of a digitizing table. One way to reduce the number of duplicate lines is to put a transparent sheet on top of the source map and to mark off each line on the transparent sheet after the line is digitized. This method can also reduce the number of missing lines.

6.4.5 Scanning

Scanning is a digitizing method that converts an analog map into a scanned file, which is then converted back to vector format through tracing (Verbyla and Chang 1997). A scanner (Figure 6.11) converts an analog map into a scanned image file in raster format. The map to be scanned is typically a black-and-white map: black lines represent map features, and white areas represent the background. The map may be a paper or Mylar map, and inked or penciled. Scanning converts the map into a binary scanned file in raster format; each pixel has a value of either 1 (map feature) or 0 (background). Map features are shown as raster lines, a series of connected pixels on the scanned file (Figure 6.12). The pixel size depends on the scanning resolution, which is often set at 300 dots per inch (dpi) or 400 dpi for digitizing. A raster line representing a thin inked line on the source map may have a width of 5 to 7 pixels (Figure 6.13).

Scribe sheets and color maps can also be scanned for digitizing. Because a scribed sheet is essentially a negative of a black-and-white map, the values of 1 and 0 are reversed in the scanned file. Expensive scanners can recognize colors. A digital raster graphic (DRG), for example, can have 13 different colors, each representing one type of map feature scanned from a USGS quadrangle map.

To complete the digitizing process, a scanned file must be vectorized. **Vectorization** turns raster lines into vector lines in a process called tracing. Tracing or vectorization involves three basic elements: line thinning, line extraction, and topological reconstruction. Tracing can be semiautomatic or manual. In semiautomatic mode, the user selects a starting point on the image map and lets the computer trace all the connecting raster lines (Figure 6.14). In manual mode, the user determines the raster line to be traced and the direction of tracing. Results of tracing depend on the robustness of the tracing algorithm that is built in the GIS package. Although no single tracing algorithm will work satisfactorily with different types of maps under different conditions, some algorithms are better than others. Examples of problems that must be solved by the tracing algorithm include: how to trace an intersection, where the width of a raster line may double or triple (Figure 6.15); how to continue when a raster line is broken or when two

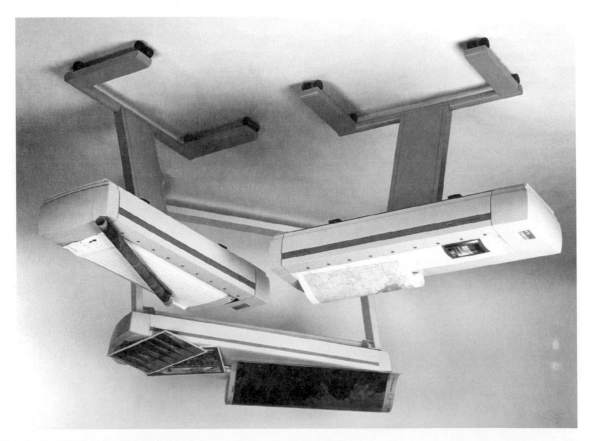

Figure 6.11
Large format drum scanners. (Courtesy of GTCO Calcomp, Inc.)

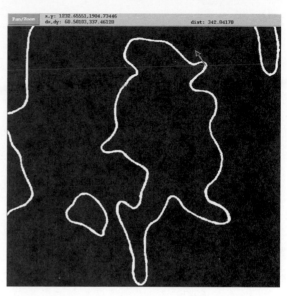

Figure 6.12
A binary scanned file: The lines are soil lines, and the black areas are the background.

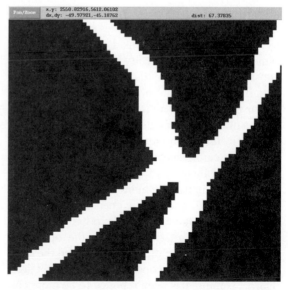

Figure 6.14
Semiautomatic tracing starts at a point (shown with an arrow) and traces all lines connected to the point.

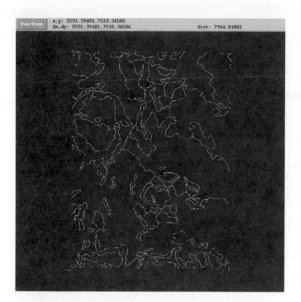

Figure 6.13
A raster line in a scanned file has a width of several pixels.

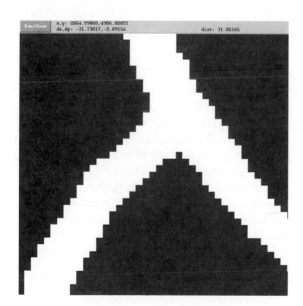

Figure 6.15
The width of a raster line doubles or triples when lines meet or intersect.

raster lines are close together; and how to separate a line from a polygon. A tracing algorithm normally uses the parameter values specified by the user in solving the problems. Box 6.4 lists the parameters that we can specify in ArcGIS.

Is scanning better than manual digitizing for data input? Large data producers apparently think so for the following reasons. First, scanning uses the machine and computer algorithm to do most of the work, thus avoiding human errors caused by fatigue or carelessness. Second, tracing has both the scanned image and vector lines on the same screen, making tracing more flexible than manual digitizing. With tracing, the operator can zoom in or out and can move around the raster image with ease. Manual digitizing requires the operator's attention on both the digitizing table and the computer monitor, and the operator can easily get tired. Third, scanning has been reported to be more cost effective than manual digitizing. The cost of scanning by a service company has dropped significantly in recent years. Scanning a black-and-white map the size of a 1:24,000 scale USGS quadrangle costs less than $10.

6.4.6 On-Screen Digitizing

On-screen digitizing, also called heads-up digitizing, is manual digitizing on the computer monitor using a data source such as a DOQ as the background. DOQs combine the image characteristics of a photograph with the geometric qualities of a map. Easily integrated in a GIS, DOQs are the ideal background for digitizing. On-screen digitizing is an efficient method for editing or updating an existing layer such as adding new trails or roads that are not on an existing layer but are on a new DOQ. Likewise, we can use the method to update new clear-cuts or burned areas in a vegetation layer.

6.4.7 Importance of Source Maps

Despite the increased availability of high-resolution remotely sensed data and GPS data, maps are still a

---

  Box 6.4 Vectorization Settings in ArcGIS

The ArcScan extension to ArcGIS is designed for converting raster data to vector features. ArcScan offers the following seven settings for batch vectorization (i.e., vectorization of an entire raster).

- **Intersection solution:** This setting determines how an intersection (i.e., a point where three or more raster lines meet) is traced. The geometrical solution preserves angles and straight lines, which are appropriate for maps showing streets, canals, and other man-made features. The median solution has nonrectilinear angles and is more appropriate for natural features such as soils, streams, and vegetation covers. The none solution does not connect lines at an intersection.
- **Maximum line width:** This setting separates lines and polygons in vectorization. Centerline vectorization applies to raster lines that are less than or equal to the maximum line width. All other lines are vectorized as polygon features.
- **Compression tolerance:** This setting determines the degree of generalization used in a vector post-processing procedure. A Douglas-Peucker algorithm (Chapter 8) is used for generalization.
- **Smoothing weight:** This setting determines the degree of smoothing in vectorization. A larger weight results in smoother line features.
- **Gap closure tolerance:** This setting determines how many pixels can be jumped over in a broken raster line.
- **Fan angle:** This setting determines the angle that is used by the gap closure function to search for raster lines when jumping over gaps.
- **Holes:** This setting determines the size of holes in a raster line that can be ignored during tracing.

dominant source for creating new GIS data. Digitizing, either manual digitizing or scanning, converts an analog map to its digital format. The accuracy of the digital map is therefore directly related to the accuracy of the source map. The digital map can be only as good or as accurate as its source map.

A variety of factors can affect the accuracy of the source map. Maps such as USGS quadrangle maps are secondary data sources because these maps have gone through the cartographic processes of compilation, generalization, and symbolization. Each of these processes can affect the accuracy of the mapped data. For example, if the compilation of the source map contains errors, these errors will be passed on to the digital map.

Paper maps generally are not good source maps because they tend to shrink and expand with changes in temperature and humidity. In even worse scenarios, GIS users may use copies of paper maps or mosaics of paper map copies for digitizing. Such source maps will not yield good results. Because of

their plastic backing, Mylar maps are much more stable than paper maps for digitizing.

The quality of line work on the source map will determine not only the accuracy of the digital map but also the operator's time and effort in digitizing and editing. The line work should be thin, continuous, and uniform, as expected from inking or scribing—never use felt-tip markers to prepare the line work. Penciled source maps may be adequate for manual digitizing but are not recommended for scanning. Scanned files are binary data files, separating only the map feature from the background. Because the contrast between penciled lines and the background (i.e., surface of paper or Mylar) is not as sharp as inked lines, we may have to adjust the scanning parameters to increase the contrast. But the adjustment often results in the scanning of erased lines and smudges, which should not be in the scanned file. To provide supplemental information on the source map, words or symbols can be drawn in orange color, which will not be scanned.

KEY CONCEPTS AND TERMS

Coordinate geometry (COGO): A branch of geometry that provides the methods for creating digital spatial data of points, lines, and polygons from survey data.

Data conversion: Conversion of geospatial data from one format to another.

Differential correction: A method that uses data from a base station to correct noise errors in GPS data.

Digital line graphs (DLGs): Digital representations of point, line, and area features from USGS quadrangle maps including contour lines, spot elevations, hydrography, boundaries, transportation, and the U.S. Public Land Survey System.

Digitizing: The process of converting data from analog to digital format.

Digitizing table: A table with a built-in electronic mesh that can sense the position of the

cursor and can transmit its x-, y-coordinates to the connected computer.

Direct translation: Use of a translator or algorithm in a GIS package to directly convert spatial data from one format to another.

Federal Geographic Data Committee (FGDC): A U.S. multiagency committee that coordinates the development of geospatial data standards.

Framework data: Data that many organizations regularly use for GIS activities.

Geospatial One-Stop: A portal established by the Federal Office of Management and Budget for accessing geospatial data.

Global positioning system (GPS) data: Longitude, latitude, and elevation data for point locations made available through a navigational satellite system and a receiver.

Metadata: Data that provide information about geospatial data.

National Elevation Dataset (NED):   A USGS program that uses a seamless system for delivering 1:24,000 scale DEMs (1:63,360 scale DEMs for Alaska).

National Land Cover Data (NLCD):   Land use/land cover data from the USGS. NLCD 1992 are based on the Thematic Mapper imagery of the early 1990s and NLCD 2001 the Landsat 7 ETM+ imagery.

Neutral format: A public format such as SDTS that can be used for data exchange.

On-screen digitizing: Manual digitizing on the computer monitor by using a data source such as a DOQ as the background.

Scanning: A digitizing method that converts an analog map into a scanned file in raster format, which can then be converted back to vector format through tracing.

Spatial Data Transfer Standard (SDTS):   Public data formats for transferring geospatial data such as DLGs and DEMs from the USGS.

SSURGO (Soil Survey Geographic):   A soil database compiled from field mapping at scales from 1:12,000 to 1:63,360 by the Natural Resources Conservation Service of the U.S. Department of Agriculture.

STATSGO (State Soil Geographic):   A soil database compiled at 1:250,000 scales by the Natural Resources Conservation Service of the U.S. Department of Agriculture.

TIGER (Topologically Integrated Geographic Encoding and Referencing):   A database prepared by the U.S. Census Bureau that contains legal and statistical area boundaries, which can be linked to the census data.

Vector product format (VPF):   A standard format, structure, and organization for large geographic databases that are based on a georelational data model.

Vectorization: The process of converting raster lines into vector lines through tracing.

REVIEW QUESTIONS

1. What kinds of data are contained in the USGS DLG files?

2. What is SSURGO?

3. Suppose you want to make a map showing the rate of population change between 1990 and 2000 by county in your state. Describe (1) the kinds of digital data you will need for the mapping project, and (2) the website(s) you will use to download the data.

4. Find the GIS data clearinghouse for your state at the Geospatial One-Stop website (http://www.geo-one-stop.gov/). Go to the clearinghouse website. Select the metadata of a data set and go over the information in each category.

5. Describe the kinds of data that are contained in a SDTS topological vector profile, point profile, and raster profile.

6. What is VPF?

7. Go to ArcGIS Desktop Help > Getting started > Data types supported in ArcGIS. The help page lists data types (data formats) developed by ESRI, other GIS companies, or government agencies that can be used in ArcGIS. List three non-ESRI data types that can be used directly in ArcGIS, and list three data types that can be used in ArcGIS through import.

8. Describe two common types of field data that can be used in a GIS project.

9. Explain how differential correction works.

10. What types of GPS data errors can be corrected by differential correction?

11. What kinds of data must exist in a text file so that the text file can be converted to a shapefile?

12. What is COGO?

13. Suppose you are asked to convert a paper map to a digital data set. What methods can you use for the task? What are the advantages and disadvantages of each method?

14. Explain the difference between point mode and stream mode for digitizing.

15. The scanning method for digitizing involves both rasterization and vectorization. Why?

16. The source map can greatly influence the quality of a digitized map. Provide an example that supports the statement.

APPLICATIONS: DATA INPUT

This applications section covers three methods for data input. Task 1 uses existing data on the Internet. Task 2 covers on-screen digitizing. Task 3 uses a table with *x*-, *y*-coordinates. There are other methods besides the ones in this section for data input. Chapter 7 covers use of a scanned file for digitizing and Chapter 17 covers address geocoding.

Task 1: Download and Process DEM and DLG from the Internet

What you need: access to the Internet and WinZip.

The first part of Task 1 guides you through the process of downloading a DEM and converting it to a grid. The second part shows you how to download and convert a DLG to a coverage. Both the DEM and the DLG are in SDTS format, and the SDTS files are in the TAR.GZ zipped format. If you do not have access to WinZip, you can skip Steps 1 to 6 and use the following folders in the Chapter 6 database to complete the task: *elevation_cp*, *hydrography_cp*, and *masterdd_cp*. Before using them, rename the folders as *elevation*, *hydrography*, and *masterdd* respectively.

1. Go to the GIS Data Depot website **http://data.geocomm.com/.** (GIS Data Depot is one of three websites for downloading USGS data.)

2. Use the following links to download the Menan Buttes, ID DEM 24K: USGS DEMs > Download DEM Data Here > Click Idaho on the map > Select Madison > Click DEM 24K > Select Menan Buttes, ID > Set up an account with GeoCommunity > Click

1704439.DEM.SDTS.TAR.GZ (10 meter) to download > Save the file to the Chapter 6 database. This DEM has a 10-meter resolution.

3. Use WinZip to open the downloaded file and save the extracted files in a folder named *elevation* in the Chapter 6 database.

4. Now you will download a DLG and a master data dictionary for the same area. Go to the USGS Geographic Data Download website (**http://edc.usgs.gov/geodata/**). Click on the 1:24K DLG, and select FTP via State. Click Idaho, and then Menan Buttes. Click on hydrography and then 1573016.HY.sdts.tar.gz. Save the file in the Chapter 6 database. Go back to the USGS Geographic Data Download web page. Click on SDTS DLGs require Master Data Dictionary. On the next page, click on 1:24K as a single tar file and save the file (00MASTERDD_LRG.SDTS.tar.gz) in the Chapter 6 database.

5. At this point, you should have 1573016.HY.sdts.tar.gz and 00MASTERDD_LRG.SDTS.tar.gz, both zipped files, in the Chapter 6 database. Before using WinZip to extract the files, you need to do two things. First, create a new folder named *hydrography* and another new folder named *masterdd*. Second, select Configuration from WinZip's Options menu and, on the Miscellaneous tab, uncheck the box next to TAR file smart CR/LF Conversion. Unless you uncheck this WinZip feature, you will get an error message in converting the SDTS profile stating that "error

reading LE01 module (record #8) / missing FLIPS 123 subfield format definition."

6. Now you can unzip the two tar files. Save the extracted files from 1573016.HY.sdts.tar.gz in the *hydrography* folder and the extracted files from 00MASTERDD_LRG.SDTS.tar.gz in the *masterdd* folder.

7. Click Show/Hide ArcToolbox Windows to open the ArcToolbox window in ArcCatalog. Double-click the Import From SDTS tool in the Coverage Tools/Conversion/To Coverage toolset. In the Import From SDTS dialog, use the browse button to navigate to the data files in the *elevation* folder. The data files all have the 1102 prefix. Double-click any of the files. The Input SDTS Transfer File Prefix in the dialog should list 1102. Change the output grid name to *menandem* and save it in the Chapter 6 folder (not the *elevation* folder). Click OK to execute the conversion.

8. Double-click the Import From SDTS tool in the Coverage Tools/Conversion/To Coverage toolset. Use the browse button to navigate to the data files in the *hydrography* folder. The data files all have the HY01 prefix. Double-click any of the files. Save the coverage as *menanhydro* in the Chapter 6 database.

9. Launch ArcMap. Change the data frame name to Task 1. Add *menandem* and the arc layer of *menanhydro* to Task 1. Change the color of *menanhydro.arc* to blue and the color ramp of *menandem* to Elevation #1. The two data sets should register spatially.

Q1. What is the elevation range (in meters) in *menandem*?

Q2. What is the coordinate system that both *menandem* and *menanhydro* are based on?

Q3. What other layers besides *arc* are contained in the *menanhydro* coverage?

Task 2: Digitize On-Screen in ArcMap

What you need: *land_dig.shp*, a background map for digitizing. *land_dig.shp* is based on UTM coordinates and measured in meters.

On-screen digitizing is technically similar to manual digitizing. The differences are as follows: you use the mouse pointer rather than the digitizer's cursor for digitizing; you use a feature or image layer as the background for digitizing; and you must zoom in and out repeatedly while digitizing. Task 2 lets you digitize several polygons off *land_dig.shp* to make a new shapefile. Task 2 assumes that *land_dig.shp* is an image, similar to a DRG or a DOQ. You will digitize the new shapefile in a "free hand" style using the image as the background.

1. Make sure that ArcCatalog is connected to the Chapter 6 folder. First create a new shapefile for digitizing. Right-click the Chapter 6 folder, point to New, and select Shapefile. In the next dialog, enter *trial1* for the name, select Polygon for the feature type, and click the Edit button for the spatial reference. Import the coordinate system of *land_dig.shp* for *trial1*.

2. Insert a data frame in ArcMap and rename it Task 2. Add *trial1* and *land_dig.shp* to Task 2. Make sure that *trial1* is on top of *land_dig* in the table of contents. Before digitizing, you need to change the symbol of the two shapefiles, define the selection layer, and set up the digitizing environment. To facilitate digitizing, you want to draw *land_dig* in red with labels and *trial1* in black. Select Properties from the context menu of *land_dig*. On the Symbology tab, click Symbol and change it to a Hollow symbol with the Outline Color in red. On the Labels tab, check the box to label features in this layer and select LAND_DIG_1 from the dropdown list for the Layer label field. Click OK to dismiss the Layer Properties dialog. Click the symbol of *trial1* in the table of contents. Select the Hollow symbol and the Outline Color of black.

3. Click the Selection tab in the table of contents. Uncheck *land_dig*. This ensures that *trial1* is the only selectable layer during digitizing. Switch back to the Display tab.

4. Click the Tools menu and make sure that Editor Toolbar is checked. (The alternative is to click the Editor Toolbar button.) Select Start

Editing from the Editor dropdown list. Make sure that the task is to Create New Feature, and the target is *trial1*. Select Options from the Editor dropdown list. On the General tab, enter 10 and select map units for the snapping tolerance. (The snapping tolerance is 10 meters because the map units of *trial1* are in meters.) Click OK. Click the Editor dropdown arrow again and select Snapping. Check the Vertex, Edge, and End boxes for *trial1* only. Close the Snapping dialog. You can use the Measure tool to see how large a snapping tolerance of 10 meters is.

5. You are ready to digitize. Zoom in the area around polygon 72. Notice that polygon 72 in *land_dig* is made of a series of lines (edges), which are connected by points (vertices). Click the Sketch Tool on the Editor toolbar. Digitize a starting point of polygon 72 by left-clicking the mouse. Use *land_dig* as a guide to digitize the other vertices. When you come back to the starting point, right-click the mouse and select Finish Sketch. The completed polygon 72 appears in cyan with an x inside it. A feature appearing in cyan is an active feature. To unselect polygon 72, click the Edit Tool and click a point outside the polygon. If you need to delete a polygon in *trial1* later, use the Edit Tool to first select and activate the polygon and then press the Delete key.

6. Digitize polygon 73. You can zoom in and out, or use other tools, any time during digitizing. Click the Sketch Tool whenever you are ready to resume digitizing.

7. Digitize polygons 74 and 75 next. The two polygons have a shared border. The strategy is to digitize the outline of both polygons first and then to cut the polygon into two. Use one end of the shared border as the starting point and digitize the outline, including a vertex at the other end of the shared border. Change the task to Cut Polygon Features. Make sure that the digitized outline is still active; if not, use the Edit Tool to select it. Click the Sketch Tool. Left-click the starting point used in digitizing the outline. Digitize

the other vertices that make up the shared border. Double-click the other end of the shared border.

8. Auto-Complete Polygon is an alternative to Cut Polygon Features in the previous step. To use this alternative method, you will digitize one of the two polygons first, switch the task to Auto-Complete Polygon, and then digitize the other polygon without going over the shared border again. But, as of July 2004, the ESRI website has a warning message about a bug with the Auto-Complete Polygon tool. Apparently, the tool does not work every time.

9. You are done with digitizing. Right-click *trial1* in the table of contents, and select Open Attribute Table. Click the first cell under Id and enter 72. Enter 73, 74, and 75 in the next three cells. (You can click the box to the left of a record and see the corresponding polygon to the record.) Close the table.

10. Select Stop Editing from the Editor dropdown list. Save the edits.

Q4. Define the snapping tolerance. (Tip: Use the Index tab in ArcGIS Desktop Help.)

Q5. Will a smaller snapping tolerance give you a more accurate digitized map? Why?

Q6. The Task dropdown menu on the Editor toolbar lists 4 categories of tasks. Which category does Auto-Complete Polygon belong to?

Task 3: Add X Y Data in ArcMap

What you need: *events.txt*, a text file containing GPS readings.

In Task 3, you will use ArcMap to create a new shapefile from *events.txt*, a text file that contains *x*-, *y*-coordinates of a series of points collected by GPS readings.

1. Insert a data frame in ArcMap and rename it Task 3. Add *events.txt* to Task 3. Select Add XY Data from the Tools menu. Make sure that *events.txt* is the table to be added as a layer. Use the dropdown list to select

REFERENCES

Anderson, J. R., E. E. Hardy, J. T. Roach, and R. E. Witmer. 1976. A Land Use and Land Cover Classification System for Use with Remote Sensor Data. U.S. Geological Survey Professional Paper 964. Washington, DC: U.S. Government Printing Office.

Guptill, S. C. 1999. Metadata and Data Catalogues. In P. A. Longley, M. F. Goodchild, D. J. Maguire, and D. W. Rhind, eds., Geographical Information Systems, 2d ed., pp. 77–92. New York: Wiley.

Kavanagh, B. F. 2003. Geomatics. Upper Saddle River, NJ: Prentice Hall.

Kennedy, M. 1996. The Global Positioning System and GIS. Ann Arbor, MI: Ann Arbor Press.

Lange, A. F., and C. Gilbert. 1999. Using GPS for GIS Data Capture. In P. A. Longley, M. F. Goodchild, D. J. Maguire, and D. W. Rhind, eds., Geographical Information Systems, 2d ed., pp. 467–79. New York: Wiley.

Moffitt, F. H., and J. D. Bossler. 1998. Surveying, 10th ed. Menlo Park, CA: Addison-Wesley.

Onsrud, H. J., and G. Rushton, eds. 1995. Sharing Geographic Information. New Brunswick, NJ: Center for Urban Policy Research.

Sperling, J. 1995. Development and Maintenance of the TIGER Database: Experiences in Spatial Data Sharing at the U.S. Bureau of the Census. In H. J. Onsrud and G. Rushton, eds., Sharing Geographic Information, pp. 377–96. New Brunswick, NJ: Center for Urban Policy Research.

Verbyla, D. L., and K. Chang. 1997. Processing Digital Images in GIS. Santa Fe, NM: OnWord Press.

EASTING for the X Field and NORTHING for the Y Field. Click the Edit button for the spatial reference of input coordinates. Select projected coordinate systems, Utm, Nad1927, and NAD 1927 UTM Zone 11N.prj. Click OK to dismiss the dialogs. events.txt Events is added to the table of contents.

2. events.txt Events can be saved as a shapefile. Right-click events.txt Events, point to Data, and select Export Data. Opt to export all features and save the output as events.shp in the Chapter 6 database.

3. You can also convert events.txt into a shapefile directly in ArcCatalog. Right-click events.txt in the Catalog tree, point to Create Feature Class, and select From XY Table. The Create Feature Class From XY Table dialog allows you to specify the X Field, the Y Field, and the output shapefile.

Challenge Task

What you need: quake.txt.

quake.txt in the Chapter 6 database contains earthquake data in northern California from January 2002 to August 2003. The quakes recorded in the file all had a magnitude of 4.0 or higher. The Northern California Earthquake Data Center maintains and catalogs the quake data (**http://quake.geo.berkeley.edu/**).

This challenge task asks you to perform two related tasks: (1) use the longitude (Lon) and latitude (Lat) readings in quake.txt to create a shapefile called quake and define its coordinate system as NAD27, and (2) download an interchange file (an .e00 file) of California counties and convert the file to a coverage called co_calif.

To download the interchange file of California counties, go to **http://casil-mirror1.ceres.ca.gov/casil/**. Select statewide data index and County Boundaries (1:24000).

Q1. How many records are in quake?

Q2. What is the maximum magnitude recorded in quake?

Q3. What coordinate system is co_calif based on?

Q4. What are the parameter values of the coordinate system for co_calif?

Q5. Were the quakes recorded in quake all on land?

GEOMETRIC TRANSFORMATION

A newly digitized map has the same measurement unit as the source map used in digitizing or scanning. If manually digitized, the map is measured in inches, same as the digitizing table. If converted from a scanned image, the map is measured in dots per inch (dpi). Obviously this newly digitized map cannot be used for data display or analysis in a GIS project. To make it usable, we must convert the newly digitized map into a projected coordinate system. This conversion is called geometric transformation, which, in this case, transforms map feature coordinates from digitizer units or dpi into projected coordinates such as UTM (Universal Transverse Mercator)

coordinates. Only through a geometric transformation, a newly digitized map can align with other layers.

Geometric transformation also applies to satellite images. Remotely sensed data are recorded in rows and columns. A geometric transformation can convert rows and columns into projected coordinates. Additionally, the transformation can correct geometric errors in the remotely sensed data, which are caused by the relative motions of a satellite (i.e., its scanners and the Earth) and uncontrolled variations in the position and altitude of the remote sensing platform. Although some of these errors (e.g., the Earth's rotation) can be removed systematically, they are typically removed through geometric transformation.

Chapter 7 has the following four sections. Section 7.1 reviews transformation methods, especially the affine transformation, which is commonly used in GIS and remote sensing. Section 7.2 examines the root mean square (RMS) error, a measure of the goodness of a geometric transformation, and how it is derived. Section 7.3 covers the interpretation of RMS errors on

117

7.1 GEOMETRIC TRANSFORMATION

Geometric transformation is the process of using a set of control points and transformation equations to register a digitized map, a satellite image, or an air photograph onto a projected coordinate system. As its definition suggests, geometric transformation is a common operation in GIS, remote sensing, and photogrammetry. But the mathematical aspects of geometric transformation are from coordinate geometry (Moffitt and Mikhail 1980).

7.1.1 Map-to-Map and Image-to-Map Transformation

A newly digitized map, either manually digitized or traced from a scanned file, is based on digitizer units. Digitizer units can be in inches or dots per inch. Geometric transformation converts the newly digitized map into projected coordinates in a process often called **map-to-map transformation. Image-to-map transformation** applies to remotely sensed data (Jensen 1996; Richards and Jia 1999). The term suggests that the transformation changes the rows and columns (i.e., the image coordinates) of a satellite image into projected coordinates. Another term describing this kind of transformation is georeferencing (Verbyla and Chang 1997; Lillesand and Kiefer 2000). A georeferenced image can register spatially with other feature or raster layers in a GIS database, as long as the coordinate system is the same.

Either map-to-map or image-to-map, a geometric transformation uses a set of control points to establish a mathematical model that relates map coordinates of one system to another or image coordinates to map coordinates. Use of control points makes the process somewhat uncertain. This is particularly true with image-to-map transformation because control points are selected directly from the original image. Misplacement of the control points can make the transformation result unacceptable.

The **root mean square (RMS)** error is a quantitative measure that can determine the quality of geometric transformation. It measures the displacement between the actual and estimated locations of the control points. If the RMS error is acceptable, then a mathematical model derived from the control points can be used for transforming the entire map or image.

A map-to-map transformation automatically creates a new map that is ready to use. An image-to-map transformation, on the other hand, requires an additional step of resampling to complete the process. Resampling fills in each pixel of the transformed image with a value that is derived from the original image.

7.1.2 Transformation Methods

Different methods have been proposed for transformation from one coordinate system to another (Taylor 1977; Moffitt and Mikhail 1980). Each method is distinguished by the geometric properties it can preserve and by the changes it allows. The effect of transformation varies from changes of position and direction, to a uniform change of scale, to changes in shape and size (Figure 7.1). The following summarizes these transformation methods and their effect on a rectangular object.

• Equiarea transformation allows rotation of the rectangle and preserves its shape and size.
• Similarity transformation allows rotation of the rectangle and preserves its shape but not size.
• **Affine transformation** allows angular distortion of the rectangle but preserves the parallelism of lines (i.e., parallel lines remain as parallel lines).
• Projective transformation allows both angular and length distortions, thus allowing the rectangle to be transformed into an irregular quadrilateral.

The above transformation methods are available in GIS packages such as ArcGIS and MGE. The general rules suggest use of the affine transformation for digitized maps. Section 7.4 deals with the resampling of pixel values for remotely sensed data after transformation.

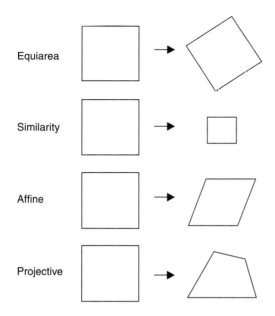

Figure 7.1
Different types of geometric transformations.

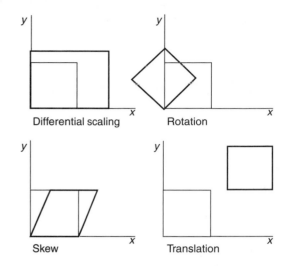

Figure 7.2
Differential scaling, rotation, skew, and translation in the affine transformation.

map-to-map or image-to-map transformations and the projective transformation for aerial photographs with relief displacement. Also available in GIS packages are general polynomial transformations that use surfaces generated from second- or higher-order polynomial equations to transform satellite images with high degrees of distortion and topographic relief displacement. The process of general polynomial transformations is commonly called warping (Jensen 1996).

7.1.3 Affine Transformation

The affine transformation allows rotation, translation, skew, and differential scaling on a rectangular object, while preserving line parallelism (Pettofrezzo 1978). Rotation rotates the object's x- and y-axis from the origin. Translation shifts its origin to a new location. Skew allows a nonperpendicularity (or affinity) between the axes, thus changing its shape to a parallelogram with a slanted direction. And differential scaling changes the scale by expanding or reducing in the x and/or y direction. Figure 7.2 shows these four transformations graphically.

Mathematically, the affine transformation is expressed as a pair of first-order polynomial equations:

(7.1)

$$X = Ax + By + C$$

(7.2)

$$Y = Dx + Ey + F$$

where x and y are the input coordinates that are given, X and Y are the output coordinates to be determined, and A, B, C, D, E, and F are the transformation coefficients. The affine transformation is also called the six-parameter transformation because it involves six estimated coefficients.

The same equations apply to both digitized maps and satellite images. But there are two differences. First, x and y represent point coordinates in a digitized map, but columns and rows in a satellite image. Second, the coefficient E is negative in the case of a satellite image. This is because the origin of a satellite image is located at the upper-left corner whereas the origin of a projected coordinate system is at the lower-left corner.

tion equations to compute the x- and y-coordinates of map features in the digitized map or pixels in the image. The third step is also called **rectification** in image analysis. The outcome from the third step is a new map or image that is based on a user-defined projected coordinate system.

7.1.4 Geometric Interpretation of Affine Transformation Coefficients

The geometric properties of the affine transformation can be interpreted from the transformation coefficients. The coefficient C represents the translation in the x direction, and F the translation in the y direction. The coefficients of A, B, D, and E are related to rotation, skew, and scaling in the following equations:

$$A = Sx \cos(t) \qquad (7.3)$$

$$B = Sy\,[k \cos(t) - \sin(t)] \qquad (7.4)$$

$$D = Sx \sin(t) \qquad (7.5)$$

$$E = Sy\,[k \sin(t) + \cos(t)] \qquad (7.6)$$

where Sx is the change of scale in x, Sy is the change of scale in y, t is the rotation angle, and k is the shear factor. The skew angle can be derived from arctan (k). The equations for A and D can be used to solve for t and, in turn, Sx. The equations for B and E can be used to solve for k and, in turn, Sy and the skew angle.

7.1.5 Control Points

Control points play a key role in determining the accuracy of an affine transformation (Bolstad et al. 1990). Because the selection of control points differs between map-to-map transformation and image-to-map transformation, the following discussion separates the two.

Operationally, an affine transformation of a digitized map or image involves three steps (Figure 7.3). First, update the x- and y-coordinates of selected control points to real-world coordinates. If real-world coordinates are not available, we can derive them by projecting the longitude and latitude values of the control points. Second, run an affine transformation on the control points and examine the RMS error. If the RMS error is higher than the expected value, select a different set of control points and rerun the affine transformation. If the RMS error is acceptable, then the six coefficients of the affine transformation estimated from the control points are used in the next step. Third, use the estimated coefficients and the transformation equations to the input features.

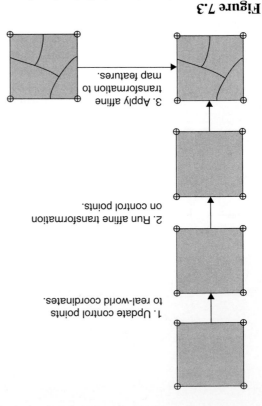

Figure 7.3

A geometric transformation typically involves three steps. Step 1 updates the control points to real-world coordinates. Step 2 uses the control points to run an affine transformation. Step 3 creates the output by applying the transformation equations to the input features.

1. Update control points to real-world coordinates.

2. Run affine transformation on control points.

3. Apply affine transformation to map features.

Box 7.1   Estimation of Transformation Coefficients

The example used here and later in Chapter 7 is a third-quadrangle soil map (one-third of a USGS 1:24,000 scale quadrangle map), which has been scanned at 300 dpi. The map has four control points marked at the corners: Tic 1 at the NW corner, Tic 2 at the NE corner, Tic 3 at the SE corner, and Tic 4 at the SW corner. X and Y denote the control points' real-world (output) coordinates in meters based on the UTM coordinate system, and x and y denote the control points' digitized (input) locations. The measurement unit of the digitized locations is 1/300 of an inch, corresponding to the scanning resolution.

The following table shows the input and output coordinates of the control points:

Tic-id	x	y	X	Y
1	465.403	2733.558	518843.844	5255910.5
2	5102.342	2744.195	528265.750	5255948.5
3	5108.498	465.302	528288.063	5251318.0
4	468.303	455.048	518858.719	5251280.0

We can solve for the transformation coefficients by using the following equation in matrix form:

$$\begin{bmatrix} C & F \\ A & D \\ B & E \end{bmatrix} = \begin{bmatrix} n & \Sigma x & \Sigma y \\ \Sigma x & \Sigma x^2 & \Sigma xy \\ \Sigma y & \Sigma xy & \Sigma y^2 \end{bmatrix}^{-1} \cdot \begin{bmatrix} \Sigma X & \Sigma Y \\ \Sigma xX & \Sigma xY \\ \Sigma yX & \Sigma yY \end{bmatrix}$$

where n is the number of control points and all other notations are the same as previously defined. The transformation coefficients derived from the equation show

$$A = 2.032, B = -0.004, C = 517909.198,$$
$$D = 0.004, E = 2.032, F = 5250353.802$$

Selection of control points for a map-to-map transformation is relatively straightforward. What we need are points with known real-world coordinates. But, if they are not available, we can use points with known longitude and latitude values and project them into real-world coordinates. A USGS 1:24,000 scale quadrangle map has 16 points with known longitude and latitude values: 12 points along the border, and 4 additional points within the quadrangle. (These 16 points divide the quadrangle into 2.5 minutes in longitude and latitude.) These 16 points are also called tics.

An affine transformation requires a minimum of three control points to estimate its six coefficients. But often four or more control points are used to reduce problems in measurement errors and to allow for a least-squares solution. After the control points are selected, they are digitized along with map fea-

tures onto the digitized map. The coordinates of these control points on the digitized map are the x, y values in Eq. (7.1) and Eq. (7.2), and the real-world coordinates of these control points are the X, Y values. Box 7.1 shows an example of using a set of four control points to derive the six coefficients. Box 7.2 shows the output from the affine transformation and the interpretation of the transformation.

Control points for an image-to-map transformation are usually called ground control points. **Ground control points (GCPs)** are points where both image coordinates (in columns and rows) and real-world coordinates can be identified. The image coordinates are the x, y values, and their corresponding real-world coordinates are the X, Y values in Eq. (7.1) and Eq. (7.2).

GCPs are selected directly from a satellite image. Therefore the selection is not as straightforward as selecting four tics for a digitized map. Ideally, GCPs are those features that show up clearly as single distinct pixels. Examples include road intersections, rock outcrops, small ponds, or distinctive features along shorelines. Georeferencing a TM (Thematic Mapper) scene may need an initial set of 20 or more GCPs. As explained in Section 7.2, some of these points are eventually removed in the transformation process. After GCPs are identified on a satellite image, their real-world coordinates can be obtained from digital maps or GPS readings.

7.2 ROOT MEAN SQUARE (RMS) ERROR

The affine transformation uses the coefficients derived from a set of control points for transforming a digitized map or a satellite image. The location of a control point on a digitized map or an image is an estimated location and can deviate from its actual location. A common measure of the goodness of the control points is the RMS error, which measures the deviation between the actual (true) and estimated (digitized) locations of the control points.

How is a RMS error derived from a digitized map? After the six coefficients have been estimated, we can use the digitized coordinates of the first control point as the inputs (i.e., the x and y values) to Eq. (7.1) and Eq. (7.2) and compute the X and Y values respectively. If the digitized control points were perfectly located, the computed X and Y values would be the same as the control point's real-world coordinates. But this is rarely the case. The deviation between the computed (estimated) X and Y values and the actual coordinates then becomes an error associated with the first control point on the output. Likewise, to derive an error associated with a control point on the input, we can use the point's real-world coordinates as the inputs and measure the deviation between the computed x and y values and the digitized coordinates.

The above procedure for deriving RMS errors also applies to GCPs used in an image-to-map transformation. Again, the difference is that rows and columns of a satellite image replace digitized coordinates.

Mathematically, the input or output for a control point is computed by:

$$\text{Error for a control point} = \sqrt{(x_{act} - x_{est})^2 + (y_{act} - y_{est})^2} \qquad (7.7)$$

where x_{act} and y_{act} are the x and y values of the actual location, and x_{est} and y_{est} are the x and y values of the estimated location.

Box 7.2 Output from an Affine Transformation

Using the data from Box 7.1, the following shows the portion of the output from the affine transformation that is related to the geometric properties of the transformation:

Scale $(X, Y) = (2.032, 2.032)$
Skew (degrees) $= (-0.014)$
Rotation (degrees) $= (0.102)$
Translation $= (517909.198, 5250353.802)$

The positive rotation angle means a rotation counterclockwise from the x-axis, and the negative skew angle means a shift clockwise from the y-axis. Both angles are very small, meaning that the change from the original rectangle to a parallelogram through the affine transformation is very slight.

Box 7.3 | RMS from an Affine Transformation

The following shows a RMS report using the data from Box 7.1.

RMS error (input, output) = (0.138, 0.281)

Tic-id	Input x Output X	Input y Output Y	X error	Y error
1	465.403	2733.558		
	518843.844	5255910.5	−0.205	−0.192
2	5102.342	2744.195		
	528265.750	5255948.5	0.205	0.192
3	5108.498	465.302		
	528288.063	5251318.0	−0.205	−0.192
4	468.303	455.048		
	518858.719	5251280.0	0.205	0.192

The output shows that the average deviation between the input and output locations of the control points is 0.281 meter based on the UTM coordinate system, or 0.00046 inch (0.138 divided by 300) based on the digitizer unit. This RMS error is well within the acceptable range. The individual X and Y errors suggest that the error is slightly lower in the y direction than the x direction and the average RMS error is equally distributed among the four control points.

The average RMS error can be computed by averaging errors from all control points:

(7.8)

$$\text{Average RMS} = \sqrt{\left(\sum_{i=1}^{n}(x_{\text{act, i}} - x_{\text{est, i}})^2 + \sum_{i=1}^{n}(y_{\text{act, i}} - y_{\text{est, i}})^2\right)\Big/n}$$

where n is the number of control points, $x_{\text{act,i}}$ and $y_{\text{act,i}}$ are the x and y values of the actual location of control point i, and $x_{\text{est,i}}$ and $y_{\text{est,i}}$ are the x and y values of the estimated location of control point i. Box 7.3 shows an example of the average RMS errors, and the output X and Y errors for each control point from an affine transformation.

To ensure the accuracy of geometric transformation, the RMS error should be within a tolerance value. The data producer defines the tolerance value, which can vary by the accuracy and the map scale or the ground resolution of the input data. A RMS error (output) of <6 meters is probably acceptable if the input map is a 1:24,000 scale USGS quadrangle map. A RMS error (input) of <1 pixel is probably acceptable for a TM scene with a ground resolution of 30 meters.

If the RMS error is within the acceptable range, then the assumption is that this same level of accuracy based on the control points can also apply to the entire map or image. But as shown later in Section 7.3, this assumption can be quite wrong in the case of newly digitized maps.

If the RMS error exceeds the established tolerance, then the control points need to be adjusted. For digitized maps, this means redigitizing the control points. For satellite images, the adjustment means removing the control points that contribute most to the RMS error and replacing them with new GCPs. Geometric transformation is therefore an iterative process of selecting control points, estimating transformation coefficients, and computing the RMS error. This process continues until a satisfactory transformation is obtained.

Figure 7.4
Inaccurate location of soil lines can result from input tic location errors. The thin lines represent correct soil lines and the thick lines incorrect soil lines. In this case, the x values of the upper two tics were increased by 0.2″ while the x values of the lower two tics were decreased by 0.2″ on a third quadrangle (15.4″ × 7.6″).

latitude readings of control points 1 and 2 (the upper two control points) are off by 10″ (e.g., 47°27′20″ instead of 47°27′30″). The RMS error from transformation would be acceptable but the soil lines would deviate from their locations on the source map (Figure 7.5). The same problem occurs if the longitude readings of control points 2 and 3 (the two control points to the right) are off by 30″ (e.g., −116°37′00″ instead of −116°37′30″). Again, this happens because the affine transformation works with parallelograms. Although we tend to take for granted the accuracy of published maps, erroneous longitude and latitude readings are quite common, especially with inset maps (maps that are smaller than the regular size) and oversized maps (maps that are larger than the regular size).

We typically use the four corner points of the source map as control points. This practice makes sense because the exact readings of longitude and latitude are usually shown at those points. Moreover, using the corner points as control points helps the process of joining the map with its adjacent maps. But the practice of using four corner control points does not preclude the use of more control points if additional points with known locations are available. As mentioned earlier, a least-squares solution is used in an affine transformation when more than three points are used as control points. Therefore, the use

7.3 INTERPRETATION OF RMS ERRORS ON DIGITIZED MAPS

If a RMS error is within the acceptable range, we usually assume that the transformation of the entire map is also acceptable. This assumption can be quite wrong, however, if gross errors are made in digitizing the control points or in inputting the longitude and latitude readings of the control points.

As an example, we can shift the locations of control points 2 and 3 (the two control points to the right) on a third quadrangle (similar to that in Box 7.1) by increasing their x values by a constant. The RMS error would remain about the same because the object formed by the four control points retains the shape of a parallelogram. But the soil lines would deviate from their locations on the source map. The same problem occurs if we increase the x values of control points 1 and 2 (the upper two control points) by a constant, and decrease the x values of control points 3 and 4 (the lower two control points) by a constant (Figure 7.4). In fact, the RMS error would be well within the tolerance value as long as the object formed by the shifted control points remains as a parallelogram.

The same problem is present in the interpretation errors of digitized map features. Suppose the longitude and latitude readings printed on paper maps are sometimes erroneous. This would lead to acceptable RMS errors but significant loca-

Figure 7.5
Incorrect location of soil lines can result from output tic location errors. The thin lines represent correct soil lines and the thick lines incorrect soil lines. In this case, the latitude readings of the upper two tics were off by 10″ (e.g., 47°27′20″ instead of 47°27′30″) on a third quadrangle.

of more control points means a better coverage of the entire map in transformation. In other situations, control points that are closer to the map features of interest should be used instead of the corner points. This ensures the location accuracy of these map features.

7.4 RESAMPLING OF PIXEL VALUES

The result of geometric transformation of a satellite image is a new image based on a projected coordinate system. But the new image has no pixel values. The pixel values must be filled through resampling. **Resampling** in this case means filling each pixel of the new image with a value or a derived value from the original image.

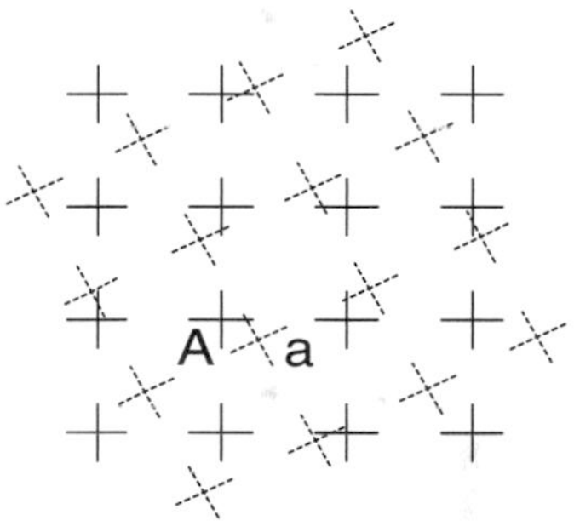

Figure 7.6
Because *a* in the original image is closest to pixel *A* in the new image, the pixel value at *a* is assigned to be the pixel value at *A* using the nearest neighbor technique.

7.4.1 Resampling Methods

Three common resampling methods are nearest neighbor, bilinear interpolation, and cubic convolution. The **nearest neighbor** resampling method fills each pixel of the new image with the nearest pixel value from the original image. For example, Figure 7.6 shows that pixel *A* in the new image will take the value of pixel *a* in the original image because it is the closest neighbor. The nearest neighbor method is computationally efficient. The

method has the additional property of preserving the original pixel values, which is important for categorical data such as land cover types and desirable for some image processing such as edge detection.

Both bilinear interpolation and cubic convolution fill the new image with distance-weighted averages of the pixel values from the original image. The **bilinear interpolation** method uses the average of the four nearest pixel values from three linear interpolations, whereas the **cubic convolution** method uses the average of the 16 nearest pixel values from five cubic polynomial interpolations (Richards and Jia 1999). Cubic convolution tends to produce a smoother output than bilinear interpolation but re-quires a longer processing time. Box 7.4 and Fig-ure 7.7 show an example of bilinear interpolation.

7.4.2 Other Uses of Resampling

Geometric transformation of satellite images is not the only operation that requires resampling. Resam-pling is needed whenever there is a change of cell location or cell size between the input raster and the output raster. For example, projecting a raster from one coordinate system to another requires resampling to fill in the cell values of the output raster. Resam-pling is also involved when a raster changes from one cell size to another (e.g., from 10 to 15 meters).

Pyramiding is a common technique for display-ing large raster data sets (Box 7.5). Resampling is the technique for building different pyramid lev-els. Regardless of its application, resampling typ-ically uses one of the three methods covered in Section 7.4.1 to produce the output raster or rasters.

Box 7.5 Pyramiding

P yramiding is a technique commonly used for dis-playing large images. GIS packages have recently adopted the technique for data display. For example, ArcGIS uses a dialog to ask if pyramiding is wanted or not before displaying a large raster. Pyramiding builds different pyramid levels to represent reduced or lower resolutions of a large raster. Because a lower-resolution raster (i.e., a pyramid level of greater than 0) requires less memory space, it can display more quickly. Therefore, when viewing the entire raster, we view it at the highest pyramid level. And, as we zoom in, we view more detailed data at a finer resolution (i.e., a pyramid level of closer to 0). Resampling is in-volved in building different pyramid levels.

Box 7.4 Computation for Bilinear Interpolation

B ilinear interpolation uses four nearest neighbors in the original image to compute a pixel value in the new image. Pixel x in Figure 7.7 represents a pixel in the new image, whose value needs to be derived from the origi-nal image. Pixel x corresponds to a location of (2.6, 2.5) in the original image. Its four nearest neighbors have the image coordinates of (2, 2), (3, 2), (2, 3), and (3, 3), and the pixel values of 10, 5, 15, and 10 respectively.

Using the bilinear interpolation method, we first perform two linear interpolations along the scan lines 2 and 3 to derive the interpolated values at a and b:

$$a = 0.6(5) + 0.4(10) = 7$$
$$b = 0.6(10) + 0.4(15) = 12$$

Then we perform the third linear interpolation be-tween a and b to derive the interpolated value at x:

$$x = 0.5(7) + 0.5(12) = 9.5$$

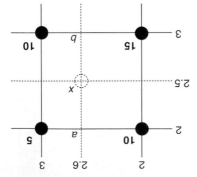

Figure 7.7

The bilinear interpolation method uses the value of the four closest pixels (black circles) in the original image to estimate the pixel value at x in the new image.

KEY CONCEPTS AND TERMS

Affine transformation: A geometric transformation method that allows rotation, translation, skew, and differential scaling on a rectangular object, while preserving line parallelism.

Bilinear interpolation: A resampling method that uses the distance-weighted average of the four nearest pixel values to estimate a new pixel value.

Cubic convolution: A resampling method that uses the distance-weighted average of the 16 nearest pixel values to estimate a new pixel value.

Geometric transformation: The process of converting a map or an image from one coordinate system to another by using a set of control points and transformation equations.

Ground control points (GCPs): Points used as control points for an image-to-map transformation.

Image-to-map transformation: One type of geometric transformation that converts the rows and columns of a satellite image into real-world coordinates.

Map-to-map transformation: One type of geometric transformation that converts a newly digitized map into real-world coordinates.

Nearest neighbor: A resampling method that uses the nearest pixel value to estimate a new pixel value.

Pyramiding: A technique that builds different pyramid levels for displaying large raster data sets at different resolutions.

Rectification: A process of applying the transformation equations to a satellite image so that the image coordinates can be converted to projected coordinates.

Resampling: A process of filling each pixel of a newly transformed image with a value or a derived value from the original image.

Root mean square (RMS) error: In geometric transformation, the RMS measures the deviation between the actual location and the estimated location of the control points.

REVIEW QUESTIONS

1. Explain map-to-map transformation.
2. Explain image-to-map transformation.
3. An image-to-map transformation is sometimes called an image-to-world transformation. Why?
4. The affine transformation allows rotation, translation, skew, and differential scaling. Describe each of these transformations.
5. Operationally, an affine transformation involves three sequential steps. What are these steps?
6. Explain the role of control points in an affine transformation.
7. How are control points selected for a map-to-map transformation?
8. How are ground control points chosen for an image-to-map transformation?
9. Define the root mean square (RMS) error in geometric transformation.
10. Explain the role of the RMS error in an affine transformation.
11. Describe a scenario in which the RMS error may not be a reliable indicator of the goodness of a map-to-map transformation.
12. Why do we have to perform the resampling of pixel values following an image-to-map transformation?
13. Describe three common resampling methods for raster data.
14. The nearest neighbor method is recommended for resampling categorical data. Why?
15. What is pyramiding?

This applications section covers three tasks. Task 1 covers the affine transformation of a scanned file. In Task 2, you will use the transformed scanned file for vectorization. As covered in Chapter 6, scanning is a popular data input method. Task 3 covers the affine transformation of a satellite image.

Task 1: Georeference and Rectify a Scanned Map

What you need: *hoytmtn.tif*, a TIFF file containing scanned soil lines.

The bi-level scanned file *hoytmtn.tif* is measured in inches. For this task, you will convert the scanned image into UTM coordinates. The conversion process involves two basic steps. First, you will georeference the image by using four control points, also called tics, which correspond to the four corner points on the original soil map. Second, you will rectify or transform the image by using the results from georeferencing. The four control points have the following longitude and latitude values in degrees-minutes-seconds (DMS):

Tic-id	Longitude	Latitude
1	−116 00 00	47 15 00
2	−115 52 30	47 15 00
3	−115 52 30	47 07 30
4	−116 00 00	47 07 30

After being projected onto the NAD 1927 UTM Zone 11N coordinate system, these four control points have the following x- and y-coordinates:

Tic-id	x	y
1	575672.2771	5233212.6163
2	585131.2232	5233341.4371
3	585331.3327	5219450.4360
4	575850.1480	5219321.5730

Now you are ready to perform the georeferencing of *hoytmtn.tif*.

1. Launch ArcMap, and rename the data frame Task 1. Add *hoytmtn.tif* to Task 1. Click the View menu, point to Toolbars, and check Georeferencing. The Georeferencing toolbar should now appear in ArcMap, and the Layer dropdown list should list *hoytmtn.tif*.

2. Zoom in *hoytmtn.tif* and locate the four control points. These control points are shown as brackets: two at the top and two at the bottom of the image. They are numbered 1 through 4 in a clockwise direction, with 1 at the upper-left corner.

3. Zoom in around the first control point. Click (Activate) the Add Control Points tool on the Georeferencing toolbar. Click the intersection point where the centerlines of the bracket meet, and then click again. A plus-sign symbol at the control point turns from green to red. Use the same procedure to add the other three control points.

4. This step is to update the coordinate values of the four control points. Click the View Link Table tool on the Georeferencing toolbar. The link table lists the four control points at the top with their X Source, Y Source, X Map, Y Map, and Residual values. The X Source and Y Source values are the coordinates on the scanned image. The X Map and Y Map values are the UTM coordinates to be entered. The link table also offers Auto Adjust, the Transformation method, and the Total RMS Error. Notice that the transformation method is 1st Order Polynomial (i.e., affine transformation). Click the first record, and enter 575672.2771 and 5233212.6163 for its X Map and Y Map values respectively. Enter the X Map and Y Map values for the other three records.

Q1. What is the total RMS error of your first trial?

Q2. What is the residual for the first record?

5. The total RMS error should be smaller than 4.0 (meters) if the control points are added

to the correct location on the image. If the RMS error is very high, highlight the record with a high residual value and delete it. Go back to the image and re-enter the control point. After you have come to an acceptable total RMS error, click OK on the Link Table dialog.

6. This step is to rectify (transform) *hoytmtn.tif*. Select Rectify from the Georeferencing dropdown menu. Take the defaults in the next dialog but save the rectified image as *rect_hoytmtn.tif*.

Task 2: Use ArcScan to Vectorize Raster Lines

What you need: *rect_hoytmtn.tif*, a rectified TIFF file from Task 1.

ArcScan is an extension to ArcGIS. To use Arc-Scan, you need to check the extension in ArcMap's Tools menu and to check the toolbar in the View menu. ArcScan can convert raster lines in a bi-level raster such as *rect_hoytmtn.tif* into line or polygon features. The output from vectorization can be saved into a shapefile or a geodatabase feature class. Task 2 is therefore an exercise for creating new spatial data from a scanned file. Scanning, vectorization, and vectorization parameters are topics already covered in Chapter 6.

Vectorization of raster lines can be challenging if a scanned image contains irregular raster lines, raster lines with gaps, and smudges. A poor-quality scanned image typically reflects the poor quality of the original map and, in some cases, the use of wrong parameter values for scanning. The scanned image you will use for this task is of excellent quality. Therefore, the result from batch vectorization should be excellent as well.

1. This step is to create a new shapefile that will store the vectorized features from *rect_hoytmtn.tif*. Right-click the Chapter 7 folder in ArcCatalog, point to New, and select Shapefile. In the Create New Shapefile dialog, enter *hoytmtn_trace.shp* for the name and polyline for the feature

type. Click the Edit button in the Spatial Reference frame. Select NAD 1927 UTM Zone 11N for the new shapefile's coordinate system.

2. Insert a new data frame in ArcMap and rename the data frame Task 2. Add *rect_hoytmtn.tif* and *hoytmtn_trace.shp* to Task 2. Select Properties from the context menu of *rect_hoytmtn.tif*. On the Symbology tab, choose Unique Values and change the symbol for the value of 0 to red. Because raster lines on *rect_hoytmtn.tif* are very thin, you do not see them at first on the monitor. Zoom in and you will see the red lines.

3. Select Start Editing from the Editor's dropdown menu. The edit mode activates the ArcScan toolbar. The Raster dropdown list should show *rect_hoytmtn.tif*.

4. This step is to set up the vectorization parameters, which are critical for batch vectorization. Select Vectorization Settings from the Vectorization menu. You can enter the parameters values in the settings dialog or choose a style with the predefined values. Click Styles. Choose Polygons and click OK in the next dialog. Click Apply and then Close to dismiss the Vectorization Settings dialog.

5. Select Generate Features from the Vectorization menu. Make sure that *hoytmtn_trace* is the layer to add the centerlines to. Notice that the tip in the dialog states that the command will generate features from the full extent of the raster. Click OK. Results of batch vectorization are now stored in *hoytmtn_trace*.

Q3. The Generate Features command adds the centerlines to *hoytmtn_trace*. Why are the lines called centerlines?

Q4. What are the other vectorization options besides batch vectorization?

6. The lower-left corner of *rect_hoytmtn.tif* has notes about the soil survey, which should be removed. Click the Select Features tool on

the standard (Tools) toolbar, select the notes, and delete them.

7. Select Stop Editing from the Editor's menu and save the edits. Check the quality of the traced soil lines in *hoytmtn_trace*. Because the scanned image is of excellent quality, the traced soil lines should also be of excellent quality.

Task 3: Perform Image-to-Map Transformation

What you need: *spot-pan.bil*, a 10-meter SPOT panchromatic satellite image; *road.shp*, a road shapefile acquired with a GPS receiver and projected onto UTM coordinates.

You will perform an image-to-map transformation in Task 3. ArcMap provides the Georeferencing toolbar that has the basic tools for georeferencing and rectifying a satellite image.

1. Insert a new data frame in ArcMap and rename the data frame Task 3. Add *spot-pan.bil* and *road.shp* to Task 3. Click the symbol for *road*, and change it to orange. Make sure that the Georeferencing toolbar is available and that the Layer on the toolbar shows *spot-pan.bil*. Click View Link Table on the Georeferencing toolbar, select all links from Task 1 in the table, and delete them.

2. You can use the Zoom tool in the Layer to see *road*, or *spot-pan.bil*, but not both. This is because they are in different coordinates. To see both of them, you must have one or more links to initially georeference *spot-pan.bil*. Figure 7.8 marks the first four recommended links. They are all at road intersections. Examine these road intersections in both *spot-pan.bil* and *road* so that you know where they are.

3. Check Auto Adjust in the Georeferencing dropdown menu. If Task 3 shows *road*, right-click *spot-pan.bil* and select Zoom to Layer. Zoom in the first road intersection in the image, click Add Control Points on the Georeferencing toolbar, and click the intersection point. Right-click *road* and select

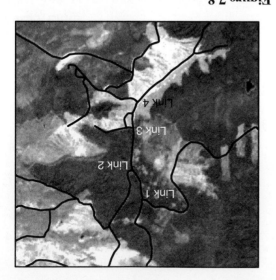

Figure 7.8
The four links to be created first.

Zoom to Layer. Zoom in the corresponding first road intersection in the layer, click the Add Control Points tool, and click the intersection point. The first link brings both the satellite image and the roads to view, but they are still far apart spatially. Repeat the same procedure to add the other three links. Each time you add a link, the Auto Adjust command uses the available links to develop a transformation.

4. Click View Link Table on the Georeferencing toolbar. The Link Table shows four records, one for each link you have added in Step 3. The X Source and Y Source values are based on the image coordinates of *spot-pan.bil*. The image has 1087 columns and 1760 rows. The X Source value corresponds to the column and the Y Source value corresponds to the row. Because the origin of the image coordinates is at the upper-left corner, the Y Source values are negative. The X Map and Y Map values are the UTM coordinates of *road*. And the Residual value shows the RMS error of the control point. The Link Table dialog also shows the transformation

method (i.e., affine transformation) and the total RMS error. You can save the link table as a text file at anytime, and you can load the file next time to continue the georeferencing process.

Q5. What is the total RMS error from the four initial links?

5. An image-to-map transformation usually requires more than four control points. At the same time, the control points should cover the extent of the study area, rather than a limited portion. For this task, try to have a total of 10 links and keep the total RMS error of less than 1 pixel or 10 meters. If a link has a large residual value, delete it and add a new one. Each time you add or delete a link, you will see a change in the total RMS error.

6. This step is to rectify *spot-pan.bil* by using a link table you have created. Select Rectify from the Georeferencing menu. The next dialog lets you specify the cell size, choose a resampling method (nearest neighbor, bilinear interpolation, or cubic convolution), and specify the output name. For this task, you can specify 10 (meters) for the cell size, nearest neighbor for the resampling method, and *rect_spot* for the output. Click Save to dismiss the dialog.

7. Now you can use *rect_spot*, a georeferenced and rectified raster, with other georeferenced data sets for the study area. To delete the control points from *rect_spot*, select Delete Control Points from the Georeferencing menu.

8. If you have difficulty in getting enough links and an acceptable RMS error, first select Delete Control Points from the Georeferencing toolbar. Click View Link Table, and load *georef.txt* from the Chapter 7 database. *georef.txt* has 10 links and a total RMS error of 9.2 meters. Then use the link table to rectify *spot-pan.bil*.

9. If the rectified raster turns out to have values ranging from 16 to 255 (and a black image), it means the value of 255 has been assigned

to the area outside the image. You can correct the problem by going through the following two steps. First, select Reclassify from the Spatial Analyst dropdown menu. In the Reclassify dialog, click Unique. Click the row with an old value of 255, change its new value to NoData, and click OK. The *Reclass of rect_spot* now has the correct value. Second, right-click *Reclass of rect_spot* and select Properties. Select Stretched in the Show frame. Change the Low Label value to 16, the High Label value to 100, and click OK. Now *Reclass of rect_spot* should look like *spot-pan.bil* except that it has been georeferenced. To save the corrected file, right-click *Reclass of rect_spot*, point to Data, and select Export Data. Then specify the output name and location.

Challenge Task

What you need: *cedarbt.tif.*

The Chapter 7 database contains *cedarbt.tif*, a bi-level scanned file of a soil map. This challenge task asks you to perform two operations. First, convert the scanned file into UTM coordinates (NAD 1927 UTM Zone 12N) and save the result into *rec_cedarbt.tif*. Second, vectorize the raster lines in *rec_cedarbt.tif* and save the result into *cedarbt_trace.shp*. There are four tics on *cedarbt.tif*. Numbered clockwise from the upper-left corner, these four tics have the following UTM coordinates:

Tic-id	x	y
1	389988.78125	4886459.5
2	399989.875	4886299.5
3	399779.1875	4872416.0
4	389757.03125	4872575.5

Q1. What is the total RMS error for the affine transformation?

Q2. What problems, if any, did you encounter in vectorizing *rec_cedarbt.tif* ?

REFERENCES

Bolstad, P. V., P. Gessler, and T. M. Lillesand. 1990. Positional Uncertainty in Manually Digitized Map Data. *International Journal of Geographical Information Systems* 4: 399–412.

Jensen, J. R. 1996. *Introductory Digital Image Processing: A Remote Sensing Perspective,* 2d ed. Upper Saddle River, NJ: Prentice Hall.

Lillesand, T. M., and R. W. Kiefer. 2000. *Remote Sensing and Image Interpretation,* 4th ed. New York: Wiley.

Moffitt, F. H., and E. M. Mikhail. 1980. *Photogrammetry,* 3rd ed. New York: Harper & Row.

Pettofrezzo, A. J. 1978. *Matrices and Transformations.* New York: Dover Publications.

Richards, J. A, and X. Jia. 1999. *Remote Sensing Digital Image Analysis: An Introduction,* 3rd ed. Berlin: Springer-Verlag.

Taylor, P. J. 1977. *Quantitative Methods in Geography: An Introduction to Spatial Analysis.* Boston: Houghton Mifflin.

Verbyla, D. L., and Chang, K. 1997. *Processing Digital Images in GIS.* Santa Fe, NM: OnWord Press.

SPATIAL DATA EDITING

Spatial data editing refers to the process of adding, deleting, and modifying features in digital layers. A major part of spatial data editing is to remove digitizing errors. Newly digitized maps, no matter how carefully prepared, always have some errors. Existing layers may be outdated or may contain errors from initial digitizing. Therefore, roads, land parcels, forest inventory, and other data require regular revision and updating. The process of updating data is basically the same as correcting errors on a newly digitized map.

There are two types of digitizing errors: location and topological errors. Location errors such as missing polygons or distorted lines relate to the geometric inaccuracies of spatial features, whereas topological errors such as dangling lines and unclosed polygons relate to the logical inconsistencies between spatial features. To correct location errors, we often have to reshape individual lines and digitize new lines. To correct topological errors, we must first learn about the topological relationships and then use a topology based GIS package to help make corrections.

Correcting digitizing errors may extend beyond individual layers. When a study area covers two or more source layers, we must match features across the layers. When two layers share some common boundaries, we must make sure that the lines are coincident. Spatial data editing can also be more than removing digitizing errors. Modification of features may take the form of simplification and smoothing.

The use of both topological and nontopological data in GIS and the introduction of the object-based data model have significantly expanded the scope of spatial data editing. In fact, determining which method to use for a given data set has become a main challenge in spatial data editing.

Chapter 8 is organized into the following six sections. Section 8.1 describes location and topological errors. Section 8.2 discusses spatial data accuracy standards in the United States, which have gone through three development phases. Section 8.3 examines topological errors with simple features, and between layers. Section 8.4 introduces topological editing using the coverage model and the geodatabase data model as examples. Section 8.5 covers nontopological or basic editing. Section 8.6 includes edgematching, line simplification, and line smoothing.

8.1 LOCATION ERRORS

Location errors refer to the geometric inaccuracies of digitized features. We can examine location errors by the data source for digitizing.

8.1.1 Location Errors Using Secondary Data Sources

If the data source for digitizing is a secondary data source such as a paper map, the evaluation of location errors typically begins by comparing the digitized map with the source map. The obvious goal in digitizing is to duplicate the source map in digital format. To determine how well the goal has been achieved, we can make a **check plot** of the digitized map at the same scale as the source map, superimpose the plot on the source map, and see how well they match.

How well should the digitized map match the source map? There are no federal standards on the threshold value. A geospatial data producer can decide on the tolerance of location error. The Natural Resources Conservation Service (NRCS), for example, stipulates that each digitized soil boundary in the Soil Survey Geographic (SSURGO) database shall be within 0.01 inch (0.254 millimeter) line width of the source map. At the scale of 1:24,000, this tolerance represents 20 feet (6 to 7 meters) on the ground.

Spatial features digitized from a source map can only be as accurate as the source map itself. A variety of factors can affect the accuracy of the source map. Perhaps the most important factor is the map scale. The accuracy of a map feature is less reliable on a 1:100,000 scale map than on a 1:24,000 scale

map. Map scale also influences the level of detail on a published map. As the map scale becomes smaller, the amount of map details decreases and the degree of line generalization increases (Monmonier 1996). For example, a meandering stream on a large-scale map becomes less sinuous on a small-scale map. Therefore we must consider the map scale when evaluating the location accuracy of digital maps.

8.1.2 Causes for Digitizing Errors

Discrepancies between digitized lines and lines on the source map may result from a variety of scenarios. Section 8.1.2 describes three common scenarios. The first is human errors in manual digitizing. Human error is not difficult to understand: when a source map has hundreds of polygons and thousands of lines, one can easily miss some lines, connect the wrong points, or digitize the same lines twice or even more times. Because of the high resolution of a digitizing table, duplicate lines will not be on top of one another but will intersect to form a series of tiny polygons.

The second scenario consists of errors in scanning and tracing. A tracing algorithm usually has problems when raster lines meet or intersect, are too close together, are too wide, or are too thin and broken (Chapter 6). Digitizing errors from tracing include collapsed lines, misshapen lines, and extra lines (Figure 8.1). Duplicate lines can also occur in tracing because semiautomatic tracing follows continuous lines even if some of the lines have already been traced. Because of multiple-pixel raster line in tracing, these duplicate lines will not be on top of one another.

The third scenario consists of errors in converting the digitized map into real-world coordinates (Chapter 7). To make a check plot at the same scale as the source map, we must convert the newly digitized map into real-world coordinates by using a set of control points. With erroneous control points, this conversion can cause discrepancies between digitized lines and source lines. Unlike seemingly random errors from the first two scenarios, discrepancies from geometric transformation often exhibit regular patterns. To correct this type of

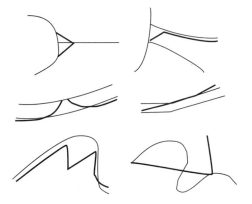

Figure 8.1
Common types of digitizing errors from tracing. The
thin lines are lines on the source map, and the thick
lines are lines from tracing.

location errors, we must redigitize control points
and rerun geometric transformation.

8.1.3 Location Errors Using Primary Data Sources

Although paper maps are still the most common
source for spatial data entry, new data entry
methods using global positioning systems (GPS)
and remote sensing imagery can bypass printed
maps and map generalization practices. The resolution of the measuring instrument determines
the accuracy of spatial data collected by GPS or
satellite images; map scale has no meaning in
this case. The spatial resolution of satellite images can range from less than 1 meter to 1 kilometer. Similarly, the spatial resolution of GPS
point data can range from several millimeters to
20 meters or higher.

8.2 SPATIAL DATA ACCURACY STANDARDS

Discussions of location errors naturally lead to the
topic of spatial data accuracy standards. As users
of spatial data, we typically do not conduct testing
of location errors but rely on published standards
in evaluating data accuracy. Spatial data accuracy

standards have evolved as maps have changed
from printed to digital format.

In the United States, the development of spatial data accuracy standards has gone through three
phases. Revised and adopted in 1947, the U.S.
National Map Accuracy Standard (NMAS) sets the
accuracy standard for published maps such as
topographic maps from the U.S. Geological Survey
(USGS) (U.S. Bureau of the Budget 1947). The
standards for horizontal accuracy require that no
more than 10 percent of the well-defined map
points tested shall be more than 1/30 inch at scales
larger than 1:20,000, and 1/50 inch at scales of
1:20,000 or smaller. This means that the threshold
value is 40 feet on the ground for 1:24,000 scale
maps and about 167 feet on the ground for
1:100,000 scale maps. But the direct linkage of the
threshold values to map scales can be problematic
in the digital age because digital spatial data can be
easily manipulated and output to any scale.

In 1990 the American Society for Photogrammetry and Remote Sensing (ASPRS) published accuracy standards for large-scale maps (American
Society for Photogrammetry and Remote Sensing
1990). The ASPRS defines the horizontal accuracy
in terms of the root mean square (RMS) error, instead of fixed threshold values. The RMS error
measures deviations between coordinate values on
a map and coordinate values from an independent
source of higher accuracy for identical points.
Examples of a higher-accuracy data source may
include digital or hard-copy map data, GPS, or
survey data. The ASPRS standards stipulate the
threshold RMS error of 16.7 feet for 1:20,000 scale
maps and 2 feet for 1:2,400 scale maps.

In 1998 the Federal Geographic Data Committee (FGDC), a committee representing 17 federal
agencies at the time, established the National Standard for Spatial Data Accuracy (NSSDA) to replace
the NMAS. The NSSDA follows the ASPRS accuracy standards, but extends to map scales smaller
than 20,000 (Federal Geographic Data Committee
1998) (Box 8.1). A major difference between the
NSSDA and NMAS or the ASPRS accuracy standards is that the NSSDA omits threshold accuracy
values that spatial data, including paper maps and

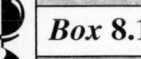

Box 8.1 **National Standard for Spatial Data Accuracy (NSSDA) Statistic**

To use the ASPRS standards or the NSSDA statistic, one must first compute the root mean square (RMS) error, which is defined by

$$RMS = \sqrt{\sum[(x_{\text{data},i} - x_{\text{check},i})^2 + (y_{\text{data},i} - y_{\text{check},i})^2]/n}$$

where $x_{\text{data},i}$ and $y_{\text{data},i}$ are the coordinates of the ith check point in the data set; $x_{\text{check},i}$ and $y_{\text{check},i}$ are the coordinates of the ith check point in the independent source of higher accuracy; n is the number of check points tested; and i is an integer ranging from 1 to n.

The NSSDA suggests that a minimum of 20 check points shall be tested. After the RMS is computed, it is multiplied by 1.7308, which represents the standard error of the mean at the 95 percent confidence level. The product is the NSSDA statistic. A handbook on how to use NSSDA to measure and report geographic data quality has been published online by the Land Management Information Center at Minnesota Planning (**http://www.lmic.state.mn.us/**).

digital data, must achieve. Instead, agencies are encouraged to establish accuracy thresholds for their products and to report the NSSDA statistic, a statistic based on the RMS error. Given the new accuracy standards, we must ask for the product specifications and determine ourselves if the level of data accuracy is acceptable for a GIS project.

Data accuracy should not be confused with data precision. Spatial data accuracy measures how close the recorded location of a spatial feature is to its ground location, whereas **data precision** measures how exactly the location is recorded. Distances may be measured with decimal digits or rounded off to the nearest meter or foot. Likewise, numbers can be stored in the computer as integers or floating points. Moreover, floating-point numbers can be single precision with 7 significant digits or double precision with up to 15 significant digits. The number of significant digits used in data recording expresses the precision of a recorded location.

8.3 TOPOLOGICAL ERRORS

Topological errors violate the topological relationships either required by a GIS package or defined by the user. The coverage model developed by ESRI, Inc. incorporates the topological relationships of connectivity, area definition, and contiguity (Chapter 3). If digitized features did not follow these relationships, they would have topological errors. The geodatabase data model from ESRI, Inc. has a total of 25 topology rules that apply to point, line, and polygon features (Chapter 4). Some of these rules relate to features within a feature class, while others relate to two or more participating feature classes. Using the geodatabase data model, we can choose the topological relationships to implement in the data sets and define the kinds of topological errors that are important to a project.

8.3.1 Topological Errors with Geometric Features

The topological relationships incorporated into the coverage model apply to the geometries of simple features in a single layer. This is also true for topological rules that apply to features in a geodatabase feature class. The following groups common topological errors with simple features by polygon, line, and point.

Topological errors with polygon features include unclosed polygons, gaps between polygons, and overlapping polygons (Figure 8.2).

A common topological error with line features is that they do not meet perfectly at a point

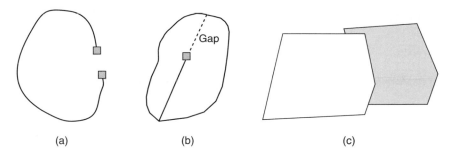

Figure 8.2

(*a*) An unclosed polygon, (*b*) a gap between two polygons, and (*c*) overlapped polygons.

(node). This type of error becomes an **undershoot** if a gap exists between lines, and an **overshoot** if a line is overextended (Figure 8.3). The result of both cases is a **dangling node** at the end of a **dangle.** Dangling nodes are, however, acceptable in special cases such as those attached to dead-end streets and small tributaries. A **pseudo node** appears along a continuous line and divides the line unnecessarily into separate lines (Figure 8.4). Some pseudo nodes are, however, acceptable. Examples include insertion of pseudo nodes at points where the attribute values of a line feature change.

The direction of a line may become a topological error. For example, a hydrologic analysis project may stipulate that all streams must follow the downstream direction and the starting point (the from-node) of a stream must be at a higher elevation than the end point (the to-node). Likewise, a traffic simulation project may require that all streets are clearly defined in terms of two-way, one-way, or dead-end (Figure 8.5).

Point features have few topological errors. The coverage model requires a label point to link a polygon to its attribute data. A polygon should therefore have one, and only one, label point. An error occurs if a polygon has zero or multiple label points. Unclosed polygons are often the cause for errors with multiple label points (Figure 8.6). When a gap exists between two polygons, they are treated topologically as a single polygon, which has an error if a label point has already been assigned to each polygon. A geodatabase may also require for some data sets that a polygon contain a point feature, such as a parcel containing an address point.

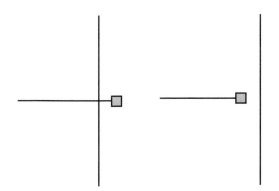

Figure 8.3

An overshoot (left) and an undershoot (right). Both types of errors result in dangling nodes.

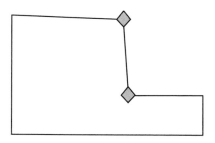

Figure 8.4

Pseudo nodes, shown by the diamond symbol, are nodes that are not located at line intersections.

8.3.2 Topological Errors Between Layers

The coverage model allows the user to work with topological errors in a single coverage. But with the new geodatabase data model, we can find and fix topological errors of spatial features between two or more layers. The layers do not have to have the same feature type. In other words, we can examine topological errors between two polygon feature classes or between a point and a polygon feature class.

A common error between two polygon layers is that their outline boundaries are not coincident (Figure 8.7). Suppose a GIS project uses a soil layer and a land-use layer for data analysis. Digitized separately, the two layers do not have coincident boundaries. If the two layers are later overlaid, the discrepancies between their boundaries become

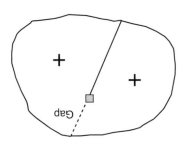

Figure 8.5
The from-node and to-node of an arc determine the arc's direction.

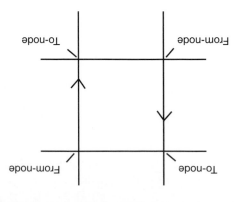

Figure 8.6
Multiple labels can be caused by unclosed polygons.

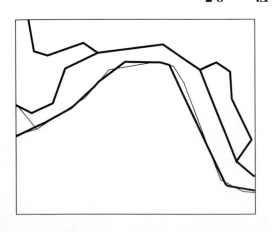

Figure 8.7
The outline boundaries of two layers, one shown in a thicker line and the other a thinner line, are not coincident at the top.

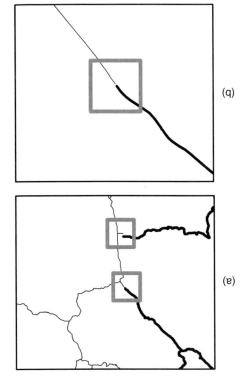

Figure 8.8
Line features from one layer do not connect perfectly with those from another layer at end points; (b) is an enlargement of the top error in (a).

small polygons, which miss either soil or land-use attributes. Similar problems can occur with individual polygon boundaries. For example, census tracts are supposed to be nested within counties and subwatersheds within watersheds. Errors occur when the larger polygons (e.g., counties) do not share boundaries with the smaller polygons (e.g., census tracts).

One type of error with two line layers occurs when line features from one layer do not connect with those from another layer at end points (Figure 8.8). For example, when we merge two highway layers from two adjacent states, we expect highways to connect perfectly across the state border. Errors can happen if highways intersect, or overlap, or have gaps. Other errors with line layers include overlapping line features (e.g., rail lines on highways) and line features not covered by another set of line features (e.g., bus routes not covered by streets).

Errors with point features can occur if they do not fall along line features in another layer. For example, gauge stations for measuring stream flow must fall along streams. Errors occur if they do

not. This is also true between section corners and the polygon boundaries of the Public Land Survey System (PLSS).

8.4 TOPOLOGICAL EDITING

Topological editing ensures that digitized spatial features follow the topological relationships that are either built into a data model or specified by the user. To perform topological editing, we must use a topology-based GIS package that can detect and display topological errors and has tools to remove them. Examples of topology-based GIS packages include ArcGIS, AutoCAD Map, and MGE (Box 8.2). They have similar capabilities in fixing topological errors with geometric features. The introduction of topology rules by ESRI, Inc. has changed the scope of topological editing considerably.

Using ArcGIS as an example, Section 8.4 covers three types of topological editing: topological editing on coverages, editing using a map topology, and editing using topology rules (Box 8.3). Although divided into three subsections, many editing concepts and procedures are similar.

 Box 8.2 | **GIS Packages for Topological Editing**

ArcGIS, AutoCAD Map, and MGE are all capable of performing topological editing with simple features. Chapter 8 references only ArcGIS, but the following list highlights the editing capabilities of the other two packages:

- AutoCAD Map: build topology, repair undershoots, snap clustered nodes, remove duplicate lines, simplify linear objects, and correct other errors.
- MGE (Base Mapper): build topology, remove redundant linear data segments, fix undershoots and overshoots, and create an intersection at a line crossing.

ArcGIS is not the only software package that uses topology rules for spatial data editing. Laser-Scan, a

company in Cambridge, England, launched an extension to Oracle9i called Radius Topology in June 2002 **(http://www.laser-scan.com/technologies/radius/ radius_topology/)**. Like the geodatabase data model, Radius Topology also implements topological relationships as rules between features and stores these rules as tables in the database. When these tables are activated, the topology rules automatically apply to spatial features in the enabled classes. An earlier version of Radius Topology was used to create the Ordnance Survey's MasterMap **(http://www.ordnancesurvey. co.uk/oswebsite/)**.

Tools for topological editing are placed in different applications and on different toolbars in ArcGIS. Chapter 8 divides these tools into three groups: coverage tools, map topology tools, and tools for topology rules.

The ArcEdit module of ArcInfo Workstation contains a large assortment of commands for editing coverages. ArcGIS 9.0 has incorporated three commands from ArcEdit into ArcToolbox: Clean and Build for constructing the topology of a coverage, and Create Labels for adding labels to a polygon coverage.

The Topology toolbar has the Map Topology button, which opens a dialog for the user to define a cluster tolerance and the feature classes to participate in a map topology. Tools for working with a map topology are the Topology Editing Tool on the Topology toolbar, and the Topology Tasks (Modify Edge and Reshape Edge) on the Editor toolbar.

Topology rules are defined through the properties dialog of a feature dataset in a geodatabase. The Topology toolbar has tools for validating, inspecting, and fixing topological errors. The options for actually fixing a topological error such as subtract features and create features are built into the topological error's context menu.

8.4.1 Topological Editing on Coverages

Topological editing on a coverage typically starts by constructing its topology. ArcGIS has the Clean command, which not only builds the topology but also applies the tolerances of dangle length and fuzzy tolerance to the entire coverage to remove some digitizing errors. The **dangle length** specifies the minimum length for dangling arcs on the output coverage. A dangling arc such as an overshoot is removed if it is shorter than the specified dangle length (Figure 8.9). The **fuzzy tolerance** specifies the minimum distance between points (vertices) and arcs on the output coverage. It is useful for removing duplicate lines within the specified tolerance (Figure 8.10). Clean also automatically inserts a node at a line intersection to enforce the topological arc-node relationship. Because it applies to an entire coverage, Clean must be used cautiously. A large dangle length can remove undershoots as well as overshoots

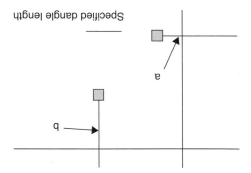

Specified dangle length

Figure 8.9
The dangle length can remove an overshoot if the overshoot, such as a, is smaller than the specified length. The overshoot b remains.

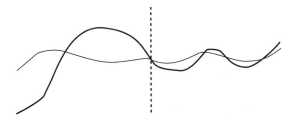

—— Specified fuzzy tolerance

Figure 8.10
The fuzzy tolerance can snap duplicate lines if the gap between the duplicate lines is smaller than the specified tolerance. In this diagram, the duplicate lines to the left of the dashed line will be snapped but not those to the right.

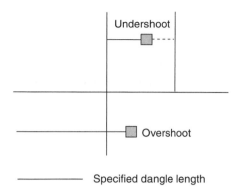

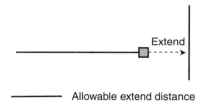

- Allowable extend distance

Figure 8.13
The allowable extend distance can remove the dangle by extending it to the line on the right.

- Specified dangle length

Figure 8.11
A large dangle length can remove the overshoots, which should be removed, and the undershoot, which should not be removed.

(Figure 8.11). Instead of being removed, undershoots should be extended to fill the gap. A large fuzzy tolerance will snap arcs that are not duplicate arcs, thus distorting the shape of map features (Figure 8.12).

For individual features rather than the entire coverage, ArcGIS has commands that can snap points and lines within specified tolerances. Nodesnap can snap nodes, Arcsnap can snap the ends of arcs to existing arcs, and Extend can extend dangling arcs to intersect existing arcs. Again, these commands must be used cautiously.

A large nodesnap tolerance, for example, can snap wrong nodes. Moreover, when a node is snapped to a new location within the nodesnap tolerance, the node moves the arc segment that is attached to it, thus altering the shape of the arc.

We can combine the above snapping commands with basic editing tools such as delete, move, add, split, unsplit, and flip, which are also available in ArcGIS, to fix many types of digitizing errors. Some examples are listed in the following:

- Dangles: One can remove an undershoot by extending the dangling arc to meet with the target arc at a new node (Figure 8.13), and remove an overshoot by deleting the extension (Figure 8.14). In general, it is easier to identify and remove an overshoot than an undershoot.
- Duplicate Arcs: One solution is to carefully select extra arcs and delete them, and the other is to delete all duplicate arcs within a box and redigitize.
- Wrong Arc Directions: One can alter the direction of an arc by flipping the arc, thus changing the relative position of the beginning and end nodes.

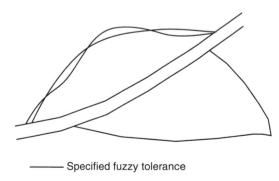

- Specified fuzzy tolerance

Figure 8.12
A large fuzzy tolerance can remove duplicate lines (top), which should be removed, as well as features such as a small stream channel (middle), which should not be removed.

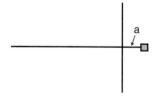

Figure 8.14
To remove the overshoot a, first select it and then delete it.

- Reshaping Arcs: To reshape an arc, one can move, add, or delete points (vertices) that make up the arc.

- Label Errors: One can add new label points with proper IDs for missing labels in a polygon coverage. If a polygon has multiple labels, the extra labels can be selected and deleted. Multiple labels may also indicate an unclosed polygon.

- Pseudo Nodes: One can remove a pseudo node by first setting the two arcs on each side of the node to have the same ID value, before unsplitting them (Figure 8.15).

Figure 8.15
To remove the pseudo node, select *a* and *b*, assign the same ID value to both, and unsplit them.

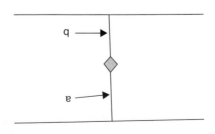

8.4.2 Editing Using Map Topology

A **map topology** is a temporary set of topological relationships between the parts of features that are supposed to be coincident. For example, a map topology can be built between a land-use layer and a soil layer so that their outlines are coincident. A map topology can also be built between a stream layer and a county layer so that, when a stream serves as the county boundary, they are coincident. Layers participating in a map topology can be shapefiles or geodatabase feature classes, but not coverages. (Shapefiles are nontopological but can participate in a temporary map topology that applies only to coincident features.)

Coincident features in a map topology are defined by a specified cluster tolerance. The **cluster tolerance** can snap vertices and lines (edges) if they fall within the specified tolerance. Functionally, the cluster tolerance is similar to the fuzzy tolerance for

editing coverages. And, like the fuzzy tolerance, a cluster tolerance should not be set too large. A large cluster tolerance can unintentionally alter the shape of lines and polygons. A good strategy is to use a small cluster tolerance and to change it only to deal with more severe but localized errors.

To edit with a map topology, we create a map topology, specify the participating feature classes, and define a cluster tolerance. Then we use the editing tools in ArcGIS to force the geometries of the participating feature classes to be coincident. (Task 2 in the applications section uses a map topology for editing.)

8.4.3 Editing Using Topology Rules

The geodatabase data model has a total of 25 topology rules for point, line, and area features. Editing with a topology rule involves three basic steps. The first step creates a new topology by defining the participating feature classes, the ranks for each feature class, the topology rule(s), and a cluster tolerance. The rank determines the relative importance of a feature class in topological editing. Suppose a topology includes two feature classes of soil and vegetation cover. If the soil feature class is deemed to be more accurate of the two, it is assigned a higher rank, thus ensuring that its polygon boundaries will not be moved as much as the other feature class.

The second step is validation of topology. This step evaluates the topology rule and creates errors indicating those features that have violated the topology rule. At the same time, the edges and vertices of features in the participating feature classes are snapped together if they fall within the specified cluster tolerance. The snapping uses the ranking of the feature classes: features of a lower-rank feature class are moved more than features of a higher-rank feature class. Validation of topology is similar to "cleaning" a coverage, and the cluster tolerance plays the role of the fuzzy tolerance. Validation results are saved into a topology layer, which is used in the third step for fixing errors and for accepting errors as exceptions (e.g., acceptable dangling nodes). The geodatabase data model provides a set of tools for fixing topological errors. For example, if the study area boundaries from two participating feature classes are not coincident and create small polygons between them, we can opt to subtract (i.e., delete) these polygons, or create new polygons, or modify their boundaries until they are coincident. (Tasks 3 and 4 in the applications section use some of these tools in fixing errors between layers.)

Similar to editing coverages, the steps of creating a topology, validating the topology rule, and fixing topological errors may have to be repeated before we are satisfied with the topological accuracy of a feature class or between two or more feature classes.

8.5 NONTOPOLOGICAL EDITING

Nontopological editing refers to a variety of basic editing operations that can modify simple features and can create new features from existing features (Box 8.4). Like topological editing, many of these basic operations also use snapping tolerances for snapping points and lines and the line and polygon sketches to edit features. The difference is that the basic operations do not involve topology as defined in the coverage model, a map topology, or a geodatabase.

8.5.1 Edit Existing Features

The following summarizes nontopological or basic editing operations on existing features.

- Extend/Trim Lines: One can extend or trim a line to meet a target line.
- Delete/Move Features: One can delete or move one or more selected features, which may be points, lines, or polygons. Because each polygon in nontopological data is a unit, separate from other polygons, moving a polygon means placing the polygon on top of an existing polygon while creating a void area in its original location (Figure 8.16).
- Reshaping Features: One can alter the shape of a line by moving, deleting, or adding points (vertices) on the line (Figure 8.17). The same method can be used to reshape a polygon. But if the reshaping is intended for a polygon and its connected polygons, one must use a topology tool so that, when a

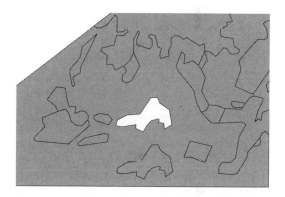

Figure 8.16
After a polygon of a shapefile is moved, a void area appears in its location.

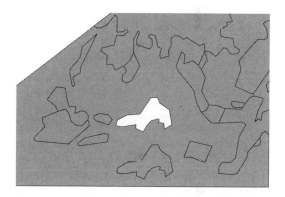

| *Box* 8.4 | **Basic Editing in ArcGIS** |

Most basic editing tools are available on the Editor toolbar. The Editor dropdown menu offers such tools as Move, Merge, Union, and Buffer. The Task dropdown menu groups tools by task. For example, the Modify Tasks group has such tools as Reshape Feature, Cut Polygon Features, and Modify Feature. Other editing tools are located on the Advanced Editing toolbar. They include Copy Feature, Extend, Trim, and Explode Multipart Feature. Many of these basic editing tools are limited to shapefiles and geodatabase feature classes.

8.5.2 Create Features from Existing Features

The following summarizes nontopological operations that can create new features from existing features.

- **Split Lines and Polygons:** One can sketch a new line that crosses an existing line to split the line, or sketch a split line through a polygon to split the polygon (Figure 8.18).

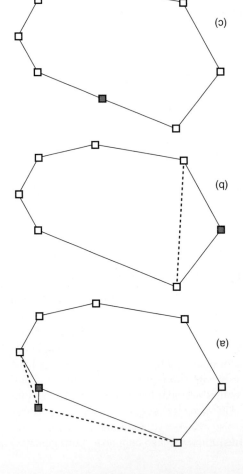

Figure 8.17

Reshape a line by moving a vertex (*a*), deleting a vertex (*b*), and adding a vertex (*c*).

- **Merge Features:** One can group selected line or polygon features into one feature (Figure 8.19). If the merged features are not spatially adjacent, they form a feature with multiple parts (e.g., multipart polygons).

- **Buffer Features:** One can create a buffer around a line or polygon feature at a specified distance.

- **Union Features:** One can combine features from different layers into one feature. This operation differs from the merge operation because it works with different layers rather than a single layer.

- **Intersect Features:** One can create a new feature from the intersection of overlapped features in different layers.

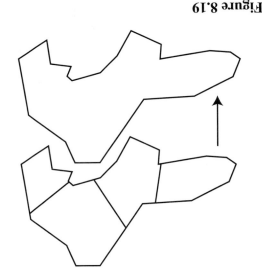

Figure 8.19

Merge four selected polygons into one.

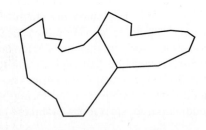

Figure 8.18

Sketch a line across the polygon boundary to split the polygon into two.

| Box 8.5 | **Edgematching and Advanced Editing in ArcGIS** |

There are two options for performing edgematching in ArcGIS. The first option is the Edgematch command, a menu-driven command, in ArcEdit. The Spatial Adjustment toolbar is the second option. This toolbar has a dropdown menu for the user to specify adjustment data, method, and properties. It also has tools for selecting links in a window area or individual links for edgematching. Both ArcInfo Workstation and ArcGIS Desktop have tools for line simplification and line smoothing. In ArcGIS Desktop, these tools are available on the Advanced Editing toolbar as well as in ArcToolbox.

8.6 OTHER EDITING OPERATIONS

There are editing operations that cannot be easily classified as either topological editing or nontopological editing. This section covers two such operations: edgematching and manipulation of line features (Box 8.5).

8.6.1 Edgematching

Edgematching matches lines along the edge of a layer to lines of an adjacent layer so that the lines are continuous across the border between the layers (Figure 8.20). For example, a regional highway layer may consist of several state highway layers that have been digitized and edited separately. These individual highway layers must be edgematched to remove errors between them and eventually joined to make the final single layer. Errors between layers are often very small (Figure 8.21). But unless they are removed, the regional highway layer cannot be used for such operation as shortest path analysis.

Edgematching involves a source layer and a target layer. Features, typically vertices, on the source layer are moved to match those on the target layer. A snapping tolerance can assist in snapping vertices (and lines) from the two layers. Edgematching can be performed on one pair of vertices, or multiple pairs, at a time. After edgematching is complete, the source layer and the

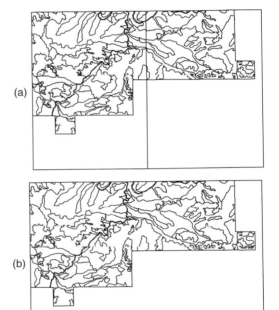

Figure 8.20
Edgematching matches the lines of two adjacent layers (*a*) so that the lines are continuous across the border (*b*).

target layer can be merged into a single layer and the border separating the two layers (e.g., the state boundary) can be dissolved. (Task 5 in the applications section covers the working details of edgematching.)

from the 1:100,000 scale source map is displayed at the 1:1,000,000 scale, lines become jumbled and

8.6.2 Line Simplification and Smoothing

Line simplification refers to the process of simplifying or generalizing a line by removing some of its points. Line simplification is a common practice in map display (Robinson et al. 1995) and a common topic under generalization in digital cartography and GIS (McMaster and Shea 1992; Weibel and Dutton 1999). When a map digitized from the 1:100,000 scale source map is displayed at the 1:1,000,000 scale, lines become jumbled and

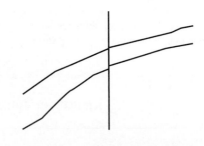

Figure 8.21
Mismatches of lines from two adjacent layers are only visible after zooming in.

fuzzy because of the reduced map space. Line simplification can also be important for GIS analysis that uses every point that makes up a line. One example is buffering, which measures a buffer distance from each point along a line (Chapter 12). Lines with too many points do not necessarily improve the result of analysis but will require more processing time.

The **Douglas-Peucker algorithm** is a well-known algorithm for line simplification (Douglas and Peucker 1973). The algorithm works line by line and with a specified tolerance. The algorithm starts by connecting the end points of a line with a trend line (Figure 8.22). The deviation of each intermediate point from the trend line is calculated. If there are deviations larger than the tolerance, then the point with the largest deviation is connected to the end points of the original line to form new trend lines (Figure 8.22a). Using these new trend lines, the algorithm again calculates the deviation of each intermediate point. This process continues until no deviations exceed the tolerance. The result is a simplified line that connects the trend

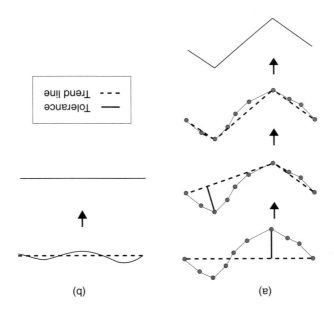

(a)   (b)

Tolerance ———
Trend line - - - -

Figure 8.22
The Douglas-Peucker line simplification algorithm is an iterative process, which requires use of a tolerance, trend lines, and calculation of deviations of vertices to the trend line. See Section 8.6.2 for explanation.

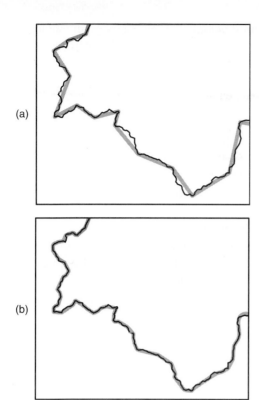

(a)

(b)

Figure 8.23
Result of line simplification can differ depending on
the algorithm used: the Douglas-Peucker algorithm
(*a*) and the bend-simplify algorithm (*b*).

lines. But if the initial deviations are all smaller
than the tolerance, the simplified line is the straight
line connecting the end points (Figure 8.22*b*).

One shortcoming of the point-removing
Douglas-Peucker algorithm is that the simplified

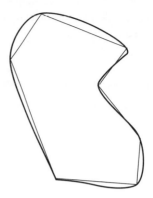

Figure 8.24
Line smoothing smoothes a line by generating new
vertices mathematically and adding them to the line.

line often has sharp angles. An alternative is to
use an algorithm that dissects a line into a series
of bends, calculates the geometric properties of
each bend, and removes those bends that are considered insignificant (Wang 1996). By emphasiz-
ing the shape of a line, this new algorithm tends
to produce simplified lines with better cartographic quality than the Douglas-Peucker algorithm (Figure 8.23).

Line smoothing also adds new points to lines,
but the location of new points is generated by
mathematical functions (e.g., Bezier curves). Line
smoothing is perhaps most important for data display. Lines derived from computer processing
such as isolines on a precipitation map are sometimes jagged and unappealing. These lines can be
smoothed for data display purposes. Figure 8.24
shows an example of line smoothing using splines.

KEY CONCEPTS AND TERMS

Check plot: A plot of a digitized map for error
checking.

Cluster tolerance: A tolerance for snapping
points and lines. It is functionally similar to a
fuzzy tolerance.

Dangle length: A tolerance that specifies the
minimum length for dangling arcs on an output
coverage.

Dangle: An arc that has the same polygon on
both its left and right sides and a dangling node at
the end of the arc.

Dangling node: A node at the end of an arc that is not connected to other arcs.

Data precision: A measure of how exactly data such as the location data of x- and y-coordinates are recorded.

Douglas-Peucker algorithm: A computer algorithm for line simplification.

Edgematching: An editing operation that matches lines along the edge of a layer to lines of an adjacent layer.

Fuzzy tolerance: A tolerance that specifies the minimum distance between vertices on a coverage.

Line simplification: The process of simplifying or generalizing a line by removing some of the line's points.

Line smoothing: The process of smoothing a line by adding new points, which are typically generated by a mathematical function such as spline, to the line.

Location errors: Errors related to the location of map features such as missing lines or missing polygons.

Map topology: A temporary set of topological relationships between coincident parts of simple features between layers.

Nontopological editing: Editing on nontopological data.

Overshoot: One type of digitizing error that results in an overextended arc.

Pseudo node: A node appearing along a continuous arc.

Topological editing: Editing on topological data to make sure that they follow the required topological relationships.

Topological errors: Errors related to the topology of map features such as dangling arcs and missing or multiple labels.

Undershoot: One type of digitizing error that results in a gap between arcs.

REVIEW QUESTIONS

1. Explain the difference between location errors and topological errors.

2. What are the primary data sources for digitizing?

3. A digitized map from a secondary data source such as a USGS quadrangle map is subject to more location errors than a primary data source. Why?

4. Although the U.S. National Map Accuracy Standard adopted in 1947 is still printed on USGS quadrangle maps, the standard is not really applicable to GIS data. Why?

5. According to the new National Standard for Spatial Data Accuracy, a geospatial data producer is encouraged to report a RMS statistic associated with a data set. In general terms, how does one interpret and use the RMS statistic?

6. Suppose a point location is recorded as (575729.0, 5228382) in data set 1 and (575729.64, 5228382.11) in data set 2. Which data set has a higher data precision? In practical terms, what does the difference in data precision in this case mean?

7. The ArcGIS Desktop Help has a poster illustrating topology rules in the geodatabase data model (ArcGIS Desktop Help > Editing in ArcMap > Editing Topology > Topology rules). View the poster. Can you think of an example (other than those on the poster) that can use the polygon rule of "Must be covered by feature class of"?

8. Give an example (other than those on the poster) that can use the polygon rule of "Must not overlap with."

9. Give an example (other than those on the poster) that can use the line rule of "Must not intersect or touch interior."

10. Use a diagram to illustrate how a large nodesnap for editing can alter the shape of line features.

11. Use a diagram to illustrate how a large cluster tolerance for editing can alter the shape of line features.

12. Explain the difference between a dangling node and a pseudo node.

13. What is a map topology?

14. Describe the three basic steps in using a topology rule.

15. Some nontopological editing operations can create features from existing features. Give two examples of such operations.

16. Edgematching requires a source layer and a target layer. Explain the difference between these two types of layers.

17. The Douglas-Peucker algorithm typically produces simplified lines with sharp angles. Why?

APPLICATIONS: SPATIAL DATA EDITING

This applications section covers five tasks. Task 1 lets you use basic editing tools on a shapefile. Task 2 asks you to use a map topology and a cluster tolerance to fix digitizing errors between two shapefiles. You will use topology rules in Tasks 3 and 4: fixing dangles in Task 3 and fixing outline boundaries in Task 4. Task 5 is an edgematching operation using the Spatial Adjustment toolbar.

Task 1: Edit a Shapefile

What you need: *editmap2.shp* and *editmap3.shp*.

Task 1 covers three basic editing operations for shapefiles: merging polygons, splitting a polygon, and reshaping the polygon boundary. While working with *editmap2.shp*, you will use *editmap3.shp* as a reference, which shows how *editmap2.shp* will look after editing.

1. Start ArcCatalog and make connection to the Chapter 8 database. Launch ArcMap. Change the name of the data frame to Task 1. Add *editmap3.shp* and *editmap2.shp* to Task 1. To edit *editmap2* by using *editmap3* as a guide, they must be shown in different outline symbols. Select Properties from the context menu of *editmap2*. On the Symbology tab, change the symbol to Hollow with the Outline Color in black. On the Labels tab,

check the box to label features in this layer and select LANDED_ID to be the label field. Click the symbol of *editmap3* in the table of contents. Choose the Hollow symbol and the Outline Color of red. On the Selection tab in the table of contents, uncheck *editmap3*. Switch to the Display tab.

2. Make sure that Editor toolbar is checked. Click the Editor dropdown arrow and choose Start Editing. Make sure that the target layer is *editmap2*. The first operation is to merge polygons 74 and 75. Click the Edit Tool. Click inside polygon 75, and then click inside polygon 74 while pressing the Shift key. The two polygons are highlighted in cyan. Click the Editor dropdown arrow and choose Merge. In the next dialog, choose the top feature and click OK to dismiss the dialog. Polygons 74 and 75 are merged into one with the label of 75.

Q1. List other editing operations beside Merge that are on the Editor menu.

3. The second operation is to cut polygon 71. Click the Task dropdown arrow and choose Cut Polygon Features. Zoom in the area around polygon 71. Use the Edit Tool to select polygon 71. Click the Sketch Tool. To cut a polygon, the cut line should cross over

Task 2: Use Cluster Tolerance to Fix Digitizing Errors Between Two Shapefiles

What you need: *land_dig.shp*, a reference shape-file; and *trial_dig.shp*, a shapefile digitized off *land_dig.shp*.

There are discrepancies between *land_dig.shp* and *trial_dig.shp* due to digitizing errors. This task uses a cluster tolerance to force the boundaries of *trial_dig.shp* to be coincident with those of *land_dig.shp*. Both *land_dig.shp* and *trial_dig.shp* are measured in meters and in UTM coordinates.

1. Insert a new data frame and rename it Task 2. Add *land_dig.shp* and *trial_dig.shp* to Task 2. Display the shapefiles with the outline symbol of different colors. Label *land_dig* with the field of LAND_DIG_1. On the Selection tab in the table of contents, uncheck *land_dig*. Zoom in and use the Measure tool to check the discrepancies between the two shapefiles. Most discrepancies are smaller than 1 meter.

2. The first step is to create a map topology between the two shapefiles. Click the View menu and make sure that both the Editor and Topology toolbars are checked. Select Start Editing from the Editor's dropdown menu. Make sure that the task is Modify Edge and the target is *trial_dig*. Click the Map Topology tool on the Topology toolbar. In the next dialog, select both *land_dig* and *trial_dig* to participate in the map topology and enter 1 (meter) for the Cluster Tolerance. Click OK to dismiss the dialog.

3. *trial_dig* has five polygons, of which three are isolated polygons and two are spatially adjacent. Start editing with the isolated polygon in the lower right. Zoom in the area around the polygon. Click the Topology Edit Tool on the Topology toolbar. Then use the

7. Select Stop Editing from the Editor dropdown list, and save the edits.

Q3. Can you suggest another option for completing the third operation?

---

the polygon boundary. Left-click the mouse where you want the cut line to start, click each vertex that makes up the cut line, and double-click the end vertex. Polygon 71 is cut into two, each labeled 71.

Q2. What other modify tasks beside Cut Polygon Features are on the Task menu?

4. The third operation is to reshape polygon 73 by extending its southern border in the form of a rectangle. Because polygon 73 shares the border (i.e., edge) with polygon 59, you need to use a map topology to modify the border. Click the Editor's dropdown arrow, point to More Editing Tools, and check Topology. Click the Map Topology tool on the Topology toolbar. In the next dialog, check the *editmap2* box. Change the Task to the topology task of Modify Edge. Click the Topology Edit Tool on the Topology toolbar, and then double-click on the southern edge of polygon 73. Now the outline of polygon 73 turns magenta with vertices in dark green and the end point in red.

5. The strategy in reshaping the polygon is to add three new vertices and to drag the vertices to form the new shape. Move the mouse pointer to about the midpoint of the southern border, right-click the point, and select Insert Vertex. A new vertex appears. Position the mouse pointer over the new vertex (vertex 1). When the pointer changes to a four-headed arrow, drag it to where the new border is going to be (use *editmap3* as a guide), and release the mouse button. (The original border of polygon 73 remains in place as a reference. It will disappear when you click anywhere outside polygon 73.)

6. Next, add another vertex (vertex 2) along the line that connects vertex 1 and the original SE corner of polygon 73. Position the mouse pointer over vertex 2. When the pointer changes, drag it to the SE corner of the new boundary. Do the same to form the SW corner of the new boundary. After you have modified the edge, right-click the edge and select Finish Sketch.

mouse pointer to double-click the boundary of the polygon. The boundary turns into an edit sketch with green squares representing vertices and a red square representing a node. Place the mouse pointer over a vertex until it changes to a four-headed arrow. Right-click on the arrow, and select Move from the context menu. Hit Enter to dismiss the next dialog. (You are using the specified cluster tolerance to snap the vertices and edges.) Click any point outside the polygon to unselect its boundary. The polygon now should coincide perfectly with its corresponding polygon in *land_dig*. Move to the other polygons in *trial_dig* and follow the same procedure to fix digitizing errors.

4. All discrepancies except one (polygon 76) are fixed between *trial_dig* and *land_dig*. The remaining discrepancy is larger than the specified cluster tolerance (1 meter). Rather than using a larger cluster tolerance, which may result in distorted features, you will use the basic editing operations to fix the discrepancy. Use the Edit Tool to double-click the boundary with the discrepancy. When the boundary turns into an edit sketch, you can drag a vertex to meet the target line.

5. After you have completed editing all five polygons, select Stop Editing from the Editor's dropdown menu and save the edits.

Q4. If you had entered 4 meters for the cluster tolerance in Step 2, what would have happened to *trial_dig.shp*?

Task 3: Use Topology Rule to Fix Dangles

What you need: *idroads.shp*, a shapefile of Idaho roads; *mtroads_idtm.shp*, a shapefile of Montana roads projected onto the same coordinate system as *idroads.shp*; and *Merge_result.shp*, a shapefile with merged roads from Idaho and Montana.

The two road shapefiles downloaded from the Internet are not perfectly connected across the state border. Therefore *Merge_result.shp* contains gaps. Unless the gaps are removed, *Merge_result.shp* cannot be used for network applications such as finding the shortest path. This task asks you to use a topology rule to symbolize where the gaps are and then use the editing tools to fix the gaps.

1. The first step for this task is to prepare a personal geodatabase and a feature dataset, and to import *Merge_result.shp* as a feature class to the feature dataset. Right-click the Chapter 8 database in ArcCatalog, point to New, and select Personal Geodatabase. Rename the geodatabase *MergeRoads.mdb*. Right-click *MergeRoads.mdb*, point to New, and select Feature Dataset. Enter *Merge* for the Name of the feature dataset. Click the Edit button in the Spatial Reference frame. On the Coordinate System tab, import the coordinate system from *idroads.shp* for the feature dataset. On the X/Y Domain tab, change the Precision to 100. (Changing the precision from the default value of 488 to 100 will result in a slightly larger cluster tolerance.) Click OK to dismiss the dialogs. Right-click *Merge,* point to Import, and select Feature Class (single). In the next dialog, select *Merge_result.shp* for the input features and enter *Merge_result* for the output feature class name. Click OK to import the shapefile.

2. This step is to build a new topology. Right-click *Merge,* point to New, and select Topology. Click Next in the first two panels. Check the box next to *Merge_result* in the third. Click Next in the fourth panel. Click the Add Rule button in the fifth panel. Select "Must Not Have Dangles" from the Rule dropdown list in the Add Rule dialog and click OK. Click Next and then Finish to complete the setup of the topology rule. After the new topology has been created, click Yes to validate it.

Q5. Each rule has a description in the Add Rule dialog. What is the rule description for "Must Not Have Dangles"?

Q6. What is the rule description for "Must Not Have Pseudos"?

3. The validation results are saved in a topology layer called *Merge_Topology* in the *Merge* feature dataset. Select Properties from the context menu of *Merge_Topology*. The Topology Properties dialog has four tabs. The General, Feature Classes, and Rules tabs show the definition of the topology rule. Click the Errors tab and then Generate Summary. The summary report shows 102 errors, meaning that *Merge_result* has 102 dangling nodes. The Preview tab in ArcCatalog shows these dangling nodes in red. Most of them are the end points along the border of the two-state area, which are acceptable dangling nodes. Those nodes along the border between the two states are the ones that need inspection and, if necessary, fixing.

4. Insert a new data frame in ArcMap, and rename the data frame Task 3. Add the *Merge* feature dataset to Task 3. Also add *idroads.shp* and *mtroads_idtm.shp* to Task 3. The two shapefiles can be used as references in inspecting and fixing errors. Use different colors to display *Merge_result*, *idroads*, and *mtroads_idtm* so that you can easily distinguish between them. On the Selection tab, uncheck *idroads* and *mtroads_idtm*.

5. Now you are ready to inspect and fix errors in *Merge_result*. Make sure that both the Editor toolbar and the Topology toolbar are available. Select Start Editing from the Editor menu. Select *MergeRoads.mdb* as the database to edit data from. There are eight places where the roads connect across the Montana-Idaho border. These places are shown with point errors. Zoom in the area around the first crossing near the top of the map until you see a pair of dangles, separated by a distance of about 5.5 meters. (Use the Measure tool on the standard toolbar to measure the distance.) Click the Fix Topology Error Tool on the Topology toolbar, and then click a red square. The red square turns black after being selected. Click the Error Inspector on the Topology toolbar. A report appears and shows the error type (Must Not Have Dangles). Use the Fix Topology Error Tool and right-click on the black square. The context menu has the tools of Snap, Extend, and Trim to fix errors. Select Snap, and a Snap Tolerance box appears. Enter 6 (meters) in the box. The two squares are snapped together into one square. Right-click the square again and select Snap. The square should now disappear. Remember that you have access to the Undo and Redo tools on the Edit menu as well as the standard toolbar.

6. The second point error, when zoomed in, shows a gap of 1.25 meters. There are at least two ways to fix the error. The first option is to use the Snap command of the Fix Topology Error Tool by applying a snap tolerance of at least 1.25. Here you will use the second option, which uses the regular editing tools. First set up the editing environment. Select Snapping from the Editor's menu, and check the boxes for Vertex and End for *Merge_result*. Next select Options from the Editor's menu. On the General tab, enter 10 for the snapping tolerance. Make sure that the Task is Create New Feature and the Target is *Merge_result*. Click the Sketch Tool. Right-click the square on the left, point to Snap to Feature, and select Endpoint. Right-click the square on the right, point to Snap to Feature, and select Endpoint. Press F2 to finish sketching, or right-click the square and select Finish Sketch. Now the gap is bridged with a new line segment. Click Validate Topology in Current Extent on the Topology toolbar. The square symbols disappear, meaning that the point error no longer exists.

7. With the exception of the last point error (the easternmost point error, along US20 in *mtroads_idtm*), you can use the above two options to fix the rest of point errors. The last point error shows the roads from the two shapefiles run parallel to each other over a distance of 50 meters. You will fix the error

using a different method. Click the Edit Tool on the Editor toolbar. Double-click the road from the Idaho side. The road turns cyan, and the dark green squares show the vertices that shape the road. Place the mouse pointer over the end point of the road until it changes to a four-headed arrow. Drag the end point to meet with the end point of the road from the Montana side. Click any point off the road to unselect it. Click Validate Topology in Current Extent on the Topology toolbar to make sure that the error has been fixed.

8. After all point errors representing misconnections of roads across the state border have been fixed, select Stop Editing from the Editor's menu and save the edits.

Task 4: Use Topology Rule to Ensure That Two Polygon Layers Cover Each Other

What you need: *landuse.shp* and *soils.shp*, two polygon shapefiles based on UTM coordinates.

Digitized from different source maps, the outlines of the two shapefiles are not completely coincident. This task shows you how to use a topology rule to symbolize the discrepancies between the two shapefiles and use the editing tools to fix the discrepancies.

1. Similar to Task 3, the first step is to prepare a personal geodatabase and a feature dataset, and to import *landuse.shp* and *soils.shp* as feature classes to the feature dataset. Right-click the Chapter 8 folder in ArcCatalog, point to New, and select Personal Geodatabase. Rename the geodatabase *Land.mdb*. Right-click *Land.mdb*, point to New, and select Feature Dataset. Enter *LandSoil* for the Name of the feature dataset. Click Edit in the Spatial Reference frame. On the Coordinate System tab, import the coordinate system from *landuse.shp* for the feature dataset. On the X/Y Domain tab, change the Precision to 100. Click OK to dismiss the dialogs. Right-click *LandSoil*, point to Import, and select Feature Class

(multiple). In the next dialog, add *landuse.shp* and *soils.shp* as the input features.

2. Next build a new topology. Right-click *LandSoil*, point to New, and select Topology. Click Next in the first two panels. In the third panel, check both *landuse* and *soils* to participate in the topology. The fourth panel allows you to set ranks for the participating feature classes. Features in the feature class with a higher rank are less likely to move. Click Next because the editing operations for this task are not affected by the ranks. Click the Add Rule button in the fifth panel. Select *landuse* from the top dropdown list, select "Must Cover Each Other" from the Rule dropdown list, and select *soils* from the bottom dropdown list. Click OK to dismiss the Add Rule dialog. Click Next and then Finish to complete the setup of the topology rule. After the new topology has been created, click Yes to validate it.

Q7. What is the rule description for "Must Cover Each Other"?

3. *LandSoil_Topology* contains the validation results. The Preview tab shows areas not covered by each other.

4. Insert a new data frame in ArcMap, and rename the data frame Task 4. Add the *LandSoil* feature dataset to Task 4. Use the outline symbols of different colors to display *landuse,* and *soils*. Zoom in the area errors. Most deviations between the two feature classes are within 1 meter.

5. Select Start Editing from the Editor's menu. Click the Fix Topology Error Tool on the Topology toolbar, and drag a box to select every area error. All area errors turn black. Right-click a black area and select Subtract. The Subtract command removes areas that are not common to both feature classes. In other words, Subtract makes sure that, after editing, every part of the area extent covered by *LandSoil* will have attribute data from both feature classes.

6. There are at least three other options to fix the errors. First, you can use the Create Feature command instead of Subtract. The Create Feature command creates new features for each area formed between the two sets of boundaries. The obvious problem with this option is that attributes must be gathered for the new features. Second, you can select one or more area errors at a time and choose either Subtract or Create Feature to fix the selected area errors. The problem with attributes for the new features still remains. Third, you can use the editing tools such as Modify Feature to fix area errors. The third option is most time consuming and requires knowledge as to how to reshape the boundaries. To see how each of these options works, you can click the Undo button between trials.

7. Select Stop Editing from the Editor dropdown menu. Save the edits.

Task 5: Perform Edgematching

What you need: *hoytmtn* and *mrblemtn*, two soil feature classes in *edgematch.mdb* (a personal geodatabase), that need to be edgematched.

Both *hoytmtn* and *mrblemtn* are based on the Universal Transverse Mercator (UTM) coordinate system and measured in meters. The task assumes that *mrblemtn* is more accurate than *hoytmtn*. Therefore, vertices (and lines) in *hoytmtn* are moved in the editing process to match those in *mrblemtn*. In ArcMap, *hoytmtn* is called the source layer and *mrblemtn* the target layer.

1. Insert a new data frame and rename it Task 5. Add *hoytmtn* and *mrblemtn* from the *edgematch.mdb* to Task 5. To differentiate the source layer from the target layer, change *hoytmtn* to an outline symbol in red and *mrblemtn* to an outline symbol in black. Click the View menu, point to Toolbars, and make sure that both the Editor toolbar and the Spatial Adjustment toolbar are checked.

2. First set the snapping environment and options. Select Start Editing from the Editor's dropdown menu. Select Snapping from the same menu, check the Vertex box for *hoytmtn*, and close the dialog. Then select Options from the Editor's menu to open the Editing Options dialog. On the General tab, enter 10 (meters) for the snapping tolerance.

3. Next set the parameters using the Spatial Adjustment menu. Select Set Adjust Data from the Spatial Adjustment dropdown menu. Select the option for adjusting all features in *hoytmtn* but not *mrblemtn*. Click the Spatial Adjustment dropdown arrow, point to Adjustment Methods, and check Edge Snap. Select Options from the Spatial Adjustment dropdown menu to open the Adjustment Properties dialog. On the General tab, click on the Options button in the Adjustment method frame and make sure that the method is Smooth. On the Edge Match tab, first select *hoytmtn* for the source layer and *mrblemtn* for the target layer. Next check Use Attributes. The two feature classes have the same attribute fields that can be used to assist the edgematching operation.

Q8. The Options button in the Adjustment method frame shows the methods of Smooth and Line. What is the difference between the two methods? (Tip: Right-click on Smooth and read the description.)

4. You are ready for edgematching. Zoom in the top area between *hoytmtn* and *mrblemtn* so that you can see the first pair of unmatched soil lines. Click the Edge Match tool on the Spatial Adjustment toolbar, and drag a box around the gap. A link should now connect the two end points: the link is shown as an arrow and the end points as squares. Select Preview Window from the Spatial Adjustment menu. The Adjustment Preview Window shows how the soil lines would connect if the adjustment were made.

5. The Edge Match tool can work with more than one link at a time. Click the Full Extent button in ArcMap. Make sure that the Edge Match tool is activated. Drag a box so that the box covers the remaining unmatched soil

lines and the bottom border. There should be 8 links. To see what each adjustment is like, zoom in a link area. Right-click the Adjustment Preview Window and select Track Data Frame Extent. The Adjustment Preview Window lets you preview each adjustment. Click the View Link Table tool on the Spatial Adjustment toolbar. The Link Table should have 9 records (8 new links plus the one from Step 4). Each record shows the shift of a vertex in *hoytmtn* to its new location through adjustment.

6. If all links are set, select Adjust from the Spatial Adjustment dropdown menu. At this point, the edgematching operation is complete. Select Stop Editing from the Editor menu and save the edits.

7. After the two feature classes are edgematched, you can use the Append tool in the Data Management Tools/General toolset to merge them into one polygon feature class and then use the Dissolve tool in the Data Management Tools/Generalization toolset to dissolve the artificial boundary between the two input feature classes. Both

Append and Dissolve are covered in Chapter 12 on vector data analysis.

Challenge Task

What you need: *idroads.shp*, *wyroads.shp*, and *idwyroads.shp*.

The Chapter 8 database contains *idroads.shp*, a major road shapefile for Idaho; *wyroads.shp*, a major road shapefile for Wyoming; and *idwyroads.shp*, a merged road shapefile of Idaho and Wyoming. All three shapefiles are projected onto the Idaho Transverse Mercator (IDTM) coordinate system and measured in meters. This challenge task asks you to examine and fix gaps that are smaller than 200 meters on roads across the Idaho and Wyoming border. There are at least two options for completing the challenge task.

Option 1: Treat the challenge task as an edgematching task. Import *idroads.shp* and *wyroads.shp* as line feature classes to a personal geodatabase, and perform the edgematching operation.

Option 2: Use a topology rule in the geodatabase data model to fix dangles in *idwyroads.shp*.

REFERENCES

American Society for Photogrammetry and Remote Sensing. 1990. ASPRS Accuracy Standards for Large-Scale Maps. *Photogrammetric Engineering and Remote Sensing* 56: 1068–70.

Douglas, D. H., and T. K. Peucker. 1973. Algorithms for the Reduction of the Number of Points Required to Represent a Digitized Line or Its Caricature. *The Canadian Cartographer* 10: 110–22.

Federal Geographic Data Committee. 1998. Part 3: *National Standard for Spatial Data Accuracy, Geospatial*

Positioning Accuracy Standards, FGDC-STD-007.3-1998. Washington, DC: Federal Geographic Data Committee.

McMaster, R. B., and K. S. Shea. 1992. *Generalization in Digital Cartography*. Washington, DC: Association of American Geographers.

Monmonier, M. 1996. *How to Lie with Maps,* 2d ed. Chicago: Chicago University Press.

Robinson, A. H., J. L. Morrison, P. C. Muehrcke, A. J. Kimerling, and S. C. Guptill. 1995. *Elements of Cartography,* 6th ed. New York: Wiley.

U.S. Bureau of the Budget. 1947. *United States National Map Accuracy Standards*. Washington, DC: U.S. Bureau of the Budget.

Wang, Z. 1996. Manual versus Automated Line Generalization. *GIS/LIS '96 Proceedings*: 94–106.

Weibel, R., and G. Dutton. 1999. Generalizing Spatial Data and Dealing with Multiple Representations. In P. A. Longley, M. F. Goodchild, D. J. MaGuire, and D. W. Rhind, eds., *Geographical Information Systems,* 2d ed., pp. 125–55. New York: Wiley.

CHAPTER 9

ATTRIBUTE DATA INPUT AND MANAGEMENT

GIS involves both spatial and attribute data: spatial data relate to the geometry of spatial features, and attribute data describe the characteristics of the spatial features. Figure 9.1, for example, shows attribute data such as street name, address ranges, and zip codes that are associated with each street segment in the TIGER/Line files. Without the attributes, the TIGER/Line files will be of limited use.

The difference between spatial and attribute data is well defined with vector-based features such as the TIGER/Line files. The georelational data model (e.g., a coverage) stores spatial data and attribute data separately and links the two by the feature ID (Figure 9.2). The two data sets are synchronized so that they can be queried, analyzed, and displayed in unison. The object-based data model (e.g., a geodatabase) combines both spatial and attribute data in a single system. Each spatial feature has a unique object ID and an attribute to store its geometry (Figure 9.3). Although the two data models handle the storage of spatial data differently, both operate in the same relational database environment. Therefore, materials covered in Chapter 9 can apply to both vector-based data models.

The raster data model presents a different scenario in terms of spatial data and attribute data. The cell value corresponds to the attribute data of a spatial phenomenon at the cell location. But only an integer raster has a value attribute table. And, unlike a vector-based attribute table, a value attribute table lists cell values and their frequencies, rather than cell by cell (Figure 9.4). If the cell value does represent some spatial unit such as the county FIPS (U.S. Federal Information Processing Standards) code, we can use the value attribute table to store county-level data and use the raster to display these county-level data. But the raster is always associated with the cell value (i.e., the FIPS

FEDIRP	FENAME	FETYPE	FRADDL	TOADDL	FRADDR	TOADDR	ZIPL	ZIPR
N	4th	St	6729	7199	6758	7198	83815	83815

Figure 9.1

Each street segment in the TIGER/Line files has a set of associated attributes. These attributes include street name, address ranges on the left side and the right side, as well as zip codes on both sides.

Record	Soil-ID	Area	Perimeter
1	1	106.39	495.86
2	2	8310.84	508,382.38
3	3	554.11	13,829.50
4	4	531.83	19,000.03
5	5	673.88	23,931 47

Figure 9.2

As an example of the georelational data model, the soil coverage uses Soil-ID to link the spatial and attribute data.

ObjectID	Shape	Shape_Length	Shape_Area
1	Polygon	106.39	495.86
2	Polygon	8310.84	508,382.38
3	Polygon	554.11	13,829.50
4	Polygon	531.83	19,000.03
5	Polygon	673.88	23,931.47

Figure 9.3

The object-based data model uses the Shape field to store the geometry of soil polygons. The table therefore contains both spatial and attribute data.

ObjectID	Value	Count
0	160,101	142
1	160,102	1580
2	160,203	460
3	170,101	692
4	170,102	1417

Figure 9.4

A value attribute table lists the attributes of value and count. The Value field stores the cell values, and the Count field stores the number of cells in the raster.

With emphasis on vector data, Chapter 9 is divided into the following four sections. Section 9.1 provides an overview of attribute data in GIS. Section 9.2 discusses the relational model, data normalization, type of data relationship, and joining and relating tables. Sections 9.3 covers attribute data entry including field definition, method, and verification. Section 9.4 discusses manipulation of fields and creation of new attribute data from existing attributes.

9.1 ATTRIBUTE DATA IN GIS

Attribute data are stored in tables. An attribute table is organized by row and column. Each row represents a spatial feature, each column describes a characteristic, and the intersection of a column and a row shows the value of a particular characteristic

code) for data query and analysis. This association between a raster and its designated cell value separates the raster data model from the vector data model and makes attribute data management a much less important topic for raster data than vector data.

A feature attribute table consists of rows and columns. Each row represents a spatial feature, and each column represents a property or characteristic of the spatial feature for a particular feature (Figure 9.5). A row is also called a **record** or a tuple, and a column is also called a **field** or an item.

Figure 9.5

A feature attribute table consists of rows and columns. Each row represents a spatial feature, and each column represents a property or characteristic of the spatial feature.

Label-ID	pH	Depth	Fertility
1	6.8	12	High
2	4.5	4.8	Low

Row → / Column ↑

9.1.1 Type of Attribute Table

There are two types of attribute tables. The first type is called the **feature attribute table,** which has access to the spatial data. Every vector data set must have a feature attribute table. In the case of the georelational data model, the feature attribute table uses the feature ID to link to the feature's geometry. In the case of the object-based data model, the feature attribute table has a field that stores the feature's geometry. Feature attribute tables also have default fields that summarize the feature geometries such as length for line features, and area and perimeter for area features.

A feature attribute table may be the only table needed if a data set has only several attributes. But this is often not the case. For example, a soil map unit can have over 80 estimated physical and chemical properties, interpretations, and performance data. To store all these attributes in a feature attribute table will require many repetitive entries, a process that wastes both time and computer memory. Moreover, the table will be extremely difficult to use and update. This is why we need the second type of attribute table.

This second type of attribute table is nonspatial, meaning that the table does not have direct access to the geometry of features but has a field that can link the table to the feature attribute table whenever necessary (Box 9.1). Tables of nonspatial data may exist as delimited text files, dBASE files, Excel files, or files managed by database software packages such as Oracle, Informix, SYBASE, SQL Server, and IBM DB2.

9.1.2 Database Management

The presence of feature attribute and nonspatial data tables means that a GIS requires a **database management system (DBMS)** to manage these data tables. A DBMS is a set of computer programs for managing an integrated and shared database (Laurini and Thompson 1992). A DBMS provides tools for data input, search, retrieval, manipulation, and output. Most commercial GIS packages include database management tools for local databases. ArcGIS Desktop uses Microsoft Access, ArcInfo Workstation uses INFO, AutoCAD Map uses VISION, and MGE uses database management tools from Oracle and Informix.

The use of a DBMS has other advantages beyond its GIS applications. Often a GIS is part of an enterprisewide information system, and attribute

---

Box 9.1   Nonspatial Data Tables

The term nonspatial data table can be confusing at times. A table with soil attributes is considered nonspatial, for example, if it uses a key such as the map unit symbol to access soil polygons (i.e., spatial data). A soil feature attribute table is spatial because it has direct access to soil polygons. Besides feature attribute tables, there are other types of tables that have access to spatial data. Examples include tables with x- and y-coordinates, tables with street addresses, and tables with route measures.

data needed for the GIS may reside in various departments of the same organization. Therefore, the GIS must function within the overall information system and interact with other information technologies.

The object-based data model such as the geodatabase has actually blurred the boundary between a GIS and a DBMS. A geodatabase stores both spatial and attribute data as tables in a single database. Except for the inclusion of a field for geometry, a geodatabase is basically the same as a database used in business or marketing. This has led some authors to refer to GIS as spatial database management systems (e.g., Shekhar and Chawla 2003).

Besides database management tools for managing local databases, many GIS packages also have database connection capabilities to access remote databases. This is important for GIS users who routinely access data from centralized databases. For example, GIS users at a ranger district office may regularly retrieve data maintained at the headquarter office of a national forest. This scenario represents a client-server distributed database system. A client (e.g., a district office user) sends a request to the server, retrieves data from the server, and processes the data on the local computer.

9.1.3 Type of Attribute Data

One method for classifying attribute data is by data type. The data type determines how an attribute is stored in a GIS. Depending on the GIS package, the available data types can vary. Common data types are number, text (or character), date, and binary large object (BLOB). Data types for numbers include integer (for numbers without decimal digits) and float (for numbers with decimal digits). Moreover, depending on the designated computer memory, an integer can be short or long and a float can be single precision or double precision. BLOB stores images, multimedia, and the geometry of spatial features as long sequences of binary numbers.

Another method is to define attribute data by measurement scale. The measurement scale concept groups attribute data into nominal, ordinal, interval, and ratio data, with increasing degrees of sophistication (Stevens 1946; Chang 1978). **Nominal data** describe different kinds or different categories of data such as land-use types or soil types. **Ordinal data** differentiate data by a ranking relationship. For example, soil erosion may be ordered from severe, moderate, to light. **Interval data** have known intervals between values. For example, a temperature reading of 70° F is warmer than 60° F by 10° F. **Ratio data** are the same as interval data except that ratio data are based on a meaningful, or absolute, zero value. Population densities are an example of ratio data, because a density of 0 is an absolute zero. In GIS applications, the four measurement scales may be grouped into two general categories: **categorical data** include nominal and ordinal scales, and **numeric data** include interval and ratio scales (Box 9.2).

 Box 9.2   **Categorical and Numeric Data**

Categorical data refer to data measured at nominal or ordinal scales, and numeric data refer to data measured at interval or ratio scales. These two types of data are clearly different. But GIS projects, especially those dealing with suitability analysis, commonly assign numeric scores to categorical data and use these scores in computation (Chrisman 2001). For example, a suitability analysis for a housing development may assign different scores to different soil types and then combine these scores with scores from other variables for computing the total suitability score. Assigning scores to categorical data requires information that is not in the base data. Scores in this case represent *interpreted* data, which are based on expert judgments.

Data type and measurement scale are obviously related. Character strings are appropriate for nominal and ordinal data. Integers and floats are appropriate for interval and ratio data, depending on whether decimal digits are included or not. But there are exceptions. For example, a study may classify the potential for groundwater contamination as high, medium, and low, but enter the information as numeric data using a lookup table. The lookup table may show 1 for low, 2 for medium, and 3 for high. In this case, the numbers are simply the numeric codes for the ordered categorical data. Because a GIS package has commands for conversion between data types, we must pay attention to the nature of attribute data before using them in analysis.

9.2 THE RELATIONAL MODEL

A database is a collection of interrelated tables in digital format. There are at least four types of database designs that have been proposed in the literature: flat file, hierarchical, network, and relational (Figure 9.6) (Laurini and Thompson 1992; Worboys 1995).

A **flat file** contains all data in a large table. A feature attribute table is like a flat file. Another example is a spreadsheet with data only. A **hierarchical database** organizes its data at different levels and uses only the one-to-many association between levels. The simple example in Figure 9.6 shows the hierarchical levels of zoning, parcel, and owner. Based on the one-to-many association, each level is divided into different branches. A **network database** builds connections across tables, as shown by the linkages between the tables in Figure 9.6. A common problem with both the hierarchical and the network database designs is that the linkages between tables must be known in advance and built into the computer code. This requirement tends to make a complicated and inflexible database. GIS vendors typically use the relational model for database management (Codd 1970; Codd 1990; Date 1995). A **relational database** is a collection of tables, also called relations, which can be connected to each other by keys. A **primary key** represents one or more attributes whose values

can uniquely identify a record in a table. Its counterpart in another table for the purpose of linkage is called a **foreign key.** Thus a key common to two tables can establish connections between corresponding records in the tables. In Figure 9.6, the key connecting zoning and parcel is the zone code and the key connecting parcel and owner is the PIN (parcel ID number). When used together, the keys can relate zoning and owner.

Compared to other database designs, a relational database is simple and flexible. It has two distinctive advantages. First, each table in the database can be prepared, maintained, and edited separately from other tables. This feature is important because, with the increased popularity of GIS technology, more data are being recorded and managed in spatial units. Second, the tables can remain separate until a query or an analysis requires that attribute data from different tables be linked together. Because the need for linking tables is often temporary, a relational database is efficient for both data management and data processing.

9.2.1 MUIR: A Relational Database Example

The Natural Resources Conservation Service (NRCS) produces the **Soil Survey Geographic (SSURGO) database** nationwide. The NRCS collects SSURGO data from field mapping and archives the data in 7.5-minute quadrangle units for a soil survey area. A soil survey area may consist of a county, multiple counties, or parts of multiple counties. The SSURGO database represents the most detailed level of soil mapping by the NRCS in the United States. In the 1990s, the NRCS developed a national **Map Unit Interpretation Record (MUIR)** attribute database to link to SSURGO data. Containing about 88 estimated soil physical and chemical properties, interpretations, and performance data in a series of tables, this comprehensive attribute database is available on the Internet (**http://soils.usda.gov/**).

The NRCS structures the MUIR database by soil survey area, map units within the survey area, and map unit components within the map unit. Four

(a) Flat file

PIN	Owner	Zoning
P101	Wang	Residential (1)
P101	Chang	Residential (1)
P102	Smith	Commercial (2)
P102	Jones	Commercial (2)
P103	Costello	Commercial (2)
P104	Smith	Residential (1)

(b) Hierarchical

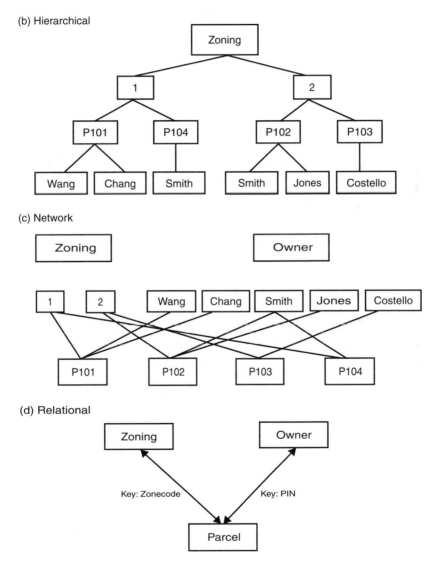

(c) Network

(d) Relational

Figure 9.6
Four types of database design: (*a*) flat file, (*b*) hierarchical, (*c*) network, and (*d*) relational.

keys are essential in relating tables in the database: state code, soil survey area ID, map unit symbol, and sequence number. The first two keys define the state and the area of the soil survey. The map unit symbol is a numeric code for each soil polygon in the survey. The sequence number key indicates the components of a soil map unit, because a unit may consist of a single dominant soil or two or more soil components. Additionally, various other keys are used to relate soil tables to lookup tables. Examples of lookup tables include soil classification and the scientific and common names of plants.

The sheer size of the MUIR database can be overwhelming at first. But the database is not difficult to use if we have a proper understanding of the relational model and the keys used in relating MUIR tables. Chapter 9 uses the MUIR database to illustrate the type of relationship between tables in Section 9.2.3. Chapter 11 uses the database as an example for data exploration.

9.2.2 Normalization

Preparing a relational database such as the MUIR must follow certain rules. An important rule is called normalization. **Normalization** is a process of decomposition, taking a table with all the attributes of a table and breaking it down to small tables while maintaining the necessary linkages between them (Vetter 1987). Normalization is designed to achieve the following objectives:

- To avoid redundant data in tables that waste space in the database and may cause data integrity problems.

- To ensure that attribute data in separate tables can be maintained and updated separately and can be linked whenever necessary.

- To facilitate a distributed database.

An example of normalization is offered here. Table 9.1 shows attribute data for a parcel map. The table contains redundant data: owner addresses are repeated for Smith and residential and commercial zoning are entered twice. The table also contains uneven records: depending on the parcel, the fields of owner and owner address can have either one or two values. An unnormalized table such as Table 9.1 cannot be easily managed or edited. To begin with, it is difficult to define the fields of owner and owner address and to store their values. If the ownership changes, the table must be updated with all the attribute data. The same difficulty applies to such operations as adding or deleting values.

Table 9.2 represents the first step in normalization. Often called the first normal form, Table 9.2 no longer has multiple values in its cells but the problem of data redundancy has increased. P101 and P102 are repeated twice except for changes of the owner and the owner address. Smith's address is repeated twice. And the zoning descriptions of residential and commercial are repeated three times each. Also, identifying the owner address is not possible with PIN alone but requires a compound key of PIN and owner.

Figure 9.7 represents the second step in normalization. In place of Table 9.2 are three small tables of parcel, owner, and address. PIN is the key relating the parcel and owner tables. Owner name

TABLE 9.1	An Unnormalized Table					
PIN	**Owner**	**Owner address**	**Sale date**	**Acres**	**Zone code**	**Zoning**
P101	Wang	101 Oak St	1-10-98	1.0	1	Residential
	Chang	200 Maple St				
P102	Smith	300 Spruce Rd	10-6-68	3.0	2	Commercial
	Jones	105 Ash St				
P103	Costello	206 Elm St	3-7-97	2.5	2	Commercial
P104	Smith	300 Spruce Rd	7-30-78	1.0	1	Residential

TABLE 9.2	First Step in Normalization					
PIN	**Owner**	**Owner address**	**Sale date**	**Acres**	**Zone code**	**Zoning**
P101	Wang	101 Oak St	1-10-98	1.0	1	Residential
P101	Chang	200 Maple St	1-10-98	1.0	1	Residential
P102	Smith	300 Spruce Rd	10-6-68	3.0	2	Commercial
P102	Jones	105 Ash St	10-6-68	3.0	2	Commercial
P103	Costello	206 Elm St	3-7-97	2.5	2	Commercial
P104	Smith	300 Spruce Rd	7-30-78	1.0	1	Residential

Parcel table

PIN	Sale date	Acres	Zone code	Zoning
P101	1-10-98	1.0	1	Residential
P102	10-6-68	3.0	2	Commercial
P103	3-7-97	2.5	2	Commercial
P104	7-30-78	1.0	1	Residential

Owner table

PIN	Owner name
P101	Wang
P101	Chang
P102	Smith
P102	Jones
P103	Costello
P104	Smith

Address table

Owner name	Owner address
Wang	101 Oak St
Chang	200 Maple St
Jones	105 Ash St
Smith	300 Spruce Rd
Costello	206 Elm St

Figure 9.7
Separate tables from the second step in normalization. The keys relating the tables are highlighted.

is the key relating the address and owner tables. The relationship between the parcel and address tables can be established through the keys of PIN and owner name. The only problem with the second normal form is data redundancy with the fields of zone code and zoning.

The final step in normalization with the above example is summarized in Figure 9.8. A new table, zone, is created to take care of the remaining data redundancy problem with zoning. Zone code is the key relating the parcel and zone tables. Unnormalized data in Table 9.1 are now fully normalized.

Figure 9.8

Separate tables after normalization. The keys relating the tables are highlighted.

Parcel table

PIN	Sale date	Acres	Zone code
P101	1-10-98	1.0	1
P102	10-6-68	3.0	2
P103	3-7-97	2.5	2
P104	7-30-78	1.0	1

Owner table

PIN	Owner name
P101	Wang
P101	Chang
P102	Smith
P102	Jones
P103	Costello
P104	Smith

Address table

Owner name	Owner address
Wang	101 Oak St
Chang	200 Maple St
Jones	105 Ash St
Smith	300 Spruce Rd
Costello	206 Elm St

Zone table

Zone code	Zoning
1	Residential
2	Commercial

Although it can achieve objectives consistent with the relational model, normalization does have a major drawback of slowing down data access. To find the addresses of parcel owners, for example, we must link three tables (parcel, owner, and address) and employ two keys (PIN and owner name). One way to increase the performance in data access is to reduce the level of normalization by, for example, removing the address table and including the addresses in the owner table. Therefore, normalization should be maintained in the conceptual design of a database but performance and other factors should be considered in its physical design.

9.2.3 Types of Relationships

A relational database may contain four types of relationships between tables, or more precisely, between records in tables: one-to-one, one-to-many, many-to-one, and many-to-many (Figure 9.9). The **one-to-one relationship** means that one and only one record in a table is related to one and only one record in another table. The **one-to-many relationship** means that one record in a table may be related to many records in another table. For example, the street address of an apartment complex may include several households. The **many-to-one relationship** means that many records in a

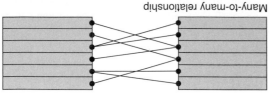
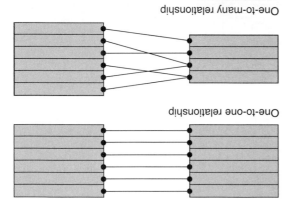

Figure 9.9

Four types of data relationship between tables: one-to-one, one-to-many, many-to-one, and many-to-many.

One-to-one relationship

One-to-many relationship

Many-to-one relationship

Many-to-many relationship

Primary key
↑

Record	Soil-ID	Area	Perimeter
1	1	106.39	495.86
2	2	8310.84	508,382.38
3	3	554.11	13,829.50
4	4	531.83	19,000.03
5	5	673.88	23,931.47

Foreign key
↑

Soil-ID	suit	soilcode
1	1	ld3
2	3	Sg
3	1	ld3
4	1	ld3
5	2	Ns1

Feature attribute table ◄────── Table to be joined

Figure 9.10
Primary key and foreign key provide the linkage to join the table on the right to the feature attribute table on the left.

table may be related to one record in another table. For example, several households may share the same street address. The **many-to-many relationship** means that many records in a table may be related to many records in another table. For example, a timber stand can grow more than one species and a species can grow in more than one stand.

To explain these relationships, especially one-to-many and many-to-one, the designation of the base table can be helpful. For example, if the purpose is to join attribute data from another table to a feature attribute table, then the feature attribute table is the base table and the other table is the table to be joined (Figure 9.10). (In ArcGIS, the tables for linking are called the *origin* and the *destination*.) The feature attribute table has the primary key and the other table has the foreign key. Often the designation of the base table depends on the storage of data and the information sought. This is illustrated in the following two examples.

The first example refers to the four normalized tables of parcel, owner, address, and zone in Figure 9.8. Suppose the question is to find who owns a selected parcel. To answer this question, we can treat the parcel table as the base table and the owner table as the table to be linked. The relationship between the tables is one-to-many: one record in the parcel table may correspond to more than one record in the owner table.

Suppose the question is now changed to find land parcels owned by a selected owner. The owner table becomes the base table and the parcel

table is the table to be linked. The relationship is many-to-one: more than one record in the owner table may correspond to one record in the parcel table. The same is true between the parcel table and the zone table. If the question is to find the zoning code for a selected parcel, it is a many-to-one relationship. If the question is to find land parcels that are zoned commercial, it is a one-to-many relationship.

The second example relates to the MUIR database. One approach to better understanding the MUIR database is to sort out the relationship between tables. Many relationships are either one-to-many or many-to-one. For example, the many-to-one relationship exists between the map unit components table and the soil classification table because different soil components may be grouped into the same soil class (Figure 9.11). On the other hand, the one-to-many relationship applies to the map unit components table and the soil profile layers table because a soil component may have multiple profile layers (Figure 9.12).

Besides the obvious effect in linking tables, the type of relationship can also influence how data are displayed. Suppose the task is to display the parcel ownership. If the relationship between the parcel and ownership tables is one-to-one, each parcel can be displayed with a unique symbol. If the relationship is many-to-one, one or more parcels can be displayed with one symbol. But if the relationship is one-to-many, data display becomes a problem because it is not right to show a multiowner parcel with an owner who happens to

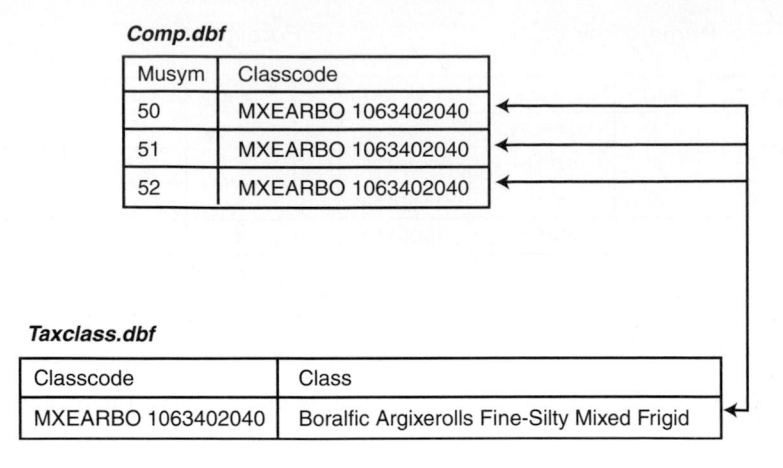

Figure 9.11
This example of a many-to-one relationship in the MUIR database relates three map unit components in *comp.dbf* to the same soil class in the *taxclass.dbf*.

Comp.dbf

Muid	Seqnum	Musym
610001	1	1

Layer.dbf

Muid	Seqnum	Layernum	Laydepl	Laydeph
610001	1	1	0	7
610001	1	2	7	18
610001	1	3	18	58
610001	1	4	58	62

Figure 9.12
This example of a one-to-many relationship in the MUIR database relates one map unit component in *comp.dbf* to four layers in *layer.dbf*.

be the first on the list. (One solution to the problem is to use a separate symbol design for the category of multiowner parcels.)

9.2.4 Join and Relate Tables

A relational database allows attribute data in separate tables to be linked whenever necessary. Two common operations for linking tables are join and relate. A **join** operation brings together two tables by using a key that is common to both tables. A typical example of join is to join attribute data from one or more nonspatial data tables to a feature attribute table. The joined table and attributes can then be used for data query or analysis.

Join is usually recommended for a one-to-one or many-to-one relationship. Given a one-to-one relationship, two tables are joined by record. Given a many-to-one relationship, many records in the base table have the same value from a record in the other table. Join is inappropriate with the one-to-many relationship, because only the first matching record value from the other table is assigned to a record in the base table.

A **relate** operation temporarily connects two tables but keeps the tables physically separate. We can connect three or more tables simultaneously by first establishing relates between tables in pairs. A Windows-based GIS package is particularly useful for working with relates because it allows the viewing of multiple tables. One advantage of relates is that they are appropriate for all four types of relationships. This is important for data exploration and data query because a relational database such as the MUIR is likely to include different types of relationships. But relates tend to slow down data access, especially if the data reside in a remote database.

9.3 ATTRIBUTE DATA ENTRY

Attribute data entry is like digitizing from a paper map. The process requires the setup of the fields to be entered, choice of a digitizing method, and verification of attribute values.

9.3.1 Field Definition

The first step in attribute data entry is to define each field in the table. A field definition usually includes the field name, data width, data type, and number of decimal digits. The width refers to the number of spaces to be reserved for a field. The width should be large enough for the largest number, including the sign, or the longest string in the data. The data type must follow data types allowed in the GIS package. The number of decimal digits is part of the definition for the float data type. (In ArcGIS, the precision defines the number of digits, and the scale defines the number of decimal digits, for the float data type.)

The field definition can be confusing at times. For example, the soil map unit in the MUIR database is defined as text, although the map unit is coded as numbers such as 610001, 610002, and so on. These are ID numbers to identify soil map units. We cannot perform computations with these soil map unit numbers.

9.3.2 Methods of Data Entry

Suppose a map has 4000 polygons, each with 50 fields of attribute data. This could require entering 200,000 values. How to reduce time and effort in attribute data entry is of interest to any GIS user.

Similar to finding existing spatial data, it is best to determine if an agency has already entered attribute data in digital format. If yes, we can simply import the digital data file into a GIS. The data format is important for importing. Most GIS packages can import delimited text files and dBASE files as well as data from other database manage-

ment systems. If attribute data files do not exist, then typing is the only option. But the amount of typing can vary depending on which method or command is used. For example, an editing command in a GIS package works with one record at a time. One way to save time is to follow the MUIR database design and to take advantage of keys and lookup tables.

For map unit symbols or feature IDs, it is better to enter them directly in a GIS. This is because we can select a feature in the view window, see where the feature is located in the base map, and enter its symbol or ID in a dialog. But for nonspatial data, it is better to use a wordprocessing (e.g., Notepad) or spreadsheet (e.g., Excel) package. These packages offer cut-and-paste, find-and-replace, and other functions that are helpful to a typist, especially a poor typist. A GIS package typically does not have these options.

9.3.3 Attribute Data Verification

Attribute data verification consists of two parts. The first part is to make sure that attribute data are properly linked to spatial data: the label or feature ID should be unique and should not contain null (empty) values. The second is to verify the accuracy of attribute data. Data verification is difficult because inaccuracies can be attributed to a large number of factors including observation errors, outdated data, and data entry errors.

There are at least two traditional methods for checking data entry errors. First, attribute data can be printed for manual verification. This is like using check plots to check the accuracy of spatial data. Second, computer programs can be written to verify data accuracy. For example, a computer program can be written to catch logical errors in soil attribute data such as data type and numeric range. The same program may also test attribute data against a master set of valid attributes to catch logical errors such as having incompatible components in the same soil map unit.

The geodatabase data model has introduced a set of validation rules (Zeiler 1999). Directly related to attribute data verification is the rule about attribute domains, which groups objects into classes by a valid range of values or a valid set of values for an attribute. Suppose the field zoning has the value of 1 for residential, 2 for commercial, and 3 for industrial for a land parcel feature class. This set of zoning values can be enforced whenever the field zoning is edited. Therefore, if a zoning value of 9 is entered, the value will be flagged or rejected because it is outside the valid set of values. Similar constraint using a valid numeric range instead of a valid set of values can be applied to lot size or height of building. Task 1 of the applications section uses an attribute domain to ensure the accuracy of data entry.

9.4 MANIPULATION OF FIELDS AND ATTRIBUTE DATA

Database management tasks involve tables as well as fields and field values in tables. Field management includes adding or deleting fields and creating new attributes through classification and computation of existing attribute data.

9.4.1 Add and Delete Fields

We regularly download data from the Internet for GIS projects. Often the downloaded data set contains far more attributes than what we need. It is a good idea to delete those fields that are not needed.

This not only reduces confusion in using the data set but also saves computer time for data processing. Deleting a field is straightforward. The process requires specifying an attribute table and the field in the table to be deleted. Obviously the table involved in the process cannot be used by another program or application.

Adding a field is a required step prior to classification or computation of attribute data. The new field is designed to receive the result of classification or computation. To add a field, we must define the field in the same way as for attribute data entry.

The object-based data model considers fields as properties of a geospatial data set. Therefore, adding or deleting a field can be performed through either the data set's or the data set's feature attribute table. This makes it more flexible in terms of manipulating fields (Box 9.3).

9.4.2 Classification of Attribute Data

New attribute data can be created from existing data through data classification. Based on an attribute or attributes, data classification reduces a data set to a small number of classes. For example, elevations may be grouped into < 500, 500 to 1000, and so on. Another example is to classify by elevation and slope: class 1 has < 500 in elevation and 0 to 10% slope, class 2 has < 500 in elevation and 10 to 20% slope, and so on.

Box 9.3 Add and Delete Fields in ArcGIS Desktop

P erhaps because adding or deleting a field is such a common operation in GIS, the operation can be performed in a number of places in ArcGIS Desktop. There are two ways to add or delete a field in ArcCatalog: through the properties dialog of a data set, or through the table view on the preview tab. To add or delete a field in ArcMap, we can use the options menu of an attribute table to add a field, and use the context menu of a field to delete it. ArcToolbox, which is available in both ArcCatalog and ArcMap, has four tools (add field, assign default to field, calculate field, and delete field) in the Data Management Tools/Fields toolset that work with fields. Regardless of which method is used, the table on which the operation is performed should not be used by another application or program.

Operationally, creating new attribute data by classification involves three steps: defining a new field for saving the classification result, selecting a data subset through query, and assigning a value to the selected data subset. The second and third steps are repeated until all records are classified and assigned new field values, unless a computer program is written to automate the procedure (see Task 5 in the applications section). The main advantage of data classification is that it simplifies a data set so that the new data set can be more easily used in GIS analysis or modeling.

9.4.3 Computation of Attribute Data

New attribute data can also be created from existing data through computation. Operationally, it involves two steps: defining a new field, and computing the new field values from the values of an existing attribute or attributes. The computation is through a formula, which can be coded manually or using a dialog with controls for various mathematical functions.

A simple example of computation is to create a new attribute called feet, for a trail map that is measured in meters. This new attribute can be computed by "length" $\times$ 3.28, where length is an existing attribute. Another example is to create a new attribute that measures the quality of wildlife habitat by evaluating the existing attributes of slope, aspect, and elevation. The task first requires a scoring system for each variable. Once the scoring system is in place, we can compute the index value, which measures the quality of wildlife habitat, by summing the scores of slope, aspect, and elevation. In some cases, different weights may be assigned to different variables. For example, if elevation is three times as important as slope and aspect, then we can compute the index value by using the following equation: slope score + aspect score + 3 $\times$ elevation score.

KEY CONCEPTS AND TERMS

Categorical data: Data that are measured at a nominal or an ordinal scale.

Database management system (DBMS): A set of computer programs for managing an integrated and shared database for such tasks as data input, search, retrieval, manipulation, and output.

Feature attribute table: An attribute table that has access to the spatial data of features.

Field: A column in a table that describes an attribute of a spatial feature.

Flat file: A database that contains all data in a large table.

Foreign key: One or more attributes that can uniquely identify a record in a table that is to be linked to a base table.

Hierarchical database: A database that is organized at different levels and uses the one-to-many association between levels.

Interval data: Data with known intervals between values, such as temperature readings.

Join: A relational database operation that brings together two tables by using a common key to both tables.

Many-to-many relationship: One type of data relationship, in which many records in a table are related to many records in another table.

Many-to-one relationship: One type of data relationship, in which many records in a table are related to one record in another table.

Map Unit Interpretation Record (MUIR): An attribute database linked to SSURGO data from the Natural Resources Conservation Service.

Network database: A database that is based on the built-in connections across tables.

Nominal data: Data that show different kinds or different categories, such as land-use types or soil types.

Ratio data: Data with known intervals between values and are based on a meaningful zero value, such as population densities.

Record: A row in a table that represents a spatial feature.

Relate: A relational database operation that temporarily connects two tables by using a common key to both tables.

Relational database: A database that consists of a collection of tables and uses keys to connect the tables.

Soil Survey Geographic (SSURGO) database: A database maintained by the Natural Resources Conservation Service, which archives soil survey data in 7.5-minute quadrangle units.

Normalization: The process of taking a table with all the attribute data and breaking it down into small tables while maintaining the necessary linkages between them in a relational database.

Numeric data: Data that are measured at an interval or a ratio scale.

One-to-many relationship: One type of data relationship, in which one record in a table is related to many records in another table.

One-to-one relationship: One type of data relationship, in which one record in a table is related to one and only one record in another table.

Ordinal data: Data that are ranked, such as large, medium, and small cities.

Primary key: One or more attributes that can uniquely identify a record in a base table.

REVIEW QUESTIONS

1. What is a feature attribute table?

2. Provide an example of a nonspatial attribute table.

3. What is a distributed database system? Can you think of an example that uses a distributed database system?

4. Describe the four types of attribute data by measurement scale.

5. Can you convert ordinal data into interval data? Why, or why not?

6. Define a relational database.

7. Explain the advantages of a relational database.

8. Sometimes it is difficult to tell a primary key from a foreign key. Take a look at the parcel table and the owner table in Figure 9.8. Which field in the parcel table is the primary key (i.e., an attribute that can uniquely identify a record in the table)? Which field in the owner table is the foreign key (i.e., an attribute designed for linking the owner table to the parcel table)?

9. Identify the primary key and the foreign key for linking the zone table to the parcel table in Figure 9.8.

10. A fully normalized database may slow down data access. To increase the performance in data access, for example, one may remove the address table in Figure 9.8. How would the database look if the address table were removed?

11. Provide a real-world example (other than those of Chapter 9) of a one-to-one relationship.

12. Explain the similarity, as well as the difference, between a join operation and a relate operation.

13. Suppose you have downloaded a GIS data set. The data set has the length measure in meters instead of feet. Describe the steps you will follow to add a length measure in feet to the data set.

14. Suppose you have downloaded a GIS data set. The feature attribute table has a field that contains values such as 12, 13, and so on. How can you find out if these values represent numbers or text strings using ArcGIS?

15. Describe two ways of creating new attributes from the existing attributes in a data set.

APPLICATIONS: ATTRIBUTE DATA INPUT AND MANAGEMENT

This applications section has six tasks. Task 1 covers attribute data entry using a geodatabase feature class. Tasks 2 and 3 cover joining tables and relating tables respectively. Tasks 4 and 5 show how to create new attributes through data classification. Task 4 uses the conventional method of repeatedly selecting a data subset and assigning a class value. Task 5, on the other hand, uses a Visual Basic script to automate the procedure. Task 6 shows how to create new attributes through data computation.

Task 1: Enter Attribute Data of a Geodatabase Feature Class

What you need: *landat.shp*, a polygon shapefile with 19 records.

In Task 1, you will learn how to enter attribute data using a geodatabase feature class and a domain. The domain and its coded values can restrict values to be entered, thus avoiding data entry errors.

1. Start ArcCatalog, and make connection to the Chapter 9 database. First, create a personal geodatabase. Right-click the Chapter 9 database in the Catalog tree, point to New, and select Personal Geodatabase. Rename the new personal geodatabase *land.mdb*.

2. This step adds *landat.shp* as a feature class to *land.mdb*. Right-click *land.mdb*, point to Import, and select Feature Class (single). Use the browse button or the drag-and-drop method to add *landat.shp* as the input features. Name the output feature class *landat*. Click OK to dismiss the dialog.

3. Now create a domain for the geodatabase. Select Properties from the context menu of *land.mdb*. The Domains tab of the Database Properties dialog shows Domain Name, Domain Properties, and Coded Values. You will work with all three frames. Click the first cell under Domain Name, and enter lucodevalue. Click the cell next to Field Type, and select Short Integer. Click the cell next to Domain Type, and select Coded Values. Click the first cell under Code and enter 100. Click

the cell next to 100 under Description, and enter 100-urban. Enter 200, 300, 400, 500, 600, and 700 following 100 under Code and enter their respective descriptions of 200-agriculture, 300-brushland, 400-forestland, 500-water, 600-wetland, and 700-barren. Click Apply and OK to dismiss the Database Properties dialog.

4. This step is to add a new field to *landat* and to specify the field's domain. Right-click *landat* and select Properties. On the Fields tab, click the first empty cell under Field Name and enter lucode. Click the cell next to lucode and select Short Integer. Click the cell next to Domain in the Field Properties frame and select lucodevalue. Click Apply and OK to dismiss the Properties dialog.

Q1. List the data types available for a new field.

5. Launch ArcMap. Rename the data frame Task 1 and add *landat* to Task 1. Open the attribute table of *landat*. lucode appears with Null values in the last field of the table.

6. Click the Editor Toolbar button to open the toolbar. Click the Editor dropdown arrow and select Start Editing. Right-click the field of LANDAT-ID and select Sort Ascending. Now you are ready to enter the lucode values. Click the first cell under lucode and select forestland (400). Enter the rest of the lucode values according to the table below.

Landat-ID	Lucode	Landat-ID	Lucode
59	400	69	300
60	200	70	200
61	400	71	300
62	200	72	300
63	200	73	300
64	300	74	300
65	200	75	200
66	300	76	300
67	300	77	300
68	200		

Q2. Describe in your own words how the domain of coded values ensure the accuracy of the attribute data that you entered in Step 6.

7. When you finish entering the lucode values, select Stop Editing from the Editor dropdown list. Save the edits.

Task 2: Join Tables in ArcMap

What you need: *wp.shp*, a forest stand shapefile, and *wpdata.dbf*, an attribute data file that contains vegetation and land-type data.

Task 2 asks you to join to a dBASE file a feature attribute table. A join operation combines attribute data from different tables into a single table, making it possible to use all attribute data in query, classification, or computation.

1. Insert a new data frame in ArcMap and rename it Task 2. Add *wp.shp* and *wpdata.dbf* to Task 2.

2. Open the attribute table of *wp*. The field ID will be used as the primary key in joining tables. Open *wpdata*. *wpdata* contains attributes such as ORIGIN, AS, and ELEV for *wp*. The field ID will be used as the key. Close both the *wp* attribute table and *wpdata*.

3. Now join *wpdata* to the attribute table of *wp*. Right-click *wp*, point to Joins and Relates, and select Join. At the top of the Join Data dialog, opt to join attributes from a table. Then, select ID in the first dropdown list, *wpdata* in the second list, and ID in the third list. Click OK to join the tables. Click Yes to create an index for the join field. Open the attribute table of *wp* to see the expanded table.

Task 3: Relate Tables in ArcMap

What you need: *wp.shp*, *wpdata.dbf*, and *wpact.dbf*. The first two are the same as in Task 2. *wpact.dbf* contains additional activity records. In Task 3, you will establish two relates among three tables.

1. Select Data Frame from the Insert menu in ArcMap. Rename the new data frame Tasks 3-6. Add *wp.shp*, *wpdata.dbf*, and *wpact.dbf* to Tasks 3-6.

2. Check the keys that can be used in relating tables. The field ID should appear in *wp*'s attribute table, *wpact*, and *wpdata*. Close the attribute table, *wpact*, and *wpdata*.

3. The first relate is between *wp* and *wpdata*. Right-click *wp*, point to Joins and Relates, and select Relate. In the Relate dialog, select ID in the first dropdown list, *wpdata* in the second list, and ID in the third list, and accept Relate1 as the relate name.

4. The second relate is between *wpdata* and *wpact*. Right-click *wpdata*, point to Joins and Relates, and select Relate. In the Relate dialog, select ID in the first dropdown list, *wpact* in the second dropdown list, and ID in the third list, and enter Relate2 as the relate name.

5. The three tables are now related. Right-click *wpdata* and select Open. Click the Options dropdown arrow and choose Select by Attributes. In the next dialog, create a new selection by entering the following SQL statement in the expression box: "ORIGIN" > 0 AND "ORIGIN" <= 1900. Click Apply. Click Selected at the bottom of the table so that only selected records are shown.

6. To see which *wp* records in the *wp* attribute table are related to the selected records in *wpdata*, go through the following steps. Click the Options dropdown arrow in the *wpdata* table, point to Related Tables, and click Relate1:*wp*. The *wp* attribute table shows the related records. And the *wp* layer shows where those selected records are located.

7. You can follow the same procedure as in Step 6 to see which records in *wpact* are related to those selected polygons in *wp*.

Q3. How many records in *wpact* are selected in Step 7?

Task 4: Create New Attribute by Data Classification

What you need: *wpdata.dbf.*

Task 4 demonstrates how the existing attribute data can be used for data classification and creation of a new attribute.

1. First click Clear Selected Features in the Selection menu in ArcMap to clear the selection. Click Show/Hide ArcToolbox Window to open ArcToolbox. Double-click the Add Field tool in the Data Management Tools/Fields toolset. Select *wpdata* for the input table, enter ELEVZONE for the field name, select SHORT for the type, and click OK.

2. Open *wpdata* in Tasks 3-6. ELEVZONE appears in *wpdata* with 0s. Click the Options dropdown arrow and choose Select by Attributes. Make sure that the selection method is to create a new selection. Enter the following SQL statement in the expression box: "ELEV" > 0 AND "ELEV" <= 40. Click Apply. Click Selected at the bottom of the table so that only selected records are shown. These selected records fall within the first class of ELEVZONE. Right-click the field ELEVZONE and select Calculate Values. Click Yes to proceed in the message box. Enter 1 in the expression box of the Field Calculator dialog, and click OK. The selected records in *wpdata* are now populated with the value of 1, meaning that they all fall within class 1.

3. Go back to the Select by Attributes dialog. Make sure that the method is to create a new selection. Enter the SQL statement: "ELEV" > 40 AND "ELEV" <= 45. Click Apply. Follow the same procedure above to calculate the ELEVZONE value of 2 for the selected records.

4. Repeat the same procedure to select the remaining two classes of 46–50 and >50, and to calculate their ELEVZONE values of 3 and 4, respectively.

Q4. How many records have the ELEVZONE value of 4?

Task 5: Use Advanced Method for Attribute Data Classification

What you need: *wpdata.dbf.*

You have classified ELEVZONE in *wpdata.dbf* by repeating the procedure of selecting a data subset and calculating the class value in Task 4. Task 5 shows how to use a Visual Basic script and the advanced option to calculate the ELEVZONE values all at once.

1. Open *wpdata* in Tasks 3-6. ELEVZONE should appear in the table with values calculated in Task 4. If necessary, clear selected records in *wpdata* by selecting Clear Selection from Options' dropdown menu. Right-click ELEVZONE and select Calculate Values. Click Yes to proceed in the message box.

2. Check Advanced in the Field Calculator dialog. Next create a Visual Basic script to perform the classification. The first line of your script declares an integer variable called intclass, which stores the classification value. Enter the following statements in the first box under "Pre-Logic VBA Script Code":

```
Dim intclass as integer
If [ELEV] > 0 and [ELEV] <= 40 Then
intclass = 1
ElseIf [ELEV] > 40 and [ELEV] <= 45 Then
intclass = 2
ElseIf [ELEV] > 45 and [ELEV] <= 50 Then
intclass = 3
ElseIf [ELEV] > 50 Then
intclass = 4
End If
```

3. Enter intclass in the second box under "ELEVZONE =". In other words, ELEVZONE will be assigned the value from intclass. Click OK to dismiss the Field Calculator. ELEVZONE is now populated with values calculated by the Visual Basic code. They should be the same as those in Task 4.

Task 6: Create New Attribute by Data Computation

What you need: *wp.shp* and *wpdata.dbf*.

You have created a new field from data classification in Tasks 4 and 5. Another common method for creating new fields is computation. Task 6 shows how a new field can be created and computed from existing attribute data.

1. Double-click the Add Field tool. Select *wp* for the input table, enter ACRES for the field name, select DOUBLE for the field type, enter 11 for the field precision, enter 4 for the field scale, and click OK.

2. Open the attribute table of *wp*. The new field ACRES appears in the table with 0s. Right-click ACRES to select Calculate Values. Click Yes in the message box. In the Field Calculator dialog, enter the following expression in the box below ACRES =: [AREA] / 1000000 × 247.11. Click OK. The field ACRES now shows the polygons in acres.

Q5. How large is FID = 10 in acres?

Challenge Task

What you need: *bailecor_id.shp*, a shapefile showing Bailey's ecoregions in Idaho. The data set is projected onto the Idaho Transverse Mercator coordinate system and measured in meters.

This challenge task asks you to add a field to *bailecor_id* that shows the number of acres for each ecoregion in Idaho.

Q1. How many acres does the Owyhee Uplands Section cover?

Q2. How many acres does the Snake River Basalts Section cover?

REFERENCES

Chang, K. 1978. Measurement Scales in Cartography. *The American Cartographer* 5: 57–64.

Chrisman, N. 2001. *Exploring Geographic Information Systems,* 2d ed. New York: Wiley.

Codd, E. F. 1970. A Relational Model for Large Shared Data Banks. *Communications of the Association for Computing Machinery* 13: 377–87.

Codd, E. F. 1990. *The Relational Model for Database Management,* Version 2. Reading, MA: Addison-Wesley.

Date, C. J. 1995. *An Introduction to Database Systems*. Reading, MA: Addison-Wesley.

Laurini, R., and D. Thompson. 1992. *Fundamentals of Spatial Information Systems*. London: Academic Press.

Shekhar, S. and S. Chawla. 2003. *Spatial Databases: A Tour*. Upper Saddle River, NJ: Prentice Hall.

Stevens, S. S. 1946. On the Theory of Scales of Measurement. *Science* 103: 677–80.

Vetter, M. 1987. *Strategy for Data Modelling*. New York: Wiley.

Worboys, M. F. 1995. *GIS: A Computing Perspective*. London: Taylor & Francis.

Zeiler, M. 1999. *Modeling Our World: The ESRI Guide to Geodatabase Design*. Redlands, CA: ESRI Press.

10

DATA DISPLAY AND CARTOGRAPHY

CHAPTER OUTLINE

10.1 Cartographic Symbolization

10.2 Types of Maps

10.3 Typography

10.4 Map Design

10.5 Map Production

Maps are an interface to GIS (Kraak and Ormeling 1996). We view, query, and analyze maps. We plot maps to examine results of query and analysis. And we produce maps for presentation and reports. As a visual tool, maps are most effective in communicating geospatial data, whether the emphasis is on the location or the distribution pattern of the data.

Common map elements are the title, body, legend, north arrow, scale bar, acknowledgment, and neatline/map border (Figure 10.1). Other elements include the graticule or grid, name of map projection, inset or location map, and data quality information.

Together, these elements bring the map information to the map reader. The map body is the most important part of a map because it contains the map information. Other elements of the map support the communication process. For example, the title suggests the subject matter and the legend relates map symbols to the mapped data. In practical terms, mapmaking may be described as a process of assembling map elements.

Data display is one area that commercial GIS packages have greatly improved in recent years. Desktop GIS packages with the graphical user interface are excellent for data display for two reasons. First, the mapmaker can simply point and click the graphic icons to construct a map. In comparison, a command-line driven package requires the user to become familiar with many commands and their parameters before making a map. Second, desktop GIS packages have incorporated some design options such as symbol choices and color schemes into menu selections.

For a novice mapmaker, these easy-to-use GIS packages and their "default options" can result

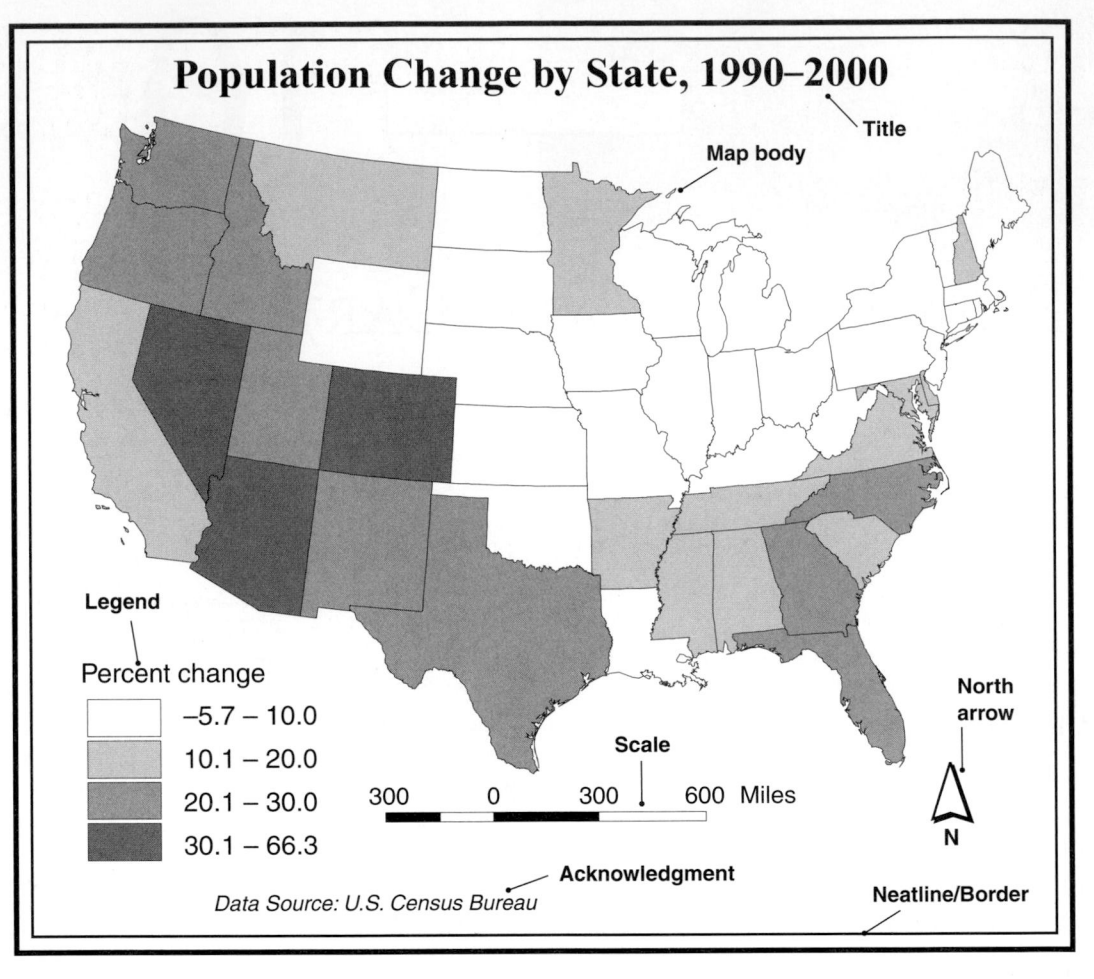

Figure 10.1
Common map elements.

in maps of questionable quality. Mapmaking should be guided by a clear idea of map design and map communication. A well-designed map can help the mapmaker communicate geospatial information to the map reader, whereas a poorly designed map can confuse the map reader and even distort the information intended by the mapmaker.

Chapter 10 emphasizes maps for presentation and reports. Maps for data exploration and 3-D visualization are covered in Chapters 11 and 14

respectively. Chapter 10 is divided into the following five sections. Section 10.1 discusses cartographic symbolization including the data–symbol relationship, use of color, and data classification. Section 10.2 considers different types of maps by function and by map symbol. Section 10.3 provides an overview of typography, selection of type variations, and placement of text. Section 10.4 covers map design and the design elements of layout and visual hierarchy. Section 10.5 examines issues related to map production.

10.1 CARTOGRAPHIC SYMBOLIZATION

Cartography is the making and study of maps in all their aspects (Robinson et al. 1995). A basic element of cartography is symbolization, the use of different map symbols to represent spatial features.

10.1.1 Spatial Features and Map Symbols

Spatial features are characterized by their locations and attributes. To display a spatial feature on a map, we use a map symbol to indicate the feature's location and a visual variable, or visual variables, with the symbol to show the feature's attribute data. For example, a thick line in red may represent an interstate highway and a thin line in black may represent a state highway. The line symbol shows the location of the highway in both cases, but the line width and color—two visual variables with the line symbol—separate the interstate from the state highway. Choosing the appropriate map symbol and visual variables is therefore the main concern for data display and mapmaking (Robinson et al. 1995; Dent 1999; Slocum et al. 2004).

Choice of map symbol is simple for raster data: the map symbol applies to cells whether the spatial feature to be depicted is a point, line, or area. Choice of map symbol for vector data depends on the feature type (Figure 10.2). The general rule is to use point symbols for point features, line symbols for line features, and area symbols for area features. But this general rule does not apply to volumetric data or aggregate data. There are no volumetric symbols for data such as elevation, temperature, and precipitation. Instead, 3-D surfaces and isolines are often used to map volumetric data. Aggregate data such as county populations are data reported at an aggregate level. A common approach is to assign aggregate data to the center of each county and display the data using point symbols.

Visual variables for data display include hue, value, chroma, size, texture, shape, and pattern (Figure 10.3). Choice of visual variable depends on the type of data to be displayed. The measurement scale is commonly used for classifying attribute data (Chapter 9). Size and texture (the spacing of symbol markings) are more appropriate for displaying ordinal, interval, and ratio data. For example, a map may use different-sized circles to represent different-sized cities. Shape and pattern

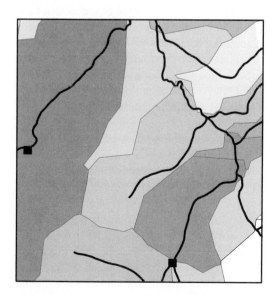

Figure 10.2

This map uses area symbols for watersheds, a line symbol for streams, and a point symbol for gauge stations.

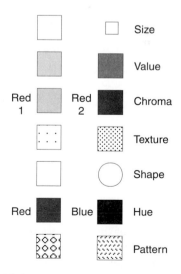

Figure 10.3

Visual variables in cartographic symbolization.

are more appropriate for displaying nominal data. For example, a map may use different area patterns to show different land-use types. Use of hue, value, and chroma as visual variables is covered in Section 10.1.2. Most GIS packages organize the selection of visual variables into palettes so that the user can easily select the variables that make up the map symbols. Some map packages also allow custom pattern designs.

The choice of visual variables is limited in the case of raster data. The visual variables of shape and size do not apply to raster data because of the use of cells. Using texture or pattern is also difficult with small cells. Display of raster data is therefore limited to different colors in most cases.

10.1.2 Use of Color

Because color adds a special appeal to a map, mapmakers will choose color maps over black and white maps whenever possible. But color is probably the most misused visual variable according to published critiques of computer-generated maps (Monmonier 1996). Use of color in mapmaking must begin with an understanding of the visual dimensions of hue, value, and chroma.

Hue is the quality that distinguishes one color from another, such as red from blue. Hue can also be defined as the dominant wavelength of light making up a color. We tend to relate different hues with different kinds of data. **Value** is the lightness or darkness of a color, with black at the lower end and white at the higher end. We generally perceive darker symbols on a map as being more important, or more in terms of magnitude (Robinson et al. 1995). Also called saturation or intensity, **chroma** refers to the richness, or brilliance, of a color. A fully saturated color is pure, whereas a low saturation approaches gray. We generally associate higher-intensity symbols with greater visual importance.

The first rule of thumb in use of color is simple: hue is a visual variable better suited for qualitative (nominal) data, whereas value and chroma are better suited for quantitative (ordinal, interval, and ratio) data.

Qualitative mapping is relatively easy. It is not difficult to find 12 or 15 distinctive hues for a map. If a map requires more symbols, we can add pattern or text to hue to make up more map symbols.

Quantitative mapping has received much more attention than qualitative mapping. Over the years, cartographers have suggested general color schemes that combine value and chroma for displaying quantitative data (Cuff 1972; Mersey 1990; Brewer 1994; Robinson et al. 1995). A basic premise among these color schemes is that the map reader can easily perceive the progression from low to high values (Antes and Chang 1990). The following is a summary of these color schemes.

- The **single hue scheme**. This color scheme uses a single hue but varies the combination of value and chroma to produce a sequential color scheme such as from light red to dark red. It is a simple but effective option for displaying quantitative data (Cuff 1972).
- The **hue and value scheme**. This color scheme progresses from a light value of one hue to a darker value of a different hue. Examples are yellow to dark red and yellow to dark blue. Mersey (1990) finds that color sequences incorporating both regular hue and value variations outperform other color schemes on the recall or recognition of general map information.
- The **diverging or double-ended scheme**. This color scheme uses graduated colors between two dominant colors. For example, a diverging scheme may progress from dark blue to light blue and then from light red to dark red. The diverging color scheme is a natural choice for displaying data with positive and negative values, or increases and decreases. But Brewer et al. (1997) report that the diverging scheme is in fact better than other color schemes for the map reader to retrieve information from quantitative maps that do not necessarily include positive and negative values.

Divergent color schemes are found in many maps of Census 2000 demographic data prepared by Brewer (2001) **(http://www.census.gov/population/ www/cen2000/atlas.html).**

- The part spectral scheme. This color scheme uses adjacent colors of the visible spectrum to show variations in magnitude. Examples of this color scheme include yellow to orange to red and yellow to green to blue.
- The full spectral scheme. This color scheme uses all colors in the visible spectrum. A conventional use of the full spectral scheme is found in elevation maps. Cartographers usually do not recommend this option for mapping other quantitative data because there is no logical sequence between hues.

10.1.3 Data Classification

Data classification involves the use of a classification method and a number of classes for aggregating data and map features. A GIS package typically offers different data classification methods. The following summarizes five commonly used methods.

- Equal interval. This classification method divides the range of data values into equal intervals.
- Equal frequency. Also called quantile, this classification method divides the total number of data values by the number of classes and ensures that each class contains the same number of data values.
- Mean and standard deviation. This classification method sets the class breaks at units of the standard deviation (0.5, 1.0, etc.) above or below the mean.
- Natural breaks. Also called the Jenks optimization method, this classification method optimizes the grouping of data values (Slocum et al. 2004). The method uses a computing algorithm to minimize differences between data values in the same class and to maximize differences between classes.

- User defined. This method lets the user choose the appropriate or meaningful class breaks. For example, in mapping rates of population change by state, the user may choose zero or the national average as a class break.

By changing the classification method, the number of classes, or both, the same data can produce different-looking maps and different spatial patterns. This is why mapmakers usually experiment with data classification before deciding on a classification scheme for the final map. Although the decision is ultimately subjective, it should be made under the guidance of map purpose and map communication.

10.2 TYPES OF MAPS

Cartographers classify maps by function and by symbolization. By function, maps can be general reference or thematic. The **general reference map** is used for general purposes. For example, a U.S. Geological Survey (USGS) quadrangle map shows a variety of spatial features, including boundaries, hydrology, transportation, contour lines, settlements, and land covers. The **thematic map** is also called the special purpose map, because its main objective is to show the distribution pattern of a theme, such as the distribution of population densities by county in a state.

By map symbol, maps can be qualitative or quantitative. A qualitative map uses visual variables that are appropriate for portraying qualitative data, whereas a quantitative map uses visual variables that are appropriate for communicating quantitative data. The following describes several common types of quantitative maps (Figure 10.4) (Robinson et al. 1995; Dent 1999; Slocum et al. 2004).

The **dot map** uses uniform point symbols to show spatial data, with each symbol representing a unit value. One-to-one dot mapping uses the unit value of one, such as one dot representing one crime location. But in most cases, it is one-to-many dot mapping and the unit value is greater than one.

The placement of dots becomes a major considera-tion in one-to-many dot mapping (Box 10.1).

The **choropleth map** symbolizes, with shad-ing, derived data based on administrative units (Box 10.2). An example is a map showing average household income by county. The derived data are usually classified prior to mapping and are sym-bolized using a color scheme for quantitative data. Therefore, the appearance of the choropleth map can be greatly affected by data classification. Cartographers often make several versions of the choropleth map from the same data and choose

Figure 10.4
Six common types of quantitative maps.

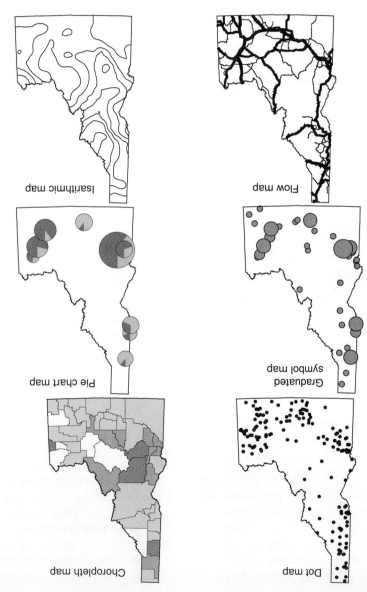

Choropleth map

Pie chart map

Isarithmic map

Dot map

Graduated symbol map

Flow map

Box 10.1 | **Locating Dots on a Dot Map**

Suppose a county has a population of 5000, and a dot on a dot map represents 500 persons. Where should the 10 dots be placed within the county? Ideally the dots should be placed at the locations of populated places. But, unless additional data are incorporated, most GIS packages including ArcMap use a random method in placing dots. A dot map with a random placement of dots is useful for comparing relative dot densities in different parts of the map. But it should not be used for locating data. One way to improve the accuracy of dot maps is to base them on the smallest administrative unit possible. Another way is to exclude areas such as water bodies that should not have dots. ArcMap allows use of mask areas for this purpose.

Box 10.2 | **Mapping Derived and Absolute Values**

Cartographers distinguish between absolute and derived values in mapping (Chang 1978). Absolute values are magnitudes or raw data such as county populations, whereas derived values are normalized values such as county population densities (derived from dividing the county population by the area of the county). County population densities are independent of the county size. Therefore, two counties with equal populations but different sizes will have different population densities, and thus different symbols, on a choropleth map. If choropleth maps are used for mapping absolute values such as county populations, size differences among counties can severely distort map comparisons (Monmonier 1996). Cartographers recommend graduated symbols for mapping absolute values.

one—typically one with a good spatial organization of classes—for final map production.

The **dasymetric map** is a variation of the simple choropleth map. By using statistics and additional information, the dasymetric map delineates areas of homogeneous values, rather than follows administrative boundaries (Robinson et al. 1995) (Figure 10.5). Dasymetric mapping used to be a time-consuming task, but the analytical functions of a GIS have simplified the mapping procedure (Holloway et al. 1999).

A GIS package such as ArcGIS uses the term **graduated color map** to cover the choropleth and dasymetric maps because both map types use a graduated color scheme to show the variation in spatial data.

The **graduated symbol map** uses different-sized symbols such as circles, squares, or triangles to represent different ranges of values. For example, we can use graduated symbols to represent cities of different population ranges. Two important issues to this map type are the range of sizes and the discernible difference between sizes. Both issues are obviously related to the number of graduated symbols on a map.

A **proportional symbol map** is a map that uses a specific symbol size for each numeric value rather than a range of values. Therefore, one circle size may represent a population of 10,000, another 15,000, and so on.

The **chart map** uses either pie charts or bar charts. A variation of the graduated circle, the pie chart can display two sets of quantitative data: the circle size can be made proportional to a value such as a county population, and the subdivisions can show the makeup of the value, such as the racial composition of the county population. Bar charts use vertical bars and their height to represent quantitative data. Bar charts are particularly useful for comparing data side by side.

The **flow map** displays different quantities of flow data such as traffic volume and stream flow by varying the line symbol width. Similar to the graduated symbols, the flow symbols usually represent ranges of values.

The **isarithmic map** uses a system of isolines to represent a surface. Each isoline connects points of equal value. GIS users often use the isarithmic map to display the terrain (Chapter 14) and the statistical surface created by spatial interpolation (Chapter 16).

GIS has introduced a new classification of maps based on vector and raster data. Maps prepared from vector data are the same as traditional maps using point, line, and area symbols. Most of Chapter 10 applies to the display of vector data. Maps prepared from raster data, although they may look like traditional maps, are cell-based (Figure 10.6). But raster data can also be classified as either quantitative or qualitative. The color schemes for qualitative and quantitative maps described in Section 10.1.2 can therefore be applied to mapping raster data as well.

10.3 TYPOGRAPHY

A map cannot be understood without text on it. Text is needed for almost every map element. Mapmakers treat text as a map symbol because, like point, line, or area symbols, text can have many type variations. Using type variations to create a pleasing and coherent map is therefore part of the mapmaking process.

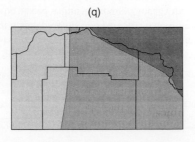

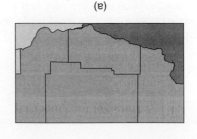

(a)

(b)

Figure 10.5
Map symbols follow the boundaries in the choropleth map (*a*) but not the dasymetric map (*b*).

Figure 10.6
Map showing raster-based elevation data. Cells with higher elevations have darker shades.

Times New Roman

Tahoma

Figure 10.7
Times New Roman is a serif typeface, and Tahoma
is a sans serif typeface.

10.3.1 Type Variations

Type can vary in typeface and form. **Typeface**
refers to the design character of the type. Two
main groups of typefaces are **serif** (with serif)
and **sans serif** (without serif) (Figure 10.7). Serifs are small, finishing touches at the ends of line
strokes, which tend to make running text in newspapers and books easier to read. Compared to
serif types, sans serif types appear simpler and
bolder. Although rarely used in books or other
text-intensive materials, a sans serif type stands
up well on maps with complex map symbols and
remains legible even in small sizes. Sans serif
types have an additional advantage in mapmaking
because many of them come in a wide range of
type variations.

   Type form variations include **type weight**
(bold, regular, or light), **type width** (condensed
or extended), upright versus slanted (or roman
versus italic), and uppercase versus lowercase (Figure 10.8). A **font** is a complete set of all variants of
a given typeface. Fonts on a computer are those
loaded from the printer manufacturer and software
packages. They are usually enough for mapping
purposes. If necessary, additional fonts can be
imported to a GIS package.

   Type can also vary in size and color. Type
size measures the height of a letter in **points,** with
72 points to an inch. Printed letters look smaller
than what their point sizes suggest. The point size
is supposed to be measured from a metal type
block, which must accommodate the lowest point
of the descender (such as p or g) to the highest
part of the ascender (such as d or b). But no letters extend to the very edge of the block. Type
color is the color of letters. In addition to color,
letters may also appear with drop shadow, halo,
or fill pattern.

Helvetica Normal

Helvetica Italic

Helvetica Bold

Helvetica Bold-Italic

Times Roman Normal

Times Roman Italic

Times Roman Bold

Times Roman Bold-Italic

Figure 10.8
Type variations in weight and roman versus italic.

10.3.2 Selection of Type Variations

Type variations for text symbols can function in
the same way as visual variables for map symbols. How does one choose type variations for a
map? A practical guideline is to group text symbols into qualitative and quantitative classes. Text
symbols representing qualitative classes such as
names of streams, mountains, parks, and so on can
vary in typeface, color, and upright versus italic.
In contrast, text symbols representing quantitative classes such as names of different-sized
cities can vary in type size, weight, and uppercase
versus lowercase. Grouping text symbols into
classes simplifies the process of selecting type
variations.

   Besides classification, cartographers also recommend legibility, harmony, and conventions for
type selection (Dent 1999). Legibility is difficult to
control on a map because it can be influenced by
not only type variations but also placement of text
and contrast between text and the background
symbol. As GIS users, we have the additional
problem of having to design a map on a computer
monitor and to print it on a larger plot. Experimen-
tation may be the only way to ensure type legibility in all parts of the map.

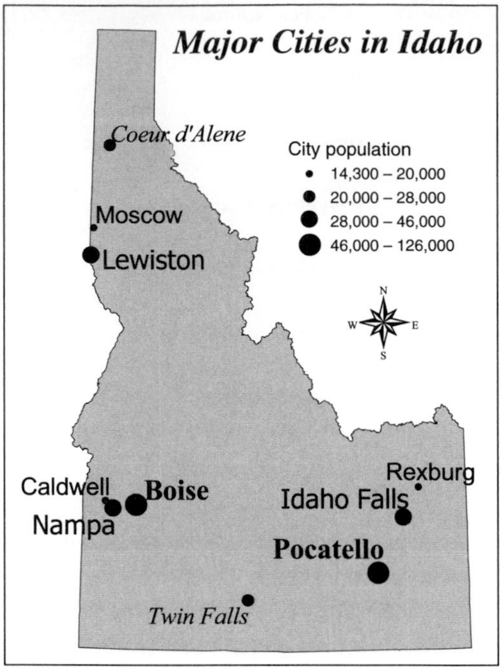

Type legibility should be balanced with harmony. As a means to communicate the map content, text should be legible but should not draw too much attention. Mapmakers can generally achieve harmony by adopting only one or two typefaces on a map (Figure 10.9). For example, many mapmakers use a sans serif type in the body of a map and a serif type for the map's title and legend. Use of conventions can also lend support for harmony. Conventions in type selection include italic for names of water features, uppercase and letter spacing for names of administrative units, and variations in type size and form for names of different-sized cities.

Figure 10.9
The look of the map is not harmonious because of too many typefaces.

Major Cities in Idaho

City population
- • 14,300 – 20,000
- ● 20,000 – 28,000
- ⬤ 28,000 – 46,000
- ⬤ 46,000 – 126,000

Coeur d'Alene · Moscow · Lewiston · Rexburg · Idaho Falls · Pocatello · Twin Falls · Caldwell · Nampa · Boise

10.3.3 Placement of Text in the Map Body

Section 10.3.3 focuses on the placement of text in the map body. Text elements in the map body, also called labels, are directly associated with the spatial features. In most cases, these labels are names of the spatial features. But they can also be some attribute values such as contour readings or precipitation amounts. Other text elements on a map such as the title and the legend are not tied to any specific locations. Instead, the placement of these text elements (i.e., graphic elements) is related to the layout of the map (Section 10.4.1).

As a general rule, a label should be placed to show the location or the area extent of the named spatial feature. Cartographers recommend placing the name of a point feature to the upper right of its symbol, the name of a line feature in a block and parallel to the course of the feature, and the name of an area feature to indicate its area extent. Other general rules suggest aligning labels with either the map border or lines of latitude, and placing labels entirely on land or on water.

Implementing labeling algorithms in a GIS package is no easy task (Mower 1993; Chirié 2000). Automated name placement presents several difficult problems to the computer programmer: names must be legible, names cannot overlap other names, names must be clear as to their intended referent symbols, and name placement must follow cartographic conventions. These problems worsen at smaller map scales as competition for the map space intensifies between names. We should not expect that labeling be completely automated. Usually, some interactive editing is needed to improve the final map's appearance. For this reason many GIS packages offer more than one method of labeling.

As an example, ArcGIS offers interactive and dynamic labeling. Interactive labeling works with one label at a time. If the placement does not work out well, the label can be moved immediately. Interactive labeling is ideal if the number of labels is small or if the location of labels must be exact. Dynamic labeling is probably the method of choice for most users because it can automatically label all or selected features. Using dynamic labeling, we can prioritize options for text placement and for solving potential conflicts (Box 10.3). For example, we can choose to place a line label in a block and parallel to the course of the line feature. We can also set rules to prioritize labels competing for the same map space. By default, ArcGIS does not allow overlapped labels. This constraint, which is sensible, can impact the placement of labels and may require adjustment of some labels (Figure 10.10).

Box 10.3   Options for Dynamic Labeling

When we choose dynamic labeling, we basically let the computer take over the labeling task. But the computer needs instructions for placement methods and for solving potential problems in labeling. ArcGIS uses the Placement Properties dialog to gather instructions from the user. The dialog has two tabs. The Placement tab deals with the placement method and duplicate labels. Different placement methods are offered for each feature type. For example, ArcGIS offers 36 options for placing a label relative to its point feature. The default placement method is usually what is recommended in the cartographic literature. Duplicate labels such as the same street name for a series of street segments can be either removed or reduced. The Conflict Detection tab handles overlapped labels and the overlapping of labels and feature symbols. ArcGIS uses a weighting scheme to determine the relative importance of a layer and its features. The Conflict Detection tab also has a buffer option that can provide a buffer around each label. We cannot always expect a "perfect" job from dynamic labeling. In most cases, we need to adjust some labels individually.

Figure 10.10

Dynamic labeling of major cities in the United States. The initial result is good but not totally satisfactory. Philadelphia is missing. Labels of San Antonio, Indianapolis, and Baltimore overlap slightly with point symbols. San Francisco is too close to San Jose.

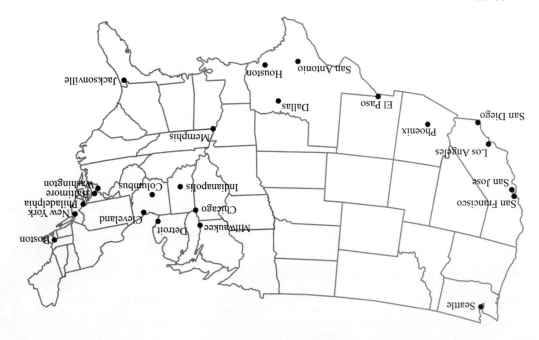

Figure 10.11

A revised version of Figure 10.10. Philadelphia is added to the map, and several city names are moved individually to be closer to their point symbols.

Dynamic labels cannot be selected or adjusted individually. But they can be first converted to text elements so that these elements can be individually moved and changed in the same way as interactive labels (Figure 10.11). One way to take care of labels in a truly congested area is to use a leader line to link a label to its feature (Figure 10.12).

Perhaps the most difficult task in labeling is to label names of streams. The general rule states that the name should follow the course of the stream, either above or below it. Both interactive labeling and dynamic labeling can curve the name if it appears along a curvy part of the stream. But the appearance of the curved name depends on the smoothness of the corresponding stream segment, the length of the stream segment, and the length of the name. Placing every name in its correct position the first time is nearly impossible (Figure 10.13). Problem names must be removed and relabeled using the **spline text** tool, which can align a text string along a curved line that is digitized on-screen (Figure 10.14).

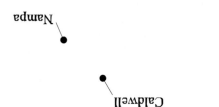

Figure 10.12

A leader line connects a point symbol to its label.

10.4 MAP DESIGN

Like graphic design, **map design** is a visual plan to achieve a goal. The purpose of map design is to enhance map communication, which is particularly important for thematic maps. A well-designed map is balanced, coherent, ordered, and interesting to look at, whereas a poorly designed map is confusing and disoriented (Antes et al. 1985). Map design is both an art and science. There may not be clear-cut right or wrong designs for maps, but

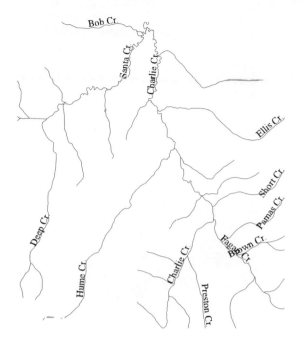

Figure 10.13
Dynamic labeling of streams may not work for every label. Brown Cr. overlaps with Fagan Cr., and Pamas Cr. and Short Cr. do not follow the course of the creek.

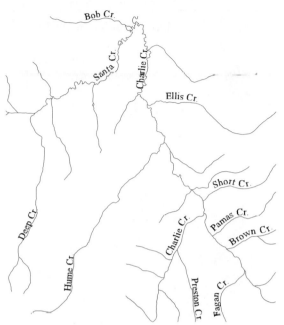

Figure 10.14
Problem labels in Figure 10.13 are redrawn with the spline text tool.

there are better or more effective maps and worse or less effective maps.

Map design overlaps with the field of graphic arts, and many map design principles have their origin in visual perception. Cartographers usually study map design from the perspectives of layout and visual hierarchy.

10.4.1 Layout

Layout, or planar organization, deals with the arrangement and composition of various map elements on a map. Major concerns with layout are focus, order, and balance. A thematic map should have a clear focus, which is usually the map body or a part of the map body. To draw the map reader's attention, the focal element should be placed near the optical center, just above the map's geometric center. The focal element should be differentiated from other map elements by contrast in line width, texture, value, detail, and color.

After viewing the focal element, the reader should be directed to the rest of the map in an ordered fashion. For example, the legend and the title are probably the next elements that the viewer needs to look at after the map body. To smooth the transition, the mapmaker should clearly place the legend and the title on the map, with perhaps a box around the legend and a larger type size for the title to draw attention to them (Figure 10.15).

A finished map should look balanced. It should not give the map reader an impression that the map "looks" heavier on the top, bottom, or the side (Figure 10.16). But balance does not suggest the breaking down of the map elements and placing them, almost mechanically, in every part of the map. Although in that case the elements would be in balance, the map would be disorganized and confusing. Mapmakers therefore should deal with balance within the context of organization and map communication.

Cartographers used to use thumbnail sketches to experiment with balance on a map. Now they

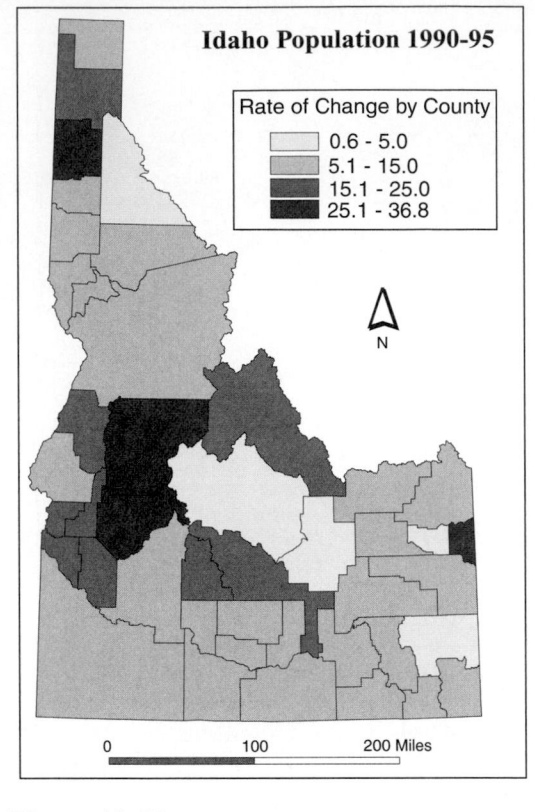

Figure 10.15
Use a box around the legend to draw the map reader's
attention to it.

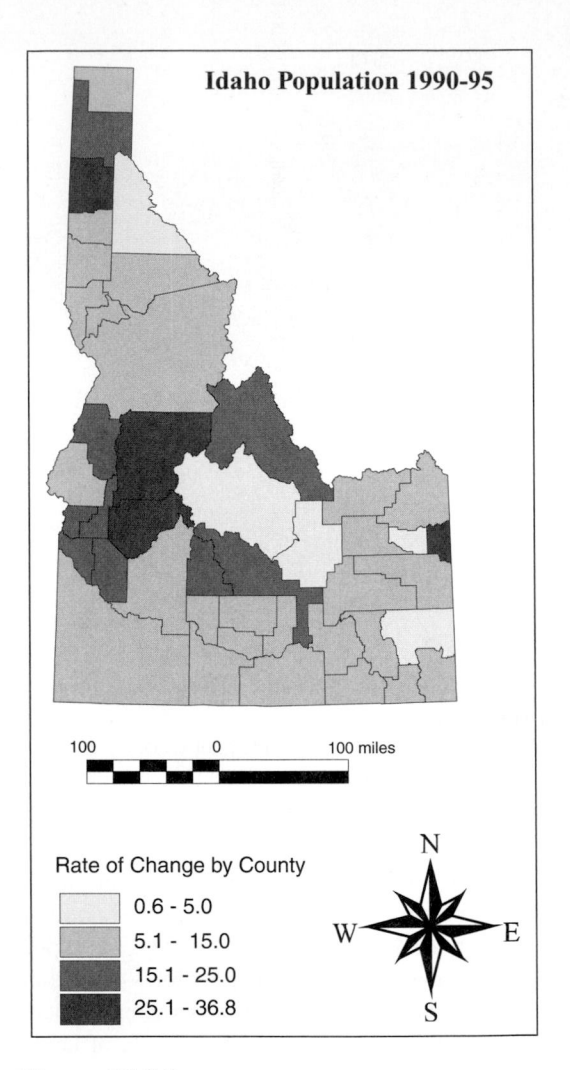

Figure 10.16
A poorly balanced map.

use computers to manipulate map elements on a
layout page. ArcGIS, for example, offers two basic
methods for layout design. The first method is to
use a layout template. These templates are grouped
into general, industry, USA, and world. Each
group has a list of choices. For example, the layout templates for the United States include USA,
conterminous USA, and five different regions of
the country. Figure 10.17 shows the basic structure
of the conterminous USA layout template. The
idea of using a layout template is to use a built-in
design option to quickly compose a map.

The second option is to open a layout page
and add map elements one at a time. ArcGIS offers
the following map elements: title, text, neatline,
legend, north arrow, scale bar, scale text, and the
graticule (Box 10.4). The user can choose one of
these map elements, place it on the layout page,
and manipulate it graphically. For example, the
user can change the type design of the title, enlarge
or reduce it, and move it around the layout. A layout design created using the second option can be
saved as a template for future use.

Regardless of the method used in layout
design, the legend deserves special consideration.
The legend includes descriptions of all the layers
that make up a map. For example, a map showing

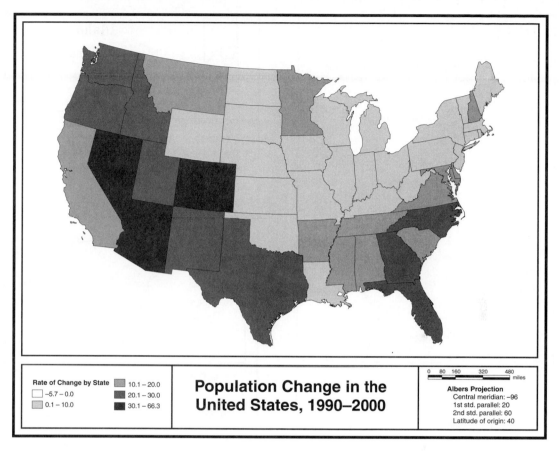

Figure 10.17
The basic structure of the conterminous USA layout template in ArcMap.

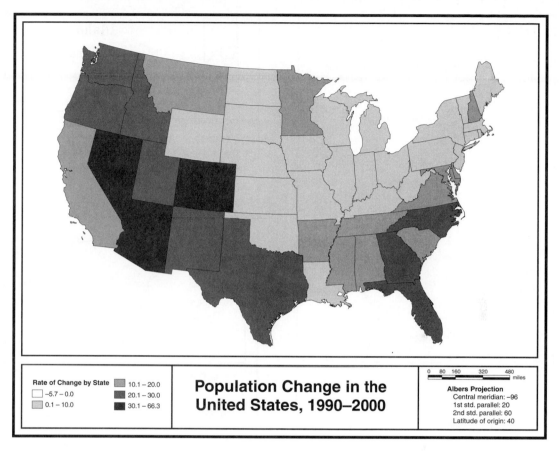

Box 10.4 | **Wizards for Adding Map Elements**

ArcMap provides wizards and selectors for adding map elements to a map. The Grids and Graticules wizard is one of them. The graticule refers to lines or tics of longitude and latitude. The grid can be a measured grid or a reference grid. A measured grid shows lines or tics of projected coordinates. A reference grid divides a map into a grid for spatial index-ing. The Grids and Graticules wizard guides the user through a series of dialogs to choose the intervals, symbol design, and placement option. Wizards are useful for the beginner but can be tedious and limiting for experienced users. We can choose to turn off the wizard and to work directly with the main property dialog of a map element.

different classes of cities and roads in a state in layout design.

Figure 10.18a

A lengthy legend is confusing and can create a problem in layout design.

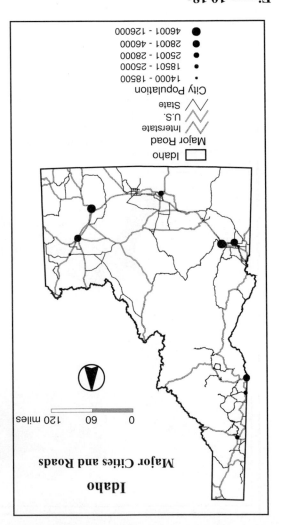

requires a minimum of three layers: one for the cities, one for the roads, and one for the state boundary. As default, these descriptions are placed together as a single graphic element, which can become quite lengthy with multiple layers. A lengthy legend presents a problem in balancing a layout design (Figure 10.18a). The solution is to break the legend into two or more columns and to remove useless legend descriptions such as the description of an outline symbol (Figure 10.18b).

10.4.2 Visual Hierarchy

Visual hierarchy is the process of developing a visual plan to introduce the 3-D effect or depth to maps (Figure 10.19). Mapmakers create the visual hierarchy by placing map elements at different visual levels according to their importance to the map's purpose. The most important element should be at the top of the hierarchy and should appear closest to the map reader. The least important element should be at the bottom. A thematic map may consist of three or more levels in a visual hierarchy. The concept of visual hierarchy is an extension of the **figure–ground relationship** in visual

Figure 10.18b

The lengthy legend in Figure 10.18a is separated into two parts. Also, the unnecessary outline symbol is removed from the legend.

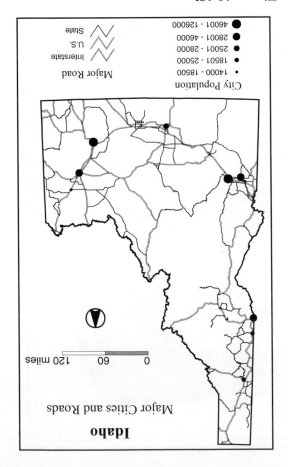

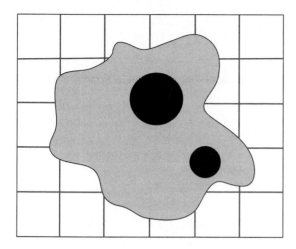

Figure 10.19
A visual hierarchy example. The two black circles are on top (closest to the map reader), followed by the gray polygon. The grid, the least important, is on the bottom.

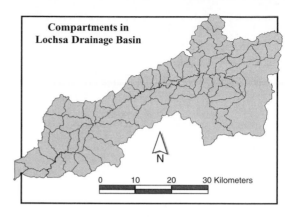

Figure 10.20
The interposition effect in map design.

perception (Arnheim 1965). The figure is more important visually, appears closer to the viewer, has form and shape, has more impressive color, and has meaning. The ground is the background. Cartographers have adopted the depth cues for developing the figure–ground relationship in map design.

Probably the simplest and yet most effective principle in creating a visual hierarchy is interposition or superimposition (Dent 1999). **Interposition** uses the incomplete outline of an object to make it appear behind another. Examples of interposition abound in maps, especially in newspapers and magazines. Continents on a map look more important or occupy a higher level in visual hierarchy if the lines of longitude and latitude stop at the coast. A map title, a legend, or an inset map looks more prominent if it lies within a box, with or without the drop shadow. When the map body is deliberately placed on top of the neatline around a map, the map body will stand out more (Figure 10.20). Because interposition is so easy to use, it can be overused or misused. A map looks confusing if several of its elements compete for the map reader's attention (Figure 10.21).

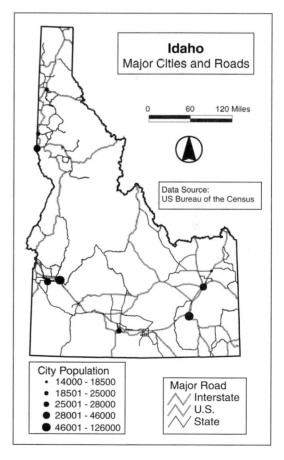

Figure 10.21
A map looks confusing if it uses too many boxes to highlight individual elements.

Subdivisional organization is a map design principle that groups map symbols at the primary and secondary levels according to the intended visual hierarchy (Robinson et al. 1995). Each primary symbol is given a distinctive hue, and the differentiation among the secondary symbols is based on color variation, pattern, or texture. For example, all tropical climates on a climate map are shown in red, and different tropical climates (e.g., wet equatorial climate, monsoon climate, and wet–dry tropical climate) are distinguished by different shades of red. Subdivisional organization is most useful for maps with many map symbols, such as climate, soil, geology, and vegetation maps.

Contrast is a basic element in map design, important to layout as well as to visual hierarchy. Contrast in size or width can make a state outline look more important than county boundaries and larger cities look more important than smaller ones (Figure 10.22). Contrast in color can separate the figure from the ground. Cartographers often use a warm color (e.g., orange to red) for the figure and a cool color (e.g., blue) for the ground. Contrast in texture can also differentiate between the figure and the ground because the area containing more details or a greater amount of texture tends to stand out on a map. Like the use of interposition, too much contrast can create a confusing map appearance. For instance, if bright red and green are used side by side as area symbols on a map, they appear to vibrate.

A tool that many GIS packages offer for data display is called transparency, which controls the percentage of a layer that is transparent. Transparency can be useful in creating a visual hierarchy by "toning down" the symbols of a background layer. Suppose we want to superimpose a layer showing major cities on top of a layer showing rates of population change by county. We can apply transparency to the county layer so that the city layer will stand out more.

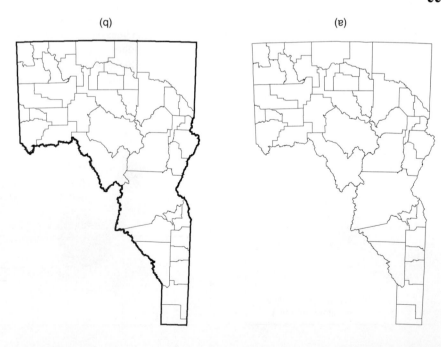

(a)

(b)

Figure 10.22
Contrast is missing in (a), whereas the line contrast makes the state outline look more important than the county boundaries in (b).

Box 10.5 | **Working with Soft-Copy Maps**

When completed, a soft-copy map can be either printed or exported. Printing a map from the computer screen requires the software interface (often called the driver) between the operating system and the print device. ArcMap uses Enhanced Metafiles (EMF) as the default and offers two additional choices of Post-Script (PS) and ArcPress. EMF files are native to the Windows operating system for printing graphics. PS, developed by Adobe Systems Inc., in the 1980s, is an industry standard for high-quality printing. Developed by ESRI, Inc., ArcPress is PS-based and useful for printing maps containing raster data sets, images, or complex map symbols.

One must specify an export format to export a computer-generated map for other uses. ArcMap, for example, offers JPEG (Joint Photographic Experts Group) for the Internet, TIFF (tagged image file format), AI (Adobe Illustrator format), BMP (bitmap), and CGM (Computer Graphics Metafile) for other graphics packages, and EMF, EPS (Encapsulated PostScript), AI, and PDF (portable document format) for printing and publishing.

Offset printing is the standard method for printing a large number of copies of a map. A typical procedure to prepare a computer-generated map for offset printing involves the following steps. First, the map is exported to separate CMYK PostScript files in a process called color separation. CMYK stands for the four process colors of cyan, magenta, yellow, and black, commonly used in offset printing. Second, these files are processed through an image setter to produce high-resolution plates or film negatives. Third, if the output from the image setter consists of film negatives, the negatives are used to make plates. Finally, offset printing runs the plates through the press to print color maps.

10.5 MAP PRODUCTION

GIS users design and make maps on the computer screen. These soft-copy maps can be used in a variety of ways. They can be printed, exported for use on the Internet, used in overhead computer projection systems, exported to other software packages, or further processed for publishing (Box 10.5).

Map production is a complex topic. We are often surprised that color symbols from the color printers do not exactly match those on the computer screen. This discrepancy results from the use of different media and color models.

Data display on the computer screen uses either **CRT (cathode-ray tube)** or **LCD (liquid crystal display).** Personal computers use CRT whereas laptop or portable computers use LCD. A CRT screen has a built-in fine mesh of pixels, and each pixel has colored dots called phosphors. When struck by electrons from an electron gun, a dot lights up. An LCD screen, also called a flat-panel display, uses two sheets of polarizing materials with a liquid crystal solution between them. Each pixel on an LCD flat-panel display can be turned on or off independently. LCDs are thinner and lighter and consume less power than CRTs. LCD uses the digital signal instead of an analog display interface such as VGA (video graphics array). As a result, LCD displays are flicker-free and can produce sharper, undistorted screen images than CRT displays. But LCD screens are more expensive and LCD technologies continue to evolve.

Using either a CRT or an LCD screen, a color symbol we see on the screen is made of pixels, and the color of each pixel is a mixture of **RGB** (red, green, and blue). The intensity of each primary color in a mixture determines its color. The number of intensity levels each primary color can have depends on the number of bit-planes assigned to the electron gun in a CRT screen or the variation of the voltage applied in an LCD screen. Typically, the intensity of each primary color can range over 256 shades. Combining the three primary colors produces a possible palette of 16.8 million colors (256 × 256 × 256).

Many GIS packages offer the RGB color model for color specification. But color mixtures of RGB are not intuitive (Figure 10.23). It is difficult to perceive that a mixture of red and green at full intensity is a yellow color. This is why other color models have been developed to specify colors that are based on the visual perception of hue, value, and

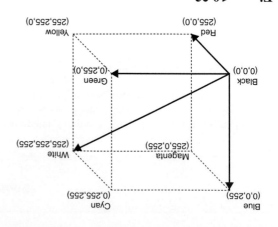

Figure 10.23
The RGB (red, green, and blue) color model.

chroma. ArcGIS, for example, has the **HSV** (hue/saturation/value) color model in addition to the RGB color model for specifying custom colors.

Printed color maps differ from color displays on a computer screen in two ways: color maps reflect rather than emit light; and the creation of colors on color maps is a subtractive rather than additive process. The three primary subtractive colors are cyan, magenta, and yellow. In printing, these three primary colors plus black form the four process colors of **CMYK.**

Color symbols are produced on a printed map in much the same way as on a computer screen. In place of pixels are color dots, and in place of varying light intensities or voltages are percentages of area covered by color dots. A deep orange color on a printed map may represent a combination of 60% magenta and 80% yellow, whereas a light orange color may represent a combination of 30% magenta and 90% yellow. To match a color symbol on the computer screen with a color symbol on the printed map requires a translation from the RGB color model to the CMYK color model. As yet there is no exact translation between the two, and therefore the color map looks different when printed (Slocum et al. 2004). The International Color Consortium, a consortium of over 70 companies and organizations worldwide, has

***Box 10.6* A Web Tool for Making Color Maps**

A free Web tool for making color maps is available at **http://www.personal.psu.edu/c/a/cab38/ColorBrewerBeta.html.** Funded initially by the National Science Foundation's Digital Government Program, the website is maintained by Cindy Brewer of Pennsylvania State University. The tool offers three main types of color schemes: sequential, diverging, and qualitative. One can select a color scheme and see how the color scheme looks on a sample choropleth map. One can also add point and line symbols to the sample map and change the colors for the map border and background. Then, for each color selected, the tool shows its color values in terms of CMYK, RGB, and other models. And, for each color scheme selected, the tool rates its potential uses including photocopy in black and white, CRT display, color printing, LCD projector, and laptop LCD display.

been working since 1993 on a color management system that can be used across different platforms and media (**http://www.color.org/**). Until such a color management system is developed, we must experiment with colors on different media.

Map production, especially production of color maps, can be a challenging task to GIS users. Box 10.6 describes a free Web tool that can help GIS users choose color symbols that are appropriate for a particular mode of map production.

KEY CONCEPTS AND TERMS

Cartography: The making and study of maps in all their aspects.

CRT (cathode-ray tube) screen: A display device for a personal computer that uses electron guns and color dots.

Chart map: A map that uses charts such as pie charts or bar charts as map symbols.

Choropleth map: A map that applies shading symbols to data or statistics collected for enumeration units such as counties or states.

Chroma: The richness or brilliance of a color. Also called *saturation* or *intensity.*

CMYK: A color model in which colors are specified by the four process colors of cyan (C), magenta (M), yellow (Y), and black (K).

Contrast: A basic element in map design that enhances the look of a map or the figure–ground relationship by varying the size, width, color, and texture of map symbols.

Dasymetric map: A map that uses statistics and additional information to delineate areas of homogeneous values, rather than following administrative boundaries.

Dot map: A map that uses uniform point symbols to show spatial data, with each symbol representing a unit value.

Figure–ground relationship: A tendency in visual perception to separate more important objects (figures) in a visual field from the background (ground).

Flow map: A map that displays different quantities of flow data by varying the width of the line symbol.

Font: A complete set of all variants of a given typeface.

General reference map: One type of map used for general purposes such as the USGS topographic map.

Graduated color map: A map that uses a progressive color scheme such as light red to dark red to show the variation in spatial data.

Graduated symbol map: A map that uses different-sized symbols such as circles, squares, or triangles to represent different magnitudes.

HSV: A color model in which colors are specified by their hue (H), saturation (S), and value (V).

Hue: The quality that distinguishes one color from another, such as red from blue. Hue is the dominant wavelength of light.

Interposition: A tendency for an object to appear behind another because of its incomplete outline.

Isarithmic map: A map that uses a system of isolines to represent a surface.

Layout: The arrangement and composition of map elements on a map.

LCD (liquid crystal display) screen: A display device on a laptop or portable computer that uses electric charge through a liquid crystal solution between two sheets of polarizing materials. Also called *flat-panel display.*

Map design: The process of developing a visual plan to achieve the map's purpose.

Point: Measurement unit of type, with 72 points to an inch.

Proportional symbol map: A map that uses a specific-sized symbol for each numeric value.

RGB: A color model in which colors are specified by their red (R), green (G), and blue (B) components.

Sans serif: Without serifs.

Serif: Small, finishing touches added to the ends of line strokes.

Spline text: A text string aligned along a curved line.

Subdivisional organization: A map design principle that groups map symbols at the primary and secondary levels according to the intended visual hierarchy.

Thematic map: One type of map that emphasizes the spatial distribution of a theme, such as a map that shows the distribution of population densities by county.

Transparency: A display tool that controls the percentage of a layer that is transparent.

Typeface: A particular style or design of type.

Type weight: Relative blackness of a type such as bold, regular, or light.

Type width: Relative width of a type such as condensed or extended.

Value: The lightness or darkness of a color.

Visual hierarchy: The process of developing a visual plan to introduce the 3-D effect or depth to maps.

Review Questions

1. What are the common elements on a map for presentation?
2. Why is it important to pay attention to map design?
3. Mapmakers apply visual variables to map symbols. What are visual variables?
4. Name common visual variables for data display.
5. Describe the three visual dimensions of color.
6. Use an example to describe a "hue and value" color scheme.
7. Use an example to describe a "diverging" color scheme.
8. How does a general reference map differ from a thematic map?
9. Define the choropleth map.
10. ArcMap offers the display options of graduated colors and graduated symbols. How do these two options differ?
11. Suppose you are asked to redo Figure 10.9. Provide a list of type designs, including typeface, form, and size, that you will use for the four classes of cities.
12. What are the general rules for achieving harmony with text on a map?
13. ArcGIS offers interactive labeling and dynamic labeling for the placement of text. What are the advantages and disadvantages of each labeling method?
14. Figure 10.17 shows a layout template available in ArcMap for the conterminous USA. Will you consider using the layout template for future projects? Why, or why not?
15. What is visual hierarchy in map design? How is the hierarchy related to the map purpose?
16. Figure 10.20 shows an example of using interposition in map design. Does the map achieve the intended 3-D effect?

17. What is subdivisional organization in map design? Can you think of an example, other than the climate map example in Chapter 10, to which you can apply the principle?

18. Explain why color symbols from a color printer do not exactly match those on the computer screen.

19. Define the RGB and CMYK color models.

APPLICATIONS: DATA DISPLAY AND CARTOGRAPHY

This applications section consists of three tasks. Task 1 guides you through the process of making a choropleth map. Task 2 involves type design, graduated symbols, color combination, and the highway shield symbols. Task 3 focuses on the placement of text. Because a layout in ArcMap will include all data frames, you must exit ArcMap at the end of each task to preserve the layout design. Making maps for presentation can be tedious. You must be patient and willing to experiment.

Task 1: Make a Choropleth Map

What you need: *us.shp*, a shapefile showing population change by state in the United States between 1990 and 2000. The shapefile is projected onto the Albers equal-area conic projection and is measured in meters.

Choropleth maps display statistics by administrative unit. For Task 1 you will map the rate of population change between 1990 and 2000 by state. The map consists of the following elements: a map of the conterminous United States and a scale bar, a map of Alaska and a scale bar, a map of Hawaii and a scale bar, a title, a legend, a north arrow, a data source statement, a map projection statement, and a neatline around all elements. The basic layout of the map is as follows. The map page is 11″ (width) × 8.5″ (height), or letter size, with a landscape orientation. One-third of the map on the left, from top to bottom, has the title, map of Alaska, and map of Hawaii. Two-thirds of the map on the right, from top to bottom, has the map of the conterminous United States and all the other elements.

1. Start ArcCatalog, and make connection to the Chapter 10 database. Launch ArcMap.

Maximize the view window of ArcMap. Add *us.shp* to the new data frame, and rename the data frame Conterminous. Use the Zoom In tool to zoom in on the lower 48 states.

2. This step symbolizes the rate of population change by state. Right-click *us* and select Properties. On the Symbology tab, click Quantities and select Graduated colors. Click the Value dropdown arrow and choose ZCHANGE (rate of population change from 1990 to 2000). Cartographers recommend use of round numbers and logical breaks such as 0 in data classification. Click the first cell under Range and enter 0. The new range should read −5.7 − 0.0. Enter 10, 20, and 30 for the next three cells, and click the empty space below the cells to unselect. Next, change the color scheme for ZCHANGE. Right-click the Color Ramp box and uncheck Graphic View. Click the dropdown arrow and choose Yellow to Green to Dark Blue. The first class from −5.7 to 0.0 is shown in yellow, and the other classes are shown in green to dark blue. Click OK to dismiss the Layer Properties dialog.

3. The Alaska map is next. Insert a new data frame and rename it Alaska. Add *us.shp* to Alaska. Use the Zoom In tool to focus on Alaska. Follow the same procedure in Step 2, or use the Import button in the Layer Properties dialog, to display ZCHANGE.

4. The Hawaii map is next. Select Data Frame from the Insert menu. Rename the new data frame Hawaii. Add *us.shp* to Hawaii. Use the Zoom In tool to focus on Hawaii. Display the map with ZCHANGE.

5. The table of contents in ArcMap now has three data frames: Conterminous, Alaska, and Hawaii. Select Layout View from the View menu. Click the Zoom Whole Page button. Select Page and Print Setup from the File menu. Check Landscape for the page orientation. Make sure that the page size is 11.5″ × 8.5″, and click OK.

6. The three data frames are stacked up on the layout. You will rearrange the data frames according to the basic layout plan. Click the Select Elements button. Click the Conterminous data frame. When the handles around the data frame appear, move the data frame to the upper-right side of the layout so that it occupies about two-thirds of the layout in both width and height and shows only the lower 48 states. Click the Alaska data frame, and move it to the center left of the layout. Click the Hawaii data frame, and place it below the Alaska data frame.

7. Now you will add a scale bar to each data frame. Begin with the Conterminous data frame. Select Scale Bar from the Insert menu. Click the selection of Alternating Scale Bar 1, and then Properties. The Scale Bar Selector dialog has three tabs: Scale and Units, Numbers and Marks, and Format. On the Scale and Units tab, start with the middle part of the dialog: select Adjust width when resizing, choose Kilometers for division units, and enter km for the label. Now work with the upper half of the dialog: enter 1000 (km) for the division value, select 2 for the number of divisions, select 0 for the number of subdivisions, and opt not to show one division before zero. On the Numbers and Marks tab, select divisions from the frequency dropdown list and Above bar from the position dropdown list. On the Format tab, select Times New Roman from the font dropdown list. Click OK to dismiss the dialogs. The scale bar appears in the map with the handles. Move the scale bar to the lower-left corner of the Conterminous data frame. The scale bar should have the divisions of 1000 and 2000 kilometers. (You can set the Zoom Control on the Layout toolbar at 100% and then use the Pan tool to check the scale bar.) Separate scale bars are needed for the other two data frames. Click the Alaska data frame. Add its scale bar by using 500 kilometers for the division value. Click the Hawaii data frame. Add its scale bar by using 100 kilometers for the division value.

Q1. Explain in your own words the number of divisions and the number of subdivisions on a scale bar.

Q2. In Step 7, you have chosen the option of "Adjust width when resizing." What does this option mean?

8. So far, you have completed the data frames of the map. The map must also have the title, legend, and other elements. Select Title from the Insert menu. An Enter Map Title box appears on the layout. Click outside the box. When the outline of the box is shown in cyan, double-click it. A Properties dialog appears with two tabs. On the Text tab, enter two lines in the text box: "Population Change" in the first line, and "by State, 1990–2000" in the second line. Click Change Symbol. The Symbol Selector dialog lets you choose color, font, size, and style. Select black, Bookman Old Style (or another serif type), 20, and B (bold) respectively. Click OK to dismiss the dialogs. Move the title to the upper left of the layout above the Alaska data frame.

9. The legend is next. Because all three data frames use the same legend, it does not matter which data frame is used. Click ZCHANGE of the active data frame, and click it one more time so that ZCHANGE is highlighted in a box. Delete ZCHANGE. (Unless ZCHANGE is removed in the table of contents, it will show up in the layout as a confusing legend descriptor.) Select Legend from the Insert menu. The Legend Wizard uses five

panels. In the first panel, make sure *us* is the layer to be included in the legend. The second panel lets you enter the legend title and its type design. Delete Legend in the Legend Title box, and enter "Rate of Population Change (%)." Then choose 14 for the size, choose Times New Roman for the font, and uncheck B (Bold). Skip the third and fourth panels, and click Finish in the fifth panel. Move the legend to the right of the Hawaii data frame.

10. A north arrow is next. Select North Arrow from the Insert menu. Choose ESRI North 6, a simple north arrow from the selector, and click OK. Move the north arrow to the upper right of the legend.

11. Next is the data source. Select Text from the Insert menu. An Enter Text box appears on the layout. Click outside the box. When the outline of the box is shown in cyan, double-click it. A Properties dialog appears with two tabs. On the Text tab, enter "Data Source: US Census 2000" in the Text box. Click on Change Symbol. Select Times New Roman for the font and 14 for the size. Click OK in both dialogs. Move the data source statement below the north arrow in the lower right of the layout.

12. Follow the same procedure as for the data source to add a text statement about the map projection. Enter Albers Equal-area Conic Projection in the text box, and change the symbol to Times New Roman with a size of 10. Move the projection statement below the data source.

13. Finally, add a neatline to the layout. Select Neatline from the Insert menu. Check to place around all elements, select Double, Graded from the Border dropdown list, and select Sand from the Background dropdown list. Click OK.

14. The layout is now complete. If you want to rearrange a map element, select the element and then move it to a new location. You can also enlarge or reduce a map element by using its handles or properties.

15. If your PC is connected to a color printer, you can print the map directly by selecting Print in the File menu. There are two other options on the File menu: save the map as an ArcMap document, or export the map as a graphic file (e.g., PostScript, JPEG, TIFF, etc.). Exit ArcMap.

Q3. In Task 1, why did you have to prepare three data frames (i.e., Conterminous, Alaska, and Hawaii)?

Task 2: Use Graduated Symbols, Line Symbols, Highway Shield Symbols, and Text Symbols

What you need: *idlcity.shp*, a shapefile showing the 10 largest cities in Idaho; *idhwy.shp*, a shapefile showing interstate and U.S. highways in Idaho; and *idoutl.shp*, an outline map of Idaho.

Task 2 lets you experiment with type design, use graduated symbols, and try the highway shield symbols in ArcMap. Task 2 also involves choice of color symbols and map design in general.

1. Make sure that ArcCatalog is still connected to the Chapter 10 database. Launch ArcMap. Rename the data frame Task 2, and add *idlcity.shp*, *idhwy.shp*, and *idoutl.shp* to Task 2. Select Page and Print Setup from the File menu. Make sure that the page has a width of 8.5 (inches), a height of 11 (inches), and a portrait orientation.

2. The only decision to be made about *idoutl* is its color, the background color for the state, which should be contrasted with the point symbols for *idlcity* and the highway symbols. Click the symbol for *idoutl* in the table of contents. Select Grey.

3. Select Properties from the context menu of *idhwy*. On the Symbology tab, select Categories and Unique values for the show option and select ROUTE_DESC from the Value Field dropdown list. Click Add All Values at the bottom. Interstate and U.S. appear as the values. Uncheck all other values. Double-click the Symbol next to Interstate and select the Freeway symbol in the Symbol Selector

checked in the Label Classes frame. Click OK to dismiss the dialog.

7. All city names should now appear in the map. But it is difficult to judge the quality of labeling in Data View. You must switch to Layout View to see how the labels will appear on a plot. Select Layout View from the View menu. Select 100% from the Zoom Control list on the Layout toolbar. Use the Pan tool to see how the labels will appear on an 8.5-by-11-inch plot.

8. All labels except Nampa are well placed. But to alter the label position of Nampa, you have to convert labels to annotation. Right-click *idlity* and check Convert Labels to Annotation. In the next dialog, select to store annotation in the map, rather than in the database. Click Convert. (The conversion will take a while to complete.) To move the label for Nampa, click the Select Elements tool on the standard toolbar, click Nampa to select it, and then move the label to below its point symbol. (Nampa is between Boise and Caldwell. You can also use the Identify tool to check which city is Nampa.)

9. The last part of this task is to label the interstates and U.S. highways with the highway shield symbols. Switch to Data View. Right-click *idhwy*, and select Properties. On the Labels tab, make sure that the Label Field is MINOR1, which lists the highway number. Then click Symbol, select the U.S. Interstate HWY shield from the Symbol Selector dialog. Click Placement Properties in the Layer Properties dialog. On the Placement tab, check Horizontal for the orientation. Click OK to dismiss the dialogs. You are ready to plot the interstate shield symbols. Click the New Text (A) dropdown arrow on the Drawing toolbar and choose the Label tool. Opt to place label at position clicked. Move the Label tool over an interstate in the map. When a number appears in a box, left-click the mouse. (The

box. Double-click the Symbol next to U.S. Select the Major Road symbol but change its color to Ginger Pink. Click OK in both dialogs.

Q4. In Step 3, you were asked to use the Category option for displaying major roads. Can you use the Quantities/Graduated Symbols option instead? How will it differ from the Category option?

4. Select Properties from the context menu of *idlity*. On the Symbology tab, select the show option of Quantities and Graduated Symbols and select POPULATION for the Value field. Next change the number of classes from 5 to 3. Change the Range values by entering 28000 in the first class and 46000 in the second class. Click Template, and choose Solar Yellow for the color.

5. Labeling the cities is next. Click the View menu, point to Toolbars, and check Labeling to open the Labeling toolbar. Click the Label Manager button on the Labeling toolbar. In the Label Manager dialog, click *idlity* in the Label Classes frame and click the Add button in the Add label classes from symbology categories frame. Click Yes to overwrite the existing labeling classes. Expand *idlity* in the Label Classes frame. You should see the three categories by population.

6. Click the first symbology category (14302 – 28000). Make sure that the label field is CITY_NAME. Select Century Gothic (or another sans serif type) and 10 (size) for the text symbol. Click the SQL Query button. Change the first part of the query expression from "POPULATION" > 14302 to "POPULATION" >= 14302. Unless the change is made, the label for the city with the population of 14302 (Rexburg) will not appear. Click the second symbology category (28001 – 46000). Select Century Gothic and 12 for the text symbol. Click the third symbology category (46001 – 125659). Select Century Gothic, 12, and B (bold) for the text symbol. Make sure that the text *idlity* is

highway number may vary along the same interstate because the interstate has multiple numbers such as 90 and 10, or 80 and 30.)

10. Follow the same procedure in Step 9 but use the U.S. Route HWY shield to label U.S. highways. Switch to Layout View and make sure that the highway shield symbols are labeled appropriately. Because you have placed the highway shield symbols interactively, these symbols can be individually adjusted.

11. To complete the layout, you must add the title, legend, and other map elements. Switch to the layout view. Start with the title. Select Title from the Insert menu. When Enter Map Title appears in a box, click outside the box. When the outline of the box is shown in cyan, double-click it. On the Text tab of the Properties dialog, enter "Idaho Cities and Highways" in the text box. Click Change Symbol. Select Bookman Old Style (or another serif type), 18, and B for the text symbol. Move the title to the upper right of the layout.

12. Next is the legend. But before plotting the legend, you want to remove the layer names of *idlcity* and *idhwy*. Click *idlcity* in the table of contents, click it again, and delete it. Follow the same procedure to delete *idhwy*. You also want to change the field names of POPULATION and ROUTE_DESC to more descriptive subtitles. Click POPULATION in the table of contents, click it again, and change it to City Population. Change ROUTE_DESC to Highway Type.

13. Select Legend from the Insert menu. By default, the legend includes all layers from the map. Because you have removed the layer names of *idlcity* and *idhwy*, they appear as blank lines. *idoutl* shows the outline of Idaho and does not have to be included in the legend. You can remove *idoutl* from the legend by clicking *idoutl* in the Legend Items box and then the left arrow button. Click Next. In the second panel, highlight

Legend in the Legend Title box and delete it. (If you want to keep the word Legend on the map, do not delete it.) Skip the next two panels, and click Finish in the fifth panel. Move the legend to the upper right of the layout below the title.

14. A scale bar is next. Select Scale Bar from the Insert menu. Click Alternating Scale Bar 1, and then click Properties. On the Scale and Units tab, first select Adjust width when resizing and select Miles for division units. Then enter the division value of 50 (miles), select 2 for the number of divisions, and select 0 for the number of subdivisions. On the Numbers and Marks tab, select divisions from the Frequency dropdown list. On the Format tab, select Times New Roman from the Font dropdown list. Click OK to dismiss the dialogs. The scale bar appears in the map. Use the handles to place the scale bar below the legend.

15. A north arrow is next. Select North Arrow from the Insert menu. Choose ESRI North 6, a simple north arrow from the selector, and click OK. Place the north arrow below the scale bar.

16. Finally, change the design of the data frame. Right-click Task 2, and select Properties. Click the Frame tab. Select Double, Graded from the Border dropdown list. Click OK.

17. You can print the map directly, save the map as an ArcMap document, or export the map as a graphic file. Exit ArcMap.

Task 3: Label Streams

What you need: *charlie.shp*, a shapefile showing Santa Creek and its tributaries in north Idaho.

Task 3 lets you try the dynamic labeling method in ArcMap. Although the method can label all features on a map and remove duplicate names, it requires adjustments on some individual labels and overlapped labels. Therefore, Task 3 also requires you to use the Spline Text tool.

1. Launch ArcMap. Rename the data frame Task 3, and add *charlie.shp* to Task 3. Select Page and Print Setup from the File menu. Enter 5 (inches) for Width and 5 (inches) for Height. Click OK to dismiss the dialog.

2. Click the View menu, point to Toolbars, and check Labeling to open the Labeling toolbar. Click the Label Manager button on the Labeling toolbar. In the Label Manager dialog, expand *charlie* in the Label Classes frame and click Default. Make sure that the label field is NAME. Select Times New Roman, 10, and I for the text symbol. Notice that the default placement properties include a parallel orientation and an above position. Click Properties. The Placement tab repeats more or less the same information as in the Label Manager dialog. The Conflict Detection tab lists label weight, feature weight, and buffer for resolving the potential problem of overlapped labels. Check *charlie* in the Label Classes frame. Click OK to dismiss the dialogs.

Q5. List the position options available for line features.

3. Switch to the layout view. Click the Zoom Whole Page button. Use the control handles to fit the data frame within the specified page size. Select 100% from the Zoom Control dropdown list, and use the Pan tool to check the labeling of stream names. The result is generally satisfactory. But you may want to change the position of some labels such as Fagan Cr., Pamas Cr., and Short Cr. Consult Figure 10.14 for possible label changes.

4. Dynamic labeling, which is what you have done up to this point, does not allow individual labels to be selected and modified. To fix the placement of individual labels, you must convert labels to annotation. Right-click *charlie*, and select Convert Labels to Annotation. Select to save annotation in the map. Click Convert. Close the overflow annotation window.

5. To make sure that the annotation you add to the map has the same look as other labels,

you must specify the drawing symbol options. Click the Drawing dropdown arrow, point to Active Annotation Target, and check charlie anno. Click the Drawing arrow again, and select Default Symbol Properties. Click Text Symbol. In the Symbol Selector dialog, select Times New Roman, 10, and I. Click OK to dismiss the dialogs.

6. Switch to Data View. The following instructions use Fagan Cr. as an example. Zoom in the lower right of the map. Use the Select Elements tool to select Fagan Cr., and delete it. Click the New Text (A) dropdown arrow on the Drawing toolbar and choose the New Spline Text tool. Move the mouse pointer to below the junction of Brown Cr. and Fagan Cr. Click along the course of the stream, and double-click to end the spline. Enter Fagan Cr. in the Text box. Fagan Cr. appears along the clicked positions. You can use the same procedure to change other labels.

Challenge Task

What you need: *country.shp*, a world shapefile that has attributes on population and area of over 200 countries.

This challenge task asks you to map the population density distribution of the world.

1. Use POP_CNTRY (population) and SQMI_CNTRY (area in square miles) in *country.shp* to create a population density field. Name the field POP_DEN and calculate the field value by [POP_CNTRY] / [SQMI_CNTRY]. One of the records in *country.shp* has a population of −99999. Change its density value to 0.

2. Classify POP_DEN into 5 classes by using class breaks of your choice.

3. Prepare a layout of the map, complete with a title ("Population Density Map of the World"), a legend (with a legend description of "Persons per Square Mile"), and a neatline around the map.

REFERENCES

Antes, J. R., and K. Chang. 1990. An Empirical Analysis of the Design Principles for Quantitative and Qualitative Symbols. *Cartography and Geographic Information Systems* 17: 271–77.

Antes, J. R., K. Chang, and C. Mullis. 1985. The Visual Effects of Map Design: An Eye Movement Analysis. *The American Cartographer* 12: 143–55.

Arnheim, R. 1965. *Art and Visual Perception.* Berkeley, CA: University of California Press.

Brewer, C. A. 1994. Color Use Guidelines for Mapping and Visualization. In A. M. MacEachren and D. R. F. Taylor, eds., *Visualization in Modern Cartography,* pp. 123–47. Oxford, England: Pergamon Press.

Brewer, C. A. 2001. Reflections on Mapping Census 2000. *Cartography and Geographic Information Science* 28: 213–36.

Brewer, C. A., A. M. MacEachren, L. W. Pickle, and D. Herrmann. 1997. Mapping Mortality: Evaluating Color Schemes for Choropleth Maps. *Annals of the Association of American Geographers* 87: 411–38.

Chang, K. 1978. Measurement Scales in Cartography. *The American Cartographer* 5: 57–64.

Chirié, F. 2000. Automated Name Placement with High Cartographic Quality: City Street Maps. *Cartography and Geographic Information Science* 27: 101–10.

Cuff, D. J. 1972. Value versus Chroma in Color Schemes on Quantitative Maps. *Canadian Cartographer* 9: 134–40.

Dent, B. D. 1999. *Cartography: Thematic Map Design,* 5th ed. Dubuque, IA: WCB/ McGraw-Hill.

Holloway, S., J. Schumacher, and R. L. Redmond. 1999. People and Place: Dasymetric Mapping Using ARC/INFO. In S. Morain, ed., *GIS Solutions in Natural Resource Management: Balancing the Technical–Political Equation,* pp. 283–91, Santa Fe, NM: OnWord Press.

Kraak, M. J., and F. J. Ormeling. 1996. *Cartography: Visualization of Spatial Data.* Harlow, England: Longman.

Mersey, J. E. 1990. Colour and Thematic Map Design: The Role of Colour Scheme and Map Complexity in Choropleth Map Communication. *Cartographica* 27(3): 1–157.

Monmonier, M. 1996. *How to Lie with Maps,* 2d ed. Chicago: Chicago University Press.

Mower, J. E. 1993. Automated Feature and Name Placement on Parallel Computers. *Cartography and Geographic Information Systems* 20: 69–82.

Robinson, A. H., J. L. Morrison, P. C. Muehrcke, A. J. Kimerling, and S. C. Guptill. 1995. *Elements of Cartography,* 6th ed. New York: Wiley.

Slocum, T. A., R. B. McMaster, F. C. Kessler, and H. H. Howard. 2004. *Thematic Cartography and Geographic Visualization,* 2d ed. Upper Saddle River, NJ: Prentice Hall.

DATA EXPLORATION

CHAPTER 11

Starting data analysis in a GIS project can be overwhelming. The GIS database may have dozens of layers and hundreds of attributes. Where do you begin? What attributes do you look for? What data relationships are there? One way to ease into the analysis phase is data exploration. Centered on the original data, data exploration allows you to examine the general trends in the data, to take a close look at data subsets, and to focus on possible relationships between data sets. The purpose of data exploration is to better understand the data and to provide a starting point in formulating research questions and hypotheses.

An important component of data exploration consists of interactive and dynamically linked visual tools. Maps (both vector- and raster-based), graphs, and tables are displayed in multiple windows and dynamically linked so that selecting records from a table will automatically highlight the corresponding features in a graph and a map. Data exploration allows data to be viewed from different perspectives, making it easier for information processing and synthesis. Windows-based GIS packages, which can work with maps, graphs, and tables in different windows simultaneously, are well suited for data exploration.

Chapter 11 is organized into the following five sections. Section 11.1 discusses elements of data exploration. Sections 11.2 and 11.3 cover methods for exploring vector data: Section 11.2 focuses on attribute data query, and Section 11.3 concentrates on spatial data query and combination of attribute and spatial data queries. Section 11.4 turns to raster data query. Section 11.5 deals with geographic visualization and map-based data manipulation. Chapter 11 frequently refers to maps. But unlike maps in Chapter 10, which are designed for data display and presentation, maps in Chapter 11 serve as a tool for data exploration.

11.1 DATA EXPLORATION

Statisticians have traditionally used descriptive statistics and graphs to explore data structure and to discover data patterns. Since the 1970s, data exploration has expanded to include exploratory data analysis and dynamic graphics. **Exploratory data analysis** involves the use of a variety of graphic techniques for examining data. It is the first step in statistical analysis, serving as a precursor to more formal and structured data analysis (Tukey 1977; Tufte 1983). **Dynamic graphics** enhances exploratory data analysis by letting the user directly manipulate data points in charts and diagrams (Cleveland and McGill 1988; Cleveland 1993). These charts and diagrams are available in multiple and dynamically linked windows.

More recently, scientists have adopted the term **data visualization** to describe the use of a variety of exploratory techniques and graphics in understanding and gaining insight into the data. Buja et al. (1996) divide data visualization activities into two areas: rendering and manipulation. Rendering deals with the decision on what to show in a graphic plot and what type of plot to make. Manipulation refers to how to operate on individual plots and how to organize multiple plots. As for the purpose of data visualization activities, Buja et al. (1996) identify three fundamental tasks: finding Gestalt, posing queries, and making comparisons. Finding Gestalt means finding patterns and properties in the data set. Posing queries means exploring data characteristics in more detail by examining data subsets. And making comparison refers to comparisons between variables or between data subsets.

11.1.1 Descriptive Statistics

Descriptive statistics summarize the values of a data set and are therefore an integral part of data exploration. Descriptive statistics can be viewed in a table or incorporated into graphics. They include the following:

- Range: the difference between the minimum and maximum values.
- Median: the midpoint value, or the 50th percentile.
- First quartile: the 25th percentile.
- Third quartile: the 75th percentile.
- Mean: the average of data values. The mean can be calculated by

$$\sum_{i=1}^{n} x_i \Big/ n,$$ where x_i is the ith value and n is the number of values.

- Variance: the average of the squared deviations of each data value about the mean. The variance can be calculated by

$$\sum_{i=1}^{n} (x_i - \text{mean})^2 \Big/ n$$

- Standard deviation: the square root of the variance.
- Z score: a standardized score that can be computed by $(x - \text{mean})/s$, where s is the standard deviation.

GIS packages typically offer descriptive statistics in a menu selection that can apply to a numeric field. ArcGIS, for example, offers the statistics of minimum, maximum, range, mean, and standard deviation. Box 11.1 includes descriptive statistics of the rate of population change by state in the United States between 1990 and 2000. This data set is frequently used as an example in Chapter 11.

11.1.2 Graphs

Section 11.1.2 covers the types of graphs that are useful for data exploration. A line graph displays data as a line. The line graph example in Figure 11.1 shows the rate of population change in the United States along the y-axis and the state along the x-axis. Notice a couple of "peaks" in the line graph.

A bar chart, also called a histogram, groups data into equal intervals and uses bars to show the

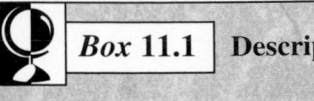

Box 11.1 | Descriptive Statistics

The following table shows, in the ascending order, the rate of population change by state in the United States from 1990 to 2000. The data set exhibits a skewness toward the higher end.

DC	–5.7	VT	8.2	AL	10.1	NM	20.1
ND	0.5	NE	8.4	MS	10.5	OR	20.4
WV	0.8	KS	8.5	MD	10.8	WA	21.1
PA	3.4	SD	8.5	NH	11.4	NC	21.4
CT	3.6	IL	8.6	MN	12.4	TX	22.8
ME	3.8	NJ	8.6	MT	12.9	FL	23.5
RI	4.5	WY	8.9	AR	13.7	GA	26.4
OH	4.7	HI	9.3	CA	13.8	ID	28.5
IA	5.4	MO	9.3	AK	14.0	UT	29.6
MA	5.5	KY	9.6	VA	14.4	CO	30.6
NY	5.5	WI	9.6	SC	15.1	AZ	40.0
LA	5.9	IN	9.7	TN	16.7	NV	66.3
MI	6.9	OK	9.7	DE	17.6		

The descriptive statistics of the data set are as follows:

- Mean: 13.45
- Median: 9.7
- Range: 72.0
- 1st quartile: 7.55, between MI (6.9) and VT (8.2)
- 3rd quartile: 17.15, between TN (16.7) and DE (17.6)
- Standard deviation: 11.38
- Z score for Nevada (66.3): 4.64

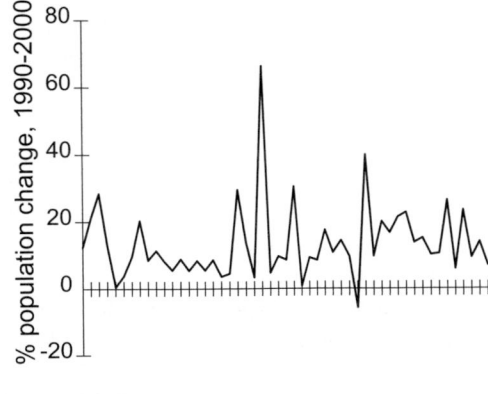

Figure 11.1
A line graph.

number or frequency of values falling within each class. A bar chart may have vertical bars or horizontal bars. Figure 11.2 uses a vertical bar chart to group rates of population change in the United States into six classes. Notice one bar at the high end of the histogram.

A cumulative distribution graph is one type of line graph that plots the ordered data values against the cumulative distribution values. The cumulative distribution value for the ith ordered value is typically calculated as $(i - 0.5)/n$, where n is the number of values. This computational formula converts the values of a data set to within the range of 0.0 to 1.0. Figure 11.3 shows a cumulative distribution graph.

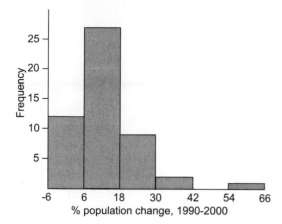

Figure 11.2
A histogram (bar chart).

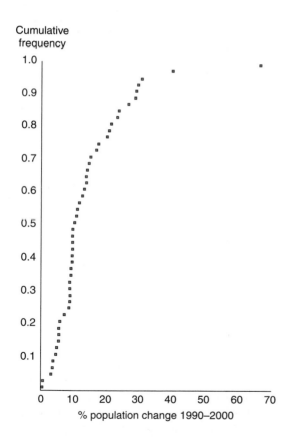

Figure 11.3
A cumulative distribution graph.

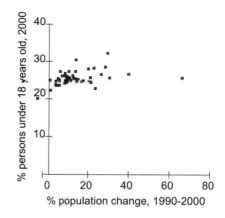

Figure 11.4
A scatterplot plotting percent persons 18 years old in 2000 against percent population change, 1990–2000. A weak positive relationship is present between the two variables.

A scatterplot uses markings to plot the values of two variables along the *x*- and *y*-axis. Figure 11.4 plots percent population change 1990–2000 against percent persons under 18 years old in 2000 by state in the United States. The scatterplot suggests a weak positive relationship between the two variables.

Bubble plots are a variation of scatterplots. Instead of using constant symbols as in a scatterplot, a bubble plot has varying-sized bubbles that are made proportional to the value of a third variable. Figure 11.5 is a variation of Figure 11.4: the additional variable shown by the bubble size is the state population in 2000. As an illustration, Figure 11.5 only shows states in the Mountain region, one of the nine regions defined by the U.S. Census Bureau.

Boxplots, also called the "box and whisker" plots, summarize the distribution of five statistics from a data set: the minimum, first quartile, median, third quartile, and maximum. By examining the position of the statistics in a boxplot, we can tell if the distribution of data values is symmetric or skewed and if there are unusual data points (i.e., outliers). Figure 11.6 shows a boxplot based on the rate of population change in the United States. This data set is clearly skewed toward the higher end. Figure 11.7 summarizes three basic types of data sets in terms of the distribution of data values.

Boxplots are therefore useful for comparisons between different data sets.

Some graphs are more specialized. Quantile–quantile plots, also called QQ plots, compare the cumulative distribution of a data set with that of some theoretical distribution such as the normal distribution, a bell-shaped frequency distribution. The points in a QQ plot fall along a straight line if the data set follows the theoretical distribution.

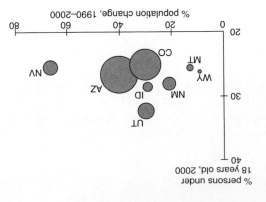

Figure 11.5

A bubble plot showing percent population change 1990–2000, percent persons under 18 years old in 2000, and state population in 2000.

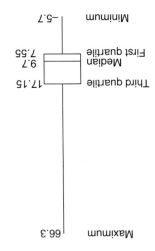

Figure 11.6

A boxplot based on the percent of population change, 1990–2000, data set.

Most GIS packages provide tools for making graphs and charts. ArcGIS, for example, has graphic tools in ArcMap and the Geostatistical Analyst extension (Box 11.2). Commercial statistical analysis packages such as SAS, SPSS, SYSTAT, S-PLUS (Insightful), and Excel also offer tools for making graphs and charts. For some projects, it may be beneficial to export data from a GIS to a statistical analysis package for data exploration (Scott 1994).

Some graphs are designed for spatial data. Figure 11.9, for example, shows a plot of spatial data values by raising a bar at each point location so that the height of the bar is proportionate to its value. This kind of plot allows the user to see the general trends among the data values in both the *x*-dimension (east–west) and *y*-dimension (north–south).

Figure 11.8 plots the rate of population change against the standardized value from a normal distribution. It shows that the data set is not normally distributed. The main departure occurs at the two highest values, which are also highlighted in previous graphs.

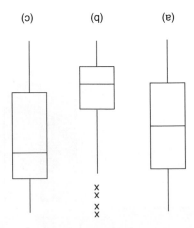

Figure 11.7

Boxplot (*a*) suggests that the data values follow a normal distribution. Boxplot (*b*) shows a positively skewed distribution with a higher concentration of data values near the high end. The x's in (*b*) may represent outliers, which are more than 1.5 box lengths from the end of the box. Boxplot (*c*) shows a negatively skewed distribution with a higher concentration of data values near the low end.

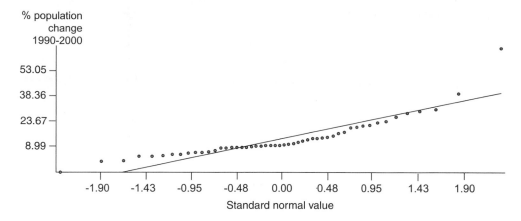

Figure 11.8
A QQ plot plotting percent population change, 1990–2000 against the standardized value from a normal distribution.

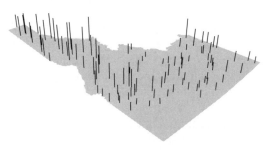

Figure 11.9
A 3-D plot showing the annual precipitation at 105 weather stations in Idaho. A north to south decreasing trend is apparent in the plot.

11.1.3 Dynamic Graphics

When graphs from Section 11.1.2 are displayed in multiple and dynamically linked windows, they become dynamic graphs. We can directly manipulate data points in dynamic graphs. For example, we can pose a query in one window and get the response in other windows, all in the same visual field. By viewing selected data points highlighted in multiple windows, we can hypothesize any patterns or relationships that may exist in the data. This is why multiple linked views have been described as the optimal framework for posing queries about data (Buja et al. 1996).

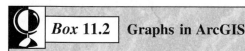

Box 11.2 | **Graphs in ArcGIS**

ArcMap has a tool menu for creating graph types of line graphs, bar charts, scatterplots, bubble plots, polar graphs, and hi-low-close graphs. The first four graph types are covered in the Chapter 11 text. A polar graph displays angular data (in degrees or radians) versus radial distances. The angular data are displayed on a circular angle axis, and the radial distances are extended from the center of the circle. A hi-low-close graph displays a range of *y* values at each *x* value. It is often used to show the high, low, opening, and closing of stock values, but it can also be used for other types of data such as air pollutant amounts at midmorning, noon, and midafternoon.

The Geostatistical Analyst extension has menu selections for creating histograms, QQ plots, 3-D plots, variograms, and other graphs for geostatistical analysis. The variogram is a scatterplot of semivariances, which is useful for detecting spatial dependency between data points. Chapter 16 covers use of variograms and other graphs for geostatistical analysis.

Brushing is a common method for manipulating dynamic graphs (Becker and Cleveland 1987). For example, we can graphically select a subset of points from a scatterplot, and view related data points in other graphics. Brushing can be extended to maps (Monmonier 1989). Figure 11.10 illustrates a brushing example that links a scatterplot and a map. Many GIS packages including ArcGIS have implemented brushing as part of the graphical user interface.

Other methods that can be used for manipulating dynamic graphics include rotation, deletion, and transformation of data points. Rotation of a 3-D plot lets the viewer see the plot from different perspectives. Deletions of data points (e.g., outliers) and data transformations (e.g., logarithmic transformation) are both useful for uncovering data relationships.

11.1.4 Data Exploration and GIS

Data exploration in GIS is similar to exploratory data analysis in statistics; it allows the user to explore data structure and to focus on a data subset of interest. But there are at least two important differences. First, data exploration in GIS involves both spatial data and attribute data. Spatial data are new in data exploration. A number of researchers have in fact suggested the term exploratory *spatial data analysis* and advocated the linking of data exploration to spatial analysis and, if necessary, to software specifically written for spatial analysis (Walker and Moore 1988; Haslett et al. 1990;

Batty and Xie 1994; Anselin 1999; Andrienko et al. 2001; Haining 2003).

Second, the media for data exploration in GIS naturally include maps and map features. Maps add a spatial perspective to data exploration. When we work with soil conditions, for example, we want to know not only how much of the study area is rated poor but also where those poor soils are distributed. To recognize the importance of maps in data exploration, some researchers have advocated geographic or cartographic visualization (MacEachren 1995; MacEachren et al. 1998; Crampton 2002). Geographic visualization emphasizes map-based manipulations as data visualization activities.

Chapter 11 incorporates the above differences into its organization. It has separate sections on attribute data query, spatial data query, and geographic visualization, but all aimed at data exploration.

11.2 ATTRIBUTE DATA QUERY

Attribute data query retrieves a data subset by working with attribute data. The selected data subset can be simultaneously examined in the table, displayed in charts, and linked to the highlighted features in the map. The selected data subset can also be saved for further processing.

Attribute data query requires the use of expressions, which must be interpretable by a GIS or

Figure 11.10

The scatterplot on the left is dynamically linked to the map on the right. The "brushing" of two data points in the scatterplot highlights the corresponding states (Washington and New Mexico) on the map.

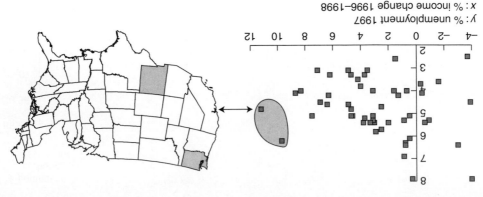

x: % income change 1996–1998
y: % unemployment 1997

Box 11.3 Query Methods in ArcGIS

Depending on the application, query expressions can vary in ArcGIS. The main difference is between the desktop application of ArcMap, which uses SQL, and ArcInfo Workstation's TABLES module, which uses INFO. Query operations in ArcMap can also vary. Query operations use SQL if the data set is a geodatabase feature class but a limited version of SQL if the data set is a coverage or a shapefile. Therefore, a geodatabase feature class allows greater flexibility and more query operations than a coverage or a shapefile. Additionally, the syntax can be different. For example, the fields in a query expression have square brackets around them if the fields belong to a geodatabase feature class but have double quotes if the fields belong to a coverage or a shapefile.

a database management system. The structure of these expressions varies from one system to another, although the general concept is the same. ArcGIS, for example, uses SQL (Structured Query Language) for query expressions (Box 11.3).

11.2.1 SQL (Structured Query Language)

SQL is a data query language designed for relational databases. IBM developed SQL in the 1970s, and many commercial database management systems such as Oracle, Informix, DB2, Access, and Microsoft SQL Server have since adopted the query language. A new development is to extend SQL to object-oriented database management systems and to spatial data (Shekhar and Chawla 2003).

To use SQL to access a database, we must follow the structure (i.e., syntax) of the query language. The basic syntax of SQL, with the keywords in italic, is

select <attribute list>

from <relation>

where <condition>

The *select* keyword selects field(s) from the database, the *from* keyword selects table(s) from the database, and the *where* keyword specifies the condition or criterion for data query. The following shows three examples of using SQL to query the tables in Figure 11.11. The Parcel table has the fields of PIN (string type), Sale_date (date type), Acres (double type), Zone_code (integer type), and Zoning

PIN	Owner_name
P101	Wang
P101	Chang
P102	Smith
P102	Jones
P103	Costello
P104	Smith

Relation 1: Owner

PIN	Sale_date	Acres	Zone_code	Zoning
P101	1-10-98	1.0	1	Residential
P102	10-6-68	3.0	2	Commercial
P103	3-7-97	2.5	2	Commercial
P104	7-30-78	1.0	1	Residential

Relation 2: Parcel

Figure 11.11
PIN (parcel ID number) relates the Owner and Parcel tables and allows use of SQL with both tables.

(string type), with the data type in parentheses. The Owner table has the fields of PIN (string type) and Owner_name (string type).

The first example is a simple SQL statement that queries the sale date of the parcel coded P101:

select Parcel.Sale_date
from Parcel
where Parcel.PIN = 'P101'

The prefix of Parcel in Parcel.Sale_date and Parcel.PIN indicate that the fields are from the Parcel table.

The second example queries parcels that are larger than 2 acres and are zoned commercial:

select Parcel.PIN
from Parcel
where Parcel.Acres > 2 AND
Parcel.Zone_code = 2

The fields used in the expression are all present in the Parcel table.

The third example queries the sale date of the parcel owned by Costello:

select Parcel.Sale_date
from Parcel, Owner
where Parcel.PIN = Owner.PIN AND
Owner_name = 'Costello'

This query involves two tables, which are joined first before the query. The *where* clause consists of two parts: the first part states that Parcel.PIN and Owner.PIN are the keys for the join operation, and the second part is the actual query expression.

SQL can be used to query a local database or an external database. But there are a couple of procedural differences. First, a GIS package may have already prepared the keywords of *select, from,* and *where* in the dialog for querying a local database. Therefore, we only have to enter the *where* clause (commonly called the query expression) in the dialog box. Second, an attribute query dialog in a GIS package is designed for a single table. If a query involves two tables, such as the third example above, they must be joined first. Therefore, the

dialog does not show the prefix to a field (e.g., Parcel in Parcel.PIN) if data query involves only one table. The prefixes only appear in the dialog if two or more tables have been joined for data query.

11.2.2 Query Expressions

Query expressions, or the *where* conditions, consist of Boolean expressions and connectors. A simple **Boolean expression** contains two operands and a logical operator. For example, Parcel.PIN = 'P101' is an expression in which PIN and P101 are operands and = is a logical operator. In this example, PIN is the name of a field, P101 is the field value used in the query, and the expression selects the record that has the PIN value of P101. Operands may be a field, a number, or a string. Logical operators may be equal to (=), greater than (>), less than (<), greater than or equal to (>=), less than or equal to (<=), or not equal to (<>).

Boolean expressions may contain calculations that involve operands and the arithmetic operators of $+$, $-$, $\times$, and $/$. Suppose length is a field measured in feet. We can use the expression, "length" $\times$ 0.3048 > 100, to find those records that have the length value of greater than 100 meters. Longer calculations, such as "length" $\times$ 0.3048 $-$ 50 > 100, evaluate the $\times$ and $/$ operators first from left to right and then the $+$ and $-$ operators. We can use parentheses to change the order of evaluation. For example, to subtract 50 from length before multiplication by 0.3048, we can use the following expression: ("length" $-$ 50) $\times$ 0.3048 > 100.

Boolean connectors are AND, OR, XOR, and NOT, which are used to connect two or more expressions in a query statement. A previous example uses AND to connect two expressions: Parcel.Acres > 2 AND Parcel.Zone_code = 2. Records selected from the statement must satisfy both Parcel.Acres > 2 and Parcel.Zone_code = 2. If the connector is changed to OR in the example, then records that satisfy either one or both of the expressions are selected. If the connector is changed to XOR, then records that satisfy one and only one of the expressions are selected. (XOR is functionally opposite to AND.) The connector NOT negates an expression so that a true expression is changed to false and vice versa. The statement, NOT Parcel.Acres > 2 AND Parcel.Zone_code = 2, for example, selects those parcels that are not larger than 2 acres and are zoned commercial.

Boolean connectors of NOT, AND, and OR are actually keywords used in the operations of Complement, Intersect, and Union on sets in probability. These operations are illustrated in Figure 11.12, with A and B representing two subsets of a universal set.

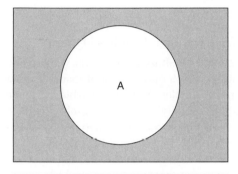

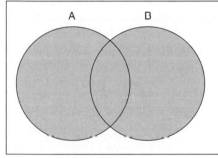

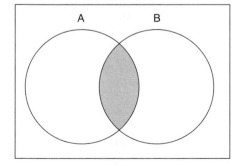

Figure 11.12
The shaded portion represents the complement of data subset A (top), the union of data subsets A and B (middle), and the intersection of A and B (bottom).

- The Complement of A contains elements of the universal set that do NOT belong to A.
- The Union of A and B is the set of elements that belong to A OR B.
- The Intersect of A and B is the set of elements that belong to both A AND B.

11.2.3 Type of Operation

Attribute data query begins with a complete data set. A basic query operation is to select a subset and divide the data set into two groups: one containing selected records and the other unselected records. Given a selected data subset, three types of operations can act on it: add more records to the subset, remove records from the subset, and select a smaller subset (Figure 11.13). Operations can also be performed between the selected and unselected subsets. We can switch between the selected and the unselected subsets, or we can clear the selection by bringing back all records.

These different types of operations allow greater flexibility in data query. For example, instead of using an expression of Parcel.Acres > 2 AND Parcel.Zone_code = 2, we can first use Parcel.Acres > 2 to select a subset and then use Parcel.Zone_code = 2 to select a subset from the previously selected subset. Although this example may be trivial, the combination of query expressions and operations can be quite useful for examining various data subsets, as shown in Section 11.2.4.

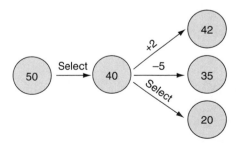

Figure 11.13
Three types of operations may be performed on the selected subset of 40 records: add more records to the subset (+2), remove records from the subset (−5), or select a smaller subset (20).

11.2.4 Examples of Query Operations

The following examples show different query operations using data from Table 11.1, which has 10 records and 3 fields.

Example 1: Select a Data Subset and Then Add More Records to It

[Create a new selection] "cost" >= 5 AND "soiltype" = 'Ns1'

0 of 10 records selected

[Add to current selection] "soiltype" = 'N3'

3 of 10 records selected

Example 2: Select a Data Subset and Then Switch Selection

[Create a new selection] "cost" >= 5 AND "soiltype" = 'Tn4' AND "area" >= 300

2 of 10 records selected

[Switch Selection]

8 of 10 records selected

Example 3: Select a Data Subset and Then Select a Smaller Subset from It

[Create a new selection] "cost" > 8 OR "area" > 400

4 of 10 records selected

[Select from current selection] "soiltype" = 'Ns1'

2 of 10 records selected

11.2.5 Relational Database Query

Relational database query works with a relational database, which may consist of many separate but interrelated tables. A query of a table in a relational database not only selects a data subset in the table but also selects records related to the subset in other tables. This feature is desirable in data exploration because it allows the user to examine related data characteristics from multiple linked tables.

To use a relational database, we must be familiar with the overall structure of the database, the designation of keys in relating tables, and a data dictionary listing and describing the fields in each table. For data query involving two or more tables, we can choose to either join or relate the tables. A join operation combines attributes from two or more tables into a single table. A relate operation dynamically links the tables but keeps the tables separate. When a record in one table is selected, the link will automatically select and highlight the corresponding record or records in the related tables. As discussed in Chapter 9, an important consideration in choosing a join or relate operation is the type of data relationship between tables. A join operation is appropriate for the one-to-one or many-to-one relationship but inappropriate for the one-to-many or many-to-many relationship. A relate operation, on the other hand, can be used with all four types of relationships.

The National Map Unit Interpretation Record (MUIR) database is a relational database developed by the Natural Resources Conservation Service

TABLE 11.1	A Data Set for Query Operation Examples				
Cost	Soiltype	Area	Cost	Soiltype	Area
1	Ns1	500	6	Tn4	300
2	Ns1	500	7	Tn4	200
3	Ns1	400	8	Ns3	200
4	Tn4	400	6	Ns3	100
5	Tn4	300	10	Ns3	100

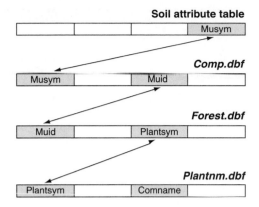

Figure 11.14
The keys relating three dBASE files in the MUIR database and the soil attribute table.

(NRCS) in the 1990s. The database contains over 80 estimated soil properties, interpretations, and performance data for each polygon on a soil map. Suppose we ask the following question: What types of plants, in their common names, are found where annual flooding frequency is rated as either frequent or occasional? To answer the question, we need the following three MUIR tables: the map unit components table or *comp.dbf*, which contains data on annual flood frequency; the woodland native plants table or *forest.dbf*, which has data on forest types; and a lookup table called *plantnm. dbf*, which has common plant names. The next step is to find the keys that can link the three tables (Figure 11.14). Finally, we link the attribute table of the soil map to *comp.dbf* by using musym (map unit symbol) as the key.

After the tables are related, we can issue the following query statement to *comp.dbf*: "anflood" = 'frequent' OR "anflood" = 'occas'. Evaluation of the query expression selects records in *comp.dbf* that meet the criteria, the corresponding records in *forest.dbf*, *plantnm.dbf*, and the feature attribute table, and the corresponding soil polygons in the map. This dynamic selection is possible because the tables are interrelated in the MUIR database and are dynamically linked to the map. A detailed description of relational database query is included in Task 4 of the applications section.

11.3 SPATIAL DATA QUERY

Spatial data query refers to the process of retrieving a data subset from a layer by working directly with features. We may select features using a cursor, a graphic, or the spatial relationship between features. As the geographic interface to the database, spatial data query complements attribute data query in data exploration.

Similar to attribute data query, results of spatial data query can be simultaneously inspected in the map, linked to the highlighted records in the table, and displayed in charts. They can also be saved as a new data set for further processing.

11.3.1 Feature Selection by Cursor

The simplest spatial data query is to select a feature by pointing at it or to select features by dragging a box around them.

11.3.2 Feature Selection by Graphic

This query method uses a graphic such as a circle, a box, a line, or a polygon to select features that fall inside or are intersected by the graphic object (Figure 11.15). We can draw the graphic for selection

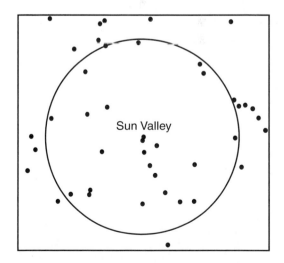

Figure 11.15
Select features by a circle centered at Sun Valley.

by using the mouse pointer in real time. Examples of query by graphic include selecting restaurants within a 1-mile radius of a hotel, selecting land parcels that intersect a proposed highway, and finding owners of land parcels within a proposed nature reserve.

11.3.3 Feature Selection by Spatial Relationship

This query method selects features based on their spatial relationships to other features. Features to be selected may be in the same layer as features for selection. Or, more commonly, they are in different layers. An example of the first type of query is to find roadside rest areas within a radius of 50 miles of a selected rest area; in this case, features to be selected and for selection are in the same layer. An example of the second type of query is to find rest areas within each county. Two layers are required for this query: one layer showing county boundaries and the other roadside rest areas.

Spatial relationships used in query include the following:

• **Containment**—selects features that fall completely within features for selection. Examples include finding schools within a selected county, and finding state parks within a selected state.

• **Intersect**—selects features that intersect features for selection. Examples include selecting land parcels that intersect a proposed road, and finding urban areas that intersect an active fault line.

• **Proximity**—selects features that are within a specified distance of features for selection. Examples include finding state parks within 10 miles of an interstate highway, and finding pet shops within 1 mile of selected streets. If features to be selected and features for selection share common boundaries and if the specified distance is 0, then proximity becomes **adjacency**. Examples of spatial adjacency include selecting land parcels that are adjacent to a flood zone, and finding vacant lots that are adjacent to a new theme park.

Box 11.4 shows the types of spatial relationships that can be used for spatial data query in ArcMap.

11.3.4 Combining Attribute and Spatial Data Queries

So far we have approached data exploration through attribute data query or spatial data query. In many cases data exploration requires both types of queries. For example, both are needed to find

---

Box 11.4 Expressions of Spatial Relationship in ArcMap

ArcMap handles feature selection by spatial relationship through the Select By Location dialog. The dialog requires the user to specify one or more layers whose features will be selected, and a layer whose features to be selected for selection. Eleven expressions of spatial relationships connect features to be selected and used for selection. These expressions may be subdivided by the relationships of containment, intersect, and proximity/adjacency:

• Containment: ''are completely within,'' ''completely contain,'' ''have their center in,'' ''contain,'' and ''are contained by.''

• Intersect: ''intersect'' and ''are crossed by the outline of.''

• Proximity/adjacency: ''are within a distance of,'' ''share a line segment with,'' ''touch the boundary of,'' and ''are identical to.''

A complete query expression in the Select By Location dialog may read as follows: ''I want to select features from *city* that are completely within the features in *quake*,'' where *city* is a city layer and *quake* is an earthquake-prone layer.

gas stations that are within 1 mile of a freeway exit in southern California and have an annual revenue exceeding $2 million each. Assuming that the layers of gas stations and freeway exits are available, there are at least two ways to solve the question.

1. Locate all freeway exits in the study area, and draw a circle around each exit with a 1-mile radius. Select gas stations within the circles through spatial data query. Then use attribute data query to find gas stations that have annual revenues exceeding $2 million.
2. Locate all gas stations in the study area, and select those stations with annual revenues exceeding $2 million through attribute data query. Next, use spatial data query to narrow the selection of gas stations to those within 1 mile of a freeway exit.

The first option queries spatial data and then attribute data. The process is reversed with the second option. Assuming that there are many more gas stations than freeway exits, the first option may be a better option, especially if the gas station map must be linked to other attribute tables for getting the revenue data.

The combination of spatial and attribute data queries opens wide the possibilities of data exploration. Some GIS users might even consider this kind of data exploration as data analysis because that is what they need to do to solve most of their routine tasks.

11.4 RASTER DATA QUERY

Although the concept and even some methods for data query are basically the same between raster data and vector data, there are enough practical differences to warrant a separate section on raster data query.

11.4.1 Query by Cell Value

The cell value in a raster typically represents a specific attribute value (e.g., land-use type, elevation value, etc.) at the cell location. Therefore, the operand in raster data query is the raster itself rather than a field as in the case of vector data query. This is illustrated in the following examples.

A raster data query uses a Boolean statement to separate cells that satisfy the query statement from cells that do not. The expression, [road] = 1, queries a road raster that has the cell value of 1. The operand [road] refers to the raster and the operand 1 refers to a cell value, which may represent the interstate category. This next expression, [elevation] > 1243.26, queries a floating-point elevation raster that has the cell value greater than 1243.26. Again the operand [elevation] refers to the raster itself. Because a floating-point elevation raster contains continuous values, querying a specific value is not likely to find any cell in the raster.

Raster data query can also use the Boolean connectors of AND, OR, and NOT to string together separate expressions. A compound statement with separate expressions usually applies to multiple rasters, which may be integer, or floating point, or a mix of both types. For example, the statement, ([slope] = 2) AND ([aspect] = 1), selects cells that have the value of 2 (e.g., 10–20% slope) in the slope raster, and 1 (e.g., north aspect) in the aspect raster (Figure 11.16). Those cells that satisfy the statement have the cell value of 1 on the output, while other cells have the cell value of 0.

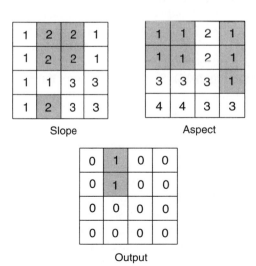

Figure 11.16
Raster data query involving two rasters: slope = 2 and aspect = 1. Selected cells are coded 1 and others 0 in the output raster.

Querying multiple rasters directly is unique with raster data. For vector data, all attributes to be used in a compound expression must be from a table or a joint attribute table. Another difference is that a GIS package such as ArcGIS has dialogs specifically designed for vector data query but does not have them for raster data query. The user interface for raster data query is often mixed with raster data analysis.

11.4.2 Query by Select Features

We can query a raster by using features such as points, circles, boxes, or polygons. The query returns an output raster with values for cells that correspond to the point locations or fall within the features for selection. Other cells on the output carry no data. Again, this type of raster data query shares the same user interface as data analysis. Chapter 13 provides more information on this topic.

11.5 GEOGRAPHIC VISUALIZATION

Geographic visualization, sometimes called cartographic visualization, refers to the use of maps for setting up a context for processing visual information and for formulating research questions or hypotheses (MacEachren 1995; MacEachren et al. 1998). Geographic visualization therefore has the same objective and the same types of interactivities as exploratory data analysis (Andrienko et al. 2002; Crampton 2002). Section 11.5 covers three methods that are map-based: data classification, spatial aggregation, and map comparison.

11.5.1 Data Classification

Data classification can be a tool for data exploration, especially if the classification is based on descriptive statistics. Suppose we want to explore rates of unemployment by state in the United States. To get a preliminary look at the data, we may place rates of unemployment into classes of above and below the national average (Figure 11.17a). Although generalized, the map divides the country into contiguous regions, which may suggest some regional factors for explaining unemployment.

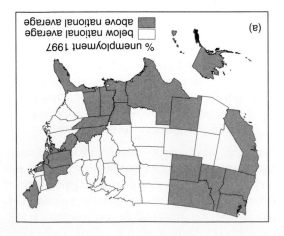

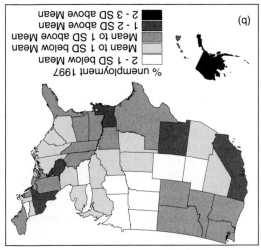

Figure 11.17

Two classification schemes: above or below the national average (a), and mean and standard deviation (SD) (b).

To isolate those states that are way above or below the national average, we can classify rates of unemployment by using the mean and standard deviation method (Figure 11.17b). We can now focus our attention on states that are, for example, more than 1 standard deviation above the mean.

Classified maps can be linked with tables, graphs, and statistics for more data exploration activities (Egbert and Slocum 1992). For example, we can link the maps in Figure 11.17 with a table showing percent change in median household income and find out whether states that have lower unemployment rates tend to have higher rates of income growth, and vice versa.

11.5.2 Spatial Aggregation

Spatial aggregation is functionally similar to data classification except that it groups data spatially. Figure 11.18 shows percent population change in the United States by state and by region. Used by the U.S. Census Bureau for data collection, regions are spatial aggregates of states. As shown in Figure 11.18*b*, a map by region gives a more general view of population growth in the country than a map by state does. Other geographic levels used by the U.S. Census Bureau are county, census tract, block group, and block. Because these levels of geographic units form a hierarchical order, we can explore the effect of spatial scaling by examining data at different spatial scales.

If distance is the primary factor in a study, we can aggregate spatial data by distance measures from points, lines, or areas. An urban area, for example, may be aggregated into distance zones away from the city center or from its streets (Batty and Xie 1994). Unlike the geography of census, these distance zones require additional data processing such as buffering and areal interpolation (Chapter 12).

Spatial aggregation for raster data means aggregating cells of the input raster to produce a coarser-resolution raster. For example, we can aggregate cells of a raster by a factor of 3. Each cell in the output raster corresponds to a 3-by-3 matrix in the input raster, and the cell value is a computed statistic such as mean, median, minimum, maximum, or sum from the nine input cell values.

11.5.3 Map Comparison

Map comparison can help a GIS user sort out the relationship between different maps. For example, the display of wildlife locations on a vegetation layer may reveal the association between the wildlife species and the distribution of vegetation covers (Figure 11.19).

If the maps to be compared consist of only point or line features, they can be coded in differ-

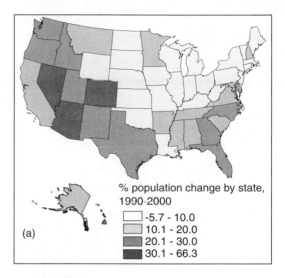

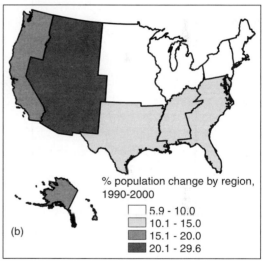

Figure 11.18
Two levels of spatial aggregation: by state (*a*), and by region (*b*).

ent colors and superimposed on one another in a single view. But this process becomes difficult if they include polygon features or raster data. One option is to use transparency as a visual variable (Chapter 10). A semitransparent layer allows another layer to show through. For example, to compare two raster layers, we can display one layer in a color scheme and the other in semitransparent

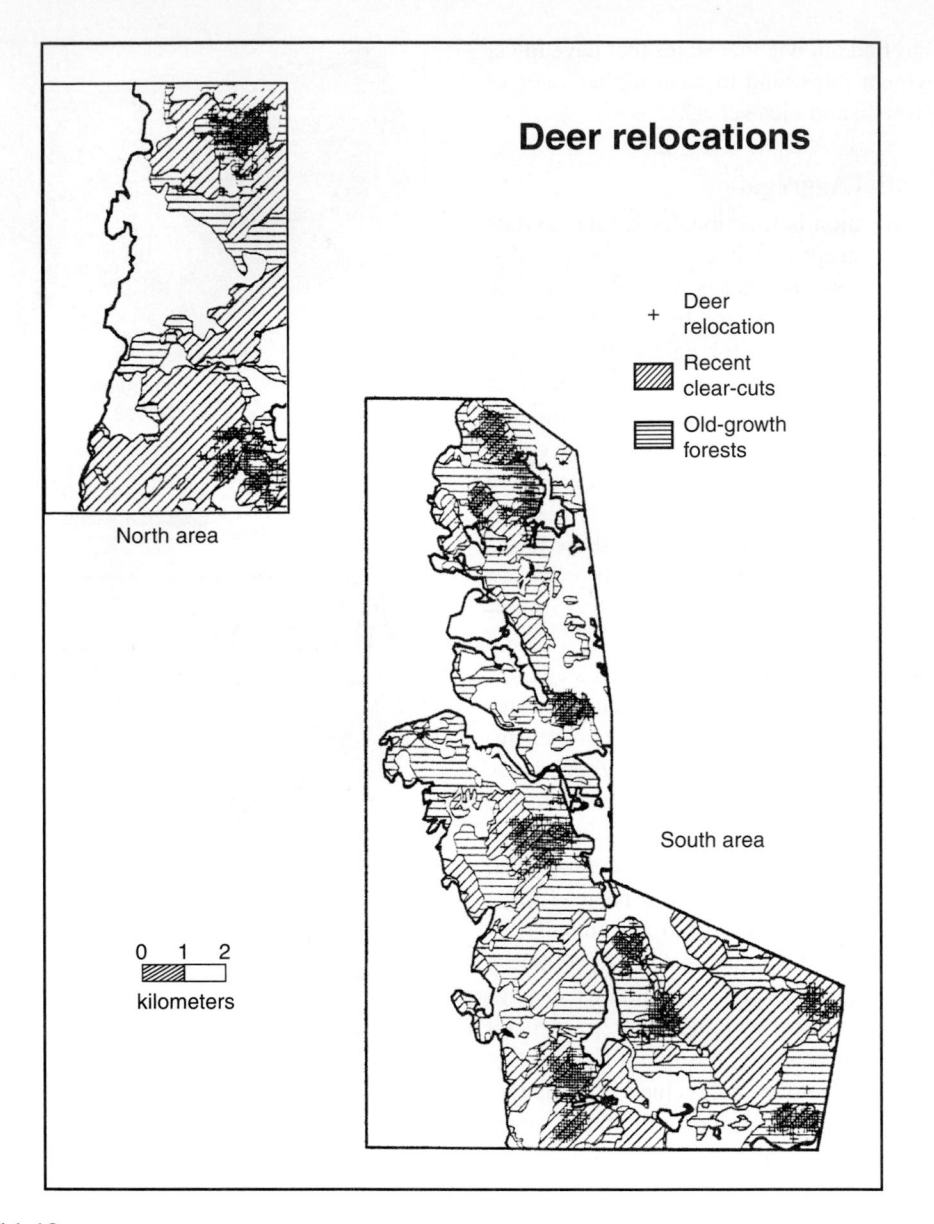

Figure 11.19
An example of map comparison. Deer relocations tend to be concentrated along the clear-cut/old forest edge.

shades of gray. The gray shades simply darken the color symbols and do not produce confusing color mixtures. Another example is to use transparency for displaying temporal changes such as land cover change between 1990 and 2000. Because one layer is semitransparent, we can follow the area extent of a land cover type from both years. But it is difficult to apply transparency to more than two layers.

There are three other options for comparing polygon or raster layers. The first option is to place all polygon and raster layers, along with other point and line layers, onto the screen but to turn on and off polygon and raster layers so that only one of them is viewed at a time. Used by many websites for interactive mapping, this option is designed for casual users.

The second option is to use a set of adjacent views, one for each polygon or raster layer, similar to the use of small multiples (Tufte 1983) or a scatterplot matrix (Cleveland 1993). This option is especially useful for observing temporal changes. For example, land cover maps of 1980, 1990, and 2000 for a study area can be placed in separate views so that the user can observe the changes over time. With maps for many points in time, one can even run map animation showing continuous changes (DiBiase et al. 1992; Weber and Buttenfield 1993; Peterson 1995).

The third option is to use map symbols that can show multiple data sets. A bar chart placed in each polygon can have two or more bars representing, for example, amounts of timber harvest from two or more years for comparison. Another example is the bivariate choropleth map (Meyer et al. 1975). Figure 11.20, for example, classifies the rate of unemployment and the rate of income change into either > or <= the national average and then shows the combinations of the two variables on a single map. To be readable, a bivariate

Figure 11.20

A bivariate map: rate of unemployment in 1997, either above or below the national average; and rate of income change, 1996–1998, either above or below the national average.

map cannot have too many classes for each variable. Otherwise, the lack of logical progression between the symbols can become a problem (Olson 1981).

KEY CONCEPTS AND TERMS

Adjacency: A spatial relationship that can be used to select features that share common boundaries.

Attribute data query: The process of retrieving data by working with attributes.

Boolean connector: A keyword such as AND, OR, XOR, or NOT that is used to construct compound expressions.

Boolean expression: A combination of a field, a value, and a logical operator, such as "class" = 2,

from which an evaluation of True or False is derived.

Brushing: A data exploration technique for selecting and highlighting a data subset in multiple views.

Containment: A spatial relationship that can be used in data query to select features that fall within specified features.

Data visualization: The process of using a variety of exploratory techniques and graphics in understanding and gaining insight into the data.

Dynamic graphics: A data exploration method, in which the user can directly manipulate data points in charts and diagrams that are displayed in multiple and dynamically linked windows.

Exploratory data analysis: Use of a variety of techniques such as charts, diagrams, and scatterplots to examine data as the first step in statistical analysis.

Geographic visualization: A method that uses maps to set up a context for visual information processing and for formulating research questions or hypotheses.

Intersect: A spatial relationship that can be used in data query to select features that intersect specified features.

Proximity: A spatial relationship that can be used to select features within a distance of specified features.

Relational database query: Query in a relational database, which not only selects a data subset in a table but also selects records related to the subset in other tables.

Spatial data query: The process of retrieving data by working with spatial features.

Structured Query Language (SQL): A data query and manipulation language designed for relational databases.

REVIEW QUESTIONS

1. Give an example of data exploration from your own experience.

2. Download % population change by county for your state between 1990 and 2000. (You can get the data from the Census Bureau's website, **http://www.census.gov/**, or from the GIS data clearinghouse website in your state.) Compute the median, first quartile, third quartile, mean, and standard deviation using the county data.

3. Use the county data and descriptive statistics from Question 2 to draw a boxplot. What kind of data distribution does the boxplot show?

4. Among the graphics presented in Section 11.1.2, which are designed for multivariate (i.e., two or more variables) visualization?

5. Figure 11.4 exhibits a weak positive relationship between % population change, 1990–2000 and % persons under 18 years old, 2000. What does a positive relationship mean in this case?

6. Describe brushing as a technique for data exploration.

7. Refer to Figure 11.11, and write a SQL statement to query the owner name of parcel P104.

8. Refer to Figure 11.11, and write a SQL statement to query the owner name of parcel P103 OR parcel P104.

9. Refer to Table 11.1, and fill in the blank for each of the following query operations:

[Create a new selection] "cost" > 8
_____ of 10 records selected

[Add to current selection] "soiltype" = 'N3' OR "soiltype" = 'Ns1'
_____ of 10 records selected

[Select from current selection] "area" > 400
_____ of 10 records selected

[Switch Selection]
_____ of 10 records selected

10. Refer to Box 11.4, and describe an example of using "intersect" for spatial data query.

11. Refer to Box 11.4, and describe an example of using "are contained by" for spatial data query.

12. You are given two digital maps of New York City: one shows landmarks, and the other

shows restaurants. One of the attributes of the restaurant layer lists the type of food (e.g., Japanese, Italian, etc.). Suppose you want to find a Japanese restaurant within 2 miles of Times Square. Describe the steps you will follow to complete the task.

13. Can you think of another solution for Question 12?

14. Refer to Figure 11.16. If the query statement is ([slope] = 1) AND ([aspect] = 3), how many cells in the output will have the cell value of 1?

15. Describe an example of using spatial aggregation for geographic visualization.

16. Describe an example of using map comparison for data exploration.

APPLICATIONS: DATA EXPLORATION

This applications section covers six tasks. Tasks 1 and 2 present an overview of data exploration in ArcMap: you will perform "select feature by location" in Task 1 and "select feature by graphic" in Task 2. In Task 3, you will query a joint attribute table; examine the query results using a magnifier window; and bookmark the area with the query results. Task 4 covers relational database query. Task 5 combines spatial and attribute data queries. Task 6 deals with raster data query.

Task 1: Select Feature by Location

What you need: *idcities.shp*, with 654 places in Idaho; and *snowsite.shp*, with 206 snow courses in Idaho and the surrounding states.

Task 1 lets you use the select feature by location method to select snow courses within 40 miles of Sun Valley, Idaho, and plot the snow station data in charts.

1. Start ArcCatalog, and make connection to the Chapter 11 database. Launch ArcMap. Add *idcities.shp* and *snowsite.shp* to Layers. Right-click Layers and select Properties. On the General tab, rename the data frame Tasks 1&2 and select Miles from the Display dropdown list.

2. Step 2 selects Sun Valley from *idcities*. Choose Select By Attributes from the Selection menu. Select *idcities* from the layer dropdown list and "Create a new selection" from the method list. Then enter in the expression box the following SQL statement: "CITY_NAME" = 'Sun Valley'.

(You can click Get Unique Values to get Sun Valley from the list.) Click Apply and close the dialog. Sun Valley is now highlighted in the map.

3. Choose Select By Location from the Selection menu. In the Select By Location dialog, choose "select features from" from the first dropdown list, check *snowsite*, choose "are within a distance of" from the second list, choose *idcities* from the third list, enter 40 miles for the distance to buffer, and click Apply. Snow courses that are within 40 miles of Sun Valley are highlighted in the map.

4. Right-click *snowsite* and select Open Attribute Table. Click Selected to show only the selected snow courses.

Q1. How many snow courses are within 40 miles of Sun Valley?

5. Now you will graph the elevation (ELEV) and maximum snow water equivalent (SWE_MAX) of the selected snow courses. Click the Tools menu, point to Graphs, and select Create. In the first panel of the Graph Wizard, click on the graph type of Scatter and the upper-left graph subtype. In the second panel, make sure that *snowsite* is the layer that contains data. Check ELEV, and uncheck all others, for the Y axis field. Check SWE_MAX only for the X axis field. In the third panel, enter Elev-SweMax for the title, uncheck Show Legend, and click Finish. A scatter

diagram of ELEV against SWE_MAX appears.

Q2. Describe the relationship between ELEV and SWE_MAX.

Task 2: Select Feature by Graphic

What you need: *idcities.shp* and *snowsite.shp*, as in Task 1.

Task 2 lets you use the select feature by graphic method to select snow courses within 40 miles of Sun Valley.

1. Make sure that the Tasks 1&2 data frame still has *idcities* and *snowsite* in ArcMap. Choose Clear Selected Features from the Selection menu to clear features selected in Task 1. Use the same procedure as in Task 1 to select Sun Valley from *idcities*.

2. This step draws a 40-mile circle around Sun Valley. Zoom in the area around Sun Valley. Click the New Rectangle dropdown arrow on the Drawing toolbar and select New Circle. Click Sun Valley in the map and drag the mouse pointer to draw a circle. (A radius reading in miles should appear in the lower left of the screen.) Right-click anywhere within the circle and select Properties. On the Size and Position tab, first click the center Anchor Point and then enter 128720 meters for the Size Width. (A width of 128720 meters is the same as a diameter of 80 miles or a radius of 40 miles.) Click OK. If the circle you have drawn is not right, you can delete it and redraw it.

3. Now you will use the circle to select features in *snowsite*. On the Selection tab in the table of contents, check *snowsite* and uncheck *idcities*. Switch back to the Display tab. Make sure that the circle is active with the handles; if not, click the circle to activate it. Click the Selection menu and click Select By Graphics. Snow courses that are within 40 miles of Sun Valley are highlighted in the map. Open the attribute table of *snowsite*. The number of the

selected snow courses should be the same as those in Task 1, unless you did not center the circle exactly at Sun Valley.

Task 3: Query Attribute Data from a Joint Attribute Table

What you need: *wp.shp*, a timber-stand shapefile; and *wpdata.dbf*, a dBASE file containing stand data.

Data query can be approached from either attribute data or spatial data. Task 3 focuses on attribute data query.

1. Insert a new data frame in ArcMap and rename it Task 3. Add *wp.shp* and *wpdata.dbf* to Task 3. Next join *wpdata* to *wp* by using ID as the key. Right-click *wp*, point to Joins and Relates, and select Join. In the Join Data dialog, opt to join attributes from a table, select ID for the field in the layer, select *wpdata* for the table, select ID for the field in the table, and click OK. Click Yes to create an index.

2. *wpdata* is now joined to the *wp* attribute table. Open the attribute table of *wp*. The table has two sets of attributes, distinguished by the prefix. Click the Options dropdown arrow and choose Select by Attributes. In the Select By Attributes dialog, make sure that the method is to create a new selection. Then enter the following SQL statement in the expression box: ''wpdata.ORIGIN'' > 0 AND ''wpdata.ORIGIN'' <= 1900. Click Apply.

Q3. How many records are selected?

3. Click Selected at the bottom of the table so that only the selected records are shown. Polygons of the selected records are also highlighted in the *wp* layer. To narrow the selected records, again choose Select By Attributes from the Options dropdown menu. In the Select By Attributes dialog, make sure that the method is to select from current selection.

Then prepare an SQL statement in the expression box that reads: "wpdata.ELEV" <= 30. Click Apply.

Q4. How many records are in the subset?

4. To take a closer look at the selected polygons in the map, click the Window menu and select Magnifier. When the magnifier window appears, click the window's title bar and drag the window over the map to see a magnified view.

5. Before moving to the next part of the task, select Clear Selection from the Options menu in the *wp* attribute table. Then choose Select By Attributes from the same menu. Enter the following SQL statement in the expression box: ("wpdata.ORIGIN" > 0 AND "wpdata.ORIGIN" <= 1900) AND "wpdata.ELEV" > 40. (The pair of parentheses is for clarity; it is not necessary to have them.) Four records are selected. The selected polygons are all near the top of the map. Use the Zoom In tool to zoom in on the selected polygons. You can bookmark the zoom-in area for future reference. Click the View menu, point to Bookmarks, and select Create. Enter *protect* for the Bookmark Name. To view the zoom-in area next time, click the View menu, point to Bookmarks, and select *protect*.

Task 4: Query Attribute Data from a Relational Database

What you need: *mosoils.shp*, a soil shapefile; *comp.dbf*, *forest.dbf*, and *plantnm.dbf*, three dBASE files from the national Map Unit Interpretation Record (MUIR) database developed by the Natural Resources Conservation Service (NRCS).

Task 4 lets you work with the MUIR database. By linking the tables in the database properly, you can explore many soil attributes in the database from any table. And, because the tables are linked to the soil map, you can also see where selected records are located.

1. Insert a new data frame in ArcMap and rename it Task 4. Add *mosoils.shp*, *comp.dbf*, *forest.dbf*, and *plantnm.dbf* to Task 4.

2. First, relate *mosoils* to *comp*. Right-click *mosoils* in the table of contents, point to Joins and Relates, and click Relate. In the Relate dialog, select MUSYM from the first dropdown list, select *comp* from the second list, select MUSYM from the third list, enter soil_comp for the relate name, and click OK.

3. Next prepare two other relates: forest_comp, relating *forest* to *comp* by using MUID as the common field; and forest_plantnm, relating *forest* to *plantnm* by using PLANTSYM as the common field.

4. The four tables (the *mosoils* attribute table, *comp*, *forest*, and *plantnm*) are now related in pairs by three relates. Right-click *comp* and select Open. Click the Options dropdown arrow and choose Select By Attributes. In the next dialog, create a new selection by entering the following SQL statement in the expression box: "ANFLOOD" = 'FREQ' OR "ANFLOOD" = 'OCCAS'. Click Apply. Click Selected at the bottom of the table so that only the selected records are shown.

Q5. How many records are selected in *comp*?

5. To see which records in the *mosoils* attribute table are related to the selected records in *comp*, go through the following steps. Click the Options dropdown arrow in the *comp* table, point to Related Tables, and click soil_comp: mosoils. The Attributes of *mosoils* table appears with the related records. And the *mosoils* map shows where those selected records are located.

6. You can follow the same procedure as in Step 5 to see which records in *forest* are related to those polygons that have frequent or occasional annual flooding.

7. Because no relate exists between *comp* and *plantnm*, you cannot open selected records in *plantnm* directly from *comp*. Instead, you can open selected records in *plantnm* from the Selected Attributes of *forest* table by using the related tables of forest_plantnm: plantnm.

Q6. How many polygons in *mosoils.shp* have a plant species with the COMNAME of "Idaho fescue"?

Task 5: Combine Spatial and Attribute Data Queries

What you need: *thermal.shp*, a shapefile with 899 thermal wells and springs; and *idroads.shp*, showing major roads in Idaho.

Task 5 assumes that you are asked by a company to locate potential sites for a hot springs resort in Idaho. You are given two criteria for selecting potential sites:

- The site must be within 2 miles of a major road.
- The temperature of the water must be greater than 60° C.

The field TYPE in *thermal.shp* uses *s* to denote springs and *w* to denote wells. The field TEMP shows the water temperature in ° C.

1. Insert a new data frame in ArcMap. Add *thermal.shp* and *idroads.shp* to the new data frame. Right-click the new data frame and select Properties. On the General tab, rename the data frame Task 5 and choose Miles from the Display dropdown list.

2. First select thermal springs and wells that are within 2 miles of major roads. Choose Select By Location from the Selection menu. Do the following in the Select By Location dialog: choose "Select features from" from the first dropdown list, check *thermal*, select "are within a distance of" from the second list, select *idroads* from the third list, and enter 2 (miles) for the distance to buffer. Click Apply. Thermal springs and wells that are within 2 miles of roads are highlighted in the map.

Q7. How many thermal springs and wells are selected?

3. Next, narrow the selection of map features by using the second criterion. Choose Select By Attributes from the Selection menu. Select *thermal* from the Layer dropdown list and "Select from current selection" from the Method list. Then enter the following SQL statement in the expression box: "TYPE" = 's'AND "TEMP" > 60. Click Apply.

4. Open the attribute table of *thermal*. Click Selected at the bottom of the attribute table so that only the selected records are shown. The selected records all have TYPE of *s* and TEMP above 60.

5. Map tips are useful for examining the water temperature of the selected hot springs. Right-click *thermal* in the table of contents and select Properties. On the Display tab, check the box to Show Map Tips (uses primary display field). On the Fields tab, select TEMP from the Primary display field dropdown list. Click OK to dismiss the Properties dialog. Click Select Elements on the standard toolbar. Move the mouse pointer to a highlighted hot spring location, and a map tip will display the water temperature of the spring.

Q8. How many hot wells and springs are within 5 kilometers of *idroads* and have temperatures above 70?

Task 6: Query Raster Data

What you need: *slope_gd*, a slope raster; and *aspect_gd*, an aspect raster.

Task 6 shows you different methods for querying a single raster or multiple rasters.

1. Select Data Frame from the Insert menu in ArcMap. Rename the new data frame Task 6, and add *slope_gd* and *aspect_gd* to Task 6.

2. Select Extension from the Tools menu and make sure that Spatial Analyst is checked. Click the View menu, point to Toolbars, and check Spatial Analyst. The Spatial Analyst toolbar appears. Select Raster Calculator from the Spatial Analyst menu. In the Raster Calculator dialog, prepare the following statement in the expression box: [slope_gd] = 2. (The = sign appears as == in the expression box.) Click Evaluate. The layer *Calculation* is added to the table of contents. Cells with the value of 1 are areas with slopes between 10 and 20 degrees.

Q9. How many cells in *Calculation* have the cell value of 1?

3. Select Raster Calculator from the Spatial Analyst menu, and prepare the following statement in the expression box: [slope_gd] = 2 AND [aspect_gd] = 4. (The word AND is shown as & in the expression box.) Click Evaluate. Cells with the value of 1 in *Calculation2* are areas with slopes between 10 and 20 degrees and the south aspect.

Q10. What percentage of area covered by the above two rasters has [slope_gd] = 3 AND [aspect_gd] = 3?

Challenge Task

What you need: *cities.shp*, a shapefile with 194 cities in Idaho; *counties.shp*, a county shapefile of Idaho; and *idroads.shp*, same as Task 5.

cities.shp has an attribute called CityChange, which shows the rate of population change between 1990 and 2000. *counties.shp* has attributes on 1990 county population (pop1990) and 2000 county population (pop2000). Add a new field to *counties.shp* and name the new field CoChange. Calculate the field values of CoChange by using the following expression: ([pop2000] − [pop1990]) × 100/[pop1990]. CoChange therefore shows the rate of population change between 1990 and 2000 at the county level.

Q1. What is the average rate of population change for cities that are located within 50 miles of Boise?

Q2. How many counties that intersect an Interstate highway have CoChange >= 30?

Q3. How many cities with CityChange >= 50 are located within counties with CoChange >= 30?

REFERENCES

Andrienko, N., G. Andrienko, A. Savinov, H. Voss, and D. Wettschereck. 2001. Exploratory Analysis of Spatial Data Using Interactive Maps and Data Mining. *Cartography and Geographic Information Science* 28: 151–65.

Andrienko, N., G. Andrienko, H. Voss, F. Bernardo, J. Hipolito, and U. Kretchmer. 2002. Testing the Usability of Interactive Maps in CommonGIS. *Cartography and Geographic Information Science* 29: 325–42.

Anselin, L. 1999. Interactive Techniques and Exploratory Spatial Data Analysis. In P. A. Longley, M. F. Goodchild, D. J. MaGuire, and D. W. Rhind, eds., *Geographical Information Systems*, 2d ed., pp. 253–66, New York: Wiley.

Batty, M., and Y. Xie. 1994. Modelling Inside GIS: Part I. Model Structures, Exploratory Spatial Data Analysis and Aggregation. *International Journal of Geographical Information Systems* 8: 291–307.

Becker, R. A., and W. S. Cleveland. 1987. Brushing Scatterplots. *Technometrics* 29: 127–42.

Buja, A., D. Cook, and D. F. Swayne. 1996. Interactive High-Dimensional Data Visualization. *Journal of Computational and Graphical Statistics* 5: 78–99.

Cleveland, W. S. 1993. *Visualizing Data*. Summit, NJ: Hobart Press.

Cleveland, W. S., and M. E. McGill, eds. 1988. *Dynamic Graphics for Statistics*. Belmont, CA: Wadsworth.

Crampton, J. W. 2002. Interactivity Types in Geographic Visualization. *Cartography and Geographic Information Science* 29: 85–98.

DiBiase, D., A. M. MacEachren, J. B. Krygier, and C. Reeves. 1992. Animation and the Role of Map Design in Scientific Visualization. *Cartography and Geographic Information Systems* 19: 201–14, 265–66.

Egbert, S. L., and T. A. Slocum. 1992. EXPLOREMAP: An Exploration System for Choropleth Maps. *Annals of the Association of American Geographers* 82: 275–88.

Haining, R. 2003. *Spatial Data Analysis: Theory and Practice.* Cambridge, England: Cambridge University Press.

Haslett, J., G. Wills, and A. Unwin. 1990. SPIDER—an Interactive Statistical Tool for the Analysis of Spatially Distributed Data. *International Journal of Geographical Information Systems 3:* 285–96.

MacEachren, A. M. 1995. *How Maps Work: Representation,*

Visualization, and Design. New York: Guilford Press.

MacEachren, A. M., F. P. Boscoe, D. Haug, and L. W. Pickle. 1998. Geographic Visualization: Designing Manipulable Maps for Exploring Temporally Varying Georeferenced Statistics. *Proceedings of the IEEE Symposium on Information Visualization,* pp. 87–94.

Meyer, M. A., F. R. Broome, and R. H. Schweitzer. 1975. Color Statistical Mapping by the U.S. Bureau of the Census. *American Cartographer 2:* 100–17.

Monmonier, M. 1989. Geographic Brushing: Enhancing Exploratory Analysis of the Scatterplot Matrix. *Geographical Analysis* 21: 81–84.

Olson, J. 1981. Spectrally Encoded Two-Variable Maps. *Annals of the Association of American Geographers* 71: 259–76.

Peterson, M. P. 1995. *Interactive and Animated Cartography.* Englewood Cliffs, NJ: Prentice Hall.

Scott, L. M. 1994. Identification of GIS Attribute Error Using Exploratory Data Analysis. *The Professional Geographer 46:* 378–86.

Shekhar, S., and S. Chawla. 2003. *Spatial Databases: A Tour.* Upper Saddle River, NJ: Prentice Hall.

Tufte, E. R. 1983. *The Visual Display of Quantitative Information.* Cheshire, CT: Graphics Press.

Tukey, J. W. 1977. *Exploratory Data Analysis.* Reading, MA: Addison-Wesley.

Walker, P. A., and D. M. Moore. 1988. SIMPLE—an Inductive Modeling and Mapping Tool for Spatially-Oriented Data. *International Journal of Geographical Information Systems 2:* 347–63.

Weber, C. R., and B. P. Butterfield. 1993. A Cartographic Animation of Average Yearly Surface Temperatures for the 48 Contiguous United States: 1897–1986. *Cartography and Geographic Information Systems* 20: 141–50.

12

VECTOR DATA ANALYSIS

CHAPTER OUTLINE

The scope of GIS analysis varies among disciplines that use GIS. GIS users in hydrology will likely emphasize the importance of terrain analysis and hydrologic modeling, whereas GIS users in wildlife management will be more interested in analytical functions dealing with wildlife point locations and their relationship to the environment. This is why GIS companies have taken two general approaches in packaging their products. One prepares a set of basic tools used by most GIS users, and the other prepares extensions designed for specific applications such as hydrologic modeling. Chapter 12 covers basic analytical tools for vector data analysis.

The vector data model uses points and their *x*-, *y*-coordinates to construct spatial features of points, lines, and polygons. Therefore vector data analysis uses the geometric objects of point, line, and polygon. And the accuracy of analysis results depends on the accuracy of these objects in terms of location and shape. Because vector data may be topology-based or nontopological, topology can also be a factor for some vector data analyses such as buffering and overlay.

As more tools are introduced in a GIS package, we must avoid confusion with use of these tools. A number of analytical tools such as Union and Intersect also appear as editing tools. Although the terms are the same, they perform different functions. As overlay tools, Union and Intersect work with both spatial and attribute data. But as editing tools, they only work with spatial data (i.e., geometries). Whenever appropriate, comments are made throughout Chapter 12 to note the differences.

Chapter 12 is grouped into the following five sections. Section 12.1 covers buffering and its applications. Section 12.2 discusses overlay, types of overlay, problems with overlay, and applications of overlay. Section 12.3 covers tools for measuring distances between points and between points and lines. Section 12.4 examines spatial statistics for pattern analysis. Section 12.5 includes tools for map manipulation.

12.1 BUFFERING

Based on the concept of proximity, **buffering** creates two areas: one area that is within a specified distance of select features and the other area that is beyond. The area that is within the specified distance is called the buffer zone.

Features for buffering may be points, lines, or areas (Figure 12.1). Buffering around points creates circular buffer zones. Buffering around lines creates a series of elongated buffer zones. And buffering around polygons creates buffer zones, extending outward from the polygon boundaries.

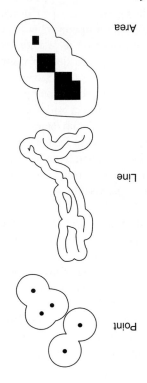

Figure 12.1
Buffering around points, lines, and areas.

12.1.1 Variations in Buffering

There are several variations in buffering from those of Figure 12.1. The buffer distance or buffer size does not have to be constant; it can vary according to the values of a given field (Figure 12.2). For example, the width of the riparian buffer can vary depending on its expected function and the intensity of adjacent land use (Box 12.1). A feature may have more than one buffer zone. As an example, a nuclear power plant may be buffered with distances of 5, 10, 15, and 20 miles, thus forming multiple rings around the plant (Figure 12.3). These buffer zones, although spaced equally from the plant, are not equal in area. The second ring from the plant in fact covers an area about three times larger than the first ring. One must consider this area difference if the buffer zones are part of an evacuation plan.

Buffering around line features does not have to be on both sides of the lines: it can be on either the left side or the right side of the line feature.

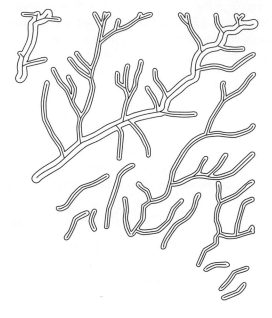

Figure 12.2
Buffering with different buffer distances.

Box 12.1 | **Riparian Buffer Width**

Riparian buffers are strips of land along the banks of rivers and streams that can filter polluted runoff and provide a transition zone between water and human land use. Riparian buffers are also complex ecosystems that can protect wildlife habitat and fisheries. Depending on what the buffer is supposed to protect or provide, the buffer width can vary. According to various reports, a width of 100 feet is necessary for filtering dissolved nutrients and pesticides from runoff. A minimum of 100 feet is usually recommended for protecting fisheries, especially cold-water fisheries. And at least 300 feet is required for protecting wildlife habitat. Many states in the United States have adopted policies of grouping riparian buffers into different classes by width.

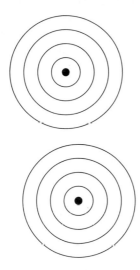

Figure 12.3
Buffering with four rings.

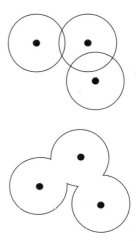

Figure 12.4
Buffer zones not dissolved (top) or dissolved (bottom).

(The left or right side is determined by the direction from the starting point to the end point of a line.) Likewise, buffer zones around polygons can be extended either outward or inward from the polygon boundaries. Boundaries of buffer zones may remain intact so that each buffer zone is a separate polygon. Or these boundaries may be dissolved so that there are no overlapped areas between buffer zones (Figure 12.4). Even the ends of buffer zones can be either rounded or flat.

Regardless of its variations, buffering uses distance measurements from select features to create the buffer zones. We must therefore know the measurement unit (e.g., meters or feet) and, if necessary, input that information prior to buffering. ArcGIS, for example, can use the map unit (i.e., the measurement unit defined for the data set) as the default distance unit or allow the user to select a different distance unit such as miles instead of feet. Because buffering uses distance measurements from spatial features, the positional accuracy of spatial features in a data set also determines the accuracy of buffer zones.

Most GIS packages offer buffering as an analysis tool. ArcGIS, for example, offers a buffer tool and a multiple ring buffer tool. Because buffering requires only distance measurements, some GIS packages including ArcGIS also offer buffering as an editing tool (Chapter 8). Such a tool can create a buffer zone around each select feature but does not have the buffering options such as dissolving overlapping buffer zone boundaries.

12.1.2 Applications of Buffering

Buffering creates a buffer zone data set, which sets the buffering operation apart from use of proximity measures for spatial data query (Chapter 11). Spatial data query using the proximity relationship can select spatial features that are located within a certain distance of other features but cannot create a buffer zone data set.

A buffer zone is often treated as a protection zone and is used for planning or regulatory purposes:

- A city ordinance may stipulate that no liquor stores or pornographic shops shall be within 1000 feet of a school or a church.
- Government regulations may set 2-mile buffer zones along streams to minimize sedimentation from logging operations.
- A national forest may restrict oil and gas well drilling within 500 feet of roads or highways.
- An urban planning agency may set aside land along the edges of streams to reduce the effects of nutrient, sediment, and pesticide runoff; to maintain shade to prevent the rise of stream temperature; and to provide shelter for wildlife and aquatic life (Thibault 1997).
- A resource agency may establish stream buffers or vegetated filter strips to protect aquatic resources from adjacent agricultural land-use practices (Castelle et al. 1994; Daniels and Gilliam 1996).

A buffer zone may be treated as a neutral zone and as a tool for conflict resolution. In controlling protesting groups, police may require protesters to be at least 300 feet from a building. Perhaps the best-known neutral zone is the demilitarized zone separating North Korea from South Korea along the 38° N parallel.

Sometimes buffer zones may represent the inclusion zones in GIS applications. For example, the siting criteria for an industrial park may stipulate that a potential site must be within 1 mile of a heavy-duty road. In this case, the 1-mile buffer zones of all heavy-duty roads become the inclusion zones.

Rather than serving as a screening device, buffer zones themselves may become the object for analysis. A forest management plan may define areas that are within 300 meters of streams as riparian zones and evaluate riparian habitat for wildlife (Iverson et al. 2001). Another example comes from urban planning in developing countries, where urban expansion typically occurs near existing urban areas and major roads. Management of future urban growth should therefore concentrate on the buffer zones of existing urban areas and major roads. Finally, buffering with multiple rings can be useful as a sampling method. A stream network can be buffered at a regular interval so that the composition and pattern of woody vegetation can be analyzed as a function of distance from the stream network (Schutt et al. 1999). One can also apply incremental banding to other studies such as land-use change around urban areas.

12.2 OVERLAY

An **overlay** operation combines the geometries and attributes of two feature layers to create the output. The geometry of the output represents the geometric intersection of features from the input layers. Figure 12.5 illustrates an overlay operation with two polygon layers. Each feature on the output contains a combination of attributes from the input layers, and this combination differs from its neighbors.

Feature layers to be overlaid must be spatially registered and based on the same coordinate system. In the case of the UTM (Universal Transverse Mercator) coordinate system or the SPC (State Plane Coordinate) system, the layers must also be in the same zone and have the same datum (e.g., NAD27 or NAD83).

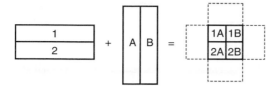

Figure 12.5
Overlay combines the geometries and attributes from two layers into a single layer. The dashed lines are for illustration only and are not included in the output.

12.2.1 Feature Type and Overlay

In practice, the first consideration for overlay is feature type. There are two groups of overlay operations. The first group uses two polygon layers as inputs. The second group uses one polygon layer and another layer, which may contain points or lines. Overlay operations can therefore be classified by feature type into point-in-polygon, line-in-polygon, and polygon-on-polygon. To distinguish the layers in the following discussion, the layer, which may be a point, line, or polygon layer, is called the input layer and the layer, which is a polygon layer, is called the overlay layer.

In a **point-in-polygon overlay** operation, the same point features in the input layer are included in the output but each point is assigned with attributes of the polygon within which it falls (Figure 12.6). For example, a point-in-polygon overlay can find the association between wildlife locations and vegetation types.

In a **line-in-polygon overlay** operation, the output contains the same line features as in the input layer but each line feature is dissected by the polygon boundaries on the overlay layer (Figure 12.7). Thus the output has more line segments than does the input layer. Each line segment on the output combines attributes from the input layer and the underlying polygon. For example, a line-in-polygon overlay can find soil data for a proposed road. The input layer includes the proposed road. The overlay layer contains soil polygons. And the output shows a dissected proposed road, each road segment having a different set of soil data from its adjacent segments.

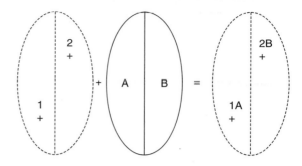

Figure 12.6
Point-in-polygon overlay. The input is a point layer. The output is also a point layer but has attribute data from the polygon layer.

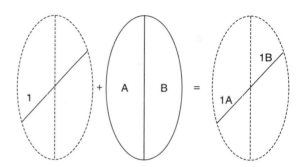

Figure 12.7
Line-in-polygon overlay. The input is a line layer. The output is also a line layer. But the output differs from the input in two aspects: the line is broken into two segments, and the line segments have attribute data from the polygon layer.

The most common overlay operation is **polygon-on-polygon,** involving two polygon layers. The output combines the polygon boundaries from the input and overlay layers to create a new set of polygons (Figure 12.8). Each new polygon carries attributes from both layers, and these attributes differ from those of adjacent polygons. For example, a polygon-on-polygon overlay can analyze the association between elevation zones and vegetation types.

two layers into a single polygon layer.

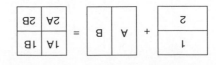

Figure 12.8
Polygon-on-polygon overlay. In the illustration, the two layers for overlay have the same area extent. The output combines the geometries and attributes from the two layers into a single polygon layer.

12.2.2 Overlay Methods

Although they may appear in different names in different GIS packages, all overlay methods are based on the Boolean connectors of AND, OR, and XOR. An overlay operation is called Intersect if it uses the AND connector. An overlay operation is called Union if it uses the OR connector. An over-lay operation that uses the XOR connector is called Symmetrical Difference or Difference. And an overlay operation is called Identity or Minus if it uses the following expression: [(input layer) AND (identity layer)] OR [(input layer) AND NOT (identity layer)]. The following ex- plains in more detail these four common overlay methods.

Union preserves all features from the inputs (Figure 12.9). The area extent of the output com- bines the area extents of both input layers. Union requires that both input layers be polygon layers.

Intersect preserves only those features that fall within the area extent common to the inputs (Figure 12.10). The input layers may contain dif- ferent feature types, although in most cases, one of them (the input layer) is a point, line, or polygon layer and the other (the overlay layer) is a polygon layer. Intersect is often a preferred method of over-

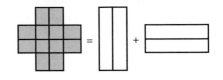

Figure 12.9
The Union method keeps all areas of the two input layers in the output.

lay because any feature on its output has attribute data from both of its inputs. For example, a forest management plan may call for an inventory of veg- etation types within riparian zones. Intersect will be a more efficient overlay method than Union in this case because the output contains only riparian zones and vegetation types within the zones.

Symmetrical Difference preserves features that fall within the area extent that is common to only one of the inputs (Figure 12.11). In other words, Symmetrical Difference is opposite to In- tersect in terms of the output's area extent. Sym- metrical Difference requires that both input layers be polygon layers.

Identity preserves features that fall within the area extent of the layer defined as the in- put layer (Figure 12.12). The other layer is called the identity layer. The input layer may contain points, lines, or polygons, and the identity layer is a polygon layer.

The choice of an overlay method becomes rel- evant only if the inputs have different area extents. If the input layers have the same area extent, then

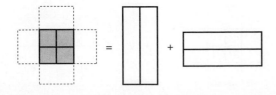

Figure 12.10
The Intersect method preserves only the area common to the two input layers in the output.

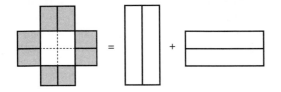

Figure 12.11
The Symmetrical Difference method preserves areas common to only one of the input layers in the output.

Box 12.2 | **Overlay Methods in ArcGIS**

The offering of overlay methods differs in ArcGIS depending on if the software in use is ArcView, the simplest version, or ArcInfo, the complete version. The Analysis Tools/Overlay toolset of ArcToolbox in ArcInfo includes Identity, Intersect, Symmetrical Difference, and Union. The same toolset in ArcView only has Intersect and Union. Another difference applies to the Intersect tool. ArcInfo accepts two or more input layers for an Intersect operation but ArcView can only have two input layers.

Because the inputs to an Intersect operation can be of different feature types, ArcInfo stipulates that the feature type of the output corresponds to the lowest dimension geometry of the inputs. For example, if the inputs consist of a point layer and two polygon layers, then the output is a point layer. But through overlay, the output point layer has attributes from the two polygon layers.

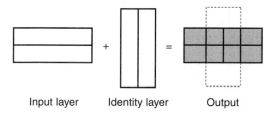

Input layer   Identity layer   Output

Figure 12.12

The Identity method produces an output that has the same extent as the input layer. But the output includes the geometry and attribute data from the identity layer.

that area extent also applies to the output. Box 12.2 describes overlay methods available in ArcGIS.

12.2.3 Overlay of Shapefiles

As the most recognizable, if not the most important, tool in GIS, overlay has a long history in GIS. Many concepts and methods developed for overlay are based on traditional vector data such as coverages from ESRI, Inc. Newer vector data such as shapefiles and geodatabase feature classes, also from ESRI, Inc., have been introduced since the 1990s. Overall, these newer vector data have not required a new way of dealing with overlay. But they have introduced some changes.

Unlike the coverage model, both shapefile and geodatabase allow polygons to have multiple com-

ponents, which may also overlap with one another. This means that overlay operations can actually be applied to a single feature layer: Union creates a new feature by combining different polygons, and Intersect creates a new feature from the area where polygons overlap. But when used with a single layer, Union and Intersect are basically editing tools for creating new features (Chapter 8). They are not overlay tools because they do not perform geometric intersections and combine attribute data from different layers.

Many shapefile users are aware of a problem with the overlay output: the area and perimeter values are not automatically updated. In fact, the output contains two sets of area and perimeter values, one from each input layer. Task 1 of the applications section shows how we can use a simple Visual Basic script to update these values. A geodatabase feature class, on the other hand, does not have the same problem because its default fields of area (shape_area) and perimeter (shape_length) are automatically updated.

12.2.4 Slivers

A common error from overlaying polygon layers is **slivers,** very small polygons along correlated or shared boundary lines (e.g., the study area boundary) of the input layers (Figure 12.13). The

in overlay.

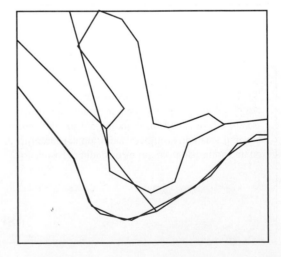

Figure 12.13
The top boundary has a series of slivers. These slivers are formed between the coastlines from the input layers in overlay.

existence of slivers often results from digitizing errors. Because of the high precision of manual digitizing or scanning, the shared boundaries on the input layers are rarely on top of one another. When the layers are overlaid, the digitized boundaries intersect to form slivers. Other causes of slivers include errors in the source map or errors in interpretation. Polygon boundaries on soil and vegetation maps are usually interpreted from field survey data, aerial photographs, and satellite images. A wrong interpretation can create erroneous polygon boundaries.

Most GIS packages incorporate some kind of tolerance in overlay operations to remove slivers. ArcGIS, for example, uses the **cluster tolerance,** which forces points and lines to be snapped together if they fall within the specified distance (Figure 12.14). The cluster tolerance is either defined by the user or based on a default value. Slivers that remain on the output of an overlay operation are those beyond the cluster tolerance. Therefore, one option to reduce the sliver problem is to increase the cluster tolerance. But because the cluster tolerance applies to the entire layer, large tolerances will likely snap shared bound-

aries as well as lines that are not shared on the input layers and create distorted features on the overlay output.

Other options for dealing with the sliver problem include data preprocessing and postprocessing. We can apply topology rules to the input layers, for example, to make sure that their shared boundaries are coincident before the overlay operation (Chapter 8). We can also apply the concept of minimum mapping unit after the overlay operation to remove slivers. The **minimum mapping unit** represents the smallest area unit that will be managed by a government agency or an organization. For example, if a national forest adopts 5 acres as its minimum mapping unit, then we can eliminate any slivers smaller than 5 acres by combining them with adjacent polygons (Section 12.5).

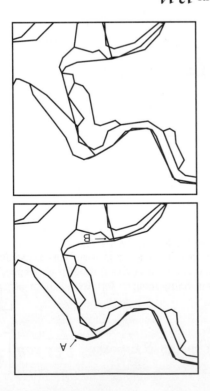

Figure 12.14
A cluster tolerance can remove many slivers along the top boundary (A) but can also snap lines that are not slivers (B).

12.2.5 Error Propagation in Overlay

Slivers are examples of errors in the inputs that can propagate to the analysis output. **Error propagation** refers to the generation of errors that are due to the inaccuracies of the input layers. Error propagation in overlay usually involves two types of error: positional and identification (MacDougall 1975; Chrisman 1987). Positional errors can be caused by the inaccuracies of boundaries that are due to digitizing or interpretation errors. Identification errors can be caused by the inaccuracies of attribute data such as the incorrect coding of polygon values. Every overlay product tends to have some combinations of positional and identification errors.

How serious can error propagation be? It depends on the number of input layers and the spatial distribution of errors in the input layers. The accuracy of an overlay output decreases as the number of input layers increases. And the accuracy decreases if the likelihood of errors occurring at the same locations in the input layers decreases.

An error propagation model proposed by Newcomer and Szajgin (1984) calculates the probability of the event that the inputs are correct on an overlay output. The model suggests that the highest accuracy that can be expected of the output is equal to that of the least accurate layer among the inputs, and the lowest accuracy is equal to:

(12.1)

$$1 - \sum_{i=1}^{n} \Pr(E_i')$$

where n is the number of input layers and $\Pr(E_i')$ is the probability that the input layer i is incorrect.

Suppose an overlay operation is conducted with three input layers and the accuracy levels of these layers are known to be 0.9, 0.8, and 0.7, respectively. According to Newcomer and Szajgin's model, we can expect the overlay output to have the highest accuracy of 0.7 and the lowest accuracy of 0.4, or $1 - (0.1 + 0.2 + 0.3)$. Newcomer and Szajgin's model illustrates the potential problem with error propagation in overlay. But it is a simple model, which, as shown in Box 12.3, can deviate significantly from the real-world operations.

12.2.6 Applications of Overlay

An overlay operation combines features and attributes from the input layers. The overlay output is useful for query and modeling purposes. Suppose an investment company is looking for a land parcel that is zoned commercial, not subject to flooding, and not more than 1 mile from a heavy-duty road. The company can first create the 1-mile road buffer and overlay the buffer zone layer with the zoning and

---

 Box 12.3 | **Error Propagation in Overlay**

Newcomer and Szajgin's model on error propagation in overlay is simple and easy to follow. But it remains largely a conceptual model. First, the model is based on square polygons, a much simpler geometry than real data sets used in overlay. Second, the model deals only with the Boolean operation of AND, that is, input layer 1 is correct AND input layer 2 is correct. The Boolean operation of OR in overlay is different because it requires that only one of the input layers be correct (i.e., input layer 1 is correct OR in-

put layer 2 is correct). Therefore, the probability of the event that the overlay output is correct actually increases as more input layers are overlaid (Veregin 1995). Third, Newcomer and Szajgin's model applies only to binary data, meaning that an input layer is either correct or incorrect. The model does not work with interval or ratio data and cannot measure the magnitude of errors. Modeling error propagation with numeric data is more difficult than binary data (Arbia et al. 1998; Heuvelink 1998).

flood plain layers. A subsequent query of the overlay output can select land parcels that satisfy the company's selection criteria. Many other examples of this type of application are included in the applications section of Chapter 12, and in Chapter 19.

A more specific application of overlay is to help solve the areal interpolation problem (Goodchild and Lam 1980). **Areal interpolation** involves transferring known data from one set of polygons (source polygons) to another (target polygons). For example, census tracts may represent source polygons with known populations in each tract from the U.S. Census Bureau, and school districts may represent target polygons with unknown population data. A common method for estimating the populations in each school district is as follows (Figure 12.15):

- Overlay the maps of census tracts and school districts.
- Query the areal proportion of each census tract that is within each school district.
- Apportion the population for each census tract to school districts according to the areal proportion.
- Sum the apportioned population from each census tract for each school district.

The above method for areal interpolation, however, assumes a uniform population distribu-

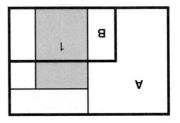

Figure 12.15

An example of areal interpolation. Thick lines represent census tracts and thin lines school districts. Census tract A has a known population of 4000 and B has 2000. The overlay result shows that the areal proportion of census tract A in school district 1 is 1/8 and the areal proportion of census tract B, 1/2. Therefore, the population in school district 1 can be estimated to be 1500, or $[(4000 \times 1/8) + (2000 \times 1/2)]$.

tion within each census tract, which is usually unrealistic. Recent studies have proposed statistical methods and use of ancillary information such as street segments for improving areal interpolation (Flowerdew and Green 1989; Goodchild et al. 1993; Flowerdew and Green 1995; Xie 1995).

12.3 DISTANCE MEASUREMENT

Distance measurement refers to measuring straight-line (euclidean) distances between features. Measurements can be made between points in a layer to points in another layer, or between each point in a layer to its nearest point or line in another layer. In both cases, distance measures are stored in a field. A GIS package such as ArcGIS may include distance measurement as an analysis tool or as a join operation between two tables (Box 12.4).

Distance measures can be used directly for data analysis. Chang et al. (1995), for example, use distance measures to test whether deer relocation points are closer to old-growth/clear-cut edges than random points located within the deer's relocation area. Fortney et al. (2000) use distance measures between home locations and medical providers to evaluate geographic access to health services.

Distance measures can also be used as inputs to data analysis. The Gravity Model, a spatial interaction model commonly used in migration studies and business applications, uses distance measures between points as the input. Pattern analysis covered in Section 12.4 also uses distance measures as inputs.

12.4 PATTERN ANALYSIS

Pattern analysis refers to the use of quantitative methods for describing and analyzing the distribution pattern of spatial features. At the general level, a pattern analysis can reveal if a distribution pattern is random, dispersed, or clustered. At the local level, a pattern analysis can detect if a distribution pattern contains local clusters of high or low values. Because pattern analysis can be a precursor to

Box 12.4 | Distance Measurement Using ArcGIS

One option for distance measurement in ArcGIS is called spatial join, or join data by location. The layer to assign data to must be a point layer. The layer to assign data from may be a point or line layer. And the spatial relationship between the two layers is nearest. The spatial join method joins attributes of the closest point or line to the appropriate record of the point attribute table. In addition, a new field called distance is added to the table, showing the distance measurement.

ArcGIS also has the Near and Point Distance tools for distance measurement. Near calculates the distance between each point in a point layer and its nearest point or line in another layer. Point Distance measures the distance between each point in a point layer and all points in another layer. Point Distance does not use the nearest relationship.

more formal and structured data analysis, some researchers have included pattern analysis as a data exploration activity (e.g., Haining 2003).

12.4.1 Nearest Neighbor Analysis

Nearest neighbor analysis uses the distance between each point and its closest neighboring point in a layer in determining if the point pattern is random, regular, or clustered. The nearest neighbor statistic is the ratio (R) of the observed average distance between nearest neighbors (d_{obs}) to the expected average for a hypothetical random distribution (d_{exp}):

(12.2)

$$R = \frac{d_{obs}}{d_{exp}}$$

The R ratio is less than 1 if the point pattern is more clustered than random, and greater than 1 if the point pattern is more dispersed than random. Nearest neighbor analysis can also produce a Z score, which indicates the likelihood that the pattern could be a result of random chance.

Figure 12.16 shows a point pattern of deer locations. The result of a nearest neighbor analysis shows an R ratio of 0.58 (more clustered than random) and a Z score of -11.4 (less than 1% likelihood that the pattern is a result of random chance).

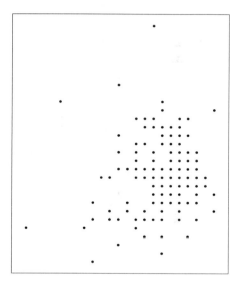

Figure 12.16
A point pattern showing deer locations.

12.4.2 Moran's I for Measuring Spatial Autocorrelation

Nearest neighbor analysis uses only distances between points as inputs. Analysis of spatial autocorrelation, on the other hand, considers both the point locations and the variation of an attribute at the locations. **Spatial autocorrelation** therefore measures the relationship among values of a variable according to the spatial arrangement of the values (Cliff and Ord 1973). The relationship may be described as highly correlated if like values are spatially close to each other, and independent or random if no pattern can be discerned from the arrangement of values. Spatial autocorrelation is also called *spatial association or spatial dependence*.

A popular measure of spatial autocorrelation is **Moran's I,** which can be computed by:

(12.3)

$$I = \frac{\sum_{i=1}^{n}\sum_{j=1}^{m} w_{ij}(x_i - \underline{x})(x_j - \underline{x})}{s^2 \sum_{i=1}^{n}\sum_{j=1}^{m} w_{ij}}$$

where x_i is the value at point i, x_j is the value at point i's neighbor j, w_{ij} is a coefficient, n is the number of points, and s^2 is the variance of x values with a mean of $\underline{x}$. The coefficient w_{ij} is the weight for measuring spatial autocorrelation. Typically, w_{ij} is defined as the inverse of the distance (d) between points i and j, or $1/d_{ij}$. Other weights such as the inverse of the distance squared can also be used.

The values Moran's I takes on are anchored at the expected value $E(I)$ for a random pattern:

(12.4)

$$E(I) = \frac{-1}{n-1}$$

$E(I)$ approaches 0 when the number of points n is large.

Moran's I is close to $E(I)$ if the pattern is random. It is greater than $E(I)$ if adjacent points tend to have similar values (i.e., spatially correlated) and less than $E(I)$ if adjacent points tend to have different values (i.e., not spatially correlated). Similar to nearest neighbor analysis, we can compute the Z score associated with a Moran's I. The Z score indicates the likelihood that the point pattern could be a result of random chance.

Figure 12.17 shows the same deer locations as Figure 12.16 but with the number of sightings as the attribute. This point distribution produces a Moran's I of 0.1, which is much higher than $E(I)$

of 0.00962, and a Z score of 12.7, which suggests that the likelihood of the pattern being a result of random chance is less than 1 percent.

The discussion has so far focused on point patterns. But Moran's I can also be applied to area patterns. Eq. (12.3) remains the same for computing the index value. The difference is the w_{ij} coefficient, which is now based on the spatial relationship between polygons. A common method is to assign 1 to w_{ij} if polygon i is adjacent to polygon j and 0 if i and j are not adjacent to each other.

Figure 12.18 shows the percentage of Latino population in Ada County, Idaho by block group. This data set produces a Moran's I of 0.05, which is higher than $E(I)$ of 0.00685, and a Z score of 6.7, which suggests that the likelihood of the pattern being a result of random chance is less than 1 percent. We can therefore conclude that adjacent block groups tend to have similar percentages of Latino population, either high or low. (With the exception of the two large but sparsely populated block groups to the south, the high percentages of Latino population are clustered near Boise.)

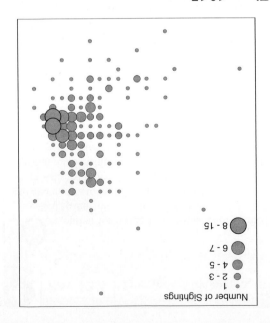

Number of Sightings
● 1
● 2 - 3
● 4 - 5
● 6 - 7
● 8 - 15

Figure 12.17
A point pattern showing deer locations and the number of sightings at each location.

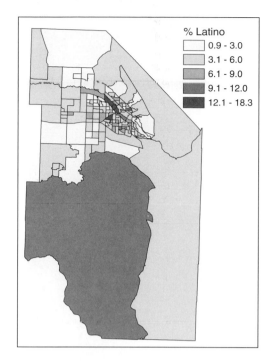

Figure 12.18
Percent Latino population by block group in Ada
County, Idaho. Boise is located in the upper center of
the map with small-sized block groups.

Recent developments in spatial statistics have
included **Local Indicators of Spatial Association
(LISA)** (Anselin 1995). A local version of
Moran's I, LISA calculates for each feature (point
or polygon) an index value and a Z score. A high
positive Z score suggests that the feature is adjacent to features of similar values, either above the
mean or below the mean. A high negative Z score
indicates that the feature is adjacent to features
of dissimilar values. Figure 12.19 shows a cluster
of highly similar values (i.e., high percentages of
Latino population) near Boise. Around the cluster
are small pockets of highly dissimilar values.

12.4.3 G-Statistic for Measuring High/Low Clustering

Moran's I, either general or local, can only detect
the presence of the clustering of similar values. It
cannot tell whether the clustering is made of high

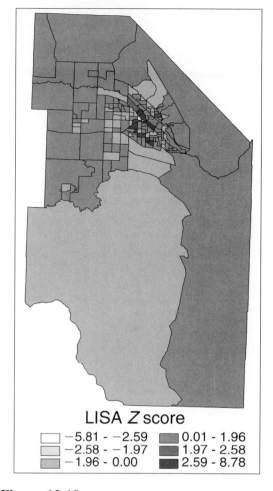

Figure 12.19
Z scores for the Local Indicators of Spatial Association
(LISA) by block group in Ada County, Idaho.

values or low values. This has led to the use of the
G-statistic, which can separate clusters of high
values from clusters of low values (Getis and Ord
1992). The general G-statistic based on a specified
distance, d, is defined as:

(12.5)

$$G(d) = \frac{\sum \sum w_{ij}(d)x_i x_j}{\sum \sum x_i x_j}, \quad i \neq j$$

where x_i is the value at location i, x_j is the value at
location j if j is within d of i, and $w_{ij}(d)$ is the spatial

weight. The weight can be based on some weighted distance (e.g., inverse distance) if i and j represent points, or 1 and 0 if i and j represent polygons.

The expected value of $G(d)$ is:

$$E(G) = \frac{\sum\sum w_{ij}(d)}{n(n-1)}$$

(12.6)

$E(G)$ is typically a very small value when n is large. A high $G(d)$ value suggests a clustering of high values, and a low $G(d)$ value suggests a clustering of low values. A Z score can be computed for a $G(d)$ to evaluate its statistical significance. The percentage of Latino population by block group in Ada County produces a general G-statistic of 0.0 and a Z score of 3.9, suggesting a spatial clustering of high values.

Similar to Moran's I, the local version of the G-statistic is also available (Ord and Getis 1995; Getis and Ord 1996). The local G-statistic, denoted by $G_i(d)$, is often described as a tool for "hot spot" analysis. A cluster of high positive Z scores suggests the presence of a cluster of high values or a hot spot. A cluster of high negative Z scores, on the other hand, suggests the presence of a cluster of low values or a cold spot. The local G-statistic also allows the use of a distance threshold d, defined as the distance beyond which no discernible increase in clustering of high or low values exists.

Figure 12.20 shows the Z scores of local G-statistics for the distribution of Latino population in Ada County. The map shows a clear hot spot in Boise but no cold spots in the county.

12.4.4 Applications of Pattern Analysis

Pattern analysis has a number of important applications. First, hot spot analysis has become a standard tool for analyzing crime locations. Standard software packages for crime mapping and analysis, such as CrimeStat (**http://www.icpsr.umich.edu/NACJD/crimestat.htm**), include local G-statistic for hot spot analysis. Spatial autocorrelation has also been applied to studies of SIDS (sudden infant death syndrome), AIDS, and housing prices (Getis and Ord 1992; Lam et al. 1996).

Second, spatial autocorrelation is useful for analyzing temporal changes of a spatial distribution. For example, we can use Moran's I or G-statistic to analyze distributions of Latino population in Ada County over time. The result may suggest a change of the distribution pattern over the past several decades.

Third, spatial autocorrelation is important for validating the use of standard statistical tests. Statistical tests typically require independent observations. If the data exhibit significant spatial autocorrelation, it should encourage the researcher

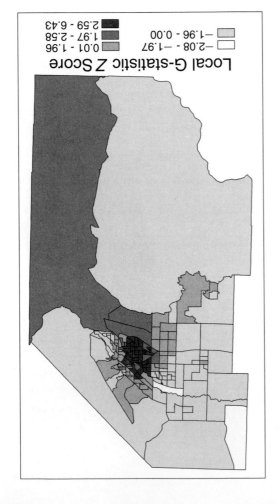

Local G-statistic Z Score

- 0.01 - 1.96
- 1.97 - 2.58
- 2.59 - 6.43
- −2.08 - −1.97
- −1.96 - 0.00

Figure 12.20

Z scores for the local G-statistics by block group in Ada County, Idaho.

to incorporate spatial dependence into the analysis (Legendre 1993).

12.5 MAP MANIPULATION

Tools are available in a GIS package for manipulating and managing maps in a database. Like buffering and overlay, these tools are considered basic GIS tools often needed for data preprocessing and data analysis. Map manipulation is easy to follow graphically, even though terms describing the various tools may differ between GIS packages. Box 12.5 summarizes map manipulation tools in ArcGIS.

Most tools covered in Section 12.5 involve two layers and are sometimes classified as overlay methods. Chapter 12, however, limits overlay to only those methods that can combine spatial and attribute data from two input layers into a single layer.

Dissolve aggregates features that have the same attribute value or values (Figure 12.21). For example, we can aggregate roads by highway number or counties by state. An important application of Dissolve is to simplify a classified polygon layer. Classification groups values of a selected attribute into classes and makes obsolete boundaries of adjacent polygons, which have different values initially but are now grouped into the same class. Dissolve can remove these unnecessary boundaries and creates a new, simpler layer with the classification results as its attribute values. Another application is to aggregate both spatial and attribute data of the input layer. For instance, to dissolve a county layer, we can choose state name as the attribute to dissolve and county population to ag-

gregate. The output is a state layer with an attribute showing the state population (i.e., the sum of county populations).

Clip creates a new layer that includes only those features of the input layer that fall within the area extent of the clip layer (Figure 12.22). Clip is

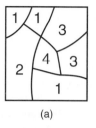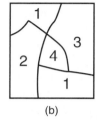

(a)   (b)

Figure 12.21
Dissolve removes boundaries of polygons that have the same attribute value in (*a*) and creates a simplified layer (*b*).

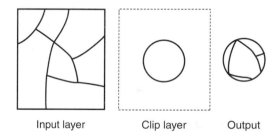

Input layer   Clip layer   Output

Figure 12.22
Clip creates an output that contains only those features of the input layer that fall within the area extent of the clip layer.

 Box **12.5**   **Map Manipulation Using ArcGIS**

Because of their diverse applications, map manipulation tools are found in different toolsets in ArcToolbox. Erase and Update are in the Analysis Tools/Overlay toolset. Clip, Select, and Split are in the Analysis Tools/Extract toolset. Append is in the Data Management Tools/General toolset. And Dissolve and Eliminate are in the Data Management Tools/Generalization toolset.

a useful tool, for example, for cutting a map acquired elsewhere to fit a study area. The input may be a point, line, or polygon layer, but the clip layer must be a polygon layer.

Append creates a new layer by piecing together two or more layers (Figure 12.23). For example, Append can put together a layer from four input layers, each corresponding to the area extent of a USGS 7.5-minute quadrangle. The output can then be used as a single layer for data query or display. But the boundaries separating the inputs still remain on the output and divide a feature into separate features if the feature crosses the boundary. Append is therefore different from edgematching (Chapter 8), which can create a seamless layer.

Select creates a new layer that contains features selected from a user-defined query expression (Figure 12.24). For example, we can create a layer showing high-canopy closure by selecting stands that have 60 to 80 percent closure from a stand layer.

Eliminate creates a new layer by removing features that meet a user-defined query expression (Figure 12.25). For example, Eliminate can implement the minimum mapping unit concept by

removing polygons that are smaller than the defined unit in a layer.

Update uses a "cut and paste" operation to replace the input layer by the update layer and its features (Figure 12.26). As the name suggests,

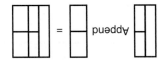

Figure 12.23

Append pieces together two adjacent layers into a single layer but does not remove the shared boundary between the layers.

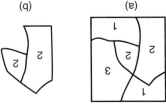

(a)

(b)

Figure 12.24

Select creates a new layer (b) with selected features from the input layer (a).

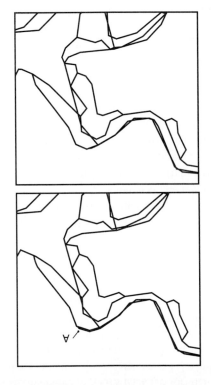

Figure 12.25

Eliminate removes some small slivers along the top boundary (A).

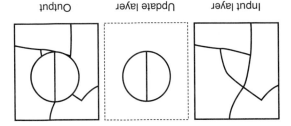

Input layer   Update layer   Output

Figure 12.26

Update replaces the input layer with the update layer and its features.

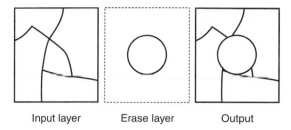

Input layer   Erase layer   Output

Figure 12.27
Erase removes features from the input layer that fall within the area extent of the erase layer.

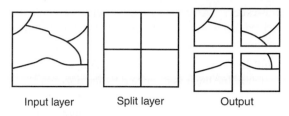

Input layer   Split layer   Output

Figure 12.28
Split uses the geometry of the split layer to divide the input layer into four separate layers.

Update is useful for updating an existing layer with new features in limited areas. It is a better option than redigitizing the entire map.

Erase removes from the input layer those features that fall within the area extent of the erase layer (Figure 12.27). Suppose a suitability analysis stipulates potential sites cannot be within 300 meters of any stream. A stream buffer layer can be used in this case as the erase layer to remove itself from further consideration.

Split divides the input layer into two or more layers (Figure 12.28). A split layer, which shows area subunits, is used as the template for dividing the input layer. For example, a national forest can split a stand layer by district so that each district office can have its own layer.

In ArcGIS, Clip and Split are also editing tools. These editing tools work with features rather than layers. For example, the editing tool of Split splits a line at a specified location or a polygon along a line sketch. The tool does not work with layers. It is therefore important that we understand the function of a tool before using it.

KEY CONCEPTS AND TERMS

Append: A GIS operation that creates a new layer by piecing together two or more layers.

Areal interpolation: A process of transferring known data from one set of polygons to another.

Buffering: A GIS operation that creates zones consisting of areas within a specified distance of select features.

Clip: A GIS operation that creates a new layer, which includes only those features of the input layer that fall within the area extent of the clip layer.

Cluster tolerance: A distance tolerance that forces points and lines to be snapped together if they fall within the specified distance.

Dissolve: A GIS operation that removes boundaries between polygons, which have the same attribute value(s).

Eliminate: A GIS operation that creates a new layer by removing features that meet a user-defined logical expression from the input layer.

Erase: A GIS operation that removes from the input layer those features that fall within the area extent of the erase layer.

Error propagation: The generation of errors in the overlay output that are due to the inaccuracies of the input layers.

G-statistic: A spatial statistic that measures the clustering of high and low values in a data set. The G-statistic can be either general or local.

Identity: An overlay method that preserves only features that fall within the area extent defined by the input layer.

Intersect: An overlay method that preserves only those features falling within the area extent common to the input layers.

Line-in-polygon overlay: A GIS operation, in which a line layer is dissected by the polygon boundaries on the overlay layer, and each line segment on the output combines attributes from the line layer and the polygon within which it falls.

Local Indicators of Spatial Association (LISA): The local version of Moran's I.

Minimum mapping unit: The smallest area unit that is managed by a government agency or an organization.

Moran's I: A statistic that measures spatial autocorrelation in a data set.

Nearest neighbor analysis: A spatial statistic that determines if a point pattern is random, regular, or clustered.

Overlay: A GIS operation that combines the geometries and attributes of the input layers to create the output.

Point-in-polygon overlay: A GIS operation, in which each point of a point layer is assigned attribute data of the polygon within which it falls.

Polygon-on-polygon overlay: A GIS operation, in which the output combines the polygon boundaries from the inputs to create a new set of polygons and each new polygon carries attributes from the inputs.

Select: A GIS operation that uses a logical expression to select features from the input layer for the output layer.

Slivers: Very small polygons found along the shared boundary of the two input layers in overlay.

Spatial autocorrelation: A spatial statistic that measures the relationship among values of a variable according to the spatial arrangement of the values. Also called *spatial association* or *spatial dependence*.

Split: A GIS operation that divides the input layer into two or more layers.

Symmetrical Difference: An overlay method that preserves features falling within the area extent that is common to only one of the input layers.

Union: A polygon-on-polygon overlay method that preserves all features from the input layers.

Update: A GIS operation that replaces the input layer by the update layer and its features.

REVIEW QUESTIONS

1. Define a buffer zone.

2. Describe *three* variations in buffering.

3. Provide an application example of buffering from your discipline.

4. Describe a point-in-polygon overlay operation.

5. A line-in-polygon overlay operation produces a line layer, which typically has more records (features) than the input line layer. Why?

6. Provide an example of a polygon-on-polygon overlay operation from your discipline.

7. Describe a scenario, in which Intersect is preferred over Union for an overlay operation.

8. Suppose the input layer shows a county and the overlay layer shows a national forest. Part of the county overlaps the national forest. We can express the output of an Intersect operation as [county] AND [national forest]. How can you express the outputs of a Union operation and an Identity operation?

9. Define slivers from an overlay operation.

10. What is a minimum mapping unit? And, how can a minimum mapping unit be used to deal with the sliver problem?

11. Although many slivers from an overlay operation represent inaccuracies in the digitized boundaries, they can also represent the inaccuracies of attribute data (i.e., identification errors). Provide an example for the latter case.

12. Both nearest neighbor analysis and Moran's I can apply to point features. How do they differ in terms of input data?

13. Explain spatial autocorrelation in your own words.

14. Both Moran's I and the G-statistic have the global (general) and local versions. How do these two versions differ in terms of pattern analysis?

15. The local G-statistic can be used as a tool for hot spot analysis. Why?

16. What does a Dissolve operation accomplish?

17. Suppose you have downloaded a vegetation map from the Internet. But the map is much larger than your study area. Describe the steps you will follow to get the vegetation map for your study area.

18. Suppose you need a map showing toxic waste sites in your county. You have downloaded a shapefile from the Environmental Protection Agency (EPA) website that shows toxic waste sites in every county of your state. What kind of operation will you use on the EPA map so that you can get only the county you need?

APPLICATIONS: VECTOR DATA ANALYSIS

This applications section has four tasks. Task 1 covers the basic tools of vector data analysis including Buffer, Overlay, and Select. Because ArcGIS does not automatically update the area and perimeter values of an overlay output in shapefile format, Task 1 also uses a Visual Basic script to calculate the area and perimeter values. Task 2 covers overlay operations with multicomponent polygons. Task 3 introduces two different options for measuring distances between point and line features. Task 4 deals with spatial autocorrelation.

Task 1: Perform Buffering and Overlay

What you need: shapefiles of *landuse*, *soils*, and *sewers*.

Task 1 simulates GIS analysis for a real-world project. The task is to find a suitable site for a new university aquaculture lab by using the following selection criteria:

- Preferred land use is brushland (i.e., LUCODE = 300 in *landuse.shp*).
- Choose soil types suitable for development (i.e., SUIT >= 2 in *soils.shp*).
- Site must be within 300 meters of sewer lines.

1. Start ArcCatalog, and make connection to the Chapter 12 database. Launch ArcMap.

Add *sewers.shp*, *soils.shp*, and *landuse.shp* to Layers, and rename Layers Task 1. All three shapefiles are measured in meters.

2. First buffer *sewers*. Click the Show/Hide ArcToolbox Window button to open the ArcToolbox window. Select Environments from the context menu of ArcToolbox, and set the Chapter 12 database to be the current workspace. Double-click the Buffer tool in the Analysis Tools/Proximity toolset. In the Buffer dialog, select *sewers* for the input features, enter *sewerbuf.shp* for the output feature class, enter 300 (meters) for the distance, select ALL for the dissolve type, and click OK. Open the attribute table of *sewerbuf*. The table has only one record for the dissolved buffer zone.

Q1. What is the definition of Side Type in the Buffer dialog?

3. Next overlay *soils*, *landuse*, and *sewerbuf*. Double-click the Intersect tool in the Analysis Tools/Overlay toolset. Select *soils*, *landuse*, and *sewerbuf* for the input features. Enter *final.shp* for the output feature class. Click OK to run the operation.

Q2. How is the cluster tolerance defined in the Intersect dialog?

Q3. How many records does *final* have?

4. The final step is to select from *final* those polygons that meet the first two criteria. Double-click the Select tool in the Analysis Tools/Extract toolset. Select *final* for the input features, name the output feature class *sites.shp*, and click the SQL button for Expression. In the Query Builder dialog, enter the following expression in the expression box: "SUIT" >= 2 AND "LUCODE" = 300. Click OK to dismiss the dialogs.

Q4. How many parcels are included in *sites*?

5. Open the attribute table of *sites*. Notice that the table contains two sets of area and perimeter. Moreover, each field contains duplicate values. This is because ArcGIS Desktop does not automatically update the area and perimeter values of the output shapefile. An easy option to get the updated values is to convert *sites.shp* to a geodatabase feature class. The feature class will have the updated values in the fields of shape_area and shape_length. For this task, you will use a Visual Basic script with four lines to perform the update.

6. Double-click the Add Field tool in the Data Management Tools/Fields toolset. Select *sites* for the input table, enter Shape_Area for the field name, select Double for the field type, enter 11 for the field precision, enter 3 for the field scale, and click OK. Use the same tool and the same field definition to add Shape_Leng as a new field to *sites*.

7. Right-click Shape_Area in the attribute table of *sites* and select Calculate Values. Click Yes to dismiss the next dialog. When the Field Calculator dialog appears, check Advanced. Enter the following VBA statements in the first text box under "Pre-Logic VBA Script Code":

```
Dim dblArea as Double
Dim pArea as IArea
```

The first line of the script declares the variable dblArea to store the updated area values. The second line declares the variable pArea, which points to the *IArea* interface. The *IArea* interface provides access to the area of a geometric object (a polygon in this case). The third line sets the pArea value to be that of Shape. And the fourth line assigns the *Area* property of pArea to dblArea. Before clicking OK to run the script, type dblArea in the second text box under "Shape_Area =".

```
Set pArea = [Shape]
dblArea = pArea.Area
```

8. The calculated area values should now appear in the attribute table. Now you are ready to calculate the perimeter values. Open the Field Calculator to calculate the field values of Shape_Leng. Enter the following VBA statements in the first text box:

```
Dim dblPerimeter as Double
Dim pCurve as ICurve
Set pCurve = [Shape]
dblPerimeter = pCurve.Length
```

Type dblPerimeter in the second text box under "Shape_Leng =". Click OK to run the script. The attribute table of *sites* should have the calculated perimeter values.

Q5. What is the Shape_Area value for FID 9 in *sites.shp*?

Task 2: Overlay Multicomponent Polygons

What you need: *boise_fire, fire1986,* and *fire1992,* three feature classes in the *regions* feature dataset of *boise_fire.mdb. boise_fire* records forest fires in the Boise National Forest from 1908 to 1996. *fire1986* fires in 1986, and *fire1992* fires in 1992. Task 2 lets you use multicomponent polygon features in overlay operations. Both *fire1986* and *fire1992* are polygon layers derived from *boise_fire.* Multicomponent polygons are similar to regions of the coverage model. An overlay operation using multicomponent polygons results in an output with fewer features (records), thus simplifying the data management task.

1. Insert a new data frame in ArcMap, and rename it Task 2. Add *regions* to Task 2. Open the attribute table of *boise_fire*. Historical fires are recorded by year in YEAR1 to YEAR6 and by name in NAME1 to NAME6. The multiple fields for year and name are necessary because a polygon can have multiple fires in the past. Open the attribute table of *fire1986*. It has only one record, although the layer actually contains 7 simple polygons. The same is true with *fire1992*.

2. First union *fire1986* and *fire1992* in an overlay operation. Double-click the Union tool in the Analysis Tools/Overlay toolset. Select *fire1986* and *fire1992* for the input features, and enter *fire_union* for the output feature class in the *regions* feature dataset. Click OK to run the operation. Open the attribute table of *fire_union*.

Q6. Explain what each record in *fire_union* represents.

3. Next intersect *fire1986* and *fire1992*. Double-click the Intersect tool in the Analysis Tools/Overlay toolset. Select *fire1986* and *fire1992* for the input features, and enter *fire_intersect* for the output feature class. Click OK to run the operation.

Q7. Explain what the single record in *fire_intersect* represents.

Task 3: Measure Distances Between Points and Lines

What you need: *deer.shp* and *edge.shp*.

Task 3 asks you to measure each deer location in *deer.shp* to its closest old-growth/clear-cut edge in *edge.shp*. There are two options to complete the task. The first option is to use the join data by location method, which is used for this task. The other option is to use the Near tool in the Analysis

Tools/Proximity toolset, which is explained at the end of the task.

1. Insert a new data frame in ArcMap, and rename it Task 3. Add *deer.shp* and *edge.shp* to Task 3.

2. Right-click *deer*, point to Joins and Relates, and select Join. Click the first dropdown arrow in the Join Data dialog, and select to join data from another layer based on spatial location. Make sure that *edge* is the layer to join to *deer*. Click the radio button stating that each point will be given all the attributes of the line that is closest to it, and a distance field showing how close that line is. Specify *deer_edge.shp* for the output shapefile. Click OK to run the operation.

3. Right-click *deer_edge* and open its attribute table. The field to the far right of the table is Distance, which lists for each deer location the distance to its closest edge.

Q8. How many deer locations are within 50 meters of their closest edge?

4. The Near tool uses a dialog to get the input features (*deer*), the near features (*edge*), and the optional search radius. The Near tool does not have the option of creating a new output data set. The result of the analysis is stored in the attribute table of *deer*, which may also include location (*x*- and *y*-coordinates of the near feature) and angle (the angle between the input and near features), in addition to distance.

Task 4: Compute General and Local G-Statistics

What you need: *adabg00.shp*, a shapefile containing block groups from Census 2000 for Ada County, Idaho.

In Task 4, you will first determine if a spatial clustering of Latino population exists in Ada

County. Then you will test if any local ''hot spots'' of Latino population exist in the county.

1. Insert a new data frame in ArcMap. Rename the new data frame Task 4, and add adabg00.shp to Task 4.

2. Right-click adabg00, and select Properties. On the Symbology tab, choose Quantities/Graduated colors to display the field values of Latino. Zoom in to the top center of the map, where Boise is located, and examine the spatial distribution of Latino population. The large block group to the southwest has a high percentage of Latino population (11%) but the block group's population is just slightly over 4600. The visual dominance of large area units is a shortcoming of the choropleth map.

Q9. What is the range of % Latino in Ada County?

3. Open ArcToolbox. You will first compute the general G-statistic. Double-click the High/Low Clustering (Getis-Ord General G) tool in the Spatial Statistics Tools/Analyzing Patterns toolset. Select adabg00 for the input feature class, select Latino for the input field, and check the box to display output graphically. Take the default for the other fields. Click OK to execute the command.

4. The graphic output shows the general G-statistic and its associated Z score. It also provides an interpretation of the result (i.e., clustering of high values) and the likelihood that the result is due to random chance (i.e., less than 1%). Close the graphic output.

5. Next you will run the local G-statistic. Double-click the Hot Spot Analysis (Getis-Ord Gi*) tool in the Spatial Statistics Tools/Mapping Clusters toolset. Select adabg00 for the input feature class. Enter local-g.shp for the output feature class in the Chapter 12 database, and specify a distance band of 5000 (meters). Click OK to execute the command.

6. Open the attribute table of local-g. The table is the same as adabg00 but has an added field of GihvDst. This new field stores the Z score of the local G-statistic for each block group.

Q10. What is the value range of GihvDst?

7. Right-click local-g, and select Properties. On the Symbology tab, choose Quantities/Graduated colors to display the field values of GihvDst. Now you can see a ''hot spot'' in Boise. Many block groups around Boise have negative Z scores. But except for one, they are not statistically significant (<−1.96). Therefore, we can conclude that there are no ''cold spots'' in the data set.

Challenge Task

What you need: lochsa.mdb, a personal geodatabase containing two feature classes for the Lochsa area of the Clearwater National Forest in Idaho.

lochsa_elk in the geodatabase has a field called USE that shows elk habitat use in summer or winter. ll_prod has a field called Prod that shows five timber productivity classes derived from the land type data, with 1 being most productive and 5 being least productive. Some polygons in ll_prod have the Prod value of −99 indicating absence of data. Also, lochsa_elk covers a larger area than ll_prod, due to the difference in data availability.

This challenge task asks you to prove the statement that ''the winter habitat area tends to have a higher area percentage of the productivity classes of 1 and 2 than the summer habitat area.'' Specifically, you are asked to calculate (1) the area percentage of the winter habitat area with the Prod value of 1 or 2, and (2) the area percentage of the summer habitat area with the Prod value of 1 or 2. The statement would be true if the percentage from (1) is higher than the percentage from (2).

Q1. What is the area percentage of the summer habitat area with the Prod value of 1 or 2?

Q2. What is the area percentage of the winter habitat area with the Prod value of 1 or 2?

REFERENCES

Anselin, L. 1995. Local Indicators of Spatial Association—LISA. *Geographical Analysis* 27: 93–116.

Arbia, G., D. A. Griffith, and R. P. Haining. 1998. Error Propagation Modeling in Raster GIS: Overlay Operations. *International Journal of Geographical Information Science* 12: 145–67.

Castelle, A. J., A. W. Johnson, and C. Conolly. 1994. Wetland and Stream Buffer Requirements: A Review. *Journal of Environmental Quality* 23: 878–82.

Chang, K., D. L. Verbyla, and J. J. Yeo. 1995. Spatial Analysis of Habitat Selection by Sitka Black-Tailed Deer in Southeast Alaska, USA. *Environmental Management* 19: 579–89.

Chrisman, N. R. 1987. The Accuracy of Map Overlays: A Reassessment. *Landscape and Urban Planning* 14: 427–39.

Cliff, A. D., and J. K. Ord. 1973. *Spatial Autocorrelation*. New York: Methuen.

Daniels, R. B., and J. W. Gilliam. 1996. Sediment and Chemical Load Reduction by Grass and Riparian Filters. *Soil Science Society of America Journal* 60: 246–51.

Flowerdew, R., and M. Green. 1989. Statistical Methods for Inference Between Incompatible Zonal Systems. In M. F. Goodchild and S. Gopal, eds., *Accuracy of Spatial Databases*, pp. 21–34. London: Taylor and Francis.

Flowerdew, R., and M. Green. 1995. Areal Interpolation and Types of Data. In S. Fotheringham and P. Rogerson, eds., *Spatial Analysis and GIS,* pp. 121–45. London: Taylor and Francis.

Fortney, J., K. Rost, and J. Warren. 2000. Comparing Alternative Methods of Measuring Geographic Access to Health Services. *Health Services & Outcomes Research Methodology* 1: 173–84.

Getis, A., and J. K. Ord. 1992. The Analysis of Spatial Association by Use of Distance Statistics. *Geographical Analysis* 24: 189–206.

Getis, A., and J. K. Ord. 1996. Local Spatial Statistics: An Overview. In P. Longley and M. Batty, eds., *Spatial Analysis: Modelling in a GIS Environment,* pp. 261–77. Cambridge, England: GeoInformation International.

Goodchild, M. F., and N. S. Lam. 1980. Areal Interpolation: A Variant of the Traditional Spatial Problem. *Geoprocessing* 1: 293–312.

Goodchild, M. F., L. Anselin, and U. Deichmann. 1993. A Framework for the Areal Interpolation of Socioeconomic Data. *Environment and Planning A* 25: 383–97.

Haining, R. 2003. *Spatial Data Analysis: Theory and Practice.* Cambridge, England: Cambridge University Press.

Heuvelink, G. B. M. 1998. *Error Propagation in Environmental Modeling with GIS*. London: Taylor and Francis.

Iverson, L. R., D. L. Szafoni, S. E. Baum, and E. A. Cook. 2001. A Riparian Wildlife Habitat Evaluation Scheme Developed Using GIS. *Environmental Management* 28: 639–54.

Lam, N. S., M. Fan, and K. Liu. 1996. Spatial-Temporal Spread of the AIDS Epidemic, 1982–1990: A Correlogram Analysis of Four Regions of the United States. *Geographical Analysis* 28: 93–107.

Legendre, P. 1993. Spatial Autocorrelation: Trouble or New Paradigm? *Ecology* 74: 1659–73.

MacDougall, E. B. 1975. The Accuracy of Map Overlays. *Landscape Planning* 2: 23–30.

Newcomer, J. A., and J. Szajgin. 1984. Accumulation of Thematic Map Errors in Digital Overlay Analysis. *The American Cartographer* 11: 58–62.

Ord, J. K., and A. Getis. 1995. Local Spatial Autocorrelation Statistics: Distributional Issues and an Application. *Geographical Analysis* 27: 286–306.

Schutt, M. J., T. J. Moser, P. J. Wigington, Jr., D. L. Stevens, Jr., L. S. McAllister, S. S. Chapman, and T. L. Ernst. 1999. Development of Landscape Metrics for Characterizing Riparian-Stream Networks. *Photogrammetric Survey and Remote Sensing* 65: 1157–67.

Thibault, P. A. 1997. Ground Cover Patterns Near Streams for Urban Land Use Categories. *Landscape and Urban Planning* 39: 37–45.

Veregin, H. 1995. Developing and Testing of an Error Propagation Model for GIS Overlay Operations. *International Journal of Geographical Information Systems* 9: 595–619.

Xie, Y. 1995. The Overlaid Network Algorithms for Areal Interpolation Problem. *Computer, Environment and Urban Systems* 19: 287–306.

CHAPTER 13

RASTER DATA ANALYSIS

The raster data model uses a regular grid to cover the space and the value in each grid cell to represent the characteristic of a spatial phenomenon at the cell location. This simple data structure of a raster with fixed cell locations not only is computationally efficient, but also facilitates a large variety of data analysis operations.

In contrast to vector data analysis, which is based on the geometric objects of point, line, and polygon, raster data analysis is based on cells and rasters. Raster data analysis can be performed at the level of individual cells, or groups of cells, or cells within an entire raster. Some raster data operations use a single raster; others use two or more rasters. An important consideration in raster data analysis is the type of cell value. Statistics such as mean and standard deviation are designed for numeric values, whereas others such as majority (the most frequent cell value) are designed for both numeric and categorical values.

Various types of data are stored in raster format. Raster data analysis, however, operates only on software-specific raster data such as ESRI grids. Therefore, to use digital elevation models (DEMs), satellite images, and other raster data in data analysis, they must be processed first and imported to software-specific raster data.

Chapter 13 covers the basic tools for raster data analysis. Section 13.1 describes the raster data analysis environment including the area for analysis and the output cell size. Sections 13.2 through 13.5 cover four common types of raster data analysis:

Box 13.1 | **How to Make an Analysis Mask**

The source of an analysis mask can be either a feature layer or a raster. An analysis mask, for example, can be the outline map of a study area (i.e., a feature layer), thus limiting the extent of raster data analysis to the study area. A raster to be used as an analysis mask must have cell values within the area of interest and no data on the outside. If necessary, reclassification can assign no data to the outside area.

Most raster data operations are available through either Spatial Analyst or ArcToolbox in ArcGIS. Spatial Analyst has the Options menu with an analysis mask option. ArcToolbox has the Environment menu with an analysis mask parameter. But as for ArcGIS 9.0, the two menus are not connected. Therefore, the analysis mask must be specified separately.

local operations, neighborhood operations, zonal operations, and physical distance measures. Section 13.6 covers operations that do not fit into the common classification of raster data analysis. Section 13.7 uses overlay and buffering as examples to compare vector- and raster-based operations.

13.1 DATA ANALYSIS ENVIRONMENT

The analysis environment refers to the area for analysis and the output cell size. The area extent for analysis may correspond to a specific raster, or an area defined by its minimum and maximum x-, y-coordinates, or a combination of rasters. Given a combination of rasters with different area extents, the area extent for analysis can be based on the union or intersect of the rasters. The union option uses an area extent that encompasses all input rasters, whereas the intersect option uses an area extent that is common to all input rasters.

An **analysis mask** can also determine the area extent for analysis. An analysis mask limits analysis to cells that do not carry the cell value of "no data." No data differs from zero. Zero is a valid cell value, whereas no data means the absence of data. For example, an elevation raster converted from a DEM often contains no-data cells along its border (Chapter 5). But in many other cases the

user enters no-data cells intentionally to limit the area extent for analysis. For example, one option to limit analysis of soil erosion to only private lands is to code public lands with no data. An analysis mask can be based on a feature layer or a raster (Box 13.1).

We can define the output cell size at any scale deemed suitable. Typically, the output cell size is set to be equal to, or larger than, the largest cell size among the input rasters. This follows the rationale that the accuracy of the output should correspond to that of the least accurate input raster. For instance, if the input cell sizes range from 10 to 30 meters, the output cell size should be 30 meters or larger. Given the specified output cell size, a GIS package uses a resampling technique to convert all input rasters to that cell size prior to data analysis. Common resampling methods are nearest neighbor, bilinear interpolation, and cubic convolution (Chapter 7).

13.2 LOCAL OPERATIONS

Constituting the core of raster data analysis, **local operations** are cell-by-cell operations. A local operation can create a new raster from either a single input raster or multiple input rasters. The cell values of the new raster are computed by a function relating the input to the output or are assigned by a classification table.

13.2.1 Local Operations with a Single Raster

Given a single raster as the input, a local operation computes each cell value in the output raster as a mathematical function of the cell value in the input raster. As shown in Figure 13.1, a large number of mathematical functions are available in a GIS package.

Converting a floating-point raster to an integer raster, for example, is a simple local operation, which uses the integer function to truncate the cell value at the decimal point on a cell-by-cell basis. Converting a slope raster measured in percent to one measured in degrees is also a local operation but requires a more complex mathematical expression. In Figure 13.2, the expression, $[slope\_d] = 57.296 \times \arctan ([slope\_p]/100)$, can convert $slope\_p$ measured in percent to $slope\_d$ measured in degrees. Because computer packages typically use radian instead of degree in trigonometric functions, the constant 57.296 $(360/2\pi, \pi = 3.1416)$ changes the angular measure to degrees.

13.2.2 Reclassification

A local operation, **reclassification** creates a new raster by classification. Reclassification is also referred to as recoding, or transforming, through lookup tables (Tomlin 1990). Two reclassification methods may be used. The first method is a one-to-one change, meaning that a cell value in the input raster is assigned a new value in the output raster. For example, irrigated cropland in a land-use raster is assigned a value of 1 in the output raster. The second method assigns a new value to a range of cell values in the input raster. For example, cells with population densities between 0 and 25 persons per square mile in a population density raster are assigned a value of 1 in the output raster and so on. An integer raster can be reclassified by either method, but a floating-point raster can only be reclassified by the second method.

Reclassification serves three main purposes. First, reclassification can create a simplified raster. For example, instead of having continuous slope values, a raster can have 1 for slopes of 0 to 10%, 2 for 10 to 20%, and so on. Second, reclassification can create a new raster that contains a unique category or value such as irrigated cropland or slopes of 10 to 20%. Third, reclassification can create a new raster that shows the ranking of cell values in the input raster. For example, a reclassified raster can show the ranking of 1 to 5, with 1 being least suitable and 5 being most suitable.

13.2.3 Local Operations with Multiple Rasters

Local operations with multiple rasters are also referred to as compositing, overlaying, or superimposing maps (Tomlin 1990). Another common term for local operations with multiple input rasters is **map algebra**. Because local operations can work with multiple rasters, they are the equivalent of vector-based overlay operations.

Figure 13.1
Arithmetic, logarithmic, trigonometric, and power functions for local operations.

Arithmetic	+, -, /, *, absolute, integer, floating-point
Logarithmic	exponentials, logarithms
Trigonometric	sin, cos, tan, arcsin, arccos, arctan
Power	square, square root, power

(a)

15.2	16.0	18.5
17.8	18.3	19.6
18.0	19.1	20.2

(b)

8.64	9.09	10.48
10.09	10.37	11.09
10.20	10.81	11.42

Figure 13.2
A local operation can convert a slope raster from percent (a) to degrees (b).

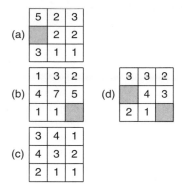

Figure 13.3

The cell value in (*d*) is the mean calculated from three input rasters (*a*, *b*, and *c*) in a local operation. The shaded cells have no data.

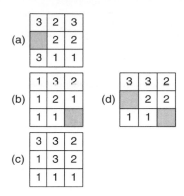

Figure 13.4

The cell value in (*d*) is the majority statistic derived from three input rasters (*a*, *b*, and *c*) in a local operation. The shaded cells have no data.

A greater variety of local operations have multiple input rasters than have a single input raster. Besides mathematical functions that can be used on individual rasters, other measures that are based on the cell values or their frequencies in the input rasters can also be derived and stored on the output raster. Some of these measures are, however, limited to rasters with numeric data.

Summary statistics, including maximum, minimum, range, sum, mean, median, and standard deviation, are measures that apply to rasters with numeric data. Figure 13.3, for example, shows a local operation that calculates the mean from three input rasters. If a cell contains no data in one of the input rasters, the cell also carries no data in the output raster.

Other measures that are suitable for rasters with numeric or categorical data are statistics such as majority, minority, and number of unique values. For each cell, a majority output raster tabulates the most frequent cell value among the input rasters, a minority raster tabulates the least frequent cell value, and a variety raster tabulates the number of different cell values. Figure 13.4, for example, shows the output with the majority statistics from three input rasters.

Some local operations do not involve statistics or computation. A local operation called Combine assigns a unique output value to each unique combination of input values. Suppose a slope raster has three cell values (0 to 20%, 20 to 40%, and greater

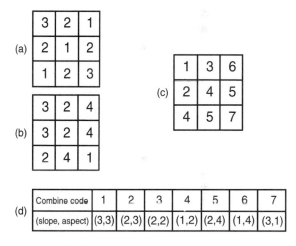

Figure 13.5

Each cell value in (*c*) represents a unique combination of cell values in (*a*) and (*b*). The combination codes and their representations are shown in (*d*).

than 40% slope), and an aspect raster has four cell values (north, east, south, and west aspects). The Combine operation creates an output raster with a value for each unique combination of slope and aspect, such as 1 for greater than 40% slope and the south aspect, 2 for (20 to 40%) slope and the south aspect, and so on (Figure 13.5).

$$A = R\,K\,L\,S\,C\,P \qquad (13.1)$$

where A is the average soil loss, R is the rainfall–runoff erosivity factor, K is the soil erodibility factor, L is the slope length factor, S is the slope steepness factor, C is the crop management factor, and P is the support practice factor. With each factor prepared as an input raster, we can multiply the rasters in a local operation to produce the output raster of average soil loss.

A study by Mladenoff et al. (1995) uses the logistic regression model for predicting favorable wolf habitat,

$$\text{logit}(p) = -6.5988 + 14.6189\,R, \qquad (13.2)$$

and

$$p = 1 / \left[1 + e^{\text{logit}(p)}\right]$$

where p is the probability of occurrence of a wolf pack, R is road density, and e is the natural exponent. logit (p) can be calculated in a local operation using a road density raster as the input. Likewise, p can be calculated in another local operation using logit (p) as the input. Chapter 19 discusses in more detail both the wolf habitat predictive model and RUSLE.

Because rasters are superimposed in local operations, error propagation can be an issue in

13.2.4 Applications of Local Operations

As the core of raster data analysis, local operations have many applications. A change detection study, for example, can use the unique combinations produced by the Combine operation to trace the change of the cell value (e.g., change of vegetation cover). But local operations are perhaps most useful for GIS models that require mathematical computation on a cell-by-cell basis.

The Revised Universal Soil Loss Equation (RUSLE) (Wischmeier and Smith 1978; Renard et al. 1997) uses six environmental factors in the equation:

Box 13.2 describes local operations available in ArcGIS and the organization of tools and commands for local operations in the software package.

Box 13.2 Local Operations in ArcGIS

B oth Spatial Analyst and ArcToolbox can perform local operations. Spatial Analyst is an extension originally developed for ArcView 3.x. ArcToolbox used to be a separate application but is now available in both ArcCatalog and ArcMap. The dialogs used in Spatial Analyst and ArcToolbox are very similar except that, for most operations, Spatial Analyst creates a temporary raster whereas ArcToolbox creates a permanent raster on disk. The major advantage of using ArcToolbox is that we can choose to use dialogs, commands, models, or scripts.

Spatial Analyst's menu includes Cell Statistics, Reclassify, and Raster Calculator. Cell Statistics can compute various statistics from multiple rasters. Reclassify performs data reclassification. Raster Calculator is a versatile tool, which can take a single

raster or multiple rasters as the input. The Raster Calculator dialog offers a variety of operators and functions: arithmetic operators, logical operators, Boolean connectors, and mathematical functions (arithmetic, logarithmic, trigonometric, and power functions). We can use these operators and functions directly to prepare a local operation statement, or use additional functions (e.g., Combine) from the Spatial Analyst functional reference in the Raster Calculator.

Spatial Analyst Tools of ArcToolbox offers local operations in three toolsets: Local, Map Algebra, and Reclass. The Local toolset includes tools such as Cell Statistics and Combine. The Map Algebra toolset has a Single Output Map Algebra tool that can use multiple input rasters. And the Reclass toolset has a number of tools for data reclassification.

interpreting the output. Raster data do not directly involve digitizing errors. (If raster data are converted from vector data, digitizing errors with vector data are carried into raster data.) Instead, the main source of errors is the quality of the cell value, which in turn can be traced to other data sources. For example, if raster data are converted from satellite images, statistics for assessing the classification accuracy of satellite images can be used to assess the quality of raster data (Congalton 1991; Veregin 1995). But these statistics are based on binary data (i.e., correctly or incorrectly classified). It is more difficult to model error propagation with interval and ratio data (Heuvelink 1998).

13.3 NEIGHBORHOOD OPERATIONS

A **neighborhood operation** involves a focal cell and a set of its surrounding cells. The surrounding cells are chosen for their distance and/or directional relationship to the focal cell. Common neighborhoods include rectangles, circles, annuluses, and wedges (Figure 13.6). A rectangle is defined by its width and height in cells, such as a 3-by-3 area centered at the focal cell. A circle extends from the focal cell with a specified radius. An annulus or doughnut-shaped neighborhood consists of the ring area between a smaller circle and a larger circle centered at the focal cell. And a wedge consists of a piece of a circle centered at the focal cell. As shown in Figure 13.6, some cells are only partially covered in the defined neighborhood. The general rule is to include a cell if the center of the cell falls within the neighborhood.

13.3.1 Neighborhood Statistics

A neighborhood operation typically uses the cell values within the neighborhood, with or without the focal cell value, in computation, and then assigns the computed value to the focal cell. To complete a neighborhood operation on a raster, the focal cell is moved from one cell to another until all cells are visited. Although a neighborhood

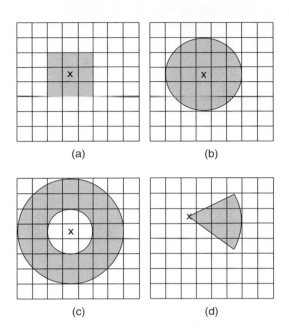

Figure 13.6

Four common neighborhood types: rectangle (*a*), circle (*b*), annulus (*c*), and wedge (*d*). The cell marked with an *x* is the focal cell.

operation works on a single raster, its process is similar to that of a local operation with multiple rasters. Instead of using cell values from different input rasters, a neighborhood operation uses the cell values from a defined neighborhood.

The output from a neighborhood operation can show summary statistics including maximum, minimum, range, sum, mean, median, and standard deviation, as well as tabulation of measures such as majority, minority, and variety. These statistics and measures are the same as those from local operations with multiple rasters.

A **block operation** is a neighborhood operation that uses a rectangle (block) and assigns the calculated value to all block cells in the output raster. Therefore, a block operation differs from a regular neighborhood operation because it does not move from cell to cell but from block to block. Box 13.3 describes neighborhood operations available in ArcGIS and their intended use.

13.3.2 Applications of Neighborhood Operations

An important application of neighborhood operations is data simplification. The moving average method, for instance, reduces the level of cell value fluctuation in the input raster (Figure 13.7). The method typically uses a 3-by-3 or a 5-by-5 rectangle as the neighborhood. As the neighborhood is moved from one focal cell to another, the average of cell values within the neighborhood is computed and assigned to the focal cell. The output raster of moving averages represents a generalization of the original cell values. Another example is a neighborhood operation that uses variety as a measure, tabulates the number of different cell values in the neighborhood, and assigns the number to the focal cell. One can use this method to show, for example, the variety of vegetation types or wildlife species in an output raster.

Neighborhood operations are common in image processing. These operations are variously called filtering, convolution, or moving window operations for spatial feature manipulation (Lillesand and Kiefer 2000). Edge enhancement, for example, can use a range filter, essentially a neighborhood operation using the range statistic (Figure 13.8). The range measures the difference between the maximum and minimum cell values within the defined neighborhood. A high range value therefore indicates the existence of an edge within the neighborhood. The opposite to edge en-

hancement is a smoothing operation that is based on the majority measure (Figure 13.9). The majority operation assigns the most frequent cell value to every cell within the neighborhood, thus creating a smoother raster than the original raster.

Another area of study that heavily depends on neighborhood operations is terrain analysis (Chapter 14). Slope, aspect, and surface curvature measures of a cell are all derived from

(a)

2	2	2	2	1
3	2	2	2	1
3	3	1	2	1
3	3	2	2	2
3	2	2	2	2

(b)

1.56	2.00	2.22
1.67	2.11	2.44
1.67	2.11	2.44

Figure 13.7
The cell values in (b) are the neighborhood means of the shaded cells in (a) using a 3-by-3 neighborhood. For example, 1.56 in the output raster is calculated from $(1 + 2 + 2 + 1 + 2 + 2 + 1 + 2 + 1)/9$.

Box 13.3   Neighborhood Operations in ArcGIS

Both Spatial Analyst and ArcToolbox can perform neighborhood operations. Spatial Analyst has a menu selection called Neighborhood Statistics. The Neighborhood Statistics dialog requires two inputs: statistic and neighborhood. The statistic options include minimum, maximum, mean, median, sum, range, standard deviation, majority, minority, and variety. The neighborhood options include rectangle, circle, doughnut, and wedge. We can also use the Raster Calculator to run additional functions such as block operations.

Spatial Analyst Tools of ArcToolbox has a Neighborhood toolset. The Focal Statistics tool has a similar design as the Neighborhood Statistics dialog in Spatial Analyst. The Neighborhood toolset also includes other tools such as Block Statistics.

(a)

200	200	110	210	210
200	200	110	210	210
150	150	100	170	170
140	140	130	160	160
140	140	130	160	160

(b)

100	110	110
100	110	110
50	70	70

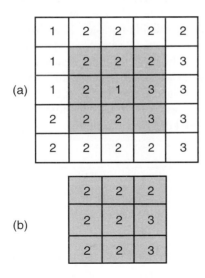

Figure 13.8
The cell values in (*b*) are the neighborhood range statistics of the shaded cells in (*a*) using a 3-by-3 neighborhood. For example, the upper-left cell in the output raster has a cell value of 100, which is calculated from (200 − 100).

(a)

1	2	2	2	2
1	2	2	2	3
1	2	1	3	3
2	2	2	3	3
2	2	2	2	3

(b)

2	2	2
2	2	3
2	2	3

Figure 13.9
The cell values in (*b*) are the neighborhood majority statistics of the shaded cells in (*a*) using a 3-by-3 neighborhood. For example, the upper-left cell in the output raster has a cell value of 2 because there are five 2s and four 1s in its neighborhood.

neighborhood operations using elevations from its adjacent neighbors.

Neighborhood operation can also be important to studies that need to select cells by their neighborhood characteristics. For example, installing a gravity sprinkler irrigation system requires information about elevation drop within a circular neighborhood of a cell. Suppose that a system requires an elevation drop of 130 feet within a distance of 0.5 mile to make it financially feasible. A neighborhood operation on an elevation raster can answer the question by using a circle with a radius of 0.5 mile as the neighborhood and (elevation) range as the statistic. A query of the output raster can show which cells meet the criterion.

Because of its ability to summarize statistics within a defined area, a neighborhood operation can be used to select sites that meet a study's specific criteria. An example is a study by Crow et al. (1999), which uses neighborhood operations to select a stratified random sample of 16 plots, representing two ownerships located within two regional ecosystems.

13.4 ZONAL OPERATIONS

A **zonal operation** works with groups of cells of same values or like features. These groups are called zones. Zones may be contiguous or noncontiguous. A contiguous zone includes cells that are spatially connected, whereas a noncontiguous zone includes separate regions of cells.

13.4.1 Zonal Statistics

A zonal operation may work with a single raster or two rasters. Given a single input raster, zonal operations measure the geometry of each zone, such as area, perimeter, thickness, and centroid (Figure 13.10). The area is the sum of the cells that fall within the zone times the cell size. The perimeter of a contiguous zone is the length of its boundary, and the perimeter of a noncontiguous zone is the sum of the length of each region. The thickness calculates the radius (in cells) of the largest circle that can be drawn within each zone. The centroid

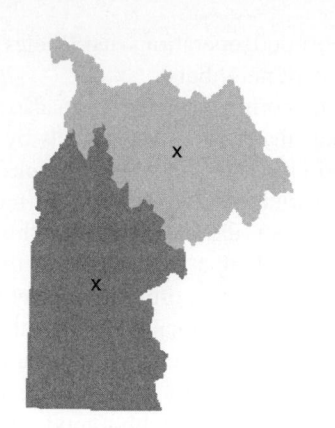

Zone	Area	Perimeter	Thickness
1	36,224	1708	77.6
2	48,268	1464	77.4

Figure 13.10
Thickness and centroid for two large watersheds (zones). Area is measured in square kilometers, and perimeter and thickness are measured in kilometers. The centroid of each zone is marked with an *x*.

Figure 13.11
The cell values in (*c*) are the zonal means derived from an input raster (*a*) and a zonal raster (*b*). For example, 2.17 is the mean of {1, 1, 2, 2, 4, 3} for zone 1.

is the geometric center of a zone located at the intersection of the major axis and the minor axis of an ellipse that best approximates the zone.

Given two rasters in a zonal operation, one input raster and one zonal raster, a zonal operation produces an output raster, which summarizes the cell values in the input raster for each zone in the zonal raster. The summary statistics and measures include area, minimum, maximum, sum, range, mean, standard deviation, median, majority, minority, and variety. (The last four measures are not available if the input raster is a floating-point raster.) Figure 13.11 shows a zonal operation of computing the mean by zone. Figure 13.11*b* is the zonal raster with three zones, Figure 13.11*a* is the input raster, and Figure 13.11*c* is the output raster. Box 13.4 describes zonal operations available in ArcGIS and their intended use.

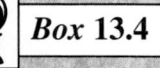 *Box 13.4* **Zonal Operations in ArcGIS**

Both Spatial Analyst and ArcToolbox can perform zonal operations. Zonal Statistics in the Spatial Analyst menu can perform zonal operations with two rasters and calculate the summary statistics of area, minimum, maximum, range, mean, standard deviation, sum, variety, majority, minority, and median. We can also use a zonal function such as Zonalarea in the Raster Calculator to calculate the zonal geometry of an input raster.

Spatial Analyst Tools of ArcToolbox has a Zonal toolset, which includes Zonal Statistics for computing the summary statistics of the input raster by zone, and Zonal Geometry for measuring the geometry of zones in an input raster.

13.4.2 Applications of Zonal Operations

Measures of zonal geometry such as area, perimeter, thickness, and centroid are particularly useful for studies of landscape ecology (Forman and Godron 1986; McGarigal and Marks 1994). Many other geometric measures can be derived from area and perimeter. For example, a shape index called the areal roundness is defined as 354 times the square root of a zone's area divided by its perimeter (Tomlin 1990). Essentially, the areal roundness compares the shape of a contiguous zone to a circle. The area roundness value approaches 100 for a circular shape and 0 for a highly distorted shape. The expression for computing areal roundness is: 354 × Sqrt(zonalarea([raster1]))/zonalperimeter ([raster1]), where raster1 is the input raster and Sqrt is the square root function.

Zonal operations with two rasters can generate useful descriptive statistics for comparison purposes. For example, to compare topographic characteristics of different soil textures, we can use a soil raster that contains the categories of sand, loam, and clay as the zonal raster and slope, aspect, and elevation as the input rasters. By running a series of zonal operations, we can summarize the slope, aspect, and elevation characteristics associated with the three soil textures.

13.5 PHYSICAL DISTANCE MEASURE OPERATIONS

In a GIS project, distances may be expressed as physical distance or cost distance. The **physical distance** measures the straight-line or euclidean distance, whereas the **cost distance** measures the cost for traversing the physical distance. The distinction between the two types of distance measures is important in real-world applications. A truck driver, for example, is more interested in the time or the fuel cost for covering a route than its physical distance. The cost distance in this case is based on not only the physical distance but also the speed limit and road condition. This section focuses on the physical distance. Chapter 18 covers the cost distance for path analysis.

Physical distance measure operations calculate straight-line distances away from cells designated as the source cells. For example, to get the distance between cells (1, 1) and (3, 3) in Figure 13.12, we can use the following formula:

$$\text{cell size} \times \sqrt{(3 - 1)^2 + (3 - 1)^2}$$

or cell size × 2.828. If the cell size were 30 meters, the distance would be 84.84 meters.

A physical distance measure operation essentially buffers the source cells with wavelike continuous distances over the entire raster (Figure 13.13). This is why physical distance measure operations are also called extended neighborhood operations (Tomlin 1990) or global (i.e., the entire raster) operations.

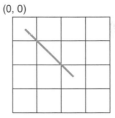

Figure 13.12
A straight-line distance is measured from a cell center to another cell center. This illustration shows the straight-line distance between cell (1, 1) and cell (3, 3).

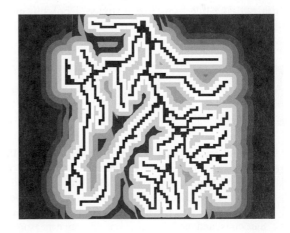

Figure 13.13
Continuous distance measures from a stream network.

Box 13.5 Distance Measure Operations in ArcGIS

Both Spatial Analyst and ArcToolbox in ArcGIS offer distance measure operations. The Distance menu of Spatial Analyst provides four selections: Straight Line Distance, Allocation, Cost Weighted, and Shortest Path. The first two use physical distance measures, and the last two use cost distance measures. Straight Line Distance creates an output raster containing continuous distance measures away from the source cells. Allocation creates a raster, in which each cell is assigned the value of its closest source cell.

Spatial Analyst Tools in ArcToolbox has a Distance toolset. The toolset includes Euclidean Allocation, Euclidean Direction, and Euclidean Distance for physical distance measures. The toolset also includes tools for cost distance measures. Chapter 18 covers cost distance measures for path analysis.

Some GIS packages including ArcGIS allow use of a vector-based data set (e.g., a shapefile) as the source in a physical distance measure operation. This option is based on the consideration of convenience because the data set is converted from vector to raster data before the operation starts.

The continuous distance raster from a physical distance measure operation can be used directly in subsequent operations. But more often it is further processed to create a specific distance zone, or a series of distance zones, from the source cells. Reclassify can convert a continuous distance raster into a raster with one or more discrete distance cells. A variation of Reclassify is an operation called **Slice**, which can divide a continuous distance raster into equal-interval or equal-area distance zones.

13.5.1 Allocation and Direction

Besides calculating straight-line distances, physical distance measure operations can also produce allocation and direction rasters (Figure 13.14). The cell value in an allocation raster corresponds to the closest source cell for the cell. The cell value in a direction raster corresponds to the direction in degrees that the cell is from the closest source cell. The direction values are based on the compass directions: 90° to the east, 180° to the south, 270° to the west, and 360° to the north. (0° is reserved for the source cell.) Box 13.5 describes distance

(a)

1.0	**2**	1.0	2.0
1.4	1.0	1.4	2.2
1.0	1.4	2.2	2.8
1	1.0	2.0	3.0

(b)

2	2	2	2
2	2	2	2
1	1	1	2
1	1	1	1

(c)

90	**2**	270	270
45	360	315	287
180	225	243	315
1	270	270	270

Figure 13.14
Based on the source cells denoted as 1 and 2, (a) shows the physical distance measures in cell units from each cell to the closest source cell; (b) shows the allocation of each cell to the closest source cell; and (c) shows the direction in degrees from each cell to the closest source cell. The cell in a dark shade (row 3, column 3) has the same distance to both source cells. Therefore, the cell can be allocated to either source cell. The direction of 243° is to the source cell 1.

measure operations available in ArcGIS and their intended use.

13.5.2 Applications of Distance Measure Operations

Like buffering around vector-based features, physical distance measure operations have many applications. For example, we can create equal-interval distance zones from a stream network or regional fault lines. Another example is the use of distance measure operations as tools for implementing a model, such as the potential nesting habitat model of greater sandhill cranes in northwestern Minnesota, developed by Herr and Queen (1993). Based on continuous distance zones measured from undisturbed vegetation, roads, buildings, and agricultural land, the model categorizes potentially suitable nesting vegetation as optimal, suboptimal, marginal, and unsuitable.

13.6 OTHER RASTER DATA OPERATIONS

Local, neighborhood, zonal, and distance measure operations discussed in Sections 13.2 to 13.5 cover the majority of raster data operations. Some raster data operations, however, do not fit well in the above classification scheme.

13.6.1 Raster Data Management

Two common raster data operations are Clip and Mosaic. **Clip** can create a new raster by using a rectangle to extract a portion of the input raster. The rectangle is defined by the minimum x, minimum y, maximum x, and maximum y that make up its area extent. **Mosaic** can piece together multiple input rasters into a single raster. If the input rasters overlap, a GIS package typically provides options for filling in the cell values in the overlapping areas. ArcGIS, for example, lets the user choose the first input raster's data or the blending of data from the input rasters for the overlapping areas. If small gaps exist between the input rasters, one option is to use the neighborhood mean operation to fill in missing values.

Clip and Mosaic are useful tools for managing raster data, especially those downloaded from the Internet. Because raster data on the Internet are usually available by USGS 1:24,000 scale quadrangle, the first step in using these data sets is to clip them, or combine them, to fit the study area for a GIS project.

13.6.2 Raster Data Extraction

Raster data extraction creates a new raster or a new field by extracting data from an existing raster. The tool for extracting raster data can be a data set, a graphic object, or a query expression. An analysis mask, which can be a raster or a polygon feature layer, is one type of data set that can be used for extracting raster data. Another type is a point feature layer. The extract-by-points operation can extract the cell values at the point locations and attach the cell values to a new field in the feature attribute table.

A graphic object for raster data extraction can be a rectangle, a set of points, a circle, or a polygon. The object is input in x-, y-coordinates. For example, a circle can be entered with a pair of x-, y-coordinates for its center and a length for its radius. Using this type of extraction method, we can extract elevation data within a radius of 50 miles from an earthquake epicenter or elevation data at a series of weather stations.

The extract-by-attribute operation creates a new raster that has cell values meeting the query expression. For example, we can create a new raster within a particular elevation zone (e.g., 900 to 1000 meters). Those cells that are outside the elevation zone carry no data on the output.

13.6.3 Raster Data Generalization

Several operations can generalize or simplify raster data. One such method is resampling, which can build different pyramid levels (different resolutions) for a large raster data set (Chapter 7). **Aggregate** is similar to a resampling technique in that it also creates an output raster that has a larger cell size (i.e., a lower resolution) than the input. But, instead of using nearest neighbor, bilinear interpolation, or cubic convolution, Aggregate calculates each output cell value as the mean, median, sum, minimum, or maximum of the input cells that fall within the output cell (Figure 13.15).

Some data generalization operations are based on zones, or groups of cells of the same value. For example, ArcGIS has a tool called RegionGroup, which identifies for each cell in the output raster the zone that the cell is connected to (Figure 13.16). We may consider RegionGroup as a classification method using both the cell values and the spatial connection of cells as the classification criteria.

Generalizing or simplifying the cell values of a raster can be useful for certain applications. For example, a raster derived from a satellite image or LIDAR (light detection and ranging) data usually has a high degree of local variations. These local

(a)   (b)

Figure 13.15
An Aggregate operation creates a lower-resolution raster (*b*) from the input (*a*). The operation uses the mean statistic and a factor of 2 (i.e., a cell in *b* covers 2-by-2 cells in *a*). For example, the cell value of 4 in (*b*) is the mean of {2, 2, 5, 7} in (*a*).

(a)   (b)

Figure 13.16
Each cell in the output (*b*) has a unique number that identifies the connected region to which it belongs in the input (*a*). For example, the connected region that has the same cell value of 3 in (*a*) has a unique number of 4 in (*b*).

variations can become unnecessary noises. To remove them, we can use Aggregate or a resampling technique.

13.7 COMPARISON OF VECTOR- AND RASTER-BASED DATA ANALYSIS

Vector data analysis and raster data analysis represent the two basic types of GIS analyses. They are treated separately because a GIS package cannot run them together in the same operation. Although some GIS packages allow use of vector data in some raster data operations (e.g., physical distance measure operations), the data are converted into raster data before the operation starts.

This section uses overlay and buffering, two most common operations in GIS, as examples to compare vector- and raster-based operations. Each GIS project is different in terms of data sources and objectives. Moreover, vector data can be easily converted into raster data and vice versa. We must therefore choose the type of data analysis that is efficient and appropriate.

13.7.1 Overlay

A local operation with multiple rasters is often compared to a vector-based overlay operation. The two operations are similar in that they both use multiple data sets as inputs. But important differences exist between them.

First, to combine the geometries and attributes from the input layers, a vector-based overlay operation must compute intersections between features and insert points at the intersections. This type of computation is not necessary for a raster-based local operation because the input rasters have the same cell size and area extent. Even if the input rasters have to be first resampled to the same cell size, the computation is still less complicated than calculating line intersections. Second, a raster-based local operation has access to various mathematical functions to create the output whereas a vector-based overlay operation only combines attributes from the input layers. Any computations with the attributes must follow the overlay operation.

Although a raster-based local operation is computationally more efficient than a vector-based overlay operation, the latter has its advantages as well. An overlay operation can combine multiple attributes from each input layer. Once combined into a layer, all attributes can be queried and analyzed individually or in combination. For example, a stand layer can have the attributes of height, crown closure, stratum, and crown diameter, and a soil layer can have the attributes of depth, texture, organic matter, and pH value. An overlay operation combines the attributes of both layers into a single layer and allows all attributes to be queried and analyzed. In contrast, each input raster in a local operation is associated with a single set of cell values (i.e., a single attribute). In other words, to query or analyze the same stand and soil attributes as above, a raster-based local operation would require one raster for each attribute. A vector-based overlay operation is therefore more efficient than a raster-based local operation if the data sets to be analyzed have a large number of attributes that share the same geometry.

13.7.2 Buffering

A vector-based buffering operation and a raster-based physical distance measure operation are similar in that they both measure distances from select features. But they differ in at least three aspects. First, a buffering operation creates buffer zones based on the specified buffer distance or distances whereas a raster-based operation creates continuous distance measures. Additional data processing is required to define buffer zones from the continuous distance measures.

Second, a buffering operation has more options than a raster-based operation. For example, a buffering operation has the option of creating separate buffer zones for each feature or a dissolved buffer zone for all features. It would be difficult to create and manipulate separate zones of distance measures using a raster-based operation.

Third, a buffering operation uses x- and y-coordinates in measuring distances whereas a raster-based operation uses rows and columns in measuring physical distances. A buffering operation can therefore create more accurate buffer zones than a raster-based operation can. The accuracy requirement can be important, for example, in implementing riparian zone management programs.

KEY CONCEPTS AND TERMS

Aggregate: A generalization operation that produces an output raster with a larger cell size (i.e., a lower resolution) than the input raster.

Analysis mask: A mask that limits raster data analysis to cells that do not carry the cell value of no data.

Block operation: A neighborhood operation that uses a rectangle (block) and assigns the calculated value to all block cells in the output raster.

Clip: A raster data operation that can create a new raster by using the extent of a rectangle to extract a portion of the input raster.

Cost distance: Distance measured by the cost or impedance of moving between cells.

Local operation: A cell-by-cell raster data operation.

Map algebra: A term referring to local operations with multiple rasters.

Mosaic: A raster data operation that can piece together multiple input rasters into a single raster.

Neighborhood operation: A raster data analysis operation that involves a focal cell and a set of its surrounding cells.

Physical distance: Straight-line distance between cells.

Physical distance measure operation: A raster data operation that calculates straight-line distances away from the source cells.

Raster data extraction: An operation that uses a data set, a graphic object, or a query expression to extract data from an existing raster.

Reclassification: A local operation that reclassifies cell values of an input raster to create a new raster.

Slice: A raster data operation that divides a continuous raster into equal-interval or equal-area classes.

Zonal operation: A raster data operation that involves groups of cells of same values or like features.

REVIEW QUESTIONS

1. How can an analysis mask save time and effort for raster data operations?

2. Why is a local operation also called a cell-by-cell operation?

3. Refer to Figure 13.3. Show the output raster if the local operation uses the minimum statistic.

4. Refer to Figure 13.4. Show the output raster if the local operation uses the variety statistic.

5. Figure 13.5c has two cells with the same value of 4. Why?

6. Neighborhood operations are also called focal operations. What is the focal cell?

7. Describe the common types of neighborhoods used in the neighborhood operation.

8. Refer to Figure 13.8. Show the output raster if the neighborhood operation uses the variety measure.

9. Refer to Figure 13.10. Show the output raster if the neighborhood operation uses the minority statistic.

10. What kinds of geometric measures can be derived from the zonal operations on a single raster?

11. A zonal operation with two rasters must define one of them as the zonal raster first. What is a zonal raster?

12. Refer to Figure 13.12. Show the output raster if the zonal operation uses the range statistic.

13. Suppose you are asked to produce a raster that shows the average precipitation in each major watershed in your state. Describe the procedure you will follow to complete the task.

14. Explain the difference between the physical distance and the cost distance.

15. What is a physical distance measure operation?

16. A government agency will most likely use a vector-based buffering operation, rather than a raster-based physical distance measure operation, to define a riparian zone. Why?

APPLICATIONS: RASTER DATA ANALYSIS

This applications section covers the basic operations of raster data analysis. Task 1 covers a local operation. Task 2 runs a local operation using the Combine function. Task 3 uses a neighborhood operation. Task 4 uses a zonal operation. And Task 5 includes the physical distance measure operation in data query. Raster data analysis typically starts by setting up an analysis environment including the area for analysis and the output cell size. However, this first step is skipped in the following tasks because no analysis mask is used and the output cell size is the same as the input cell size.

Task 1: Perform a Local Operation

What you need: *emidalat*, an elevation raster with a cell size of 30 meters.

Task 1 lets you run a local operation to convert the elevation values of *emidalat* from meters to feet.

1. Start ArcCatalog, and make connection to the Chapter 13 database. Select Properties from the context menu of *emidalat* in the Catalog tree. The Raster Dataset Properties dialog shows that *emidalat* has 186 columns, 214 rows, a cell size of 30 (meters), and a value range of 855 to 1337 (meters). Also, *emidalat* is a floating-point ESRI grid.

2. Launch ArcMap. Add *emidalat* to Layers, and rename Layers Tasks 1&3. Make sure that the Spatial Analyst extension is checked in the Tools menu and the Spatial Analyst toolbar is checked in the View menu.

3. Select Raster Calculator from the Spatial Analyst dropdown list. Enter the following expression in the Raster Calculator's expression box: [emidalat] $\times$ 3.28. Click Evaluate. *Calculation* shows *emidalat* in feet.

4. *Calculation* is a temporary raster. Right-click *Calculation* and select Make Permanent. Name the permanent raster *emidaft* and click Save.

Q1. What area percentage of *emidaft* has elevation above 4000 feet?

Task 2: Perform a Combine Operation

What you need: *slope_gd*, a slope raster with 4 slope classes; and *aspect_gd*, an aspect raster with flat areas and 4 principal directions.

Task 2 covers the use of the Combine function. Combine is a local operation that can work with two or more rasters. You can use Combine in either Spatial Analyst or ArcToolbox.

1. Select Data Frame from the Insert menu in ArcMap. Rename the new data frame Task 2, and add *slope_gd* and *aspect_gd* to Task 2.

2. First use Spatial Analyst to perform Combine. Select Raster Calculator from the Spatial Analyst dropdown list. The Raster Calculator does not show Combine in the dialog. You need to build an expression using the Combine function. Enter Combine([slope_gd], [aspect_gd]) in the Raster Calculator's expression box. Click Evaluate. *Calculation* shows a unique output value for each unique combination of input values. Open the attribute table of *Calculation* to find the unique combinations and their counts. Make *Calculation* permanent and name the output *combine*.

Q2. How many cells in *combine* have the combination of the slope class of 2 and the aspect class of 4?

3. Next use ArcToolbox to perform the Combine operation. Click the Show/Hide ArcToolbox Window button to open the ArcToolbox window. Double-click the Combine tool in the Spatial Analyst Tools/Local toolset. In the next dialog, select *aspect_gd* and *slope_gd* for the input rasters, and enter *slp_asp* for the output raster. Click OK to run the operation. The output should be exactly the same as *combine* from Step 2.

Task 3: Perform a Neighborhood Operation

What you need: *emidalat*, as in Task 1.

Task 3 asks you to run a neighborhood mean operation on *emidalat*.

1. Activate Tasks 1&3 in ArcMap. Select Neighborhood Statistics from the dropdown menu of Spatial Analyst. Select *emidalat* for the input data. Notice that the default statistic type is Mean, the default neighborhood is a 3-by-3 rectangle, and the output cell size is 30. Click OK to run the

operation. *NbrMean of emidalat* shows the neighborhood mean of emidalat.

2. Make *NbrMean of emidalat* a permanent raster, and name it *emidamean*.

Q3. What other neighborhood statistics are available in Spatial Analyst besides the mean?

3. The Focal Statistics tool in the Spatial Analyst Tools/Neighborhood toolset can also accomplish the task.

Task 5: Measure Physical Distances

What you need: *strmgd*, a raster showing streams; and *elevgd*, a raster showing elevation zones.

Task 5 asks you to locate the potential habitats of a plant species. The cell values in *strmgd* are the ID values of streams. The cell values in *elevgd* are elevation zones 1, 2, and 3. Both rasters have the cell resolution of 100 meters. The potential habitats of the plant species must meet the following criteria:

- Elevation zone 2
- Within 200 meters of streams

1. Select Data Frame from the Insert menu in ArcMap. Rename the new data frame Task 5, and add *strmgd* and *elevgd* to Task 5.

2. Click the Spatial Analyst dropdown arrow, point to Distance, and select Straight Line. In the Straight Line dialog, select *strmgd* for distance to, enter 100 (meters) for the output cell size, and opt for a temporary output raster. Click OK to run the operation.

3. *Distance to strmgd* shows continuous distance zones away from streams in *strmgd*.

4. This step is to create a new raster that shows areas within 200 meters of streams. Select Reclassify from the Spatial Analyst dropdown menu. In the Reclassify dialog, select *Distance to strmgd* for the input raster and click Classify. In the Classification dialog, first select 2 for the classes. Then click the first value under Break Values and enter 200. Click the empty space in the second Break Values frame to unselect the second cell. Click OK to dismiss the Classification dialog. Opt for a temporary output raster in the Reclassify dialog, and click OK. *Reclass of Distance to strmgd* separates areas that are within 200 meters of streams from areas that are beyond.

5. Select Raster Calculator from the Spatial Analyst dropdown menu. Enter the following query expression in the expression box: [Reclass of Distance to strmgd] = 1

Task 4: Perform a Zonal Operation

What you need: *precipgd*, a raster showing average annual precipitation in Idaho; and *hucgd*, a watershed raster.

Task 4 asks you to derive annual precipitation statistics by watershed. Both *precipgd* and *hucgd* are projected onto the same projected coordinate system and are measured in meters. The precipitation measurement unit for *precipgd* is 1/100 of an inch; for example, the cell value of 675 means 6.75 inches.

1. Select Data Frame from the Insert menu in ArcMap. Rename the new data frame Task 4, and add *precipgd* and *hucgd* to Task 4.

2. Select Zonal Statistics from the dropdown menu of Spatial Analyst. Select *hucgd* for the zone dataset, Value for the zone field, and *precipgd* for the value raster. Check the boxes to ignore NoData in calculations, to join output table to zone layer, and to chart statistic with Mean. Enter *zstats.dbf* for the output table. Click OK.

3. The chart shows the mean of *precipgd* within each zone of *hucgd*. The table shows the statistics of *precipgd*, including the minimum, maximum, range, mean, standard deviation, sum, variety, majority, minority, and median within each zone of *hucgd*.

Q4. Which watershed has the highest average annual precipitation in Idaho? Where is the watershed located?

4. The Zonal Statistics tool in the Spatial Analyst Tools/Zonal toolset can also accomplish the task.

And [elevgd] = 2. (The Raster Calculator uses == for = and & for And.) Click Evaluate. The *Calculation* layer shows areas that meet the criteria with the value of 1.

Q5. What area percentage of *Calculation* meets the selection criteria?

6. There are three other options for completing the task. First, a feature layer can replace *strmgd* as the source for the distance measure operation. Second, you can use *Distance to strmgd* directly without reclassification. In that case, the expression in the Raster Calculator dialog should be changed to: [Distance to strmgd] <= 200 And [elevgd] = 2. Third, ArcToolbox has the Euclidean Distance tool in the Spatial Analyst Tools/Distance toolset for running the distance measure operation, the Reclassify tool in the Spatial Analyst Tools/Reclass toolset for reclassification, and the Single Output Map Algebra tool in the Spatial Analyst Tools/Map Algebra toolset for raster data query.

Challenge Task

What you need: *emidalat, emidaslope,* and *emidaaspect.*

This challenge task asks you to construct a raster-based model using elevation, slope, and aspect.

1. Use the following table to reclassify *emidalat* and save the output as *emidaelev.*

Old Values	New Values
855–900	1
900–1000	2
1000–1100	3
1100–1200	4
>1200	5

2. *emidaslope* and *emidaaspect* have already been reclassified and ranked. Create a model by using the following equation: *emidaelev* + 3 × *emidaslope* + *emidaaspect.* Name the model *emidamodel.*

Q1. What is the range of cell values in *emidamodel*?

Q2. What area percentage of *emidamodel* has cell value > 20?

REFERENCES

Congalton, R. G. 1991. A Review of Assessing the Accuracy of Classification of Remotely Sensed Data. *Photogrammetric Engineering & Remote Sensing* 37: 35–46.

Crow, T. R., G. E. Host, and D. J. Mladenoff. 1999. Ownership and Ecosystem as Sources of Spatial Heterogeneity in a Forested Landscape, Wisconsin, USA. *Landscape Ecology* 14: 449–63.

Forman, R. T. T., and M. Godron. 1986. *Landscape Ecology*. New York: Wiley.

Herr, A. M., and L. P. Queen. 1993. Crane Habitat Evaluation Using GIS and Remote Sensing. *Photogrammetric Engineering & Remote Sensing* 59: 1531–38.

Heuvelink, G. B. M. 1998. *Error Propagation in Environmental Modelling with GIS*. London: Taylor and Francis.

Lillesand, T. M., and R. W. Kiefer. 2000. *Remote Sensing and Image Interpretation,* 4th ed. New York: Wiley.

McGarigal, K., and B. J. Marks. 1994. *Fragstats: Spatial Pattern Analysis Program for Quantifying Landscape Structure*. Forest Science Department: Oregon State University.

Mladenoff, D. J., T. A. Sickley, R. G. Haight, and A. P. Wydeven. 1995. A Regional Landscape Analysis and Prediction of Favorable Gray Wolf Habitat in the Northern Great Lakes Regions. *Conservation Biology* 9: 279–94.

Renard, K. G., G. R. Foster, G. A. Weesies, D. K. McCool, and D. C. Yoder (coordinators). 1997. Predicting Soil Erosion by Water: A Guide to Conservation Planning with the Revised Universal Soil Loss Equation (RUSLE). *Agricultural Handbook 703*. Washington,

DC: U.S. Department of Agriculture.

Tomlin, C. D. 1990. *Geographic Information Systems and Cartographic Modeling*. Englewood Cliffs, NJ: Prentice Hall.

Veregin, H. 1995. Developing and Testing of an Error Propagation Model for GIS Overlay Operations. *International Journal of Geographical Information Systems* 9: 595–619.

Wischmeier, W. H., and D. D. Smith. 1978. Predicting Rainfall Erosion Losses: A Guide to Conservation Planning. *Agricultural Handbook 537*. Washington, DC: U.S. Department of Agriculture.

14

TERRAIN MAPPING AND ANALYSIS

The terrain with its undulating, continuous surface is a familiar phenomenon to GIS users. The land surface has been the object of mapping and analysis for hundreds of years. Mapmakers have devised various techniques for terrain mapping such as contouring, hill shading, hypsometric tinting, and 3-D perspective views. And geomorphologists have developed measures of the land surface including slope, aspect, and surface curvature.

Terrain mapping and analysis techniques are no longer tools for specialists. GIS has made it relatively easy to incorporate them into a variety of applications. Slope and aspect play a regular role in hydrologic modeling, snow cover evaluation,

soil mapping, landslide delineation, soil erosion, and predictive mapping of vegetation communities (Lane et al. 1998; Wilson and Gallant 2000). Hill shading, perspective views, and even 3-D flyby are common features in presentations and reports.

Most GIS packages treat elevation data (z values) as attribute data at point or cell locations rather than as an additional coordinate to x- and y-coordinates as in a true 3-D model. In raster format, the z values correspond to the cell values. In vector format, the z values are stored in an attribute field or with the feature geometry. Terrain mapping and analysis can use raster data, vector data, or both as inputs. This is perhaps why GIS vendors typically group terrain mapping and analysis functions into a module or an extension, separate from the basic GIS tools.

Chapter 14 is organized into the following five sections. Section 14.1 covers two common types of input data for terrain mapping and analysis: DEM (digital elevation model) and TIN (triangulated irregular network). Section 14.2 describes different terrain mapping methods. Section 14.3 covers slope and aspect analysis using either a

DEM or a TIN. Section 14.4 focuses on deriving surface curvature from a DEM. Section 14.5 compares DEM and TIN for terrain mapping and analysis. Viewshed analysis and watershed analysis, which are closely related to terrain analysis, are covered in Chapter 15.

14.1 DATA FOR TERRAIN MAPPING AND ANALYSIS

Two common types of input data for terrain mapping and analysis are the raster-based DEM and the vector-based TIN. Although we cannot use a DEM and a TIN together in an operation, we can convert a DEM into a TIN and a TIN into a DEM.

14.1.1 DEM

A DEM represents a regular array of elevation points. Most GIS users in the United States use DEMs from the U.S. Geological Survey (USGS). Alternative sources for DEMs come from satellite images, radar data, and LIDAR (light detection and ranging) data (Chapter 5). Regardless of its origin, a point-based DEM must be converted to software-specific raster data (e.g., ESRI grid) before it can be used for terrain mapping and analysis. This conversion simply places each elevation point in a DEM at the center of a cell in an elevation raster. DEM and elevation raster can therefore be used interchangeably.

The quality of a DEM can influence the accuracy of terrain measures including slope and aspect. The USGS classifies the quality of 7.5-minute DEMs into three levels, with level 1 having the poorest quality. Using known sources such as benchmarks (vertical control points) and spot elevations as test points, the USGS calculates the root mean square (RMS) error of the DEM data. Level-1 accuracy has an RMS error target of 7 meters and a maximum RMS error of 15 meters. Level-2 accuracy has a maximum RMS error of one-half the contour interval. And level-3 accuracy has a maximum RMS error of one-third contour interval—not to exceed 7 meters. Most USGS DEMs in use are either level 1 or level 2. Level-1 DEMs are available at a spacing of 30 meters, and level-2 DEMs are available at a spacing of 30 meters or 10 meters.

Errors in level-1 USGS DEMs may be classified as either global or relative (Carter 1989). Global errors are systematic errors caused by displacements of the DEM, as evidenced by mismatching elevations along the boundaries of adjacent DEMs. We can usually correct global errors by applying geometric transformations (Chapter 7). Relative errors are local but significant errors relative to the neighboring elevations. Examples of relative errors are artificial peaks and blocks of elevations extending above the surrounding lands. Relative errors can only be corrected by editing the DEM data. Many of these global and relative errors in level-1 DEMs have been removed from level-2 DEMs.

Although not yet widely used for terrain mapping and analysis, LIDAR data have the advantage over USGS DEMs in providing high-quality DEMs with a spatial resolution of 0.5 to 2 meters and a vertical accuracy of about 15 centimeters (Flood 2001). LIDAR data are therefore ideal for studies that require detailed topographic data such as floodplain mapping, telecommunications, transportation, ecosystems, and forest structure (Hill et al. 2000; Lefsky et al. 2002; Lim et al. 2003). As of June 2004, the only LIDAR data that can be downloaded from the National Elevation Dataset website (http://gisdata.usgs.gov/ned/) are for the Puget Sound area. Bare-ground DEMs compiled from these LIDAR data have a spatial resolution of 6 feet (1.83 meters).

14.1.2 TIN

A TIN approximates the land surface with a series of nonoverlapping triangles. Elevation values (z values) along with x-, y-coordinates are stored at nodes that make up the triangles. In contrast to a DEM, a TIN is based on an irregular distribution of elevation points.

DEMs are usually the primary data source for compiling preliminary TINs through a conversion process. But a TIN can use other data sources. Additional point data may include surveyed elevation points, GPS (global positioning system) data, and LIDAR data. Line data may include contour lines and breaklines. **Breaklines** are line features that represent changes of the land surface such as streams, shorelines, ridges, and roads. And area data may include lakes and reservoirs.

Because triangles in a TIN can vary in size by the complexity of topography, not every point in a DEM needs to be used to create a TIN. Instead, the process is to select points that are more important in representing the terrain. Several algorithms for selecting significant points from a DEM have been proposed in GIS (Lee 1991; Kumler 1994). Here we examine two algorithms: VIP (very important points) and maximum z-tolerance.

To select points from a DEM, the **VIP** algorithm first converts a DEM to a raster and then evaluates the importance of each point (i.e., each cell in the elevation raster) by measuring how well its value can be estimated from the neighboring values in a 3-by-3 moving window (Chen and Guevara 1987). To assess the importance of P in Figure 14.1a, for example, VIP first estimates the elevation at P by using four pairs of its neighbors: $B–F$, $H–D$, $C–G$, and $A–E$. Figure 14.1b shows in a vertical profile how the estimate is made with $C–G$. P_e is the estimated elevation at P from the elevations at C and G (z_C and z_G), and P_h is the actual elevation at P. The difference between P_h and P_e, or d, therefore represents the elevation offset at P. But VIP uses s, which measures the distance of the perpendicular line from P_h to line $z_C z_G$, for the offset measure at P. The offset s is a better measure than d in flat or steep slope areas according to the developers of the VIP algorithm (Chen and Guevara 1987). After calculating four offset values, one for each pair of neighbors around P, VIP uses their average as an indicator of the significance of P.

VIP computes the significance value for each cell in an elevation raster. A frequency distribution of the significance values from a raster usually looks like a normal distribution, with more cells having lower significance values (Chen and Guevara 1987). The selection of points by VIP can then be based on either a desired number or a specified significance level.

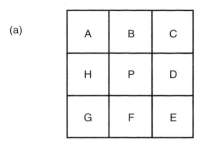

(a)

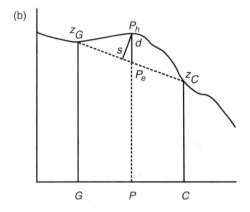

(b)

Figure 14.1

P_e is the estimated elevation, P_h is the actual elevation, and d is the offset between P_e and P_h at cell P. VIP uses s, a distance between P_h and the hypothetical surface between G and C, for measuring the significance of P.

The **maximum z-tolerance** algorithm selects points from an elevation raster to construct a TIN such that, for every point in the elevation raster, the difference between the original elevation and the estimated elevation from the TIN is within the specified maximum z-tolerance. The algorithm uses an iterative process. The process begins by constructing a candidate TIN. Then, for each triangle in the TIN, the algorithm computes the elevation difference from each point in the raster to the enclosing triangular facet. The algorithm determines the point with the largest difference. If the difference is greater than a specified z-tolerance, the algorithm flags the point for addition to the TIN. After every triangle in the current TIN is checked, a new triangulation is recomputed with

point of the DEM to be estimated (interpolated) from its neighboring nodes that make up the TIN. Each of these nodes has its x-, y-coordinates as well as a z (elevation) value. Because each triangular facet is supposed to have a constant slope and aspect, converting a TIN to a DEM can be based on local first-order polynomial interpolation. (Chapter 16 covers local polynomial interpolation as one of the spatial interpolation methods.) TIN to DEM conversion is useful for producing a DEM from LIDAR data. The process first connects LIDAR data (points) to form a TIN and then compiles the DEM by interpolating elevation points from the TIN.

14.2 TERRAIN MAPPING

Section 14.2 covers five terrain mapping techniques: contouring, vertical profiling, hill shading, hypsometric tinting, and perspective view.

14.2.1 Contouring

Contouring is the most common method for terrain mapping. **Contour lines** connect points of equal elevation, the **contour interval** represents the vertical distance between contour lines, and the **base contour** is the contour from which contouring starts. Suppose a DEM has elevation readings ranging from 743 to 1986 meters. If the base contour were set at 800 and the contour interval at 100, then contouring would create the contour lines of 800, 900, 1000, and so on.

The arrangement and pattern of contour lines reflect the topography. For example, contour lines are closely spaced in steep terrain and are curved in the upstream direction along a stream (Figure 14.3). With some training and experience in reading contour lines, we can visualize, and even judge the accuracy of, the terrain as simulated by digital data.

Automated contouring follows two basic steps: (1) detecting a contour line intersecting a raster cell or a triangle, and (2) drawing the contour line through the raster cell or triangle (Jones et al. 1986). A TIN is a good example for illustrating automated contouring because it has elevation readings for all nodes from triangulation. Given a contour line,

the selected additional points. This process continues until all points in the raster are within the specified maximum z-tolerance.

Elevation points selected by the VIP algorithm or the maximum z-tolerance algorithm, plus additional elevation points from contour lines, survey data, GPS data, or LIDAR data are connected to form a series of nonoverlapping triangles in an initial TIN. A common algorithm for connecting points is the **Delaunay triangulation** (Watson and Philip 1984; Tsai 1993). Triangles formed by the Delaunay triangulation have the following characteristics: all nodes (points) are connected to the nearest neighbors to form triangles; and triangles are as equiangular, or compact, as possible.

Depending on the software design, breaklines may be included in the initial TIN or used to modify the initial TIN. Breaklines provide the physical structure in the form of triangle edges to show changes of the land surface (Figure 14.2). If the z values for each point of a breakline are known, they can be stored in a field. If not, they can be estimated from the underlying DEM or TIN surface.

We often notice that triangles along the border of a TIN are stretched and elongated, thus distorting the landform features derived from those triangles. The cause of this irregularity stems from the sudden drop of elevation along the edge. One way to solve this problem is to include elevation points beyond the border of a study area for processing and then to clip the study area from the larger coverage. The process of building a TIN is certainly more complex than preparing an elevation raster from a DEM.

The discussion has so far dealt with the conversion of a DEM to a TIN. We can also convert a version of a DEM to a TIN. The process requires each elevation

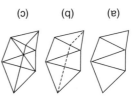

(a)

(b)

(c)

Figure 14.2

A breakline, shown as a dashed line in (b), modifies the triangles in (a) and creates new triangles in (c).

Figure 14.3
A contour line map.

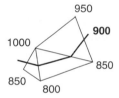

Figure 14.4
The contour line of 900 connects points that are interpolated to have the value of 900 along the triangle edges.

every triangle edge is examined to determine if the line should pass through the edge. If it does, linear interpolation, which assumes a constant gradient between the end nodes of the edge, can determine the contour line's position along the edge. After all the positions are calculated, they are connected to form the contour line (Figure 14.4). The initial contour line consists of straight-line segments, which can be smoothed by fitting a mathematical function such as splines to points that make up the line. Another way of producing smooth contour lines is to divide a triangle into a series of smaller triangles and to use these smaller triangles for contouring.

Contour lines do not intersect one another or stop in the middle of a map, although they can be close together in cases of cliffs or form closed lines in cases of depressions or isolated hills. Contour maps created from a GIS sometimes contain irregularities and even errors. Irregularities are often caused by use of large cells, whereas errors are caused by use of incorrect parameter values in the smoothing algorithm (Clarke 1995).

14.2.2 Vertical Profiling

A **vertical profile** shows changes in elevation along a line, such as a hiking trail, a road, or a stream (Figure 14.5). The manual method usually involves the following steps:

1. Draw a profile line on a contour map.
2. Mark each intersection between a contour and the profile line and record its elevation.

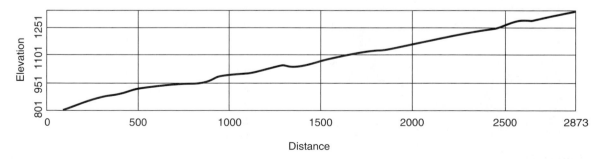

Figure 14.5
A vertical profile.

Automated profiling follows the same procedure but substitutes the contour map with an elevation raster or a TIN.

3. Raise each intersection point to a height proportional to its elevation.
4. Plot the vertical profile by connecting the elevated points.

14.2.3 Hill Shading

Also known as *shaded relief*, **hill shading** simulates how the terrain looks with the interaction between sunlight and surface features (Figure 14.6). A mountain slope directly facing incoming light will be very bright; a slope opposite to the light will be dark. Hill shading helps viewers recognize the shape of landform features. Hill shading can be mapped alone, such as the well-known example of

Figure 14.6
An example of hill shading, with the sun's azimuth at 315° (NW) and the sun's altitude at 45°.

Theilen and Pike's (1991) digital shaded-relief map of the United States (**http://www.usgs.gov/reports/misc/Misc._Investigations_Series_Maps_(I._Series)/I_2206/usa_dem.gif**). But often hill shading is the background for terrain or thematic mapping.

Hill shading used to be produced by talented artists. But the computer can now generate high-quality shaded-relief maps. Four factors control the visual effect of hill shading. The sun's azimuth is the direction of the incoming light, ranging from 0° (due north) to 360° in a clockwise direction. Typically, the default for the sun's azimuth is 315°. With the light source located above the upper-left corner of the shaded-relief map, the shadows appear to fall toward the viewer, thus avoiding the pseudoscopic effect (Box 14.1). The sun's altitude is the angle of the incoming light measured above the horizon between 0° and 90°. The other two factors are the surface's slope and aspect: slope ranges from 0° to 90° and aspect from 0° to 360° (Section 14.3).

An algorithm for hill shading uses the following equation to compute the relative radiance value for every cell in an elevation raster or for every triangle in a TIN (Eyton 1991):

$$R_f = \cos(A_f - A_s) \sin(H_f) \cos(H_s) + \cos(H_f) \sin(H_s)$$

(14.1)

where R_f is the relative radiance value of a facet (a raster cell or a triangle), A_f is the facet's aspect, A_s is the sun's azimuth, H_f is the facet's slope, and H_s is the sun's altitude. R_f ranges in value from 0 to 1. If multiplied by the constant 255, R_f can be converted to the illumination value (I_f) for display. An I_f value of 255 would result in white and an I_f value of 0 would result in black on a shaded-relief map. GIS packages such as ArcGIS use I_f for hill shading (Box 14.2).

Relative radiance is similar to another measure called the incidence value (Franklin 1987):

$$\cos(H_f) + \cos(A_f - A_s) \sin(H_f) \cot(H_s)$$

(14.2)

Box 14.1 | **The Pseudoscopic Effect**

A shaded-relief map looks right when the shadows appear to fall toward the viewer. If the shadows appear to fall away from the viewer, such as using 135° as the sun's azimuth, the hills on the map look like depressions and the depressions look like hills in an optical illusion called the pseudoscopic effect (Campbell 1984). Of course, the sun's azimuth at 315° is totally unrealistic for most parts of the Earth's surface.

Box 14.2 | **A Worked Example of Computing Relative Radiance**

Suppose a cell in an elevation raster has a slope value of 10° and an aspect value of 297° (W to NW), the sun's altitude is 65°, and the sun's azimuth is 315° (NW). The relative radiance value of the cell can be computed by:

$$R_f = \cos(297 - 315)\sin(10)\cos(65) + \cos(10)\sin(65) = 0.9623$$

The cell will appear bright with an R_f value of 0.9623.

If the sun's altitude is lowered to 25° and the sun's azimuth remains at 315°, then the cell's relative radiance value becomes:

$$R_f = \cos(297 - 315)\sin(10)\cos(25) + \cos(10)\sin(25) = 0.5658$$

The cell will appear in medium gray with an R_f value of 0.5658.

The notations in Eq. (14.2) are the same as Eq. (14.1). One can also derive the incidence value by multiplying the relative radiance value by $\sin(H_s)$. Besides producing hill shading, both relative radiance and incidence can be used in image processing as variables representing the interaction between the incoming radiation and local topography.

14.2.4 Hypsometric Tinting

Hypsometry depicts the distribution of the Earth's mass with elevation. **Hypsometric tinting,** also known as *layer tinting,* applies color symbols to different elevation zones (Figure 14.7). The use of well-chosen color symbols can help viewers see the progression in elevation, especially on a small-scale map. One can also use hypsometric tinting to highlight a particular elevation zone, which may be important, for instance, in a wildlife habitat study.

Figure 14.7
A hypsometric map. Different elevation zones are shown in different gray symbols.

14.2.5 Perspective View

Perspective views are 3-D views of the terrain: the terrain has the same appearance as viewed with an angle from an airplane (Figure 14.8). Four parameters can control the appearance of a 3-D view (Figure 14.9):

- **Viewing azimuth** is the direction from the observer to the surface, ranging from $0°$ to $360°$ in a clockwise direction.

Figure 14.8
A 3-D perspective view.

- **Viewing angle** is the angle measured from the horizon to the altitude of the observer. A viewing angle is always between $0°$ and $90°$. An angle of $90°$ provides a view from directly above the surface. And an angle of $0°$ provides a view from directly ahead of the surface. Therefore, the 3-D effect reaches its maximum as the angle approaches $0°$ and its minimum as the angle approaches $90°$.
- **Viewing distance** is the distance between the viewer and the surface. Adjustment of the viewing distance allows the surface to be viewed up close or from a distance.
- **z-scale** is the ratio between the vertical scale and the horizontal scale. Also called the *vertical exaggeration factor*, z-scale is useful for highlighting minor landform features.

Because of its visual appeal, 3-D perspective view is a display tool in many GIS packages. The 3D Analyst extension to ArcGIS, for example, provides the graphical interfaces for manipulating the viewing parameters. Using the extension, we can rotate the surface, navigate the surface, or take a close-up view of the surface. To make perspective views more realistic, we can superimpose these views with layers such as hydrographic features (Figure 14.10), land cover, vegetation, and roads in

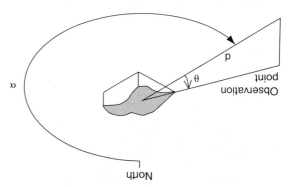

Figure 14.9
Three controlling parameters of the appearance of a 3-D view: the viewing azimuth α is measured clockwise from the north, the viewing angle θ is measured from the horizon, and the viewing distance d is measured between the observation point and the 3-D surface.

Figure 14.10
Draping of streams and shorelines on a 3-D surface.

Figure 14.11
A 3-D perspective view of elevation zones.

a process called **3-D draping.** We can also add clouds and change the color of the sky.

Typically, DEMs and TINs provide the surfaces for 3-D perspective views and 3-D draping. But as long as a feature layer has a field that stores z values or a field that can be used for calculating z values, the layer can be viewed in 3-D perspective. Figure 14.11, for example, is a 3-D view based on elevation zones. Likewise, to show building features in 3-D, we can extrude them by using the building height as the z value.

14.3 SLOPE AND ASPECT

As two most important measures of the terrain, slope and aspect used to be derived manually from a contour map. Fortunately, this practice has become rare with the use of GIS.

Slope measures the rate of change of elevation at a surface location, and **aspect** is the directional measure of slope. If we define the elevation (z) of a point on the land surface as a function of the point's position (x and y), then we can define slope (S) at the point as a function of the first-order derivatives of the surface in the x and y directions:

(14.3)

$$S = \sqrt{(\partial_z/\partial_x)^2 + (\partial_z/\partial_y)^2}$$

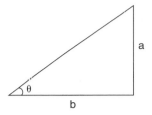

Figure 14.12
Slope θ, either measured in percent or degrees, can be calculated from the vertical distance a and the horizontal distance b.

And we can define the slope's directional angle as:

(14.4)

$$A = \arctan((\partial_z/\partial_y)/(\partial_z/\partial_x))$$

Slope may be expressed as percent slope or degree slope. Percent slope is 100 times the ratio of rise (vertical distance) over run (horizontal distance), whereas degree slope is the arc tangent of the ratio of rise over run (Figure 14.12). Aspect (A) is a directional measure in degrees. Aspect starts with 0° at the north, moves clockwise, and ends with 360° also at the north.

Aspect is a circular measure. An aspect of 10° is closer to 360° than to 30°. We often have to manipulate aspect measures before using them in data analysis. A common method is to classify aspects into the four principal directions (north, east, south, and west) or eight principal directions (north, northeast, east, southeast, south, southwest, west, and northwest) and to treat aspects as categorical data (Figure 14.13). Rather than converting aspects to categorical data, Chang and Li (2000) have proposed a method for capturing the principal direction while retaining the numeric measure. For instance, to capture the N–S principal direction, we can set 0° at north, 180° at south, and 90° at both west and east (Figure 14.14). Perhaps the most common method for converting aspect measures to linear measures is to use their sine or cosine values, which range from −1 to 1 (Zar 1984).

As basic elements for analyzing and visualizing landform characteristics, slope and aspect are important in studies of watershed units, landscape

the quantity and direction of tilt of the cell's normal vector—a directed line perpendicular to the cell (Figure 14.15). Given a normal vector (n_x, n_y, n_z), the formula for computing the cell's slope is:

(14.5)
$$\sqrt{n_x^2 + n_y^2}/n_z$$

And the formula for computing the cell's aspect is:

(14.6)
$$\arctan (n_y/n_x)$$

Different approximation methods have been proposed for calculating slope and aspect. Here we will examine three common methods. All three methods use a 3-by-3 moving window to estimate the slope and aspect of the center cell, but they differ in the number of neighboring cells used in the estimation and the weight applying to each cell.

The first method, which is attributed to Fleming and Hoffer (1979) and Ritter (1987), uses the four immediate neighbors of the center cell. The slope (S) at C_0 in Figure 14.16 can be computed by:

(14.7)
$$S = \sqrt{(e_1 - e_3)^2 + (e_4 - e_2)^2}/2d$$

where e_i are the neighboring cell values, and d is the cell size. The n_x component of the normal vector to C_0 is ($e_1 - e_3$), or the elevation difference in the x dimension. The n_y component is ($e_4 - e_2$), or the elevation difference in the y dimension. To compute the percent slope at C_0, we can multiply S by 100.

units, and morphometric measures (Moore et al. 1991). When used with other variables, slope and aspect can assist in solving problems in forest inventory estimates, soil erosion, wildlife habitat suitability, site analysis, and many other fields.

14.3.1 Computing Algorithms for Slope and Aspect Using Raster

When an elevation raster is used as the data source, slope and aspect are computed for each cell in the raster. The slope and aspect for a cell are measured by

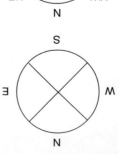

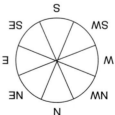

Figure 14.13
Aspect measures are often grouped into the four principal directions or eight principal directions.

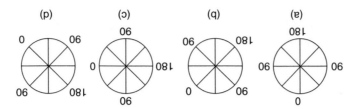

Figure 14.14
Transformation methods to capture the N–S direction (a), the NE–SW direction (b), the E–W direction (c), and the NW–SE direction (d).

Box 14.3 | Conversion of *D* to Aspect

The notations used here are the same as in Eq. (14.5) to (14.8). In the following, the text after an apostrophe is an explanatory note.

```
If S <> 0 then
    T = D × 57.296
    If nₓ = 0
        If n_y < 0 then
            Aspect = 180
        Else
```

```
            Aspect = 360
    ElseIf nₓ > 0 then
        Aspect = 90 − T
    Else 'nₓ < 0
        Aspect = 270 − T
Else 'S = 0
    Aspect = −1 'undefined aspect for flat surface
End If
```

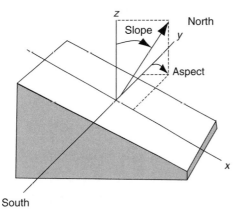

Figure 14.15
The normal vector to the cell is the directed line perpendicular to the cell. The quantity and direction of tilt of the normal vector determine the slope and aspect of the cell. (Redrawn from Hodgson, 1998, *CaGIS* 25, (3): pp. 173–185; reprinted with the permission of the American Congress on Surveying and Mapping.)

	e_2	
e_1	C_0	e_3
	e_4	

Figure 14.16
Ritter's algorithm for computing slope and aspect at C_0 uses the four immediate neighbors of C_0.

S's directional angle *D* can be computed by:

(14.8)

$$D = \arctan((e_4 - e_2)/(e_1 - e_3))$$

D is measured in radians and is with respect to the *x*-axis. Aspect, on the other hand, is measured in degrees and from a north base of 0°. Box 14.3 shows an algorithm for converting *D* to aspect (Ritter 1987; Hodgson 1998).

The second method for computing slope and aspect is called Horn's algorithm (1981), an algorithm used in ArcGIS. Horn's algorithm uses eight neighboring cells and applies a weight of 2 to the four immediate neighbors and a weight of 1 to the four corner cells. Horn's algorithm computes slope at C_0 in Figure 14.17 by:

(14.9)

$$S = \sqrt{[(e_1 + 2e_4 + e_6) - (e_3 + 2e_5 + e_8)]^2 + [(e_6 + 2e_7 + e_8) - (e_1 + 2e_2 + e_3)]^2}/8d$$

And the *D* value at C_0 is computed by:

(14.10)

$$D = \arctan([(e_6 + 2e_7 + e_8) - (e_1 + 2e_2 + e_3)]/[(e_1 + 2e_4 + e_6) - (e_3 + 2e_5 + e_8)])$$

D can be converted to aspect by using the same algorithm for the first method except that $n_x = (e_1 + 2e_4 + e_6)$ and $n_y = (e_3 + 2e_5 + e_8)$ (Box 14.4).

Box 14.4 A Worked Example of Computing Slope and Aspect Using Raster

T he following diagram shows a 3-by-3 window of an elevation raster. Elevation readings are measured in meters, and the cell size is 30 meters.

1006	1012	1017
1010	1015	1019
1012	1017	1020

This example computes the slope and aspect of the center cell using Horn's algorithm:

$$n_x = (1006 + 2 \times 1010 + 1012) - (1017 + 2 \times 1019 + 1020) = -37$$

$$n_y = (1012 + 2 \times 1017 + 1020) - (1006 + 2 \times 1012 + 1017) = 19$$

$$S = \sqrt{(-37)^2 + (19)^2}/(8 \times 30) = 0.1733$$

$$S_p = 100 \times 0.1733 = 17.33$$

$$D = \arctan(n_y/n_x) = \arctan(19/-37) = -0.4744$$

$$T = -0.4744 \times 57.296 = -27.181$$

Because $S <> 0$ and $n_x < 0$,

Aspect = $270 - (-27.181) = 297.181$.

For comparison, S_p has a value of 17.16 using the Fleming and Hoffer's algorithm and a value of 17.39 using the Sharpnack and Akin's algorithm. Aspect has a value of 299.06 using the Fleming and Hoffer's algorithm and a value of 296.56 using the Sharpnack and Akin's algorithm.

The third method called Sharpnack and Akin's algorithm (1969) also uses eight neighboring cells but applies the same weight to every cell. The formula for computing S is:

(14.11)
$$S = \sqrt{[(e_1 + e_4 + e_6) - (e_3 + e_5 + e_8)]^2 + [(e_6 + e_7 + e_8) - (e_1 + e_2 + e_3)]^2}/6d$$

And the formula for computing D is:

(14.12)
$$D = \arctan\left([(e_6 + e_7 + e_8) - (e_1 + e_2 + e_3)]/[(e_1 + e_4 + e_6) - (e_3 + e_5 + e_8)]\right)$$

e_1	e_2	e_3
e_4	C_0	e_5
e_6	e_7	e_8

Figure 14.17
Both Horn's algorithm and Sharpnack and Akin's algorithm for computing slope and aspect at C_0 use the eight neighboring cells of C_0.

14.3.2 Computing Algorithms for Slope and Aspect Using TIN

The algorithms for computing slope and aspect for a triangle in a TIN also use the bidirectional normal vector, the vector perpendicular to the triangular surface. Suppose a triangle is made of the following three nodes: $A (x_1, y_1, z_1)$, $B (x_2, y_2, z_2)$, and $C (x_3, y_3, z_3)$ (Figure 14.18). The normal vector is a cross product of vector AB, $[(x_2 - x_1)$, $(y_2 - y_1)$, $(z_2 - z_1)]$, and vector AC, $[(x_3 - x_1)$, $(y_3 - y_1)$, $(z_3 - z_1)]$. And the three components of the normal vector are:

(14.13)
$$n_x = (y_2 - y_1)(z_3 - z_1) - (y_3 - y_1)(z_2 - z_1)$$
$$n_y = (z_2 - z_1)(x_3 - x_1) - (z_3 - z_1)(x_2 - x_1)$$
$$n_z = (x_2 - x_1)(y_3 - y_1) - (x_3 - x_1)(y_2 - y_1)$$

Box 14.5 | A Worked Example of Computing Slope and Aspect Using TIN

Suppose a triangle in a TIN is made of the following nodes with their x, y, and z values measured in meters.

Node 1: $x_1 = 532260$, $y_1 = 5216909$, $z_1 = 952$

Node 2: $x_2 = 531754$, $y_2 = 5216390$, $z_2 = 869$

Node 3: $x_3 = 532260$, $y_3 = 5216309$, $z_3 = 938$

The following shows the steps for computing the slope and aspect of the triangle.

$n_x = (5216390 - 5216909)(938 - 952) - (5216309 - 5216909)(869 - 952) = -42534$

$n_y = (869 - 952)(532260 - 532260) - (938 - 952)(531754 - 532260) = -7084$

$n_z = (531754 - 532260)(5216309 - 5216909) - (532260 - 532260)(5216390 - 5216909) = 303600$

$S_p = 100 \times [\sqrt{(-42534)^2 + (-7084)^2}/303600] = 14.20$

$D = \arctan(-7084/-42534) = 0.165$

$T = 0.165 \times 57.296 = 9.454$

Because $S <> 0$ and $n_x < 0$, aspect $= 270 - T = 260.546$.

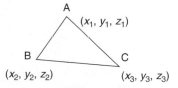

Figure 14.18

The algorithm for computing slope and aspect of a triangle in a TIN uses the x, y, and z values at the three nodes of the triangle.

The S and D values of the triangle can be derived from Eq. (14.5) and (14.6), and the D value can then be converted to the aspect measured in degrees and from a north base of $0°$ (Box 14.5).

14.3.3 Factors Influencing Slope and Aspect Measures

The accuracy of slope and aspect measures can influence the performance of models that use slope and aspect as inputs (Srinivasan and Engel 1991). Therefore, it is important to examine several factors that can influence slope and aspect measures.

The first and perhaps the most important factor is the resolution of DEM used for deriving slope and aspect. Slope and aspect layers created from the 7.5-minute DEM are expected to contain greater amounts of detail than those from the 1-degree DEM (Isaacson and Ripple 1990). Reports from experimental studies also show that the accuracy of slope and aspect estimates decreases with a decreasing DEM resolution (Chang and Tsai 1991; Gao 1998).

Figure 14.19 shows hill-shaded maps of a small area north of Tacoma, Washington created from three different DEMs. The 30-meter and 10-meter DEMs are USGS DEMs. The 1.83-meter DEM is a bare-ground DEM compiled from LIDAR data. It is not difficult to see an increase of topographic details with an increasing DEM resolution. Slope maps in Figure 14.20 are derived from the three DEMs, with darker symbols for steeper slopes. Slope measures range from $0°$ to $77.82°$ for the 30-meter DEM, $0°$ to $83.44°$ for the 10-meter DEM, and $0°$ to $88.45°$ for the 1.83-meter DEM. Details on the slope maps therefore increase with an increasing DEM resolution.

The quality of DEM can also influence slope and aspect measures. We can now extract DEMs from satellite imagery (e.g., SPOT images) on the personal computer (Chapter 5). But the quality of such DEMs varies depending on the software package and the quality of input data including ground control points. From a study comparing a USGS 7.5-minute DEM and a DEM produced from a SPOT panchromatic stereopair, Bolstad and Stowe (1994) report that slope and aspect errors for the SPOT DEM are significantly different from zero, whereas those for the USGS DEM are not.

Slope and aspect measures can vary by the computing algorithm. But there is no consensus as to which algorithm is better. Skidmore (1989) reports that Horn's algorithm and Sharpnack and

Akin's algorithm, both involving eight neighboring cells, are the best for estimating both slope and aspect in an area of moderate topography in Australia. Ritter's algorithm involving four immediate neighbors turns out to be better than other algorithms in two more recent studies: slope measures based on a synthetic surface and a study area in Tennessee by Hodgson (1998), and slope and aspect estimates using a synthetic surface and a DEM in Scotland by Jones (1998). Finally, local topography can be a factor in estimating slope and aspect. Errors in slope estimates

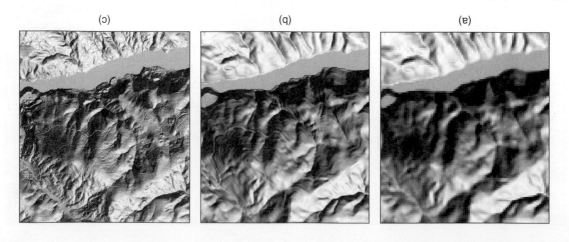

Figure 14.19

DEMs at three different resolutions: USGS 30-meter DEM (a), USGS 10-meter DEM (b), and 1.83-meter DEM derived from LIDAR data (c).

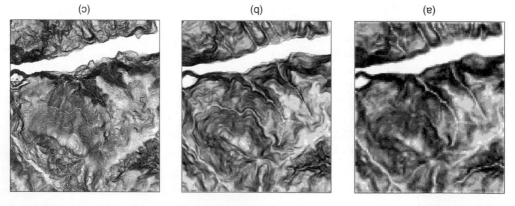

Figure 14.20

Slope layers derived from the three DEMs in Figure 14.19. The darkness of the symbol increases as the slope becomes steeper.

tend to be greater in areas of higher slopes, but errors in aspect estimates tend to be greater in areas of lower relief (Chang and Tsai 1991). Data precision problem (i.e., the rounding of elevations to the nearest whole number) can be the cause for aspect and slope errors in areas of low relief (Carter 1992; Florinsky 1998). Slope errors on steeper slopes may be due in part to difficulties in stereocorrelation in forested terrain (Bolstad and Stowe 1994).

14.4 SURFACE CURVATURE

GIS applications in hydrological studies often require computation of surface curvature to determine if the surface at a cell location is upwardly convex or concave (Gallant and Wilson 2000). Similar to slope and aspect, different algorithms are available for calculating surface curvature (Schmidt et al. 2003). A common algorithm is to fit a 3-by-3 window with a quadratic polynomial equation (Zevenbergen and Thorne 1987; Moore et al. 1991):

(14.14)

$$z = Ax^2y^2 + Bx^2y + Cxy^2 + Dx^2 + Ey^2 + Fxy + Gx + Hy + I$$

The coefficients $A-I$ can be estimated by using the elevation values in the 3-by-3 window and the raster cell size (Box 14.6). Three curvature measures can then be computed from the coefficients:

(14.15)

$$\text{Profile curvature} = -2\left[(DG^2 + EH^2 + FGH)/(G^2 + H^2)\right]$$

(14.16)

$$\text{Plan curvature} = 2\left[(DH^2 + EG^2 - FGH)/(G^2 + H^2)\right]$$

(14.17)

$$\text{Curvature} = -2\,(D + E)$$

Box 14.6 | A Worked Example of Computing Surface Curvature

1017	1010	1017
1012	1006	1019
1015	1012	1020

e_1	e_2	e_3
e_4	e_0	e_5
e_6	e_7	e_8

The diagram above represents a 3-by-3 window of an elevation raster, with a cell size of 30 meters. This example shows how to compute the profile curvature, plan curvature, and surface curvature at the center cell. The first step is to estimate the coefficients $D-H$ of the quadratic polynomial equation that fits the 3-by-3 window:

$$D = [(e_4 + e_5)/2 - e_0]/L^2$$
$$E = [(e_2 + e_7)/2 - e_0]/L^2$$
$$F = (-e_1 + e_3 + e_6 - e_8)/4L^2$$
$$G = (-e_4 + e_5)/2L$$
$$H = (e_2 - e_7)/2L$$

where e_0 to e_8 are elevation values within the 3-by-3 window according to the diagram below, and L is the cell size.

$$\text{Profile curvature} = -2\left[(DG^2 + EH^2 + FGH)/(G^2 + H^2)\right] = -0.0211$$

$$\text{Plan curvature} = 2\left[(DH^2 + EG^2 - FGH)/(G^2 + H^2)\right] = 0.0111$$

$$\text{Curvature} = -2\,(D + E) = -0.0322$$

All three measures are based on 1/100 (z units). The negative curvature value means the surface at the center cell is upwardly concave in the form of a shallow basin surrounded by higher elevations in the neighboring cells.

Profile curvature is estimated along the direction of maximum slope. Plan curvature is estimated across the direction of maximum slope. And curvature measures the difference between the two: profile curvature — plan curvature. A positive curvature value at a cell means that the surface is upwardly convex at the cell location. A negative curvature value means that the surface is upwardly concave. And a 0 value means that the surface is flat.

14.5 RASTER VERSUS TIN

We can often choose either elevation rasters or TINs for terrain mapping and analysis. GIS packages, such as ArcGIS, allow use of elevation rasters or TINs and can convert a raster to a TIN or a TIN to a raster (Box 14.7). Given the options, we might ask which data model to use. There is no easy answer to the question. Essentially, rasters and TINs differ in data flexibility and computational efficiency.

A main advantage of using a TIN lies in the flexibility with input data sources. We can construct a TIN using inputs from DEM, contour lines, GPS data, LIDAR data, and survey data. We can add elevation points to a TIN at their precise locations and add breaklines, such as streams, roads, ridgelines, and shorelines, to define surface discontinuities. In comparison, a DEM or an elevation raster cannot indicate a stream in a hilly area and the accompanying topographic characteristics if the stream width is smaller than the DEM resolution.

An elevation raster is fixed with a given cell size. We cannot add new sample points to an elevation raster to increase its surface accuracy. Assuming the production method is the same, the only way to improve the accuracy of a raster is to increase its resolution, for example, from 30 meters to 10 meters. Researchers, especially those working with small watersheds, have in fact advocated DEMs with a 10-meter resolution (Zhang and Montgomery 1994) and even higher (Gertner et al. 2002). But increasing DEM resolution is a costly operation because it requires the recompiling of elevation data and more computer memory.

Box 14.7   Terrain Mapping and Analysis Using ArcGIS

Tools for terrain mapping and analysis are available through the Spatial Analyst and 3D Analyst extensions to ArcGIS. Both extensions have a surface analysis menu with tools for contour, hillshade, slope, and aspect. 3D Analyst has the additional functionalities for creating and editing TINs and for converting between TIN, raster, and feature layer. 3D Analyst also has a 3-D viewing application called ArcScene, which can be used to prepare and manipulate perspective views, 3-D draping, and 3-D animation.

ArcGIS 9.0 has the additional option of performing terrain mapping and analysis using ArcToolbox. The Spatial Analyst Tools/Surface toolset includes tools for aspect, contour, curvature, hillshade, and slope. The 3D Analyst Tools/Conversion toolset has tools for conversion between TIN, raster, and features. These tools use dialogs very similar to those in Spatial Analyst and 3D Analyst. ArcToolbox has some new tools that are not available in the extensions. For example, the 3D Analyst Tools/TIN Surface toolset has tools for deriving aspect, contour, and slope feature layers directly from TINs. Besides the new tools, ArcToolbox has two other advantages: it is available in ArcCatalog and ArcMap, and it offers the choices of dialog, command line, model, and script for working with tools. ArcToolbox is therefore more flexible than the extensions in terms of data processing.

ArcGlobe is a new 3-D visualization and analysis application in ArcGIS 9.0. Available through the 3D Analyst extension, ArcGlobe is functionally similar to ArcScene except that it can work with large and varied data sets such as high-resolution satellite images, high-resolution DEMs, and vector data. ArcScene, on the other hand, is designed for small local projects.

Besides data flexibility, TIN is also an excellent data model for terrain mapping and 3-D display. The triangular facets of a TIN better define the land surface than an elevation raster and create a sharper image. Most GIS users seem to prefer the look of a map based on a TIN rather than an elevation raster (Kumler 1994).

Computational efficiency is the main advantage of using rasters for terrain analysis. The simple data structure makes it relatively easy to perform neighborhood operations on an elevation raster. Therefore, using an elevation raster to compute slope, aspect, surface curvature, relative radiance, and other topographic variables is fast and efficient. In contrast, the computational load using a TIN can increase significantly as the number of triangles increases. For some terrain analysis operations, a GIS package may in fact convert a TIN to an elevation raster prior to data analysis.

Finally, which data model is more accurate in measuring elevation, slope, aspect, and other topographic parameters? The answer depends on how a TIN or a DEM is made. If a TIN is made from sampling a DEM, the TIN cannot be as accurate as the full DEM. Based on a series of comparisons between DEMs and TINs derived from the VIP algorithm, the maximum z-tolerance algorithm, and contour lines, Kumler (1994) confirms that TINs are inferior to the full DEM and are not as efficient as DEMs in terrain modeling. Likewise, if a DEM is interpolated from a TIN, the DEM cannot be as accurate as the TIN.

KEY CONCEPTS AND TERMS

3-D drape: The method of superimposing thematic layers such as vegetation and roads on 3-D perspective views.

Aspect: The directional measure of slope.

Base contour: The contour from which contouring starts.

Breaklines: Line features that represent changes of the land surface such as streams, shorelines, ridges, and roads.

Contour interval: The vertical distance between contour lines.

Contour lines: Lines connect points of equal elevation.

Delaunay triangulation: An algorithm for connecting points to form triangles such that all points are connected to their nearest neighbors and triangles are as compact as possible.

Hill shading: A graphic method, which simulates how the land surface looks with the interaction between sunlight and landform features. The method is also known as *shaded relief*.

Hypsometric tinting: A mapping method, which applies color symbols to different elevation zones. The method is also known as *layer tinting*.

Maximum z-tolerance: A TIN construction algorithm, which ensures that, for each elevation point selected, the difference between the original elevation and the estimated elevation from the TIN is within the specified tolerance.

Perspective view: A graphic method that produces 3-D views of the land surface.

Slope: The rate of change of elevation at a surface location, measured as an angle in degrees or as a percentage.

Vertical profile: A chart showing changes in elevation along a line such as a hiking trail, a road, or a stream.

Viewing angle: A parameter for creating a perspective view, measured by the angle from the horizon to the altitude of the observer.

Viewing azimuth: A parameter for creating a perspective view, measured by the direction from the observer to the surface.

Viewing distance: A parameter for creating a perspective view, measured by the distance between the viewer and the surface.

VIP: An elevation point selection algorithm, which evaluates the importance of an elevation point by measuring how well its value can be estimated from the neighboring point values.

z-scale: The ratio between the vertical scale and the horizontal scale in a perspective view. Also called the *vertical exaggeration factor.*

REVIEW QUESTIONS

1. Describe the two common types of data for terrain mapping and analysis.
2. Describe level-2 DEMs from the USGS in terms of point spacing and vertical accuracy.
3. USGS DEMs are known to contain relative errors. What are relative errors?
4. List the types of data that can be used to compile an initial TIN.
5. List the types of data that can be used to modify a TIN.
6. The maximum z-tolerance algorithm is an algorithm used by ArcGIS for converting a DEM into a TIN. Explain how the algorithm works.
7. Suppose you are given a DEM to make a contour map. The elevation readings in the DEM range from 856 to 1324 meters. If you were to use 900 as the base contour and 100 as the contour interval, what contour lines would be on the map?
8. Describe factors that can influence the visual effect of hill shading.
9. Explain how the viewing azimuth, viewing angle, viewing distance, and z-scale can change a 3-D perspective view.
10. Describe in your own words (no equation) how a computing algorithm derives the slope of the center cell by using elevations of its four immediate neighbors.
11. Describe in your own words (no equation) how ArcGIS derives the slope of the center cell by using elevations of its eight surrounding neighbors.
12. What factors can influence the accuracy of slope and aspect measures from a DEM?
13. Suppose you need to convert a raster from degree slope to percent slope. How can you complete the task in ArcGIS?
14. Suppose you have a polygon layer with a field showing degree slope. What kind of data processing do you have to perform in ArcGIS so that you can use the polygon layer in percent slope?
15. What are the advantages of using an elevation raster for terrain mapping and analysis?
16. What are the advantages of using a TIN for terrain mapping and analysis?

APPLICATIONS: TERRAIN MAPPING AND ANALYSIS

This applications section includes three tasks. Task 1 lets you create a contour layer, a vertical profile, a shaded-relief layer, and a 3-D perspective view. In Task 2, you will derive a slope layer, an aspect layer, and a surface curvature layer from DEM data. Task 3 lets you build and modify a TIN. You will use Spatial Analyst, 3D Analyst, and ArcToolbox to perform terrain mapping and analysis in this section.

Task 1: Use DEM for Terrain Mapping

What you need: *plne*, an elevation raster; and *streams.shp*, a stream shapefile.

The elevation raster *plne* is imported from a USGS 7.5-minute DEM. The shapefile *streams.shp* shows major streams in the study area.

1.1 Create a contour layer

1. Start ArcCatalog, and make connection to the Chapter 14 database. Launch ArcMap. Add *plne* to Layers, and rename Layers Tasks 1&2. Make sure that both the Spatial Analyst and 3D Analyst extensions are checked in the Tools menu and their toolbars are checked in the View menu.

2. Click the Spatial Analyst dropdown arrow, point to Surface Analysis, and select Contour. In the Contour dialog, select *plne* for the input surface, enter 100 (meters) for the contour interval and 800 (meters) for the base contour, and save the output features as *ctour.shp*. Click OK to run the analysis.

3. *ctour* appears in the map. Select Properties from the context menu of *ctour*. On the Labels tab, check the box to label features in this layer and select CONTOUR for the label field. Click OK. The contour lines are now labeled. (To remove the contour labels, right-click *ctour* and uncheck Label Features.)

1.2 Create a vertical profile

1. Add *streams.shp* to Task 1. This step selects a stream for the vertical profile. Open the attribute table of *streams*. Click the Options dropdown arrow and choose Select By Attributes. Enter the following SQL statement in the expression box: "USGH_ID" = 167. Click Apply. Close the *streams* attribute table. Zoom in on the selected stream.

2. Click the Interpolate Line tool on the 3D Analyst toolbar. Use the mouse pointer to digitize points along the selected stream. Double-click the last point to finish

digitizing. A rectangle with handles appears around the digitized stream.

3. Click the Create Profile Graph tool on the 3D Analyst toolbar. A vertical profile appears with a default title and subtitle. Right-click the title bar of the graph and select Properties. The Graph Properties dialog allows you to enter a new title and subtitle and to choose other advanced design options.

Q1. What is the elevation range along the vertical profile? Does the range correspond to the readings on *ctour* from Task 1.1?

4. The digitized stream becomes a graphic element on the map. You can delete it by using the Select Elements tool to first select it. To unselect the stream, choose Clear Selected Features from the Selection menu.

1.3 Create a hillshade layer

1. Click the Spatial Analyst dropdown arrow, point to Surface Analysis, and select Hillshade. In the Hillshade dialog, select *plne* for the Input surface. Take the default values of 315 for the azimuth, 45 for the altitude, 1 for the Z factor, and 30 for the output cell size. Opt for a temporary output raster. Click OK to run the operation. *Hillshade of plne* is added to the table of contents.

2. Try different values of azimuth and altitude to see how these two parameters affect hill shading.

Q2. Does the hillshade layer look darker or lighter with a lower altitude?

1.4 Create a perspective view

1. Click the ArcScene tool on the 3D Analyst toolbar to open the ArcScene application. Add *plne* and *stream.shp* to view. By default, *plne* is displayed in a planimetric view, without the 3-D effect. Select Properties from the context menu of *plne*. On the Base Heights tab, click the radio button to obtain heights for layer from surface, and select *plne* for the surface. Click OK to dismiss the dialog.

2. *plne* is now displayed in a 3-D perspective view. The next step is to drape *streams* on the surface. Select Properties from the context menu of *streams*. On the Base Heights tab, click the radio button to obtain heights for layer from surface, and select *plne* for the surface. Click OK.

3. Using the properties of *plne* and *streams*, you can change the look of the 3-D view. For example, you can change the color scheme for displaying *plne*. Select Properties from the context menu of *plne*. On the Symbology tab, right-click the Color Ramp box and uncheck Graphic View. Click the Color Ramp dropdown arrow and select Elevation #1. Click OK. Elevation #1 uses the conventional color scheme to display the 3-D view of *plne*. Click the symbol for *streams* in the table of contents. Select the River symbol from the Symbol Selector, and click OK.

4. You can tone down the color symbols for *plne* so that *streams* can stand out more. Select Properties from the context menu of *plne*. On the Display tab, enter 40 (%) transparent and click OK.

5. Click the View menu and select Scene Properties. The Scene Properties has four tabs: General, Coordinate System, Extent, and Illumination. The General tab has options for the vertical exaggeration factor (the default is 1 or none), background color, and a check box for enabling animated rotation. The Illumination tab has options for azimuth and altitude.

6. ArcScene has standard tools to navigate, zoom in or out, center on target, and to perform other 3-D manipulations. For example, the Navigate tool allows you to rotate the 3-D surface.

7. Besides the above standard tools, ArcScene has additional toolbars for perspective views. Click the View menu, point to Toolbars, and check the boxes for 3D Effects and

Animation. The 3D Effects toolbar has tools for adjusting transparency, lighting, and shading. The Animation toolbar has tools for making animations. For example, you can save an animation as an .avi file and use it in a PowerPoint presentation.

Task 2: Derive Slope, Aspect, and Curvature from DEM

What you need: *plne*, an elevation raster, same as Task 1.

Task 2 covers slope, aspect, and surface curvature.

2.1 Derive a slope layer

1. Click Show/Hide ArcToolbox Window to open the ArcToolbox window in ArcMap. Set the Chapter 14 database as the current workspace. Double-click the Slope tool in the Spatial Analyst Tools/Surface toolset. Select *plne* for the input raster, specify *plne_slope* for the output raster, select PERCENT_RISE for the output measurement, and click OK to execute the command.

Q3. What is the range of percent slope values in *plne_slope*?

2. *plne_slope* is a continuous raster. You can divide *plne_slope* into slope classes. Double-click Reclassify in the Spatial Analyst Tools/Reclass toolset. In the Reclassify dialog, select *plne_slope* for the input raster and click on Classify. In the next dialog, change the number of classes to 5, enter 10, 20, 30, and 40 as the first four break values, and click OK. Save the output raster as *slope_reclass* in the Reclassify dialog, before running the command.

2.2 Derive an aspect layer

1. Double-click the Aspect tool in the Spatial Analyst Tools/Surface toolset. Select *plne* for the input raster, specify *plne_aspect* for the output raster, and click OK.

2. *plne_aspect* shows an aspect layer with the eight principal directions and flat area. But it is actually a continuous raster. You can verify the statement by checking the layer properties. To create an aspect raster with the eight principal directions, you need to reclassify *plne_aspect*.

3. Double-click the Reclassify tool in the Spatial Analyst Tools/Reclass toolset. Select *plne_aspect* for the input raster and click on Classify. In the Classification dialog, make sure that the number of classes is 10. Then click the first cell under Break Values and enter −1. Enter 22.5, 67.5, 112.5, 157.5, 202.5, 247.5, 292.5, 337.5, and 360 in the following nine cells. Click OK to dismiss the Classification dialog.

4. The old values in the Reclassify dialog are now updated with the break values you have entered. Now you have to change the new values. Click the first cell under new values and enter −1. Click and enter 1, 2, 3, 4, 5, 6, 7, 8, and 1 in the following 9 cells. The last cell has the value of 1 because the cell (337.5° to 360°) and the second cell (−1° to 22.5°) make up the north aspect. Change the name of the output raster to *aspect_rec* before running the command. *aspect_rec* is an integer aspect raster with the eight principal directions and flat (−1).

Q4. The value with the largest number of cells on *aspect_rec* is −1. Can you speculate why?

2.3 Derive a surface curvature layer

1. Double-click the Curvature tool in the Spatial Analyst Tools/Surface toolset. Select *plne* for the input raster, specify *plne_curv* for the output raster, and click OK.

2. A positive cell value in *plne_curv* indicates that the surface at the cell location is upwardly convex. A negative cell value indicates that the surface at the cell location is upwardly concave. The ArcGIS Desktop

Help further suggests that the curvature output value should be within −0.5 to 0.5 in a hilly area and within −4 to 4 in rugged mountains. The elevation data set *plne* is from the Priest Lake area in North Idaho, a mountainous area with steep terrain. Therefore, it is no surprise that *plne_curv* has cell values ranging from −6.89 to 6.33.

3. Right-click *plne_curv* and select Properties. On the Symbology tab, change the show type to Classified and then click on Classify. In the Classification dialog, first select 6 for the number of classes and then enter the following break values: −4, −0.5, 0, 0.5, 4, and 6.34. Return to the Properties dialog, select a diverging color ramp (e.g., red-to-yellow-to-green), and click OK.

4. Through the color symbols, you can now tell upwardly convex cells in *plne_curv* from upwardly concave cells. Priest Lake on the west side carries the symbol for the −0.5 to 0 class; its cell value is actually 0 (flat surface). Add *streams.shp* to Tasks 1&2, if necessary. Check cells along the stream channels. Many of these cells should have symbols indicating upwardly concave.

Task 3: Build and Display a TIN

What you need: *emidalat*, an elevation raster; and *emidastrm.shp*, a stream shapefile.

Task 3 shows you how to construct a TIN from an elevation raster and to modify the TIN with *emidastrm.shp* as breaklines. You will also display different features of the TIN.

1. Select Data Frame from the Insert menu in ArcMap. Rename the new data frame Task 3, and add *emidalat* and *emidastrm.shp* to Task 3.

2. Double-click the Raster to TIN tool in the 3D Analyst Tools/Conversion toolset. Select *emidalat* for the input raster, specify *emidatin* for the output TIN, and change the

Z Tolerance value to 10. Click OK to run the command.

Q5. The default Z tolerance in the Convert Raster to TIN dialog is 48.2. What happens when you change the tolerance to 10?

3. *emidatin* is an initial TIN converted from *emidalat*. This step is to modify *emidatin* with *emidastrm*, which contains streams. Double-click the Edit TIN tool in the 3D Analyst Tools/TIN Creation toolset. Select *emidatin* for the input TIN, and select *emidastrm* for the input feature class. Notice that the default for SF_type (surface feature type) is hardline. Specify *emidatin_mod* for the output TIN before clicking OK.

4. You can view *emidatin_mod* in a variety of ways. Select Properties from the context menu of *emidatin_mod*. Click the Symbology tab. Click the Add button below the Show frame. An Add Renderer scroll list appears with choices related to the display of edges, faces, or nodes that make up *emidatin_mod*. Click "Faces with the same symbol?" in the list, click Add, and click Dismiss. Uncheck all the boxes in the Show frame except Faces. Make sure that the box to show hillshade illumination effect in 2-D display is checked. Click OK on the Layer Properties dialog. With its faces in the same symbol, *emidatin_mod* can be used as a

background in the same way as a shaded-relief map for displaying map features such as streams, vegetation, and so on.

Q6. How many nodes and triangles are in *emidatin*?

Challenge Task

What you need: *lidar*, *usgs10*, and *usgs30*.

This challenge task lets you work with DEMs at three different resolutions: *lidar* at 1.83 meters, *usgs10* at 10 meters, and *usgs30* at 30 meters.

1. Insert a new data frame in ArcMap, rename the data frame Challenge. Add *lidar*, *usgs10*, and *usgs30* to Challenge.

Q1. How many rows and columns are in each DEM?

2. Create a hillshade layer from each DEM and compare the layers in terms of the coverage of topographic details.

3. Create a degree slope layer from each DEM and reclassify the slope layer into 9 classes: 0–10°, 10–20°, 20–30°, 30–40°, 40–50°, 50–60°, 60–70°, 70–80°, and 80–90°.

Q2. List the area percentage of each slope class from each DEM.

Q3. Summarize the effect of DEM resolution on the slope layers.

REFERENCES

Bolstad, P. V., and T. Stowe. 1994. An Evaluation of DEM Accuracy: Elevation, Slope, and Aspect. *Photogrammetric Engineering and Remote Sensing* 60: 1327–32.

Campbell, J. 1984. *Introductory Cartography*. Englewood Cliffs, NJ: Prentice Hall.

Carter, J. R. 1989. Relative Errors Identified in USGS Gridded DEMs. *Proceedings, AUTO-CARTO 9*, pp. 255–65.

Carter, J. R. 1992. The Effect of Data Precision on the Calculation of Slope and Aspect Using Gridded DEMs. *Cartographica* 29: 22–34.

Chang, K., and B. Tsai. 1991. The Effect of DEM Resolution on Slope and Aspect Mapping. *Cartography and Geographic Information Systems* 18: 69–77.

Chang, K., and Z. Li. 2000. Modeling Snow Accumulation with a Geographic Information System. *International Journal of Geographical Information Science* 14: 693–707.

Chen, Z. T., and J. A. Guevara. 1987. Systematic Selection of Very Important Points (VIP) from Digital Terrain Model for Constructing Triangular Irregular Networks. *Proceedings, AUTO-CARTO* 8, pp. 50–56.

Clarke, K. C. 1995. *Analytical and Computer Cartography,* 2d ed. Englewood Cliffs, NJ: Prentice Hall.

Eyton, J. R. 1991. Rate-of-Change Maps. *Cartography and Geographic Information Systems* 18: 87–103.

Fleming, M. D., and R. M. Hoffer. 1979. *Machine Processing of Landsat MSS Data and DMA Topographic Data for Forest Cover Type Mapping.* LARS Technical Report 062879. Laboratory for Applications of Remote Sensing, Purdue University, West Lafayette, IN.

Flood, M. 2001. Laser Altimetry: From Science to Commercial LIDAR Mapping. *Photogrammetric Engineering and Remote Sensing* 67: 1209–17.

Florinsky, I. V. 1998. Accuracy of Local Topographic Variables Derived from Digital Elevation Models. *International Journal of Geographical Information Systems* 12: 47–61.

Franklin, S. E. 1987. Geomorphometric Processing of Digital Elevation Models. *Computers & Geosciences* 13: 603–9.

Gallant J. C., and J. P. Wilson. 2000. Primary Topographic Attributes. In J. P. Wilson and J. C. Gallant, eds., *Terrain Analysis: Principles and Applications,* pp. 51–85. New York: Wiley.

Gao, J. 1998. Impact of Sampling Intervals on the Reliability of Topographic Variables Mapped from Raster DEMs at a Micro-Scale. *International Journal of Geographical Information Systems* 12: 875–90.

Gertner, G., G. Wang, S. Fang, and A. B. Anderson. 2002. Effect and Uncertainty of Digital Elevation Model Spatial Resolutions on Predicting the Topographical Factor for Soil Loss Estimation. *Journal of Soil and Water Conservation* 57: 164–74.

Hill, J. M, L. A. Graham, and R. J. Henry. 2000. Wide-Area Topographic Mapping and Applications Using Airborne Light Detection and Ranging (LIDAR) Technology. *Photogrammetric Engineering and Remote Sensing* 66: 908–14.

Hodgson, M. E. 1998. Comparison of Angles from Surface Slope/Aspect Algorithms. *Cartography and Geographic Information Systems* 25: 173–85.

Horn, B. K. P. 1981. Hill Shading and the Reflectance Map. *Proceedings of the IEEE* 69 (1): 14–47.

Isaacson, D. L., and W. J. Ripple. 1990. Comparison of 7.5-Minute and 1-Degree Digital Elevation Models. *Photogrammetric Engineering and Remote Sensing* 56: 1523–27.

Jones, K. H. 1998. A Comparison of Algorithms Used to Compute Hill Slope as a Property of the DEM. *Computers & Geosciences* 24: 315–23.

Jones, T. A., D. E. Hamilton, and C. R. Johnson. 1986. *Contouring Geologic Surfaces with the Computer.* New York: Van Nostrand Reinhold.

Kumler, M. P. 1994. An Intensive Comparison of Triangulated Irregular Networks (TINs) and Digital Elevation Models (DEMs). *Cartographica* 31 (2): 1–99.

Lane, S. N., K. S. Richards, and J. H. Chandler, eds., 1998. *Landform Monitoring, Modelling and Analysis.* Chichester, England: Wiley.

Lee, J. 1991. Comparison of Existing Methods for Building Triangular Irregular Network Models of Terrain from Raster Digital Elevation Models. *International Journal of Geographical Information Systems* 5: 267–85.

Lefsky, M. A., W. B. Cohen, G. G. Parker, and D. J. Harding. 2002. Lidar Remote Sensing for Ecosystem Studies. *BioScience* 52: 19–30.

Lim, K., P. Treitz, M. Wulder, B. St-Onge, and M. Flood. 2003. Lidar Remote Sensing of Forest Structure. *Progress in Physical Geography* 27: 88–106.

Moore, I. D., R. B. Grayson, and A. R. Ladson. 1991. Digital Terrain Modelling: A Review of Hydrological, Geomorphological, and Biological Applications. *Hydrological Process* 5: 3–30.

Ritter, P. 1987. A Vector-Based Slope and Aspect Generation Algorithm. *Photogrammetric Engineering and Remote Sensing* 53: 1109–11.

Schmidt, J., I. Evans, and J. Brinkmann. 2003. Comparison of Polynomial Models for Land Surface Curvature Calculation. *International Journal of Geographical Information Science* 17: 797–814.

Sharpnack, D. A., and G. Akin. 1969. An Algorithm for Computing Slope and Aspect from Elevations. *Photogrammetric Engineering* 35: 247–48.

Skidmore, A. K. 1989. A Comparison of Techniques for Calculating Gradient and Aspect from a

Gridded Digital Elevation Model. *International Journal of Geographical Information Systems* 3: 323–34.

Srinivasan, R., and B. A. Engel. 1991. Effect of Slope Prediction Methods on Slope and Erosion Estimates. *Applied Engineering in Agriculture* 7: 779–83.

Thelin, G. P., and R. J. Pike. 1991. *Landforms of the Conterminous United States: A Digital Shaded-Relief Portrayal.* Map I-2206, scale 1:3,500,000. Washington, DC: U.S. Geological Survey.

Tsai, V. J. D. 1993. Delaunay Triangulations in TIN Creation: An Overview and Linear Time Algorithm. *International Journal of Geographical Information Systems* 7: 501–24.

Watson, D. F., and G. M. Philip. 1984. Systematic Triangulations. *Computer Vision, Graphics, and Image Processing* 26: 217–23.

Wilson, J. P., and J. C. Gallant, eds. 2000. *Terrain Analysis: Principles and Applications.* New York: Wiley.

Zar, J. H. 1984. *Biostatistica Analysis,* 2d ed. Englewood Cliffs, NJ: Prentice Hall.

Zevenbergen, L. W., and C. R. Thorne. 1987. Quantitative Analysis of Land Surface Topography. *Earth Surface Processes and Landforms* 12: 47–56.

Zhang, W., and D. R. Montgomery. 1994. Digital Elevation Model Raster Size, Landscape Representation, and Hydrologic Simulations. *Water Resources Research* 30: 1019–28.

15

VIEWSHEDS AND WATERSHEDS

CHAPTER OUTLINE

Terrain analysis covers basic topographic parameters such as slope, aspect, and surface curvature (Chapter 14) as well as more specific applications. Chapter 15 focuses on two such applications: viewshed analysis and watershed analysis. A viewshed is the portion of the land surface that is visible from one or more viewpoints. A watershed is the area that drains surface water to a common outlet. Viewshed analysis can be based on a digital elevation model (DEM) or a triangulated irregular network (TIN). Delineation of watershed boundaries, on the other hand, is typically based on a DEM.

The terrain serves primarily as the backdrop in viewshed analysis. A large part of viewshed analysis involves the setup of the viewpoint and viewing parameters. The location and height of the viewpoint can change the size of a viewshed. And the viewing angle, the search distance, and even the tree height can affect the size and shape of a viewshed.

Watershed analysis traces the flow and channeling of surface water on the terrain. A basic principle in watershed analysis is that water flows in the downslope direction. In practice, flow direction is calculated by a neighborhood operation on an elevation raster, similar to the calculation of slope and aspect. Subsequent raster data operations on the flow direction raster and its derivatives complete the automatic process for delineating watersheds.

Chapter 15 is organized into six sections. Section 15.1 introduces viewshed analysis. Section 15.2 examines various parameters that can affect viewshed analysis. Section 15.3 provides an overview of viewshed applications. Section 15.4 introduces watershed analysis and the analysis procedure. Section 15.5 discusses factors that can influence the outcome of a watershed analysis. Section 15.6 covers applications of watershed analysis.

15.1 VIEWSHED ANALYSIS

A **viewshed** refers to the portion of the land surface that is visible from one or more viewpoints (Figure 15.1). The process for deriving viewsheds is called viewshed or visibility analysis. A viewshed analysis requires two input data sets. The first is usually a point layer that contains one or more

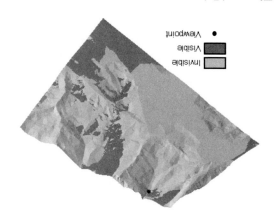

Figure 15.1
A viewshed example.

viewpoints. If a line layer is used, the viewpoints are points (vertices and nodes) that make up the linear features. The second input is a DEM (i.e., an elevation raster) or a TIN, which represents the land surface.

The basis for viewshed analysis is the **line-of-sight** operation. The line of sight, also called *sight-line*, connects the viewpoint and the target. If any land, or any object on the land, rises above the line, then the target is invisible to the viewpoint. If no land or object blocks the view, then the target is visible to the viewpoint. Rather than just marking the target point as visible or not, a GIS can display a sightline with symbols for the visible and invisible portions along the sightline.

Figure 15.2 illustrates a line-of-sight operation. Figure 15.2a shows the visible (white) and invisible (black) portions of the sightline connecting the viewpoint and the target point. Figure 15.2b shows the vertical profile along the sightline. In this case, the viewpoint is at an elevation of 994 meters on the east side of a stream. The visible portion of the sightline follows the downslope, crosses the stream at the elevation of 932 meters, and continues uphill on the west side of the stream. The ridgeline at an elevation of 1028 meters presents an obstacle and marks the beginning of the invisible portion.

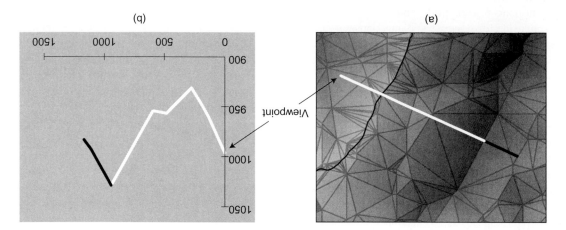

Figure 15.2
A sightline connects two points on a TIN in (*a*). The vertical profile of the sightline is depicted in (*b*). In both diagrams, the visible portion is shown in white and the invisible portion in black.

Viewshed analysis expands the line-of-sight operation to cover every possible cell or every possible TIN facet in the study area. Because viewshed analysis can be a time-consuming operation, various algorithms have been developed for computing viewsheds. Some algorithms are designed for elevation rasters, and others are for TINs. A commercial GIS package usually does not provide choices of algorithms or information on the adopted algorithm. ArcGIS, for example, can take either an elevation raster or a TIN as the data source but save the output of a viewshed analysis only in raster format. It is possible that ArcGIS converts a TIN into an elevation raster for viewshed analysis to take advantage of the computational efficiency of raster data.

Deriving a viewshed from an elevation raster follows a series of steps. First, a sightline is set up between the viewpoint and a target location. Second, a set of intermediate points is derived along the sightline. Typically, these intermediate points are chosen from the intersections between the sightline and the grid lines of the elevation raster. Third, the elevations of the intermediate points are interpolated. Finally, the computing algorithm examines the elevations of the intermediate points and determines if the target is visible or not.

The above procedure can be repeated using each cell in the elevation raster as a target (Clarke 1995). The result is a raster that classifies cells into the visible and invisible categories. An algorithm proposed by Wang et al. (1996) takes a different approach to save the computer processing time. Before running the line-of-sight operations, their algorithm first screens out invisible cells by analyzing the local surface at the viewpoint and the target.

A basic algorithm for defining viewsheds on TINs locates the viewpoint at a TIN vertex and classifies an entire triangle as either visible or invisible (Goodchild and Lee 1989; Lee 1991). A TIN triangle is visible only if all three of its edges are visible. This algorithm can save some computer processing time when compared to other algorithms that divides a TIN triangle into visible and invisible parts (De Floriani and Magillo 1994, 1999).

The output of a viewshed analysis, either using an elevation raster or a TIN as the data source, is a binary map showing visible and not visible areas. Given one viewpoint, a viewshed map has the value of 1 for visible and 0 for not visible. Given two viewpoints, a viewshed map has three possible values: 2 for visible from both points, 1 for visible from one point, and 0 for not visible (Figure 15.3).

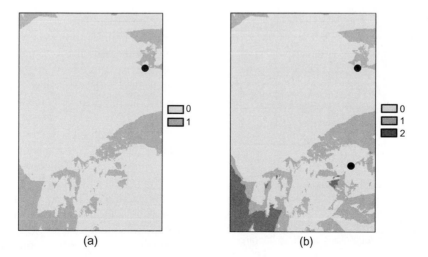

(a)   (b)

Figure 15.3

A viewshed based on one viewpoint in (a) has two categories: 0 for not visible, and 1 for visible. A viewshed based on two viewpoints in (b) has three categories: 0 for not visible, 1 for visible from one viewpoint, and 2 for visible from both viewpoints.

In other words, the number of possible values in a viewshed map is $n + 1$, where n is the number of viewpoints. A viewshed map based on two or more viewpoints is often called a **cumulative viewshed map.**

The accuracy of viewshed analysis can depend on the accuracy of DEM data and the data model (i.e., TIN versus DEM) (Fisher 1991). Even the definition of the target (e.g., center of a cell or one of the cell's corner points) can influence the analysis outcome (Fisher 1993). As shown in Figure 15.4, visibility can vary within a cell. A cell may be classified as visible or invisible depending on where the target is supposed to be located within the cell.

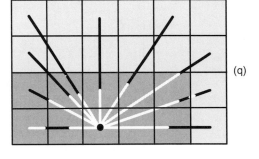

Figure 15.4
In (*a*), cells of an elevation raster are divided into visible (gray cells) and invisible (white cells) from the viewpoint located at the elevation of 918 in a viewshed analysis. The sightlines in (*b*) suggest that the analysis result can vary depending on where the target point is supposed to be located within the cell. For example, the cell at row 2, column 6 is classified as visible in (*a*). But the sightline shows that the center of the cell is actually invisible to the viewpoint.

15.2 PARAMETERS OF VIEWSHED ANALYSIS

A number of parameters can influence the result of a viewshed analysis. The first parameter is the viewpoint. A viewpoint located along a ridgeline would have a wider view than a viewpoint located in a narrow valley. There are at least two scenarios in dealing with the viewpoint in GIS. The first scenario assumes that the location of the point is fixed. If the elevation at the point is known, it can be entered directly in a field. If not, it can be estimated from an elevation raster or a TIN. ArcGIS, for example, uses bilinear interpolation (Chapter 7) to interpolate the elevation of a viewpoint from an elevation raster. The second scenario assumes that the viewpoint is to be selected. If we further assume that the objective is to gain maximum visibility, then we should select the viewpoint at a high elevation with open views. A GIS provides various tools that can help locate suitable viewpoints (Box 15.1).

After the viewpoint is determined, its elevation should be increased by the height of the observer and, in some cases, the height of a physical structure. For instance, a forest lookout station is usually 15 to 20 meters high. The height of the observation station, when added as an offset value to the elevation at the station, makes the viewpoint higher than its immediate surroundings, thus increasing its viewshed (Figure 15.5).

The second parameter is the **viewing azimuth,** which sets horizontal angle limits to the view.

A recent report suggests that the average level of agreement between GIS-predicted viewsheds and field-surveyed viewsheds is only slightly higher than 50 percent (Maloy and Dean 2001). An alternative to a binary viewshed map is to express visibility in probabilistic terms (Fisher 1996). Although currently not available from commercial GIS packages, it is possible to produce a probabilistic visibility map by incorporating an error source such as DEM errors in Monte Carlo simulations (Nackaerts et al. 1999).

Box 15.1	**Tools for Selecting Viewpoints**

A GIS package has various tools that we can use to select viewpoints that are at high elevations with open views. Tools such as contouring and hill shading can provide an overall view of the topography in a study area. Tools for data query can narrow the selection to a specific elevation range such as within 100 meters of the highest elevation in the data set. Data extraction tools can extract elevation readings from an underlying surface (i.e., a raster or a TIN) at point locations, along a line, or within a circle, a box, or a polygon. ArcGIS offers data extraction tools through Spatial Analyst, 3D Analyst, and ArcToolbox. These extraction tools are useful for narrowing the selection of a viewpoint to a small area with high elevations. It would be difficult, however, to know the exact elevation of a viewpoint because ArcGIS uses the bilinear interpolation method and four closest cell values to estimate a viewpoint's elevation from the underlying elevation raster.

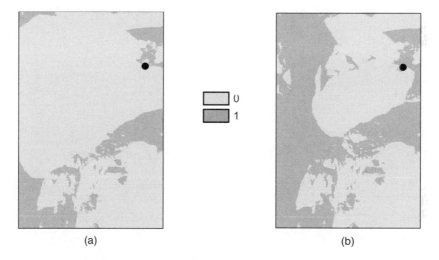

(a)   (b)

Figure 15.5
The increase of the visible areas from (*a*) to (*b*) is a direct result of adding 20 meters to the height of the viewpoint.

Figure 15.6, for example, uses a viewing angle from 0° to 180°. The default is a full 360° sweep, which is unrealistic in many instances. To simulate the view from the window of a property (e.g., a home or an office), a 90° viewing azimuth (45° either side of the perpendicular to the window) is more realistic than a 360° sweep (Lake et al. 1998).

Viewing radius is the third parameter, which sets the search distance for deriving visible areas.

Figure 15.7, for example, shows viewable areas within a radius of 8000 meters around the viewpoint. The default view distance is typically infinity. The setting of the search radius can vary with project. A 5-km radius has been used, for example, to simulate the view from an observation tower that guards the access to an ancient city (Nackaerts et al. 1999).

Other parameters include vertical viewing angle limits, the Earth's curvature, and tree height.

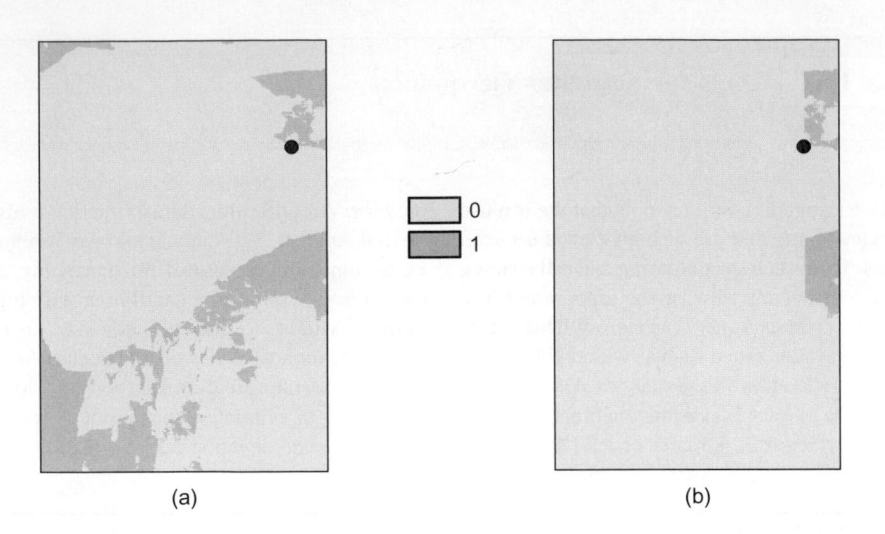

(a)   (b)

Figure 15.6
The difference in the visible areas between (a) and (b) is due to the viewing angle: 0° to 360° in (a) and 0° to 180° in (b).

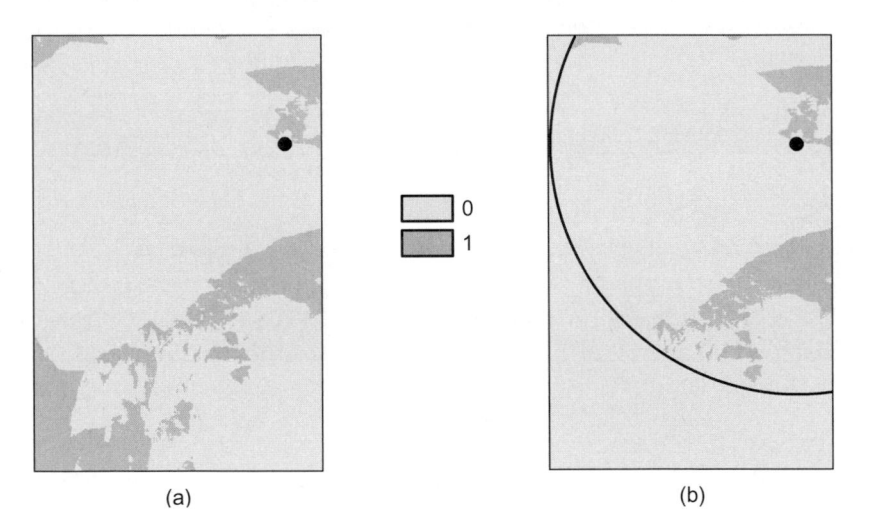

(a)   (b)

Figure 15.7
The difference in the visible areas between (a) and (b) is due to the search radius: infinity in (a) and 8000 meters from the viewpoint in (b).

Vertical viewing angles can range from 90° above the horizontal plane to −90° below. The Earth's curvature can be either ignored or corrected in deriving a viewshed. Tree height can be an important factor if a viewshed analysis is conducted within a forested land from a road or trail. In that case, estimated tree height can be added to ground elevation to create forest elevation (Wing and Johnson 2001).

A GIS package handles the parameters for viewshed analysis as attributes in the viewpoint

Box 15.2 | Setting Parameters for Viewshed Analysis

Tools for viewshed analysis are available through Spatial Analyst, 3D Analyst, and ArcToolbox in ArcGIS. But to set up the parameters for viewshed analysis, we must work with the attribute table of the viewpoint data set. Each parameter has a predefined field name. OFFSETA and OFFSETB define the heights to be added to the viewpoint and the target respectively. AZIMUTH1 and AZIMUTH2 define the view's horizontal angle limits. For example, to limit the viewing angle from 0° to 180°, we can assign 0 to AZIMUTH1 and 180 to AZIMUTH2. RADIUS1 and RADIUS2 define the search distance. For example, to limit the search distance to 2 miles from the viewpoint, we can assign 0 to RADIUS1 and 2 miles to RADIUS2. VERT1 and VERT2 define the view's vertical angle limits. The vertical angle values range from 90° to −90°. Task 1 in the applications section shows how to define and add an OFFSETA value to a viewpoint.

data set. Therefore, we must set up the parameters before running viewshed analysis. Box 15.2 describes the setup of viewshed analysis parameters in ArcGIS.

15.3 APPLICATIONS OF VIEWSHED ANALYSIS

Viewshed analysis is useful for the site selection of facilities such as forest lookout stations, wireless telephone base stations, and microwave towers for radio and television. The location of these facilities is chosen to maximize the viewable areas without having too much overlap. Viewshed analysis can help locate these facilities, especially at the preliminary stage.

Viewshed analysis can also be used for housing and resort area developments, although the objective of the analysis can differ between the developer and current residents. New facilities can easily intrude on residents in rural areas (Davidson et al. 1993). Visual intrusion and noise associated with road development can also affect property values in an urban environment (Lake et al. 1998).

Viewshed analysis is closely related to analysis of visual landscape. The U.S. Forest Service, for example, uses viewshed analysis for delineating landscape view areas and for reducing the visual impact of clear-cuts and logging activities.

A national park or a nature reserve may want to monitor and assess the impacts of land cover change on the visual landscape (Miller 2001). Viewshed analysis can also be integrated with least-cost path calculations (Chapter 18) to provide scenic paths for hikers and others (Lee and Stucky 1998). A recent study has attempted to derive measures that can quantify the perceptual characteristics of landscapes (O'Sullivan and Turner 2001).

15.4 WATERSHED ANALYSIS

Defined by topographic divides, a **watershed** is an area that drains surface water to a common outlet. A watershed is a hydrologic unit that is often used for the management and planning of natural resources. Chapter 15 uses watershed as a general term, rather than as a specific class in the Watershed Boundary Dataset that is being developed by government agencies in the United States (Box 15.3). **Watershed analysis** in Chapter 15 refers to the process of using DEMs and raster data operations to delineate watersheds and to derive topographic features such as stream networks.

Traditionally, watershed boundaries are drawn manually onto a topographic map. The person who draws the boundaries uses topographic features on the map to determine where a divide is located. Today computer programs are also used to derive watersheds from DEMs. Using computer technology, we can generate preliminary watershed boundaries in a fraction of the time needed for the traditional method.

Delineation of watersheds can take place at different spatial scales. A large watershed may cover an entire stream system and, within the watershed, there are smaller watersheds, one for each tributary in the stream system. Delineation of watersheds can also be area-based or point-based. An area-based method divides a study area into a series of watersheds, one for each stream section. A point-based method, on the other hand, derives a watershed for each select point. The select point may be an outlet, a gauge station, or a dam. Whether area- or point-based, the automated method for delineating watersheds follows a series of steps, starting with a filled DEM.

15.4.1 Filled DEM

A filled **DEM** or elevation raster is void of depressions. A depression is a cell or cells in an elevation raster that are surrounded by higher-elevation values, and thus represents an area of internal drainage. Although some depressions are real, such as quarries or glaciated potholes, many are imperfections in the DEM. Therefore depressions must be removed from an elevation raster. A common method for removing a depression is to increase its cell value to the lowest overflow point out of the sink (Jenson and Domingue 1988). The flat surface resulting from sink filling still needs to be interpreted to define the drainage flow. One approach is to impose two shallow gradients and to force flow away from higher terrain surrounding the flat surface toward the edge bordering lower terrain (Garbrecht and Martz 2000).

15.4.2 Flow Direction

A **flow direction raster** shows the direction water will flow out of each cell of a filled elevation raster. A widely used method for deriving flow direction is the D8 method. Used by ArcGIS, the D8 method assigns a cell's flow direction to one of its eight surrounding cells that has the steepest distance-weighted gradient (Figure 15.8) (O'Callaghan and Mark 1984). The method does not allow flow to be distributed to multiple cells.

The D8 method produces good results in zones of convergent flows and along well-defined valleys (Freeman 1991). But it tends to produce flow in parallel lines along principal directions (Moore 1996) and fails to represent adequately divergent flow over convex slopes and ridges (Freeman 1991). Other algorithms have been proposed to introduce a degree of randomness into the flow

---

  Box 15.3 Watershed Boundary Dataset (WBD)

T he term *watershed* in this text refers to an area that drains surface water to a common outlet. But in the Watershed Boundary Dataset (WBD), a watershed represents a specific hydrologic unit. The WBD is being developed by a number of federal agencies in the United States using a common set of guidelines documented in the Federal Standards for Delineation of Hydrologic Unit Boundaries (**http://water.usgs.gov/wicp/acwi/spatial/index.html**). The main objective of the WBD program is to add two finer levels of watershed and subwatershed to the current four levels of region, subregion, basin (formerly accounting unit), and subbasin (formerly cataloging unit). Watersheds and subwatersheds are being delineated and georeferenced to the USGS 1:24,000 scale quadrangle maps. On average, a watershed covers 40,000 to 250,000 acres and a subwatershed covers 10,000 to 40,000 acres.

1014	1011	1004
1019	1015	1007
1025	1021	1012

(a)

+1	+4	+11
−4		+8
−10	−6	+3

(b)

(c)

Figure 15.8

The flow direction of the center cell in (a) is determined by first calculating the distance-weighted gradient to each of its eight neighbors. For the four immediate neighbors, the gradient is calculated by dividing the elevation difference (b) between the center cell and the neighbor by 1. For the four corner neighbors, the gradient is calculated by dividing the elevation difference (b) by 1.414. The results show that the steepest gradient, and therefore the flow direction, is from the center cell to its right.

direction computations and to allow flow divergence (Gallant and Wilson 2000). An example is the D∞ (D Infinity) method (Tarboton 1997). The D∞ method first forms eight triangles by connecting the centers of the cell and its eight surrounding cells. It selects the triangle with the maximum downhill slope as the flow direction. The two neighboring cells that the triangle intersects receive the flow in proportion to their closeness to the aspect of the triangle. The D∞ method is available as an extension to ArcGIS.

15.4.3 Flow Accumulation

A **flow accumulation raster** tabulates for each cell the number of cells that will flow to it. The tabulation is based on the flow direction raster (Figure 15.9). With the appearance of a spanning tree (Figure 15.10), a flow accumulation raster records how many upstream cells will contribute drainage to each cell (the cell itself is not counted).

A flow accumulation raster can be interpreted in two ways. First, cells having high accumulation values generally correspond to stream channels, whereas cells having an accumulation value of zero generally correspond to ridgelines. Second, if

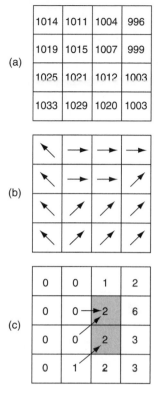

(a)

1014	1011	1004	996
1019	1015	1007	999
1025	1021	1012	1003
1033	1029	1020	1003

(b)

(c)

Figure 15.9

This illustration shows a filled elevation raster (a), a flow direction raster (b), and a flow accumulation raster (c). Both shaded cells in (c) have the same flow accumulation value of 2. The top cell receives its flow from its left and lower-left cells. The bottom cell receives its flow from its lower-left cell, which already has a flow accumulation value of 1.

multiplied by the cell size, the cell value equals the drainage area.

15.4.4 Stream Network

A stream network can be derived from a flow accumulation raster. The derivation is based on a threshold accumulation value. A threshold value of 500, for example, means that each cell of the drainage network has a minimum of 500 contributing cells. Given the same flow accumulation raster, a higher threshold value will result in a less dense

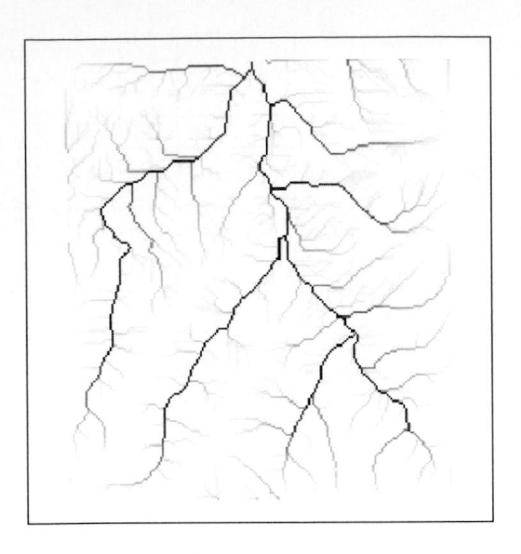

Figure 15.10
A flow accumulation raster, with darker symbols representing higher flow accumulation values.

stream network and fewer internal watersheds than a lower threshold value. Figure 15.11 illustrates the effect of the threshold value. Figure 15.11*a* shows the flow accumulation raster, Figure 15.11*b* the stream network based on a threshold value of

500 cells, and Figure 15.11*c* the stream network based on a threshold value of 100 cells.

The threshold value is a necessary input to watershed analysis. But the choice of a threshold value can be arbitrary. Ideally, the resulting stream network from a threshold value should correspond to a network obtained from traditional methods such as from high-resolution topographic maps or field mapping (Tarboton et al. 1991). Figure 15.12 shows the hydrography from the 1:24,000 scale digital line graph (DLG) for the same area as Figure 15.11. A threshold value between 100 and 500 cells seems to best capture the stream network in the area.

15.4.5 Stream Links

After a stream network is derived from a flow accumulation raster, each section of the stream raster line is assigned a unique value and is associated with a flow direction (Figure 15.13). A stream link raster therefore resembles a topology-based stream layer: the intersections or junctions are like nodes, and the stream sections between junctions are like arcs or reaches (Figure 15.14).

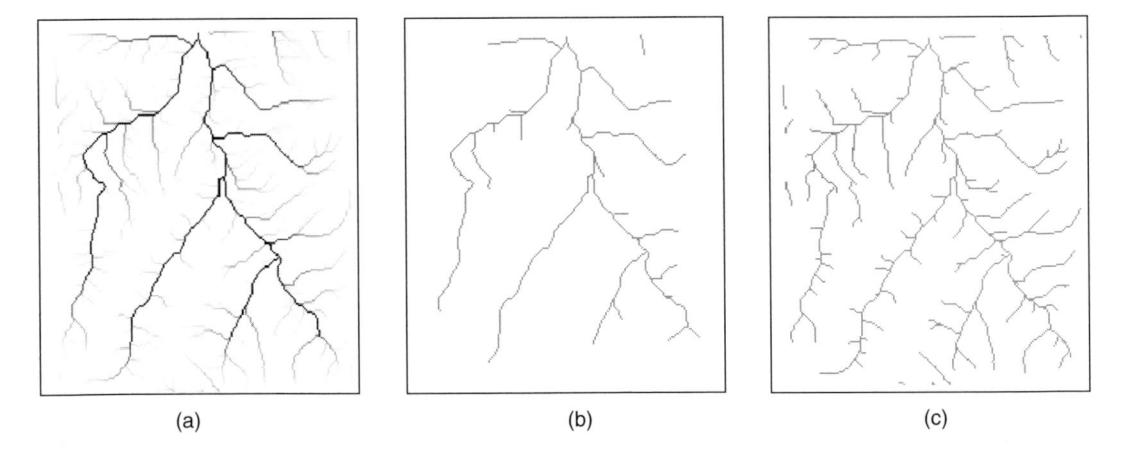

(a)   (b)   (c)

Figure 15.11
(*a*) A flow accumulation raster; (*b*) a stream network based on a threshold value of 500 cells; and (*c*) a stream network based on a threshold value of 100 cells.

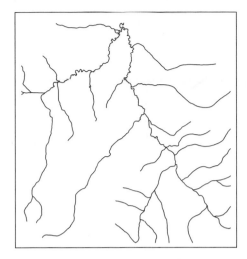

Figure 15.12
The stream network from the 1:24,000 scale digital line graph for the same area as Figure 15.11.

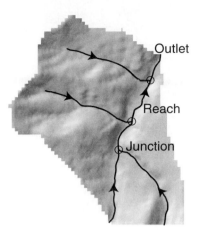

Figure 15.14
A stream link raster includes reaches, junctions, flow directions, and an outlet.

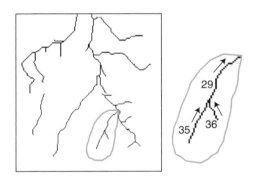

Figure 15.13
To derive the stream links, each section of the stream network is assigned a unique value and a flow direction. The inset map on the right shows three stream links.

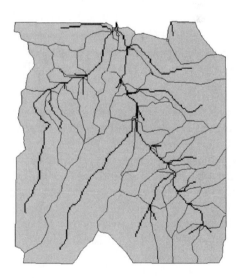

Figure 15.15
Areawide watersheds.

15.4.6 Areawide Watersheds

The final step is to delineate a watershed for each stream section (Figure 15.15). This operation uses the flow direction raster and the stream link raster as the inputs. A denser stream network (i.e., based on a smaller threshold value) will have more, but smaller, watersheds. Figure 15.15 does not cover the entire area of the original DEM. The missing areas around the rectangular border are areas that do not have flow accumulation values higher than the specified threshold value.

15.4.7 Point-Based Watersheds

Instead of deriving a watershed for each identified stream section, the task for some projects is to delineate individual watersheds based on points of interest (Figure 15.16). These points of interest may be stream gauge stations or dams. They may also correspond to surface drinking water system intake locations (Perez 2000). In watershed analysis, these points of interest are called **pour points** or *outlets*.

Delineation of individual watersheds based on pour points follows the same procedure as for delineation of areawide watersheds. The only difference is to substitute a point raster for a stream link raster. In the point raster, a cell representing a pour point must be located over a cell that is part of the stream link. If a pour point is not located directly over a stream link, it will result in a small, incomplete watershed for the outlet.

Figure 15.17 illustrates the importance of the location of a pour point. The pour point in the example represents a USGS gauge station on a river. The geographic location of the station is recorded in longitude and latitude values at the USGS website. When plotted on the stream link raster, the station is about 50 meters away from the stream. The location error results in a very small watershed for the station, as shown in Figure 15.17. By moving the station to be on the stream, it

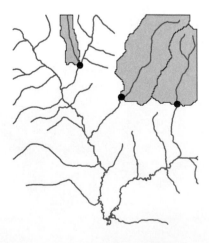

Figure 15.16
Point-based watersheds (shaded area).

results in a large watershed that actually spans beyond the length of a 1:24,000 scale quadrangle map (Figure 15.18). ArcGIS has a command that can snap a pour point to a stream cell within a user-defined search radius (Box 15.4).

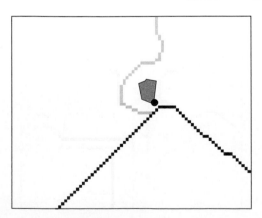

Figure 15.17
If a pour point (black circle) is not snapped to a cell with a high flow accumulation value (dark cell symbol), it usually has a small number of cells (shaded area) identified as its watershed.

Figure 15.18
When the pour point in Figure 15.17 is snapped to a cell representing a stream channel, its watershed extends to the border of a USGS 1:24,000 scale quadrangle map and beyond.

Box 15.4 | **Snapping Pour Points**

The SnapPour command, which is available through Spatial Analyst and ArcToolbox in ArcGIS, can snap pour points to the cell of highest flow accumulation within a user-defined search distance. The SnapPour operation should be considered as part of data preprocessing for delineating point-based watersheds. In many instances, pour points are digitized on screen, converted from a table with x-, y-coordinates, or selected from an existing data set. These points are rarely right on top of a computer-generated stream channel. The discrepancy can be caused by the poor data quality of the pour points, the inaccuracy of the computer-generated stream channel, or both. Task 4 of the applications section covers use of the SnapPour command.

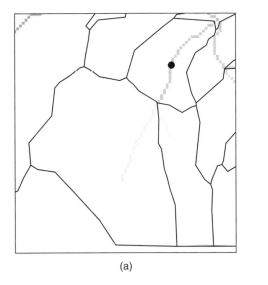
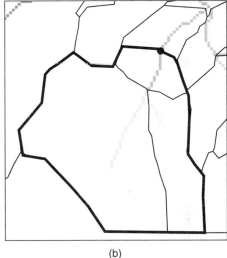

(a)   (b)

Figure 15.19
The pour point (black circle) in (*a*) is located along a stream section rather than at a junction. The watershed derived for the pour point is a merged watershed, shown in the thick black line in (*b*), which represents the upstream contributing area at the pour point.

The algorithm for deriving a point-based watershed varies according to the location of the pour point. If the pour point is located at a junction, then the watersheds (i.e., watersheds as shown in Figure 15.15) upstream from the junction are merged to form the watershed for the pour point. If the pour point is located between two junctions, then the watershed assigned to the stream section between the two junctions is divided into two, one upstream from the pour point and the other downstream (Figure 15.19). The upstream portion of the watershed is then merged with watersheds further upstream to form the watershed for the pour point.

15.5 FACTORS INFLUENCING WATERSHED ANALYSIS

Several factors can influence the outcome of a watershed analysis. As the primary data source, DEMs play a crucial role in watershed analysis. DEMs can vary in both resolution and quality. A 30-meter DEM is likely to be too coarse to provide detailed topographic features for geomorphic and hydrologic modeling. Figure 15.20 compares the topography as depicted by a 30-meter DEM versus a 10-meter DEM. Clearly the 10-meter DEM can better define the topographic features than the 30-meter DEM. Figure 15.21 shows the stream networks derived from these two types of DEMs using the same flow accumulation threshold. As expected, the 10-meter DEM yields a more detailed stream network than the 30-meter DEM. For studies of small catchments, researchers have advocated DEMs with a 10-meter resolution (Zhang and Montgomery 1994) and even higher (Gertner et al. 2002).

The quality of DEMs is also important. As described in Section 15.4, the first step in watershed analysis is to prepare a depression-free DEM. Many depressions are examples of imperfections in DEMs. Accuracy of elevation data is especially important in low-relief areas. Whenever possible, one should choose level-2 over level-1 USGS DEMs for watershed analysis.

The algorithm for deriving flow directions is another important factor. Commercial GIS packages including ArcGIS use the D8 method mainly because it is simple and efficient. The method produces good results in mountainous topography with convergent flows but does not do as well in highly variable topography with floodplains and wetlands (Liang and Mackay 2000). Figure 15.22, for example, shows that the D8 method performs well in well-defined valleys but poorly in relatively flat areas. The D8 method also ranks low in a couple of recent studies, which compares different flow routing algorithms in experimental settings (Zhou and Liu 2002; Endreny and Wood 2003).

Automated watershed delineation typically uses the stream network derived from a flow accumulation raster. But the stream network can deviate from that on a USGS 1:24,000 scale DLG, especially in areas of low relief. A method that can integrate a vector-based stream layer into the

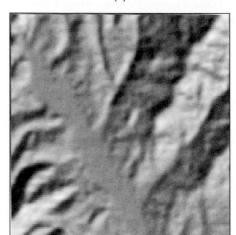
(a)

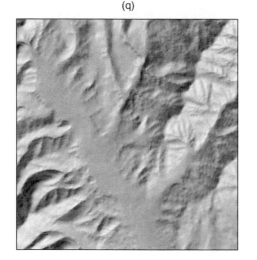
(b)

Figure 15.20
DEMs at a 30-meter resolution (a) and a 10-meter resolution (b).

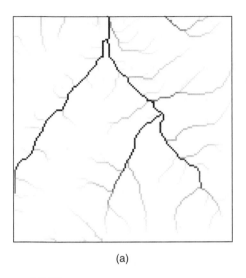
(a)

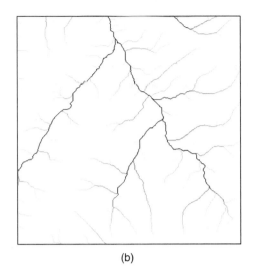
(b)

Figure 15.21
Stream networks derived from the DEMs in Figure 15.20. The stream network derived from the 30-meter DEM (*a*) has fewer details than that from the 10-meter DEM (*b*).

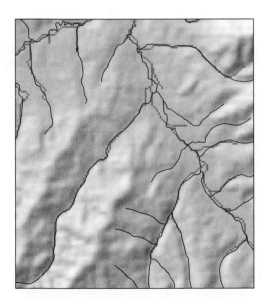

Figure 15.22
The gray raster lines represent stream segments derived using the D8 method. The thin black lines are stream segments from the 1:24,000 scale DLG. The two types of lines correspond well in well-defined valleys but poorly on the bottomlands.

process of watershed delineation is commonly referred to as "stream burning." But the method requires extensive data preprocessing, and the hydrography-enhanced DEM may produce distorted watershed boundaries (Saunders 2000).

15.6 APPLICATIONS OF WATERSHED ANALYSIS

A watershed is a hydrologic unit that is often used for the management and planning of natural resources. Therefore, an important application of watershed analysis is in the area of watershed management. **Watershed management** approaches the organization and planning of human activities on a watershed by recognizing the interrelationships among land use, soil, and water as well as the linkage between uplands and downstream areas (Brooks et al. 2003). One requirement for implementing watershed management programs is the analytical capability that can provide not only watershed boundaries but also hydrologic parameters useful for the management programs.

The Clean Water Act, introduced in the 1970s in the United States, is aimed at restoring and protecting water. Among the Act's current action plans is the call for a unified policy for ensuring a watershed approach to federal land and resource management (http://www.cleanwater.gov/ufp/). The policy's guiding principle is to use a consistent and scientific approach to manage federal lands and resources to assess, protect, and restore watersheds. Some areas of the country do not yet have delineated watersheds and, even if they do exist, they are not at a large enough map scale to be useful for many applications. This is why a multiagency effort is currently under way to create the Watershed Boundary Dataset (WBD) (http://www.ncgc.nrcs.usda.gov/branch/gdb/products/watershed/index.html). As the lead agency in this effort, the Natural Resources Conservation Service (NRCS) is responsible for certification of the WBD.

Another major application of watershed analysis is to provide the necessary inputs for hydrologic modeling. The Hydrologic Engineering Center (HEC) of the U.S. Army Corps of Engineers, for example, has a software package called Hydrologic Modeling System (HMS), which can simulate the precipitation runoff processes using different scenarios (http://www.hec.usace.army.mil/). One data set required by HMS is the basin model that includes parameter and connectivity data for such hydrologic elements as subbasin, reach, junction, source, and sink. As covered in Chapter 15, these hydrologic elements can be generated from a watershed analysis.

Flood prediction models and snowmelt runoff models are other examples that can use topographic features generated from watershed analysis. A flood prediction model requires such variables as contributing drainage area, channel slope, stream length, and basin elevation; and a snowmelt runoff model requires the snow-covered area of the watershed and its topographic features.

KEY CONCEPTS AND TERMS

Cumulative viewshed: A viewshed based on two or more viewpoints.

Filled DEM: A digital elevation model that is void of depressions.

Flow accumulation raster: A raster that shows for each cell the number of cells that will flow to it.

Flow direction raster: A raster that shows the direction water will flow out of each cell of a filled elevation raster.

Line-of-sight: A line connecting the viewpoint and the target in viewshed analysis. Also called *sightline*.

Pour points: Points used for deriving contributing watersheds.

Viewing radius: A parameter that sets the search distance for deriving a viewshed.

Viewing azimuth: A parameter that sets horizontal angle limits to the view from a viewpoint.

Viewshed: Areas of the land surface that are visible from one or more viewpoints.

Watershed: An area that drains water and other substances to a common outlet.

Watershed analysis: An analysis that involves derivation of flow direction, watershed boundaries, and stream networks.

Watershed management: A practice of managing human activities on a watershed by recognizing the interrelationships among land use, soil, and water as well as the linkage between uplands and downstream areas.

REVIEW QUESTIONS

1. Describe the two types of input data required for viewshed analysis.

2. The output from a viewshed analysis is a binary map. What does a binary map mean in this case?

3. Some researchers have advocated probabilistic visibility maps. Why?

4. What parameters can we choose for viewshed analysis?

5. Suppose you are asked by the U.S. Forest Service to run viewshed analysis along a scenic highway. You have chosen a number of points along the highway as viewpoints. You want to limit the view to within 2 miles from the highway and within a horizontal viewing angle from due west to due east. What parameters will you use? What parameter values will you specify?

6. What does the parameter of OFFSETA do for viewshed analysis in ArcGIS?

7. Besides the examples cited in Chapter 15, could you think of another viewshed application from your discipline?

8. Draw a diagram that shows the elements of watershed, topographic divide, stream section, stream junction, and outlet.

9. What is a filled DEM? Why is a filled DEM needed for a watershed analysis?

10. The example in Figure 15.8 shows an eastward flow direction. Suppose the elevation of the lower-left cell is changed from 1025 to 1028. Will the flow direction remain the same?

11. What kinds of criticisms has the D8 method received?

12. How do you interpret a flow accumulation raster (Figure 15.9)?

13. Deriving a drainage network from a flow accumulation raster requires the use of a threshold value. Explain how the threshold value can alter the outcome of the drainage network.

14. To generate areawide watersheds from a DEM, we must create several intermediate rasters. Draw a flow chart that starts with a DEM, followed by the intermediate rasters, and ends with a watershed raster.

15. A watershed identified for a pour point is often described as a merged watershed. Why?

16. Describe the effect of DEM resolution on watershed boundary delineation.

17. Describe an application example of watershed analysis in your discipline.

APPLICATIONS: VIEWSHEDS AND WATERSHEDS

This applications section includes four tasks. Task 1 covers viewshed analysis and the effect of the viewpoint's height offset on the viewshed. Task 2 creates a cumulative viewshed by using two viewpoints, one of which is added through on-screen digitizing. Task 3 covers the steps for deriving areawide watersheds from a DEM. Task 4 focuses on the derivation of point-based watersheds and the importance of snapping points of interest to the stream channel. Viewshed tools are available through Spatial Analyst, 3D Analyst, and Arc-Toolbox in ArcGIS. Watershed tools are available through Spatial Analyst and ArcToolbox. You will use Spatial Analyst for Tasks 1, 2, and 4, and ArcToolbox for Task 3.

Task 1: Perform Viewshed Analysis

What you need: *plne*, an elevation raster; and *lookout.shp*, a lookout point shapefile.

The lookout point shapefile contains a viewpoint. In Task 1, you first create a hillshade map of pine to better visualize the terrain. Next you run a viewshed analysis without specifying any parameter value. Then you add 15 meters to the height of the viewpoint to increase the viewshed coverage.

1. Launch ArcCatalog and make connection to the Chapter 15 database. Start ArcMap, and rename the data frame Tasks 1&2. Add pine and lookout.shp to Tasks 1&2. First, create a hillshade map of pine. Click the Spatial Analyst dropdown arrow, point to Surface Analysis, and select Hillshade. Select pine for the input surface and take the default values for the other parameters. Click OK to dismiss the dialog. Hillshade of pine is added to the map. Right-click Hillshade of pine and select Properties. On the Display tab, enter 30% transparent. The 30% transparency allows Hillshade of pine to be superimposed with other layers.

2. Now run a viewshed analysis. Click the Spatial Analyst dropdown arrow, point to Surface Analysis, and select Viewshed. Make sure that the input surface is pine and the observer point is from lookout. Opt for a temporary output raster. Click OK to run the viewshed analysis.

3. Viewshed of lookout separates visible areas from not visible areas. Open the attribute table of Viewshed of lookout. The table shows the cell counts for the visibility classes of 0 (not visible) and 1 (visible).

Q1. What area percentage of pine is visible from the viewpoint?

4. Suppose the viewpoint has a physical structure that adds a height of 15 meters. You can use the field of OFFSETA to include this height in viewshed analysis. Open ArcToolbox. Double-click the Add Field tool in the Data Management Tools/Fields toolset. Select lookout for the input table, enter OFFSETA for the field name, and click OK. Double-click the Calculate Field tool in the Data Management

Tools/Fields toolset. Select lookout for the input table, select OFFSETA for the field name, enter 15 for the expression, and click OK. Open the attribute table of lookout to make sure that the offset is set up correctly.

5. Follow Step 2 to run another viewshed analysis with the added height of 15 meters to the viewpoint. The result should show an increase of visible areas.

Q2. What area percentage of pine is visible from the viewpoint with the added height?

Task 2: Create a New Lookout Shapefile for Viewshed Analysis

What you need: pine and lookout.shp, same as in Task 1.

Task 2 asks you to digitize one more lookout location before running a viewshed analysis. The output from the analysis represents a cumulative viewshed.

1. Select Copy from the context menu of lookout in the table of contents. Select Paste Layer(s) from the context menu of Tasks 1&2. The copied shapefile is also named lookout. Right-click the top lookout, and select Properties. On the General tab, change the layer name from lookout to newpoints.

2. Make sure that the Editor toolbar is available. Click Editor's dropdown arrow and select Start Editing. The task is to Create New Feature and the target is newpoints.

3. Next add a new viewpoint. To find suitable viewpoint locations, you can use Hillshade of pine as a guide and the Zoom In tool for close-up looks. You can also use pine and the Identify tool to find elevation data. When you are ready to add a viewpoint, click the Sketch Tool first and then click the intended location of the point. The new viewpoint has an OFFSETA value of 0. Open the attribute table of newpoints.

Click the cell of OFFSETA for the new point and enter 15. Also, to separate the two viewpoints, enter the ID values of 1 and 2 respectively. Click the Editor menu and select Stop Editing. Save the edits. You are ready to use *newpoints* for viewshed analysis.

4. Click the Spatial Analyst dropdown arrow, point to Surface Analysis, and select Viewshed. Make sure that the input surface is *plne* and the observer points are from *newpoints*. Opt for a temporary output raster. Click OK to run the operation.

5. *Viewshed of newpoints* shows visible and not visible areas. The visible areas represent the cumulative viewshed. Portions of the viewshed are visible to only one viewpoint, whereas others are visible to both viewpoints. The attribute table of *Viewshed of newpoints* provides the cell counts of visible from one point and visible from two points.

Q3. What area percentage of *plne* is visible from *newpoints*? Report the increase in viewshed from one to two viewpoints.

6. To save *newpoints* as a shapefile, right-click *newpoints*, point to Data, and select Export Data. In the Export Data dialog, specify the path and name of the output shapefile.

Task 3: Delineate Areawide Watersheds

What you need: *emidalat*, an elevation raster; and *emidastrm.shp*, a stream shapefile.

Task 3 shows you the process of delineating areawide watersheds using an elevation raster, which is converted from a DEM, as the data source. *emidastrm.shp* serves as a reference. ArcToolbox in ArcGIS 9.0 (the ArcInfo version) has a Hydrology toolset that has tools for completing Task 3.

1. Insert a new data frame in ArcMap. Rename the new data frame Task 3, and add *emidalat* and *emidastrm.shp* to Task 3. If

necessary, click Show/Hide ArcToolbox Window to open the ArcToolbox window. Set the Chapter 15 database as the current workspace.

2. First check if there are any sinks in *emidalat*. Double-click the Flow Direction tool in the Spatial Analyst Tools/Hydrology toolset. Select *emidalat* for the input surface raster, enter *temp_flowd* for the output flow direction raster, and click OK. Double-click the Sink tool. Select *temp_flowd* for the input flow direction raster, specify *sinks* for the output raster, and click OK.

Q4. How many sinks does *emidalat* have? Describe where these sinks are located.

3. This step fills the sinks in *emidalat*. Double-click the Fill tool. Select *emidalat* for the input surface raster, specify *emidafill* for the output surface raster, and click OK.

4. You will use *emidafill* for the rest of Task 3. Double-click the Flow Direction tool. Select *emidafill* for the input surface raster, and specify *flowdirection* for the output flow direction raster. Run the command.

Q5. If a cell in *flowdirection* has a value of 64, what is the cell's flow direction? (Use the index of Flow Direction tool/command in ArcGIS Desktop Help to get the answer.)

5. Next create a flow accumulation raster. Double-click the Flow Accumulation tool. Select *flowdirection* for the input flow direction raster, enter *flowaccumu* for the output accumulation raster, and click OK.

Q6. What is the range of cell values in *flowaccumu*?

6. Next create a source raster, which will be used as the input later for watershed delineation. Creating a source raster involves two steps. First select from (or threshold) *flowaccumu* those cells that have more than 500 cells flowing into them. Double-click the Con tool in the Spatial Analyst Tools/Conditional toolset. Select

flowaccumu for the input conditional raster, enter 1 for the constant value, specify *net* for the output raster, and enter Value > 500 for the expression. (You can also click the SQL button to set the expression. Make sure to have a space before and after >.) Run the command. Second, assign a unique value to each section of *net* between junctions (intersections). Go back to the Hydrology toolset. Double-click the Stream Link tool. Select *net* for the input stream raster, select *flowdirection* for the input flow direction raster, and specify *source* for the output raster. Run the command.

7. Now you have the necessary inputs for watershed delineation. Double-click the Watershed tool. Select *flowdirection* for the input flow direction raster, select *source* for the input source raster, specify *watershed* for the output raster, and click OK. Change the symbology of *watershed* to that of unique values so that you can see individual watersheds.

Q7. How many watersheds are in *watershed*?

Q8. If the flow accumulation threshold were changed from 500 to 1000, would it increase, or decrease, the number of watersheds?

8. Using the command line option in ArcGIS 9.0, you can run the commands in Task 3 all at once. This is particularly useful if the study area is large and you do not want to wait between dialogs. To use this option, first click Show/Hide Command Line Window in ArcMap to open the command line window. Type the first command line, and press Ctrl-Enter to enter the second and subsequent lines. Assuming that the workspace is c:\chap15 and the workspace contains *emidalat*, the following lists the command lines needed to complete Task 3:

Workspace c:\chap15
FlowDirection emidalat temp_flowd
Sink temp_flowd sinks
Fill emidalat emidalfill

FlowDirection emidalfill flowdirection
FlowAccumulation flowdirection flowaccumu
Con flowaccumu 1 net # "Value > 500"
StreamLink net flowdirection source
Watershed flowdirection source watershed

After the above command lines are entered, press Enter to execute them. The bottom part of the command line window shows the progress. It also displays error messages if there are any syntax errors.

Task 4: Derive Upstream Contributing Areas at Pour Points

What you need: *Flowdirection, flowaccumu,* and *source,* all created in Task 3; and *pourpoints.shp,* a shapefile with three points.

In Task 4, you will derive a specific watershed (i.e., upstream contributing area) for each point in *pourpoints.shp.* As of ArcGIS 9.0, Spatial Analyst and ArcToolbox do not share the same data analysis environment. Therefore all operations in Task 4 are performed using Spatial Analyst. You may not get the same result if you mix operations in Spatial Analyst and ArcToolbox.

1. Insert a data frame in ArcMap and rename it Task 4. Add *flowdirection, flowaccumu, source,* and *pourpoints.shp* to Task 4.

2. First convert *pourpoints* to a raster. Select Options from the Spatial Analyst dropdown menu. On the General tab, select the Chapter 15 database for the working directory. On the Extent tab, select *flowaccumu* for the analysis extent. On the Cell Size tab, select *flowaccumu* for the analysis cell size. Click the Spatial Analyst dropdown arrow, point to Convert, and select Feature to Raster. Select *pourpoints* for the input features, enter *pourptgd* for the output raster, and click OK to convert.

3. Select Raster Calculator from the Spatial Analyst dropdown menu. Enter the following expression in the expression box:
watershed([flowdirection], [pourptgd]).

(pourptgd may appear as pourptgd – pourptgd; ignore the discrepancy.) Click Evaluate. The command creates *Calculation* as a temporary raster.

Q9. How many cells are associated with each of the pour points?

4. Zoom in on a pour point. The pour point is not right on *source*, the stream link raster created in Task 3. It is the same with the other points. This is why the pour points generated very small watersheds in Step 3. ArcGIS has a SnapPour command, which can snap a pour point to the cell with the highest flow accumulation value within a search distance. Use the Measure tool to measure the distance between the pour point and the nearby stream segment. A snap distance of 90 meters (3 cells) should place the pour points onto the stream channel.

5. Select Raster Calculator from the Spatial Analyst dropdown menu. Enter the following expression in the expression box: snappour([pourptgd], [flowaccumu], 90). Click Evaluate. The command creates *Calculation2* as a temporary grid.

6. Run the Watershed command again. Enter the following expression in the Raster Calculator's expression box: watershed([flowdirection], [Calculation2]). Click Evaluate. *Calculation3* should have many more cells for each snapped pour point.

Q10. How many cells are associated with each of the new pour points?

Challenge Task

What you need: access to the Internet.

This challenge task asks you to find the upstream contributing area for a USGS gauge station in your local area.

1. The first step is to locate a station in your local area. The USGS maintains a National Water Information System (NWIS) website, which provides a wide variety of water resource data. Go to **http://nwis. waterdata.usgs.gov/nwis.** Click on the Surface Water link and then the Streamflow link. On the Daily Streamflow for the Nation page, you can select site location by state. On the page that lists all site locations for your state, select one site (station) for the challenge task. Click on the site number. Write down the latitude and longitude readings of the station as well as the datum (e.g., NAD27).

2. Next get a 30-meter DEM for the area around the selected station. Go to the GIS Data Depot website **(http://www.geocomm.com)** to download the DEM. Unzip the downloaded data. Use the Conversion Tools/Import to Raster/DEM to Grid tool in ArcToolbox to convert the DEM to an elevation grid. Name the elevation grid *elev* and add *elev* to a new data frame in ArcMap.

3. You need to go through a couple of steps to add the station to ArcMap. Convert the latitude and longitude readings of the station from DMS (degrees-minutes-seconds) to DD (decimal degrees). Prepare a comma delimited text file for the station. The first line of the text file has the headings of ID, longitude, and latitude. The second line has the station ID (e.g., 1) and the station's longitude and latitude readings. Name the text file *gauge.txt*. Use the ADD XY Data tool in ArcMap to add *gauge.txt*. Export *gauge.txt* Events to a point shapefile and call the shapefile *gauge.shp*.

4. *gauge.shp* has an undefined coordinate system. First define its geographic coordinate according to the listed datum on the USGS website (e.g., North American Datum of 1927). Then project the shapefile by importing the coordinate system from *elev*, which is most likely the UTM (Universal Transverse Mercator) coordinate system. Name the projected output shapefile *gageutm.shp*.

5. Add *gagetm.shp* back to ArcMap. Make sure the Spatial Analyst's analysis environment (i.e., extent and cell size) is set to be the same as *elev*. Then convert *gagetm.shp* to a raster and name the raster *gagegd*.

6. Derive a flow direction raster from *elev*, and then use *gagegd* and the flow direction raster to derive the watershed for the station.

7. If the watershed from Step 6 is very small, it means that the station is not on the stream channel. Follow the same procedure as in Task 4 to snap the station to the stream channel and to run another watershed analysis.

REFERENCES

Brooks, K. N., P. F. Ffolliott, H. M. Gregersen, and L. F. DeBano. 2003. *Hydrology and the Management of Watersheds*, 3d ed. Ames, IA: Iowa State Press.

Clarke, K. C. 1995. *Analytical and Computer Cartography*, 2d ed. Englewood Cliffs, NJ: Prentice Hall.

Davidson, D. A., A. I. Watson, and P. H. Selman. 1993. An Evaluation of GIS as an Aid to the Planning of Proposed Developments in Rural Areas. In P. M. Mather, ed., *Geographical Information Handling: Research and Applications*, pp. 251–59. London: Wiley.

De Floriani, L., and P. Magillo. 1999. Intervisibility on Terrains. In P. A. Longley, M. F. Goodchild, D. J. Maguire, and D. W. Rhind, eds., *Geographical Information Systems, Vol. 1: Principles and Technical Issues*, 2d ed., pp. 543–56. New York: Wiley.

De Floriani, L., and P. Magillo. 1994. Visibility Algorithms on Triangulated Terrain Models. *International Journal of Geographical Information Systems* 8: 13–41.

Enderby, T. A., and E. F. Wood. 2003. Maximizing Spatial Congruence of Observed and DEM-Delineated Overland Flow Networks. *International Journal of Geographical Information Science* 17: 699–713.

Fisher, P. R. 1996. Extending the Applicability of Viewsheds in Landscape Planning. *Photogrammetric Engineering and Remote Sensing* 62: 1297–1302.

Fisher, P. F. 1993. Algorithm and Implementation Uncertainty in Viewshed Analysis. *International Journal of Geographical Information Systems* 7: 331–47.

Fisher, P. F. 1991. First Experiments in Viewshed Uncertainty: The Accuracy of the Viewshed Area. *Photogrammetric Engineering and Remote Sensing* 57: 1321–27.

Freeman, T. G. 1991. Calculating Catchment Area with Divergent Flow Based on a Regular Grid. *Computers and Geosciences* 17: 413–22.

Gallant, J. C., and J. P. Wilson. 2000. Primary Topographic Attributes. In J. P. Wilson and J. C. Gallant, eds., *Terrain Analysis: Principles and Applications*, pp. 51–85. New York: Wiley.

Garbrecht, J., and L. W. Martz. 2000. Digital Elevation Model Issues in Water Resources Modeling. In D. Maidment and D. Djokic, eds., *Hydrologic and Hydraulic Modeling Support with Geographic Information Systems*, pp. 1–27. Redland, CA: ESRI Press.

Gertner, G., G. Wang, S. Fang, and A. B. Anderson. 2002. Effect and Uncertainty of Digital Elevation Model Spatial Resolutions on Predicting the Topographical Factor for Soil Loss Estimation. *Journal of Soil and Water Conservation* 57: 164–74.

Goodchild, M. F., and J. Lee. 1989. Coverage Problems and Visibility Regions on Topographic Surfaces. *Annals of Operations Research* 18: 175–86.

Jenson, S. K., and J. O. Domingue. 1988. Extracting Topographic Structure from Digital Elevation Data for Geographic Information System Analysis. *Photogrammetric Engineering and Remote Sensing* 54: 1593–1600.

Lake, I. R., A. A. Lovett, I. J. Bateman, and I. H. Langford. 1998. Modelling Environmental Influences on Property Prices in an Urban Environment. *Computers, Environment and Urban Systems* 22: 121–36.

Lee, J. 1991. Analyses of Visibility Sites on Topographic Surfaces. *International Journal of Geographical Information Systems* 5: 413–29.

Lee, J., and D. Stucky. 1998. On Applying Viewshed Analysis for Determining Least-Cost Paths on Digital Elevation Models. *International Journal of Geographical Information Science* 12: 891–905.

Liang, C., and D. S. Mackay. 2000. A General Model of Watershed Extraction and Representation Using Globally Optimal Flow Paths and Up-Slope Contributing Areas. *International Journal of Geographical Information Science* 14: 337–58.

Maloy, M. A., and D. J. Dean. 2001. An Accuracy Assessment of Various GIS-Based Viewshed Delineation Techniques. *Photogrammetric Engineering and Remote Sensing* 67: 1293–98.

Miller, D. 2001. A Method for Estimating Changes in the Visibility of Land Cover. *Landscape and Urban Planning* 54: 93–106.

Moore, I. D. 1996. Hydrological Modeling and GIS. In M. F. Goodchild, L. T. Steyaert, B. O. Parks, C. Johnston, D. Maidment, M. Crane, and S. Glendinning, eds., *GIS and Environmental Modeling:*

Progress and Research Issues, pp. 143–48. Fort Collin, CO: GIS World Books.

Nackaerts, K., G. Govers, and J. Van Orshoven. 1999. Accuracy Assessment of Probabilistic Visibilities. *International Journal of Geographical Information Science* 13: 709–21.

O'Callaghan, J. F., and D. M. Mark. 1984. The Extraction of Drainage Networks from Digital Elevation Data. *Computer Vision, Graphics and Image Processing* 28: 323–44.

O'Sullivan, D., and A. Turner. 2001. Visibility Graphs and Landscape Visibility Analysis. *International Journal of Geographical Information Science* 15: 221–37.

Perez, A. 2000. Surface Water Protection Project: A Comparison of Watershed Delineation Methods in ARC/INFO and ArcView GIS. In D. Maidment and D. Djokic, eds., *Hydrologic and Hydraulic Modeling Support with Geographic Information Systems,* pp. 53–64. Redland, CA: ESRI Press.

Saunders, W. 2000. Preparation of DEMs for Use in Environmental Modeling Analysis. In D. Maidment and D. Djokic, eds., *Hydrologic and Hydraulic Modeling Support with Geographic Information*

Systems, pp. 29–51. Redland, CA: ESRI Press.

Tarboton, D. G. 1997. A New Method for the Determination of Flow Directions and Upslope Areas in Grid Digital Elevation Models. *Water Resources Research* 32: 309–19.

Tarboton, D. G., R. L. Bras, and I. Rodrigues-Iturbe. 1991. On the Extraction of Channel Networks from Digital Elevation Data. *Water Resources Research* 5: 81–100.

Wang, J. J., G. J. Robinson, and K. White. 1996. A Fast Solution to Local Viewshed Computation Using Grid-Based Digital Elevation Models. *Photogrammetric Engineering and Remote Sensing* 62: 1157–64.

Wing, M. G., and R. Johnson. 2001. Quantifying Forest Visibility with Spatial Data. *Environmental Management* 27: 411–20.

Zhang, W., and D. R. Montgomery. 1994. Digital Elevation Model Raster Size, Landscape Representation, and Hydrologic Simulations. *Water Resources Research* 30: 1019–28.

Zhou, Q., and X. Liu. 2002. Error Assessment of Grid-Based Flow Routing Algorithms Used in Hydrological Models. *International Journal of Geographical Information Science* 16: 819–42.

CHAPTER 16

SPATIAL INTERPOLATION

The terrain is one type of surface that is familiar to us. In GIS we also work with another type of surface, which may not be physically present but can be visualized in the same way as the land surface. Cartographers call this type of surface the statistical surface (Robinson et al. 1995). Examples of the statistical surface include precipitation, snow accumulation, water table, and population density.

How can one construct a statistical surface? The answer is similar to that for the land surface except that the input data are typically limited to a sample of point data. To make a precipitation map, for example, we will not find a regular array of weather stations like a digital elevation model (DEM). A process of filling in data between the sample points is therefore required.

Spatial interpolation is the process of using points with known values to estimate values at other points. Through spatial interpolation, we can estimate the precipitation value at a location with no recorded data by using known precipitation readings at nearby weather stations. In GIS applications, spatial interpolation is typically applied to a raster with estimates made for all cells. Spatial interpolation is therefore a means of creating surface data from sample points so that the surface data can be used for analysis and modeling.

Chapter 16 has five sections. Section 16.1 reviews the elements of spatial interpolation including control points and type of spatial interpolation. Section 16.2 covers global methods including trend surface and regression models. Section 16.3 covers local methods including Thiessen polygons, density estimation, inverse distance weighted, and splines. Section 16.4 examines kriging, a widely used stochastic local method. Section 16.5 discusses comparison of interpolation methods and comparison measures. Perhaps more than any other topic in GIS, spatial interpolation depends on the computing algorithm. Worked examples are included in Chapter 16 to show how spatial interpolation is carried out mathematically.

16.1 ELEMENTS OF SPATIAL INTERPOLATION

Spatial interpolation requires two basic inputs: known points and an interpolation method. The requirement of known points distinguishes spatial interpolation from isopleth mapping, which uses assigned points such as polygon centroids for interpolation (Robinson et al. 1995).

16.1.1 Control Points

Control Points are points with known values. Also called *known points*, *sample points*, or *observations*, control points provide the data necessary for the development of an interpolator (e.g., a mathematical equation) for spatial interpolation. The number and distribution of control points can greatly influence the accuracy of spatial interpolation. A basic assumption in spatial interpolation is that the value to be estimated at a point is more influenced by nearby known points than those that are farther away. To be effective for estimation, control points should be well distributed within the study area. But this ideal situation is rare in real-world applications because a study area often contains data-poor areas.

Figure 16.1 shows 130 weather stations in Idaho and 45 additional stations from the surrounding states. The map clearly shows data-poor areas in Clearwater Mountains, Salmon River Mountains, Lemhi Range, and Owyhee Mountains. These 175 stations, and their 30-year (1971–2000) average annual precipitation data, are used as sample data throughout Chapter 16. As often shown later, the data-poor areas can cause problems for spatial interpolation.

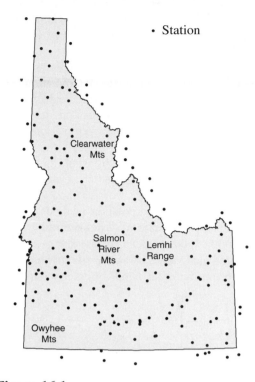

Figure 16.1
A map of 175 weather stations in and around Idaho.

16.1.2 Type of Spatial Interpolation

Spatial interpolation methods can be categorized in several ways. First, they can be grouped into global and local methods. A **global interpolation** method uses every known point available to estimate an unknown value. A **local interpolation** method, on the other hand, uses a sample of known points to estimate an unknown value. Because the difference between the two groups lies in the number of control points used in estimation, one may view the scale from global to local as a continuum.

Conceptually, a global interpolation method is designed to capture the general trend of the surface and a local interpolation method the local or short-range variation. For many phenomena, it is more efficient to estimate the unknown value at a point using a local method than a global method. Far-away points have little influence on the estimated value; in some cases, they may even distort the estimated value. A local method is also preferred because it requires much less computation than a global method does.

Second, spatial interpolation methods can be grouped into exact and inexact interpolation (Figure 16.2). **Exact interpolation** predicts a value at the point location that is the same as its known value. In other words, exact interpolation generates a surface that passes through the control points. In contrast, **inexact interpolation**, or approximate interpolation, predicts a value at the point location that differs from its known value.

Third, spatial interpolation methods may be deterministic or stochastic. A **deterministic interpolation** method provides no assessment of errors with predicted values. A **stochastic interpolation** method, on the other hand, offers assessment of

prediction errors with estimated variances. The assumption of a random process is normally required for a stochastic method.

Table 16.1 shows a classification of spatial interpolation methods covered in Chapter 16. Notice that the two global methods can also be used for local operations.

16.2 GLOBAL METHODS

Section 16.2 covers the global methods of trend surface models and regression models.

16.2.1 Trend Surface Models

An inexact interpolation method, **trend surface analysis** approximates points with known values with a polynomial equation (Davis 1986; Bailey and Gatrell 1995). The equation or the interpolator can then be used to estimate values at other points. A linear or first-order trend surface uses the equation:

$$z_{x,y} = b_0 + b_1x + b_2y \qquad (16.1)$$

where the attribute value z is a function of x and y coordinates. The b coefficients are estimated from the known points (Box 16.1). Because the trend surface model is computed by the least-squares method similar to a regression model, the "goodness of fit" of the model can be measured and tested. Also, the deviation or the residual between the observed and the estimated values can be computed for each known point.

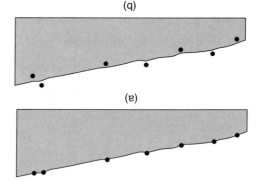

(a)

(b)

Figure 16.2
Exact interpolation (a) and inexact interpolation (b).

TABLE 16.1	A Classification of Spatial Interpolation Methods			
	Global		**Local**	
	Deterministic	**Stochastic**	**Deterministic**	**Stochastic**
	Trend surface (inexact)*	Regression (inexact)	Thiessen (exact)	Kriging (exact)
			Density estimation (inexact)	
			Inverse distance weighted (exact)	
			Splines (exact)	

*Given some required assumptions, trend surface analysis can be treated as a special case of regression analysis and thus a stochastic method (Griffith and Amrhein 1991).

Box 16.1 | A Worked Example of Trend Surface Analysis

Figure 16.3 shows five weather stations with known values around point 0 with an unknown value. The table below shows the x-, y-coordinates of the points, measured in row and column of a raster with a cell size of 2000 meters, and their known values.

Point	x	y	Value
1	69	76	20.820
2	59	64	10.910
3	75	52	10.380
4	86	73	14.600
5	88	53	10.560
0	69	67	?

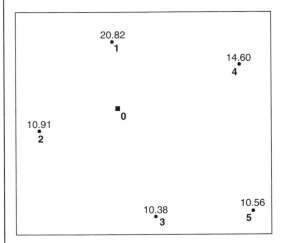

Figure 16.3
Estimation of the unknown value at point 0 from five surrounding known points.

This example shows how we can use Eq. (16.1), or a linear trend surface, to interpolate the unknown value at point 0. The least-squares method is commonly used to solve for the coefficients of b_0, b_1, and b_2 in Eq. (16.1). Therefore, the first step is to set up three normal equations, similar to those for a regression analysis:

$$\Sigma z = b_0 n + b_1 \Sigma x + b_2 \Sigma y$$
$$\Sigma xz = b_0 \Sigma x + b_1 \Sigma x^2 + b_2 \Sigma xy$$
$$\Sigma yz = b_0 \Sigma y + b_1 \Sigma xy + b_2 \Sigma y^2$$

The equations can be rewritten in matrix form as:

$$\begin{bmatrix} n & \Sigma x & \Sigma y \\ \Sigma x & \Sigma x^2 & \Sigma xy \\ \Sigma y & \Sigma xy & \Sigma y^2 \end{bmatrix} \cdot \begin{bmatrix} b_0 \\ b_1 \\ b_2 \end{bmatrix} = \begin{bmatrix} \Sigma z \\ \Sigma xz \\ \Sigma yz \end{bmatrix}$$

Using the values of the five known points, we can calculate the statistics and substitute the statistics into the equation:

$$\begin{bmatrix} 5 & 377 & 318 \\ 377 & 29007 & 23862 \\ 318 & 23862 & 20714 \end{bmatrix} \cdot \begin{bmatrix} b_0 \\ b_1 \\ b_2 \end{bmatrix} = \begin{bmatrix} 67.270 \\ 5043.650 \\ 4445.800 \end{bmatrix}$$

We can then solve the b coefficients by multiplying the inverse of the first matrix on the left (shown with four decimal digits because of the very small numbers) by the matrix on the right:

$$\begin{bmatrix} 23.2102 & -0.1631 & -0.1684 \\ -0.1631 & 0.0018 & 0.0004 \\ -0.1684 & 0.0004 & 0.0021 \end{bmatrix} \cdot \begin{bmatrix} 67.270 \\ 5043.650 \\ 4445.800 \end{bmatrix} = \begin{bmatrix} -10.094 \\ 0.020 \\ 0.347 \end{bmatrix}$$

Using the coefficients, the unknown value at point 0 can be estimated by:

$$z_0 = -10.094 + (0.020)(69) + (0.347)(67) = 14.535$$

The distribution of most natural phenomena is usually more complex than an inclined plane surface from a first-order model. Higher-order trend surface models are required to approximate more complex surfaces. A cubic or a third-order model, for example, includes hills and valleys. A cubic trend surface is based on the equation:

(16.2)

$$z_{x,y} = b_0 + b_1 x + b_2 y + b_3 x^2 + b_4 xy + b_5 y^2 + b_6 x^3 + b_7 x^2 y + b_8 xy^2 + b_9 y^3$$

A third-order trend surface requires estimation of 10 coefficients (i.e., b_i), compared to three coefficients for a first-order surface. A higher-order trend surface model therefore requires more computation than a lower-order model does. A GIS package may offer up to 12th-order trend surface models.

Figure 16.4 shows an isoline (isohyet) map derived from a third-order trend surface of annual precipitation in Idaho created from 175 data points with a cell size of 2000 meters. An isoline map is like a contour map, useful for visualization as well as measurement.

There are variations of trend surface analysis. Logistic trend surface analysis uses known points with binary data (i.e., 0 and 1) and produces a probability surface. **Local polynomial interpolation** uses a sample of known points to estimate the unknown value of a cell. Conversion of a triangulated irregular network (TIN) to a DEM (Chapter 14), for instance, uses local polynomial interpolation.

16.2.2 Regression Models

A **regression model** relates a dependent variable to a number of independent variables in a linear equation (an interpolator), which can then be used for prediction or estimation. Many regression models use nonspatial attributes and are not considered methods for spatial interpolation. But exceptions can be made for regression models that use spatial variables such as distance to a river or location-specific elevation (Burrough and McDonnell 1998; Rogerson 2001).

A watershed-level regression model for snow accumulation developed by Chang and Li (2000) is an example. The model uses snow water equivalent (SWE) as the dependent variable and location and topographic variables as the independent variables. One of their watershed models takes the form of:

$$SWE = b_0 + b_1 EASTING + b_2 SOUTHING + b_3 ELEV + b_4 PLAN1000$$

(16.3)

where EASTING and SOUTHING correspond to the column number and the row number in an elevation raster, ELEV is the elevation value, and PLAN1000 is a surface curvature measure. After the b coefficients in Eq. (16.3) are estimated using known values at snow courses, the model can be used to estimate SWE for all cells in the watershed and to produce a continuous SWE surface.

PRISM (parameter-elevation regressions on independent slopes model), a popular tool for generating precipitation estimates in mountainous regions, is another example (Daly et al. 1994). The input to PRISM consists of point measurements of climate data and a DEM, and the output is a predicted map in raster format. The estimate for each cell is based on a separate regression equation using data from nearby climate stations, which are in turn weighted by the variables of distance, elevation, vertical layer, topographic facet, and coastal proximity. The Natural Resources Conservation Service (NRCS) has used PRISM to map precipitation

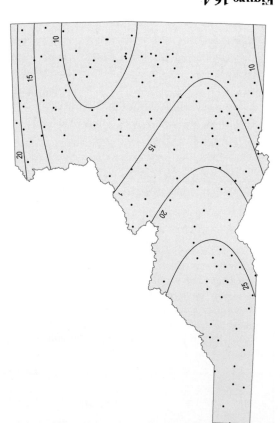

Figure 16.4
An isohyet map in inches from a third-order trend surface model. The point symbols represent known points within Idaho.

for all 50 states of the United States **(http://www.nrcs.usda.gov/technical/maps.html).**

16.3 LOCAL METHODS

Because local interpolation uses a sample of known points, it is important to know how to select a sample. The first issue in sampling is the number of points (i.e., the sample size) to be used in estimation. GIS packages typically let users specify the number of points or use a default number (e.g., 7 to 12 points). One might assume that more points would result in more accurate estimates. But the validity of this assumption depends on the distribution of known points relative to the cell to be estimated, the extent of spatial autocorrelation, and the quality of data (Yang and Hodler 2000). More points usually imply more generalized estimations.

After the number of points is determined, the next task is to search for those known points (Figure 16.5). A simple option is to use the closest known points to the point to be estimated. An alternative is to select known points within a circle, the size of which depends on the sample size. Some search options may incorporate a quadrant or octant requirement (Davis 1986). A quadrant requirement means selecting known points from each of the four quadrants around a cell to be estimated. An octant

requirement means using eight sectors. Other search options may consider the directional component by using an ellipse, with its major axis corresponding to the principal direction.

16.3.1 Thiessen Polygons

Thiessen polygons assume that any point within a polygon is closer to the polygon's known point than any other known points. Thiessen polygons were originally proposed to estimate areal averages of precipitation by making sure that any point within a polygon is closer to the polygon's weather station than any other station (Tabios and Salas 1985). Thiessen polygons, also called *Voronoi polygons,* are used in a variety of applications, especially for service area analysis of public facilities such as hospitals.

Thiessen polygons do not use an interpolator but require initial triangulation for connecting known points. Because different ways of connecting points can form different sets of triangles, the Delaunay triangulation—the same method for constructing a TIN—is often used in preparing Thiessen polygons (Davis 1986). The Delaunay triangulation ensures that each known point is connected to its nearest neighbors, and that triangles are as equilateral as possible. After triangulation, Thiessen polygons can be easily constructed by

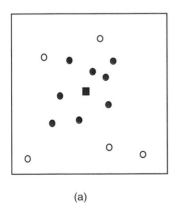
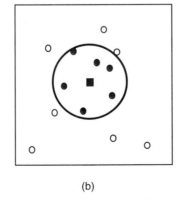
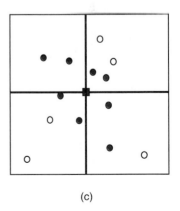

(a)   (b)   (c)

Figure 16.5

Three search methods for sample points: (*a*) find the closest points to the point to be estimated, (*b*) find points within a radius, and (*c*) find points within each quadrant.

16.3.2 Density Estimation

Density estimation measures cell densities in a raster by using a sample of known points. There are simple and kernel density estimation methods. To use the simple density estimation method, we can place a raster on a point distribution, tabulate points that fall within each cell, sum the point values, and estimate the cell's density by dividing the total point value by the cell size. Figure 16.7 shows the input and output of an example of simple density estimation. The input is a distribution of sighted

older trees (Nelson et al. 2004).

deer locations plotted with a 50-meter interval to accommodate the resolution of telemetry. Each deer location has a count value measuring how many times a deer was sighted at the location. The output is a density raster, which has a cell size of 10,000 square meters or 1 hectare and a density measure of number of sightings per hectare. A circle, rectangle, wedge, or ring based at the center of a cell may replace the cell in the calculation.

Kernel density estimation associates each known point with a kernel function for the purpose of estimation (Silverman 1986; Scott 1992; Bailey and Gatrell 1995). Expressed as a bivariate probability density function, a kernel function looks like a "bump," centering at a known point

connecting lines drawn perpendicular to the sides of each triangle at their midpoints (Figure 16.6).

Thiessen polygons are smaller in areas where points are closer together and larger in areas where points are farther apart. This size differentiation is the basis, for example, for determining the quality of public service. A large polygon means greater distances between home locations and public service providers. The size differentiation can also be used for other purposes such as predicting forest age classes, with larger polygons belonging to

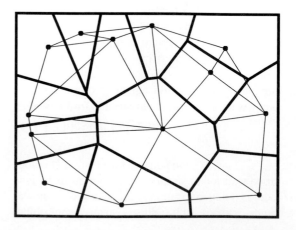

Figure 16.6

Thiessen polygons (in thicker lines) are interpolated from the known points and the Delaunay triangulation (in thinner lines).

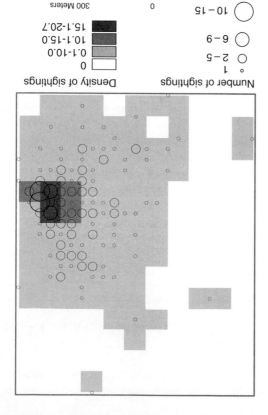

Figure 16.7

Deer sightings per hectare calculated by the simple density estimation method.

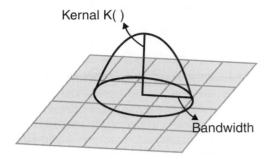

Figure 16.8
A kernel function, which represents a probability density function, looks like a "bump" above a grid.

and tapering off to 0 over a defined bandwidth or window area (Silverman 1986) (Figure 16.8). The kernel function and the bandwidth determine the shape of the bump, which in turn determines the amount of smoothing in estimation. The kernel density estimator at point x is then the sum of bumps placed at each known point x_i within the bandwidth:

(16.4)

$$\hat{f}(x) = \frac{1}{nh^d} \sum_{i=1}^{n} K(\frac{1}{h}(x - x_i))$$

where $K(\ )$ is the kernel function, h is the bandwidth, n is the number of known points within the bandwidth, and d is the data dimensionality. For two-dimensional data ($d = 2$), the kernel function is usually given by:

(16.5)

$$K(x) = 3\pi^{-1}(1 - X^TX)^2, \text{ if } X^TX < 1$$
$$K(x) = 0, \text{ otherwise}$$

By substituting Eq. (16.5) for $K(\ )$, Eq. (16.4) can be rewritten as:

(16.6)

$$\hat{f}(x) = \frac{3}{nh^2\pi} \sum_{i=1}^{n} \{1 - \frac{1}{h^2}[(x - x_i)^2 + (y - y_i)^2]\}^2$$

where π is a constant, and $(x - x_i)$ and $(y - y_i)$ are the deviations in x-, y-coordinates between

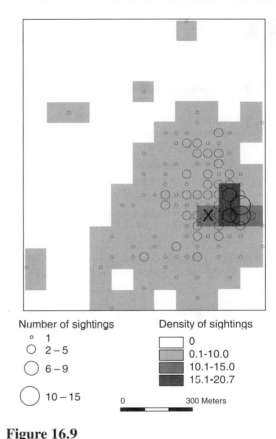

Number of sightings	
○	1
○	2–5
○	6–9
○	10–15

Density of sightings	
☐	0
	0.1-10.0
	10.1-15.0
	15.1-20.7

0   300 Meters

Figure 16.9
Deer sightings per hectare calculated by the kernel density estimation method. The letter X marks the cell, which is used as an example in Box 16.2.

point x and known point x_i that is within the bandwidth.

Using the same input as for the simple estimation method, Figure 16.9 shows the output raster from kernel density estimation. Density values in the raster are expected values rather than probabilities (Box 16.2). Kernel density estimation usually produces a smoother surface than the simple estimation method does.

16.3.3 Inverse Distance Weighted Interpolation

Inverse distance weighted (IDW) interpolation is an exact method that enforces that the estimated value of a point is influenced more by nearby known points than those farther away. The general equation for the IDW method is:

$$(16.7) \quad z_0 = \frac{\sum\limits_{i=1}^{s} z_i \frac{1}{d_i^k}}{\sum\limits_{i=1}^{s} \frac{1}{d_i^k}}$$

where z_0 is the estimated value at point 0, z_i is the z value at known point i, d_i is the distance between point i and point 0, s is the number of known points used in estimation, and k is the specified power.

The power k controls the degree of local influence. A power of 1.0 means a constant rate of change in value between points (linear interpolation). A power of 2.0 or higher suggests that the rate of change in values is higher near a known point and levels off away from it. The degree of local influence also depends on the number of known points used in estimation. A study by Zimmerman et al. (1999) shows that a smaller number of known points (6) actually produce better estimations than a larger number (12).

An important characteristic of IDW interpolation is that all predicted values are within the range of maximum and minimum values of the known points. Figure 16.10 shows an annual precipitation surface created by the IDW method with a power

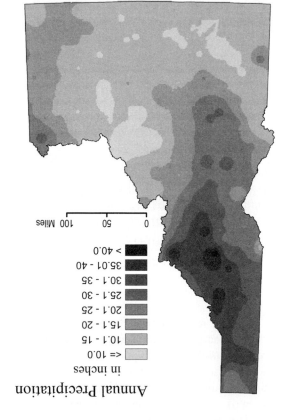

Annual Precipitation
in inches

- <= 10.0
- 10.1 - 15
- 15.1 - 20
- 20.1 - 25
- 25.1 - 30
- 30.1 - 35
- 35.01 - 40
- > 40.0

0   50   100 Miles

Figure 16.10
An annual precipitation surface created by the inverse distance squared method.

Box 16.2 A Worked Example of Kernel Density Estimation

This example shows how the value of the cell marked X in Figure 16.9 is derived. The window area is defined as a circle with a radius of 100 meters (h). Therefore, only points within the 100-meter radius of the center of the cell can influence the estimation of the cell density. Using the 10 points within the cell's neighborhood, we can compute the cell density by:

$$\frac{3}{\pi} \sum_{i=1}^{10} n_i \{1 - \frac{1}{h^2}[(x - x_i)^2 + (y - y_i)^2]\}^2$$

where n_i is the number of sightings at point i, x_i and y_i are the x-, y-coordinates of point i, and x and y are the x-, y-coordinates of the center of the cell to be estimated. Because the density is measured per 10,000 square meters or hectare, h^2 in Eq. (16.6) is canceled out. Also, because the output shows an expected value rather than a probability, n in Eq. (16.6) is not needed. The computation shows the cell density to be 11.421.

> *Box* **16.3**   **A Worked Example of Inverse Distance Weighted Estimation**

This example uses the same data set as in Box 16.1, but interpolates the unknown value at point 0 by the IDW method. The table below shows the distances in thousands of meters between point 0 and the five known points:

Between points	Distance
0,1	18.000
0,2	20.880
0,3	32.310
0,4	36.056
0,5	47.202

We can substitute the parameters in Eq. (16.7) by the known values and the distances to estimate z_0:

$$\Sigma z_i \, 1/d_i^2 = (20.820)(1/18.000)^2 +$$
$$(10.910)(1/20.880)^2 + (10.380)(1/32.310)^2 +$$
$$(14.600)(1/36.056)^2 + (10.560)(1/47.202)^2 = 0.1152$$

$$\Sigma 1/d_i^2 = (1/18.000)^2 + (1/20.880)^2 +$$
$$(1/32.310)^2 + (1/36.056)^2 + (1/47.202)^2 = 0.0076$$

$$z_0 = 0.1152/0.0076 = 15.158$$

of 2 (Box 16.3). Figure 16.11 shows an isoline map of the surface. Small, enclosed isolines are typical of IDW interpolation. The odd shape of the 10-inch isoline in the southwest corner is due to the absence of known points.

16.3.4 Thin-Plate Splines

Splines for spatial interpolation are conceptually similar to splines for line smoothing except that in spatial interpolation they apply to surfaces rather than lines. **Thin-plate splines** create a surface that passes through the control points and has the least possible change in slope at all points (Franke 1982). In other words, thin-plate splines fit the control points with a minimum curvature surface. The approximation of thin-plate splines is of the form:

(16.8)

$$Q(x, y) = \Sigma A_i d_i^2 \log d_i + a + bx + cy$$

where x and y are the x-, y-coordinates of the point to be interpolated, $d_i^2 = (x - x_i)^2 + (y - y_i)^2$, and x_i and y_i are the x-, y-coordinates of control point i. Thin-plate splines consist of two components: $(a + bx + cy)$ represents the local trend function, which has the same form as a linear or first-order

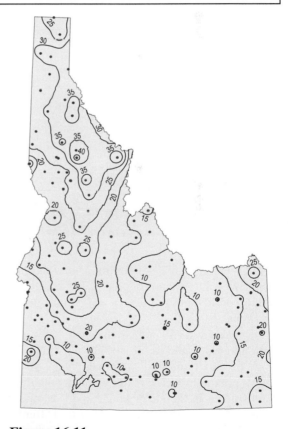

Figure 16.11

An isohyet map created by the inverse distance squared method.

Unlike the IDW method, the predicted values from thin-plate splines are not limited within the range of maximum and minimum values of the known points. In fact, a major problem with thin-plate splines is the steep gradients in data-poor areas, often referred to as overshoots. Different methods for correcting overshoots have been proposed. Thin-plate splines with tension, for example, allow the user to control the tension to be

pulled on the edges of the surface (Franke 1985; Mitas and Mitasova 1988). Other methods include regularized splines (Mitas and Mitasova 1988) and regularized splines with tension (Mitasova and Mitas 1993). All these methods belong to a diverse group called **radial basis functions (RBF)** (Box 16.4).

The **thin-plate splines with tension** method has the following form:

$$\sum_{i=1}^{n} A_i R(d_i) + a \tag{16.10}$$

where a represents the trend function, and the basis function $R(d)$ is:

$$-\frac{1}{2\pi\varphi^2}\left[\ln\left(\frac{d\varphi}{2}\right) + c + K_0(d\varphi) \right] \tag{16.11}$$

where φ is the weight to be used with the tension method. If the weight φ is set close to 0, then the approximation with tension is similar to the basic thin-plate splines method. A larger φ value reduces the stiffness of the plate and thus the range of interpolated values, with the interpolated surface resembling the shape of a membrane passing through the control points (Franke 1985). Box 16.5 shows a worked example using the thin-plate splines with tension method.

trend surface, and $d_i^2 \log d_i$ represents a basis function, which is designed to obtain minimum curvature surfaces (Watson 1992). The coefficients A_i, and a, b, and c are determined by a linear system of equations (Franke 1982):

$$\sum_{i=1}^{n} A_i d_i^2 \log d_i + a + bx + cy = f_i$$
$$\sum_{i=1}^{n} A_i = 0$$
$$\sum_{i=1}^{n} A_i x_i = 0$$
$$\sum_{i=1}^{n} A_i y_i = 0 \tag{16.9}$$

where n is the number of control points, and f_i is the known value at control point i. The estimation of the coefficients requires $n + 3$ simultaneous equations.

 Box 16.4   Radial Basis Functions

Radial basis functions (RBF) refer to a large group of interpolation methods. All of them are exact interpolators. The selection of a basis function or equation determines how the surface will fit in between the control points. Both ArcToolbox and the Spatial Analyst extension to ArcGIS offer thin-plate splines with tension and regularized splines. The Geostatistical Analyst extension to ArcGIS, on the other hand, has a menu choice of five RBF methods: thin-plate spline, spline with tension, completely regularized spline, multi-quadric function, and inverse multiquadric function. Each RBF method also has a parameter that controls the smoothness of the generated surface. Although each combination of an RBF method and a parameter value can create a new surface, the difference between the surfaces is usually small.

Box 16.5 | **A Worked Example of Thin-Plate Splines with Tension**

This example uses the same data set as in Box 16.1 but interpolates the unknown value at point 0 by splines with tension. The method first involves calculation of $R(d)$ in Eq. (16.11) using the distances between the point to be estimated and the known points, distances between the known points, and the φ value of 0.1. The following table shows the $R(d)$ values along with the distance values.

Points	0,1	0,2	0,3	0,4	0,5
Distance	18.000	20.880	32.310	36.056	47.202
$R(d)$	−7.510	−9.879	−16.831	−18.574	−22.834
Points	1,2	1,3	1,4	1,5	2,3
Distance	31.240	49.476	34.526	59.666	40.000
$R(d)$	−16.289	−23.612	−17.879	−26.591	−20.225
Points	2,4	2,5	3,4	3,5	4,5
Distance	56.920	62.032	47.412	26.076	40.200
$R(d)$	−25.843	−27.214	−22.868	−13.415	−20.305

The next step is to solve for A_i in Eq. (16.10). We can substitute the calculated $R(d)$ values into Eq. (16.10)

and rewrite the equation and the constraint about A_i in matrix form:

$$\begin{bmatrix} 1 & 0 & -16.289 & -23.612 & -17.879 & -26.591 \\ 1 & -16.289 & 0 & -20.225 & -25.843 & -27.214 \\ 1 & -23.612 & -20.225 & 0 & -22.868 & -13.415 \\ 1 & -17.879 & -25.843 & -22.868 & 0 & -20.305 \\ 1 & -26.591 & -27.214 & -13.415 & -20.305 & 0 \\ 0 & 1 & 1 & 1 & 1 & 1 \end{bmatrix} \cdot \begin{bmatrix} a \\ A_1 \\ A_2 \\ A_3 \\ A_4 \\ A_5 \end{bmatrix} = \begin{bmatrix} 20.820 \\ 10.910 \\ 10.380 \\ 14.600 \\ 10.560 \\ 0 \end{bmatrix}$$

The matrix solutions are:

$$a = 13.203 \quad A_1 = 0.396 \quad A_2 = -0.226$$
$$A_3 = -0.058 \quad A_4 = -0.047 \quad A_5 = -0.065$$

Now we can calculate the value at point 0 by:

$$z_0 = 13.203 + (0.396)(-7.510) +$$
$$(-0.226)(-9.879) + (-0.058)(-16.831) +$$
$$(-0.047)(-18.574) + (-0.065)(-22.834) = 15.795$$

Using the same data set, the estimation of z_0 by other splines is as follows: 16.350 by thin-plate splines and 15.015 by regularized splines (using the τ value of 0.1).

Thin-plate splines and their variations are recommended for smooth, continuous surfaces such as elevation and water table. Splines have also been used for interpolating mean rainfall surface (Hutchinson 1995) and land demand surface (Wickham et al. 2000). Figures 16.12 and 16.13 show annual precipitation surfaces created by the regularized splines method and the splines with tension method respectively. The isolines in both figures are smoother than those generated by the IDW method. Also noticeable is the similarity between the two sets of isolines.

16.4 KRIGING

Kriging is a geostatistical method for spatial interpolation. Kriging differs from the interpolation methods discussed so far because kriging

can assess the quality of prediction with estimated prediction errors. Originated in mining and geologic engineering in the 1950s, kriging has since been adopted in a wide variety of disciplines (Davis 1986; Isaaks and Srivastava 1989; Webster and Oliver 1990; Cressie 1991; Bailey and Gatrell 1995; Goovaerts 1997; Webster and Oliver 2001).

Kriging assumes that the spatial variation of an attribute such as changes in grade within an ore body is neither totally random (stochastic) nor deterministic. Instead, the spatial variation may consist of three components: a spatially correlated component, representing the variation of the regionalized variable; a "drift" or structure, representing a trend; and a random error term. The interpretation of these components has led to development of different kriging methods for spatial interpolation.

16.4.1 Semivariogram

Kriging uses the **semivariance** to measure the spatially correlated component, a component that is also called spatial dependence or spatial autocorrelation. As a measure of spatial autocorrelation, semivariance overlaps with Moran's I and G-statistic (Chapter 12). The semivariance is computed by:

$$(16.12)$$

$$\gamma(h) = \frac{1}{2}[z(x_i) - z(x_j)]^2$$

where $\gamma(h)$ is the semivariance between known points, x_i and x_j, separated by the distance h; and z is the attribute value.

Figure 16.14 is a semivariogram cloud, which plots $\gamma(h)$ against h for all pairs of known points in a data set. If spatial dependence does exist in a data set, known points that are close to each other are expected to have small semivariances, and known points that are farther apart are expected to have larger semivariances.

A semivariogram cloud is an important tool for investigating the spatial variability of the phenomenon under study (Gringarten and Deutsch 2001). But because it has all pairs of known points, a semivariogram cloud is difficult to manage and use. A process called **binning** is typically used in kriging to average semivariance data by distance and direction. The first part of the binning process is to group pairs of sample points into lag classes.

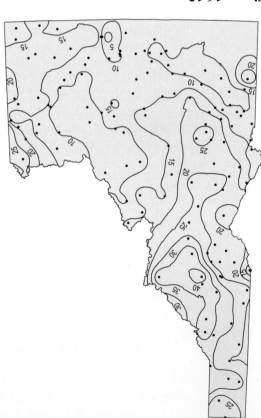

Figure 16.12

An isohyet map created by the regularized splines method.

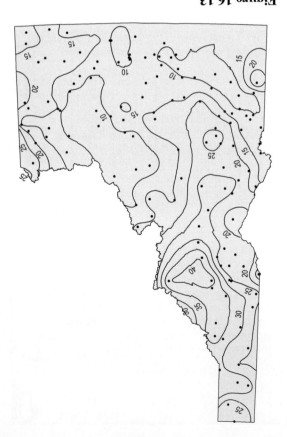

Figure 16.13

An isohyet map created by the splines with tension method.

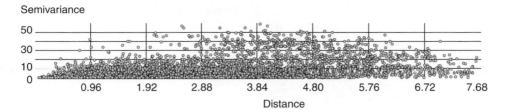

Figure 16.14
A semivariogram cloud.

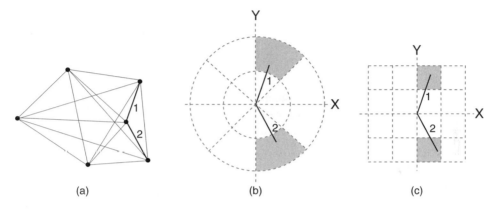

Figure 16.15
A common method for binning pairs of sample points by direction, such as 1 and 2 in (*a*), is to use the radial sector (*b*). Geostatistical Analyst to ArcGIS uses grid cells instead (*c*).

For example, if the lag size (i.e., distance interval) is 2000 meters, then pairs of points separated by less than 2000 meters are grouped into the lag class of 0–2000, pairs of points separated between 2000 and 4000 meters are grouped into the lag class of 2000–4000, and so on. The second part of the binning process is to group pairs of sample points by direction. A common method is to use radial sectors; the Geostatistical Analyst extension to ArcGIS, on the other hand, uses grid cells (Figure 16.15).

The result of the binning process is a set of bins (e.g., grid cells) that sort pairs of sample points by distance and direction. The next step is to compute the average semivariance by:

(16.13)

$$\gamma(h) = \frac{1}{2n} \sum_{i=1}^{n} [z(x_i) - z(x_i + h)]^2$$

where $\gamma(h)$ is the average semivariance between sample points separated by lag h; n is the number of pairs of sample points sorted by direction in the bin; and z is the attribute value.

A **semivariogram** plots the average semivariance against the average distance (Figure 16.16). Because of the directional component, one or more average semivariances may be plotted at the same distance. We can examine the semivariogram in Figure 16.16 by distance. If spatial dependence exists among the sample points, then pairs of points that are closer in distance will have more similar values than pairs that are farther apart. In other words, the semivariance is expected to increase as the distance increases in the presence of spatial dependence.

A semivariogram can also be examined by direction. If spatial dependence has directional differences, then the semivariance values may change more rapidly in one direction than another. **Anisotropy** is the term describing the existence of directional differences in spatial dependence (Eriksson and Siska 2000). Isotropy represents the opposite case in which spatial dependence changes with the distance but not the direction.

16.4.2 Models

A semivariogram such as Figure 16.16 may be used alone as a measure of spatial autocorrelation in the data set. But to be used as an interpolator in kriging, the semivariogram must be fitted with a mathematical function or model (Figure 16.17). The fitted semivariogram can then be used for estimating the semivariance at any given distance.

Fitting a model to a semivariogram is a difficult and often controversial task in geostatistics (Webster and Oliver 2001). One reason for the difficulty is the number of models to choose from. For example, Geostatistical Analyst offers 11 models. The other reason is the lack of a standardized procedure for comparing the models. Webster and Oliver (2001) recommend a procedure that combines visual inspection and cross-validation. Cross-validation, as discussed later in Section 16.5, is a method for comparing interpolation methods. Others suggest use of an artificial intelligent system for selecting an appropriate interpolator according to task-related knowledge and data characteristics (Jarvis et al. 2003).

Two common models for fitting semivariograms are spherical (the default model in Geostatistical Analyst) and exponential (Figure 16.18). A spherical model shows a progressive decrease of spatial dependence until some distance, beyond which spatial dependence levels off. An exponential model exhibits a less gradual pattern than a spherical model: spatial dependence decreases exponentially with increasing distance and disappears completely at an infinite distance.

A fitted semivariogram can be dissected into three possible elements: nugget, range, and sill (Figure 16.19). The **nugget** is the semivariance at the distance of 0, representing measurement error, or microscale variation, or both. The **range** is the distance at which the semivariance starts to level off. In other words, the range corresponds to the

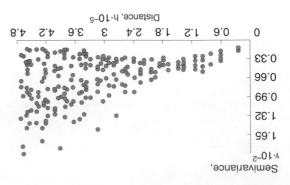

Figure 16.16

A semivariogram after binning.

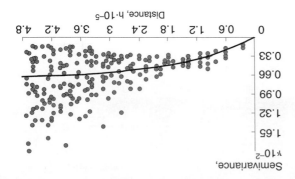

Figure 16.17

Fitting a semivariogram with a mathematical function or a model.

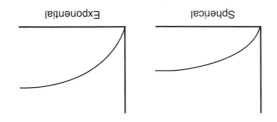

Figure 16.18

Two common models for fitting semivariograms: spherical and exponential.

Semivariance

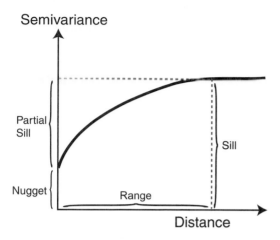

Figure 16.19
Nugget, range, sill, and partial sill.

spatially correlated portion of the semivariogram. Beyond the range, the semivariance becomes a relatively constant value. The semivariance, at which the leveling takes place, is called the **sill.** The sill comprises two components: the nugget and the **partial sill.** To put it another way, the partial sill is the difference between the sill and the nugget.

16.4.3 Ordinary Kriging

Assuming the absence of a drift, **ordinary kriging** focuses on the spatially correlated component and uses the fitted semivariogram directly for interpolation. The general equation for estimating the z value at a point is:

(16.14)

$$z_0 = \sum_{i=1}^{s} z_x W_x$$

where z_0 is the estimated value, z_x is the known value at point x, W_x is the weight associated with point x, and s is the number of sample points used in estimation. The weights can be derived from solving a set of simultaneous equations. For example, the following equations are needed for a point (0) to be estimated from three known points (1, 2, 3):

(16.15)

$$W_1\gamma(h_{11}) + W_2\gamma(h_{12}) + W_3\gamma(h_{13}) + \lambda = \gamma(h_{10})$$
$$W_1\gamma(h_{21}) + W_2\gamma(h_{22}) + W_3\gamma(h_{23}) + \lambda = \gamma(h_{20})$$
$$W_1\gamma(h_{31}) + W_2\gamma(h_{32}) + W_3\gamma(h_{33}) + \lambda = \gamma(h_{30})$$
$$W_1 + W_2 + W_3 + 0 = 1.0$$

where $\gamma(h_{ij})$ is the semivariance between known points i and j, $\gamma(h_{i0})$ is the semivariance between the ith known point and the point to be estimated, and λ is a Lagrange multiplier, which is added to ensure the minimum possible estimation error. The above equations can be rewritten in matrix form:

$$\begin{bmatrix} \gamma(h_{11}) & \gamma(h_{12}) & \gamma(h_{13}) & 1 \\ \gamma(h_{21}) & \gamma(h_{22}) & \gamma(h_{23}) & 1 \\ \gamma(h_{31}) & \gamma(h_{32}) & \gamma(h_{33}) & 1 \\ 1 & 1 & 1 & 0 \end{bmatrix} \cdot \begin{bmatrix} W_1 \\ W_2 \\ W_3 \\ \lambda \end{bmatrix} = \begin{bmatrix} \gamma(h_{10}) \\ \gamma(h_{20}) \\ \gamma(h_{30}) \\ 1 \end{bmatrix}$$

Once the weights are solved, Eq. (16.14) can be used to estimate z_0

$$z_0 = z_1W_1 + z_2W_2 + z_3W_3$$

The above example shows that weights used in kriging involve not only the semivariances between the point to be estimated and the known points but also those between the known points. This differs from the IDW method, which uses only weights applicable to the point to be estimated and the known points. Another important difference between kriging and other local methods is that kriging produces a variance measure for each estimated point to indicate the reliability of the estimation. For the above example, the variance estimation can be calculated by:

(16.16)

$$s^2 = W_1\gamma(h_{10}) + W_2\gamma(h_{20}) + W_3\gamma(h_{30}) + \lambda$$

Figure 16.20 shows an annual precipitation surface created by ordinary kriging with the exponential model. Figure 16.21 shows the distribution of the standard error of the predicted surface. As expected, the standard error is highest in data-poor areas. A worked example of ordinary kriging is included in Box 16.6.

16.4.4 Universal Kriging

Universal kriging assumes that the spatial variation in z values has a drift or a trend in addition to the spatial correlation between the sample points. Typically, universal kriging incorporates a first-order (plane surface) or a second-order (quadratic surface) polynomial in the kriging process. A first-order polynomial is:

(16.17)
$$M = b_1 x_i + b_2 y_i$$

where M is the drift, x_i and y_i are the x-, y-coordinates of sampled point i, and b_1 and b_2 are the drift coefficients. A second-order polynomial is:

(16.18)
$$M = b_1 x_i + b_2 y_i + b_3 x_i^2 + b_4 x_i y_i + b_5 y_i^2$$

Higher-order polynomials are usually not recommended for two reasons. First, kriging is performed on the residuals after the trend is removed. A higher-order polynomial will leave little variation in the residuals for assessing uncertainty. Second, a higher-order polynomial means a larger number of the b_i coefficients, which must be estimated along with the weights, and a larger set of simultaneous equations to be solved.

Figure 16.22 shows an annual precipitation surface created by universal kriging with the linear (first-order) drift, and Figure 16.23 shows the

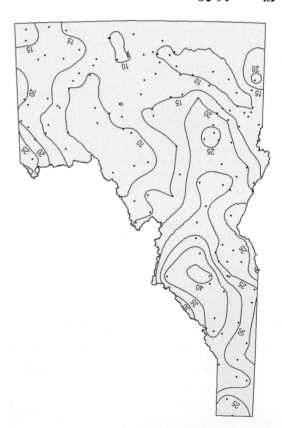

Figure 16.20
An isohyet map created by ordinary kriging with the exponential model.

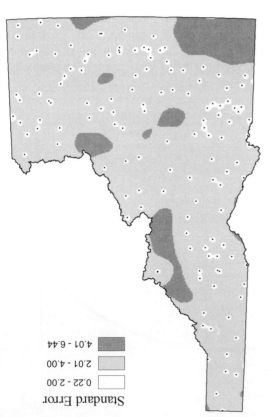

Standard Error

- ☐ 0.22 - 2.00
- ▨ 2.01 - 4.00
- ▨ 4.01 - 6.44

Figure 16.21
Standard errors of the annual precipitation surface in Figure 16.20.

Box 16.6 | A Worked Example of Ordinary Kriging Estimation

This worked example uses ordinary kriging for spatial interpolation. To keep the computation simpler, the semivariogram is fitted with the linear model, which is defined by:

$$\gamma(h) = C_0 + C(h/a), 0 < h <= a$$
$$\gamma(h) = C_0 + C, h > a$$
$$\gamma(0) = 0$$

where $\gamma(h)$ is the semivariance at distance h, C_0 is the semivariance at distance 0, a is the range, and C is the sill, or the semivariance at a. The output from ArcGIS shows:

$$C_0 = 0, C = 112.475, \text{ and } a = 458000.$$

Now we can use the model for spatial interpolation. The scenario is the same as in Box 16.1: using five points with known values to estimate an unknown value. The estimation begins by computing distances between points (in thousands of meters) and the semivariances at those distances based on the linear model:

Points ij	0,1	0,2	0,3	0,4	0,5
h_{ij}	18.000	20.880	32.310	36.056	47.202
$\gamma(h_{ij})$	4.420	5.128	7.935	8.855	11.592
Points ij	1,2	1,3	1,4	1,5	2,3
h_{ij}	31.240	49.476	34.526	59.666	40.000
$\gamma(h_{ij})$	7.672	12.150	8.479	14.653	9.823
Points ij	2,4	2,5	3,4	3,5	4,5
h_{ij}	56.920	62.032	47.412	26.076	40.200
$\gamma(h_{ij})$	13.978	15.234	11.643	6.404	9.872

Using the semivariances, we can write the simultaneous equations for solving the weights in matrix form:

$$\begin{bmatrix} 0 & 7.672 & 12.150 & 8.479 & 14.653 & 1 \\ 7.672 & 0 & 9.823 & 13.978 & 15.234 & 1 \\ 12.150 & 9.823 & 0 & 11.643 & 6.404 & 1 \\ 8.479 & 13.978 & 11.643 & 0 & 9.872 & 1 \\ 14.653 & 15.234 & 6.404 & 9.872 & 0 & 1 \\ 1 & 1 & 1 & 1 & 1 & 0 \end{bmatrix} \cdot \begin{bmatrix} W_1 \\ W_2 \\ W_3 \\ W_4 \\ W_5 \\ \lambda \end{bmatrix} = \begin{bmatrix} 4.420 \\ 5.128 \\ 7.935 \\ 8.855 \\ 11.592 \\ 1 \end{bmatrix}$$

The matrix solutions are:

$$W_1 = 0.397 \quad W_2 = 0.318 \quad W_3 = 0.182$$
$$W_4 = 0.094 \quad W_5 = 0.009 \quad \lambda = -1.161$$

Using Eq. (16.14), we can estimate the unknown value at point 0 by:

$$z_0 = (0.397)(20.820) + (0.318)(10.910) + (0.182)(10.380) + (0.094)(14.600) + (0.009)(10.560) = 15.091$$

We can also estimate the variance at point 0 by:

$$s^2 = (4.420)(0.397) + (5.128)(0.318) + (7.935)(0.182) + (8.855)(0.094) + (11.592)(0.009) - 1.161 = 4.605$$

In other words, the standard error of estimate at point 0 is 2.146.

distribution of the standard error of the predicted surface. Universal kriging produces less reliable estimates than ordinary kriging in this case. A worked example of universal kriging is included in Box 16.7.

16.4.5 Other Kriging Methods

The three basic kriging methods are ordinary kriging, universal kriging, and simple kriging. Simple

kriging assumes that the mean of the data set is known. This assumption, however, is unrealistic in most cases.

Other kriging methods include indicator kriging, disjunctive kriging, and block kriging (Bailey and Gatrell 1995; Burrough and McDonnell 1998; Webster and Oliver 2001). Indicator kriging uses binary data (i.e., 0 and 1) rather than continuous data. The interpolated values are therefore between

0 and 1, similar to probabilities. Disjunctive kriging uses a function of the attribute value for interpolation and is more complicated than other kriging methods computationally. Block kriging estimates the average value of a variable over some small area or block rather than at a point.

Cokriging uses one or more secondary variables, which are correlated with the primary variable, in interpolation. It assumes that the correlation between the variables can be used to improve the prediction of the value of the primary variable. For example, better results in precipitation interpolation have been reported by including elevation as an additional variable in cokriging

(Martinez-Cob 1996). Cokriging can be ordinary cokriging, universal cokriging, and so on, depending on the kriging method that is applied to each data set.

16.5 COMPARISON OF SPATIAL INTERPOLATION METHODS

Using the same data but different methods, we can expect to find different interpolation results. Likewise, different predicted values can occur by using the same method but different parameter values. GIS packages such as ArcGIS offer a large number of

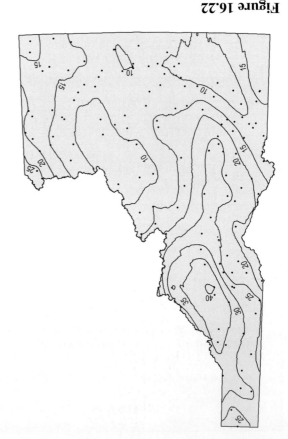

Figure 16.22
An isohyet map created by universal kriging with the linear drift and the spherical model.

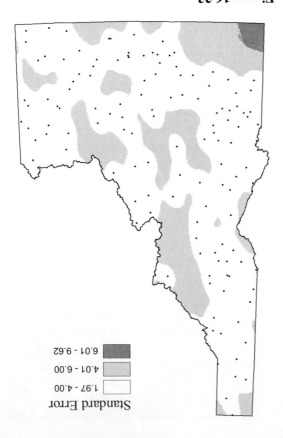

Figure 16.23
Standard errors of the annual precipitation surface in Figure 16.22.

Standard Error
1.97 - 4.00
4.01 - 6.00
6.01 - 9.62

Box 16.7 | A Worked Example of Universal Kriging Estimation

This example uses universal kriging to estimate the unknown value at point 0 (Box 16.1) and assumes that (1) the drift is linear and (2) the semivariogram is fitted with a linear model. Because of the additional drift component, this example uses eight simultaneous equations:

$$W_1\gamma(h_{11}) + W_2\gamma(h_{12}) + W_3\gamma(h_{13}) + W_4\gamma(h_{14}) + W_5\gamma(h_{15}) + \lambda + b_1x_1 + b_2y_1 = \gamma(h_{10})$$

$$W_1\gamma(h_{21}) + W_2\gamma(h_{22}) + W_3\gamma(h_{23}) + W_4\gamma(h_{24}) + W_5\gamma(h_{25}) + \lambda + b_1x_2 + b_2y_2 = \gamma(h_{20})$$

$$W_1\gamma(h_{31}) + W_2\gamma(h_{32}) + W_3\gamma(h_{33}) + W_4\gamma(h_{34}) + W_5\gamma(h_{35}) + \lambda + b_1x_3 + b_2y_3 = \gamma(h_{30})$$

$$W_1\gamma(h_{41}) + W_2\gamma(h_{42}) + W_3\gamma(h_{43}) + W_4\gamma(h_{44}) + W_5\gamma(h_{45}) + \lambda + b_1x_4 + b_2y_4 = \gamma(h_{40})$$

$$W_1\gamma(h_{51}) + W_2\gamma(h_{52}) + W_3\gamma(h_{53}) + W_4\gamma(h_{54}) + W_5\gamma(h_{55}) + \lambda + b_1x_5 + b_2y_5 = \gamma(h_{50})$$

$$W_1 + W_2 + W_3 + W_4 + W_5 + 0 + 0 + 0 = 1$$

$$W_1x_1 + W_2x_2 + W_3x_3 + W_4x_4 + W_5x_5 + 0 + 0 + 0 = x_0$$

$$W_1y_1 + W_2y_2 + W_3y_3 + W_4y_4 + W_5y_5 + 0 + 0 + 0 = y_0$$

where x_0 and y_0 are the x-, y-coordinates of the point to be estimated, and x_i and y_i are the x-, y-coordinates of known point i; otherwise, the notations are the same as in Box 16.6. The x-, y-coordinates are actu-

ally rows and columns in the output raster with a cell size of 2000 meters.

Similar to ordinary kriging, semivariance values for the equations can be derived from the semivariogram and the linear model. The next step is to rewrite the equations in matrix form:

$$\begin{bmatrix} 0 & 7.672 & 12.150 & 8.479 & 14.653 & 1 & 69 & 76 \\ 7.672 & 0 & 9.823 & 13.978 & 15.234 & 1 & 59 & 64 \\ 12.150 & 9.823 & 0 & 11.643 & 6.404 & 1 & 75 & 52 \\ 8.479 & 13.978 & 11.643 & 0 & 9.872 & 1 & 86 & 73 \\ 14.653 & 15.234 & 6.404 & 9.872 & 0 & 1 & 88 & 53 \\ 1 & 1 & 1 & 1 & 1 & 0 & 0 & 0 \\ 69 & 59 & 75 & 86 & 88 & 0 & 0 & 0 \\ 76 & 64 & 52 & 73 & 53 & 0 & 0 & 0 \end{bmatrix} \cdot \begin{bmatrix} W_1 \\ W_2 \\ W_3 \\ W_4 \\ W_5 \\ \lambda \\ b_1 \\ b_2 \end{bmatrix} = \begin{bmatrix} 4.420 \\ 5.128 \\ 7.935 \\ 8.855 \\ 11.592 \\ 1 \\ 69 \\ 67 \end{bmatrix}$$

The solutions are:

$$W_1 = 0.387 \; W_2 = 0.311 \; W_3 = 0.188 \; W_4 = 0.093$$
$$W_5 = 0.021 \; \lambda = -1.154 \; b_1 = 0.009 \; b_2 = -0.010$$

The estimated value at point 0 is:

$$z_0 = (0.387)(20.820) + (0.311)(10.910) + (0.188)(10.380) + (0.093)(14.600) + (0.021)(10.560) = 14.981$$

And, the variance at point 0 is:

$$s^2 = (4.420)(0.387) + (5.128)(0.311) + (7.935)(0.188) + (8.855)(0.093) + (11.592)(0.021) - 1.154 = 4.710$$

The standard error (s) at point 0 is 2.170. These results from universal kriging are very similar to those from ordinary kriging.

spatial interpolation methods (Box 16.8). How can we compare these methods? Section 16.5 describes the methods and useful statistics for comparison.

Figure 16.24 shows the difference of interpolated surfaces between IDW (inverse distance weighted) and ordinary kriging. The difference ranges from –8 to 3.4 inches: a negative value means that IDW has a smaller estimate than ordinary kriging, and a positive value means a reverse

pattern. Data-poor areas clearly have the largest difference (i.e., more than 3 inches, either positive or negative). This suggests the importance of having more control points in the data-poor areas. No matter which method is used, spatial interpolation can never substitute for observed data. But if the option of adding more points is not feasible, how can one tell which interpolation method, or which parameter value, is better?

Box 16.8 Spatial Interpolation Using ArcGIS

The Spatial Analyst extension to ArcGIS has menu access to density estimation (simple and kernel), inverse distance weighted, splines (splines with tension and regularized splines), and kriging (ordinary and universal). Spatial Analyst Tools in ArcToolbox offer the same tools in the Density and Interpolation toolsets. Additionally, ArcToolbox includes Trend in the Spatial Analyst Tools/Interpolation toolset and Thiessen in the Coverage Tools/Proximity toolset. Most users, however, will opt to run spatial interpolation using the Geostatistical Analyst extension.

Geostatistical Analyst's main menu has the selections of explore data, geostatistical wizard, and create subsets. The explore data selection offers histogram, semivariogram, QQ plot, and others for exploratory data analysis (Chapter 11). The geostatistical wizard offers a large variety of global and local interpolation methods including IDW, trend surface, local polynomial, radial basis function, kriging, and cokriging. The wizard also provides cross-validation measures. The create subsets selection is designed for model validation.

Comparison of local methods is usually based on statistical measures, although some studies have also suggested the importance of the visual quality of generated surfaces such as preservation of distinct spatial pattern and visual pleasantness and faithfulness (Declercq 1996; Yang and Hodler 2000). Cross-validation and validation are two common statistical techniques for comparison (Phillips et al. 1992; Garen et al. 1994; Carroll and Cressie 1996; Zimmerman et al. 1999).

Cross-validation compares the interpolation methods by repeating the following procedure for each interpolation method to be compared:

1. Remove a known point from the data set.
2. Use the remaining points to estimate the value at the point previously removed.
3. Calculate the predicted error of the estimation by comparing the estimated with the known value.

After completing the procedure for each known point, one can calculate diagnostic statistics to assess the accuracy of the interpolation method. Two common diagnostic statistics are the root mean square (RMS) and the standardized RMS:

$$RMS = \sqrt{\frac{1}{n}\sum_{i=1}^{n}(z_{i,\,act} - z_{i,\,est})^2}$$

(16.19)

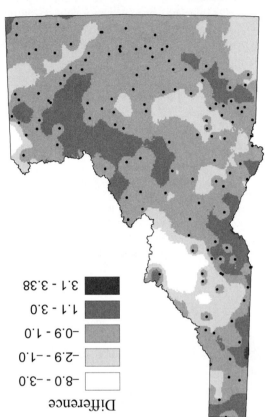

Figure 16.24
Differences between the interpolated surfaces from ordinary kriging and IDW.

Difference
- -8.0 - -3.0
- -2.9 - -1.0
- -0.9 - 1.0
- 1.1 - 3.0
- 3.1 - 3.38

(16.20)

$$\text{Standardized RMS} = \sqrt{\frac{1}{n} \sum_{i=1}^{n} \frac{(z_{i,\text{act}} - z_{i,\text{est}})^2}{s^2}}$$

$$= \frac{\text{RMS}}{s}$$

where n is the number of points, $z_{i,\text{act}}$ is the known value of point i, $z_{i,\text{est}}$ is the estimated value of point i, s^2 is the variance, and s is the standard error.

The RMS statistic is available for all exact local methods. But the standardized RMS is only available for kriging because the variance is required for the computation. The interpretation of the statistics is as follows:

- A better interpolation method should yield a smaller RMS. By extension, an optimal method should have the smallest RMS, or the smallest average deviation between the estimated and known values at sample points.
- A better kriging method should yield a smaller RMS and a standardized RMS closer to 1.

If the standardized RMS is 1, it means that the RMS statistic equals s. Therefore the estimated standard error is a reliable or valid measure of the uncertainty of predicted values.

The **validation** technique compares the interpolation methods by first dividing known points into two samples: one sample for developing the model for each interpolation method to be compared and the other sample for testing the accuracy of the models. The diagnostic statistics of RMS and standardized RMS derived from the test sample can then be used to compare the methods. Validation may not be a feasible option if the number of known points is too small to be split into two samples.

KEY CONCEPTS AND TERMS

Anisotropy: A term describing the existence of directional differences in spatial dependence.

Binning: A process used in kriging to average semivariance data by distance and direction.

Control points: Points with known values in spatial interpolation. Also called *known points, sample points,* or *observations.*

Cross-validation: A technique for comparing different interpolation methods.

Density estimation: A local interpolation method, which measures densities in a raster based on a distribution of points and point values.

Deterministic interpolation: A spatial interpolation method that provides no assessment of errors with predicted values.

Exact interpolation: An interpolation method that predicts the same value as the known value at the control point.

Global interpolation: An interpolation method that uses every control point available in estimating an unknown value.

Inexact interpolation: An interpolation method that predicts a different value from the known value at the control point.

Inverse distance weighted (IDW) interpolation: A local interpolation method, which enforces that the unknown value of a point is influenced more by nearby points than those farther away.

Kernel density estimation: A local interpolation method, which associates each known point with a kernel function in the form of a bivariate probability density function.

Kriging: A stochastic interpolation method, which assumes that the spatial variation of an attribute includes a spatially correlated component.

Local interpolation: An interpolation method that uses a sample of known points in estimating an unknown value.

Local polynomial interpolation: A local interpolation method that uses a sample of points with known values and a polynomial equation to estimate the unknown value of a point.

Nugget: The semivariance value at the distance of 0 in a semivariogram.

Ordinary kriging: A kriging method, which assumes the absence of a drift or trend and focuses on the spatially correlated component.

Partial sill: The difference between the sill and the nugget in a semivariogram.

Radial basis functions (RBF): A diverse group of methods for spatial interpolation including thin-plate splines, thin-plate splines with tension, and regularized splines.

Range: The distance at which the semivariance starts to level off in a semivariogram.

Regression model: A global interpolation method that uses a number of independent variables to estimate a dependent variable.

Semivariance: A measure of the degree of spatial dependence among points used in kriging.

Semivariogram: A diagram relating the semivariance to the distance between sample points used in kriging.

Sill: The semivariance at which the leveling starts in a semivariogram.

Spatial interpolation: The process of using points with known values to estimate unknown values at other points.

Stochastic interpolation: A spatial interpolation method that offers assessment of prediction errors with estimated variances.

Thiessen polygons: A local interpolation method, which ensures that every unsampled point within a polygon is closer to the polygon's known point than any other known points. Also called *Voronoi polygons*.

Thin-plate splines: A local interpolation method, which creates a surface passing through points with the least possible change in slope at all points.

Thin-plate splines with tension: A variation of thin-plate splines for spatial interpolation.

Trend surface analysis: A global interpolation method that uses points with known values and a polynomial equation to approximate a surface.

Universal kriging: A kriging method, which assumes that the spatial variation of an attribute has a drift or a structural component in addition to the spatial correlation between sample points.

Validation: A technique for comparing interpolation methods, which splits control points into two samples, one for developing the model and the other for testing the accuracy of the model.

REVIEW QUESTIONS

1. What is spatial interpolation?
2. What kinds of inputs are required for spatial interpolation?
3. Explain the difference between a global method and a local method.
4. How does an exact interpolation method differ from an inexact interpolation method?
5. Go to the website, **http://www.nrcs.usda.gov/technical/maps.html,** and examine the PRISM map for precipitation for your state. Derived from regression equations, the PRISM map is essentially a model and may contain unusual precipitation amounts.
6. Given a sample size of 12, illustrate the difference between a sampling method that uses the closest points and a quadrant sampling method.
7. Describe how cell densities are derived using the kernel density estimation method.
8. The power k in inverse distance weighted interpolation determines the rate of change in values from the sample points. Can you

think of a spatial phenomenon that should be interpolated with a *k* value of 2 or higher?

9. Describe how the semivariance can be used to quantify the spatial dependence in a data set.

10. Binning is a process for creating a usable semivariogram from empirical data. Describe how binning is performed.

11. A semivariogram must be fitted with a mathematical model before it can be used in kriging. Why?

12. Both IDW and kriging use weights in estimating an unknown value. Describe the difference between the two interpolation methods in terms of derivation of the weights.

13. Explain the main difference between ordinary kriging and universal kriging.

14. The root mean square (RMS) statistic is commonly used for selecting an optimal interpolation method. What does the RMS statistic measure?

15. Explain how one can use the validation technique for comparing different interpolation methods.

APPLICATIONS: SPATIAL INTERPOLATION

This applications section has five tasks. Task 1 covers trend surface analysis. Task 2 deals with kernel density estimation. Task 3 uses IDW for local interpolation. Tasks 4 and 5 cover kriging: Task 4 uses ordinary kriging and Task 5 universal kriging. Except for Task 2, you will run spatial interpolation in Geostatistical Analyst so that you can use the cross-validation statistics such as the root mean square (RMS) statistic to compare models. Geostatistical Analyst also provides more information and better user interface than Spatial Analyst or ArcToolbox for spatial interpolation.

Task 1: Use Trend Surface Model for Interpolation

What you need: *stations.shp*, a shapefile containing 175 weather stations in and around Idaho; and *idoutlgd*, an Idaho outline raster.

In Task 1 you will first explore the average annual precipitation data in *stations.shp*, before running a trend surface analysis.

1. Start ArcCatalog, and make connection to the Chapter 16 database. Launch ArcMap. Add *stations.shp* and *idoutlgd* to Layers and rename the data frame Task 1. Make sure that both the Geostatistical Analyst and Spatial Analyst extensions are checked in the Tools menu and their toolbars are checked in the View menu.

2. Click the Geostatistical Analyst dropdown arrow, point to Explore Data, and select Trend Analysis. At the bottom of the Trend Analysis dialog, click the dropdown arrow to select *stations* for the layer and ANN_PREC for the attribute.

3. Maximize the Trend Analysis dialog. The 3-D diagram shows two trend projections: The *YZ* plane dips from north to south, and the *XZ* plane dips initially from west to east and then rises slightly. The north–south trend is much stronger than the east–west trend, suggesting that the general precipitation pattern in Idaho decreases from north to south. Close the dialog.

4. Click the Geostatistical Analyst dropdown arrow, and select Geostatistical Wizard. The Step 1 panel lets you choose the input data and geostatistical method. Click the Input Data dropdown arrow and select *stations*. Click the Attribute dropdown arrow and select ANN_PREC. In the Methods frame, click Global Polynomial Interpolation.

5. The Step 2 panel lets you choose the power of the trend surface model. The Power list provides the choice from 1 to 10. Select 1 for

the power. The next panel shows scatter plots (Predicted versus Measured values, and Error versus Measured values) and statistics related to the first-order trend surface model. The RMS statistic measures the overall fit of the trend surface model. In this case, it has a value of 6.073. Click Back and change the power to 2. The RMS statistic for the power of 2 has a value of 6.085. Repeat the same procedure with other power numbers. The trend surface model with the lowest RMS statistic is the best overall model for this task. For ANN_PREC, the best overall model has the power of 5. Change the power to 5, and click Finish. Click OK in the Output Layer Information dialog.

Q1. What is the RMS statistic for the power of 5?

6. *Global Polynomial Interpolation Prediction Map* is a Geostatistical Analyst (ga) output layer and has the same area extent as *stations.* Right-click *Global Polynomial Interpolation Prediction Map* and select Properties. The Symbology tab has four Show options: Hillshade, Contours, Grid, and Filled Contours. Uncheck all Show boxes except Filled Contours, click Filled Contours, and then click on Classify. In the Classification dialog, select the Manual method and 7 classes. Then enter the class breaks of 10, 15, 20, 25, 30, and 35 between the Min and Max values. Click OK to dismiss the dialogs. The contour (isohyet) lines are color-coded.

7. To clip *Global Polynomial Interpolation Prediction Map* to fit Idaho, first convert the ga data set to a raster. Right-click *Global Polynomial Interpolation Prediction Map,* point to Data, and select Export to Raster. In the Export to Raster dialog, enter 2000 (meters) for the cell size and specify *trend5_temp* for the output raster. Click OK to export the data set. Add *trend5_temp* to the map. (Extreme cell values in *trend5_temp* are located outside the state border.)

8. Now you are ready to clip *trend5_temp.* Click the Spatial Analyst dropdown arrow and select Options. On the General tab, select

idoutlgd for the analysis mask and click OK. Select Raster Calculator from the Spatial Analyst dropdown menu. Double-click *trend5_temp* in the Layers frame so that *trend5_temp* appears in the expression box. Click Evaluate. *Calculation* is the clipped *trend5_temp.*

9. You can generate contours from *Calculation* for data visualization. Click the Spatial Analyst dropdown arrow, point to Surface Analysis, and select Contour. In the Contour dialog, make sure that *Calculation* is the input surface, enter 5 for the contour interval and 10 for the base contour, and specify *trend5Contour.shp* for the output features. Click OK. To label the contour lines, right-click *trend5Contour* and select Properties. On the Labels tab, check the box to label features in this layer, select CONTOUR from the Label Field dropdown list, and click OK. The map now shows contour labels.

10. *Calculation* is a temporary grid. You can save it as a permanent grid by going through the following steps: right-click *Calculation* and select Make Permanent, and enter the name for the permanent grid.

Task 2: Use Kernel Density Estimation Method

What you need: *deer.shp,* a point shapefile showing deer locations.

Task 2 uses the kernel density estimation method to compute the average number of deer sightings per hectare from *deer.shp.* Deer location data have a 50-meter minimum discernible distance; therefore, some locations have multiple sightings.

1. Insert a new data frame in ArcMap and rename it Task 2. Add *deer.shp* to Task 2. Select Properties from the context menu of *deer.* On the Symbology tab, select Quantities and Graduated symbols in the Show box and select SIGHTINGS from the Value dropdown list. Click OK. The map shows deer sightings at each location in graduated symbols.

Q2. What is the value range of SIGHTINGS?

2. Click Show/Hide ArcToolbox Window to open ArcToolbox. Double-click the Kernel Density tool in the Spatial Analyst Tools/ Density toolset. Select *deer* for the input point features, select SIGHTINGS for the population field, specify *kernel_d* for the output raster, enter 100 for the output cell size, enter 100 for the search radius, and select HECTARES for the area units. Click OK to run the command. *kernel_d* shows deer sighting densities computed by the kernel density estimation method.

Q3. What is the value range of deer sighting densities?

3. To view *kernel_d* on top of *deer*, you can use the transparent option. Right-click *kernel_d* and select Properties. On the Display tab, enter 30% transparent. You can now see two layers superimposed on top of one another.

Task 3: Use IDW for Interpolation

What you need: *stations.shp* and *idoutlgd*, same as in Task 1.

This task lets you create a precipitation raster using the IDW method.

1. Insert a new data frame in ArcMap and rename it Task 3. Add *stations.shp* and *idoutlgd* to Task 3. Make sure that the Geostatistical Analyst and Spatial Analyst toolbars are both available.

2. Click the Geostatistical Analyst dropdown arrow and select Geostatistical Wizard. Select *stations* for the input data and ANN_PREC for the attribute. Click Inverse Distance Weighting in the Methods frame. Click Next.

3. The Step 1 panel includes a graphic frame and a method frame for specifying IDW parameters. The default IDW method uses a power of 2, 15 neighbors (control points), and a circular area from which control points are selected. The graphic frame shows *stations* and the points and their weights (shown in percentages and color symbols)

used in deriving the estimated value for a test location. You can click the "Predict/ Estimate Value and Identify Neighbors in the Center of Ellipse" tool and then click any point within the graphic frame to see how the point's estimated value is derived.

4. The Optimize Power Value button is included in the Step 1 panel. Because a change of the power value will change the estimated value at a point location, you can click the button and ask the Geostatistical Wizard to find the optimal power value while holding other parameter values constant. Geostatistical Wizard employs the cross-validation technique to find the optimal power value. Click the Optimize Power Value button, and the Power field shows a value of 3.191. Click Next.

5. The Step 2 panel lets you examine the cross-validation results including the RMS statistic.

Q4. What is the RMS statistic by using the default parameters including the optimal power value?

Q5. Change the power to 2 and the number of neighbors to include to 10 (including at least 6). What RMS statistic do you get?

6. Set the parameters back to the default including the optimal power value. Click Finish. Click OK in the Output Layer Information dialog. You can follow the same steps as in Task 1 to convert *Inverse Distance Weighting Prediction Map* to a raster, to clip the raster by using *idoutlgd* as the analysis mask, and to create isolines from the clipped raster.

Task 4: Use Ordinary Kriging for Interpolation

What you need: *stations.shp* and *idoutlgd*.

In Task 4, you will first examine the semivariogram cloud from 175 points in *stations.shp*. Then you will run ordinary kriging on *stations.shp* to generate an interpolated precipitation raster and a standard error raster.

1. Select Data Frame from the Insert menu in ArcMap. Rename the new data frame

Tasks 4&5, and add *stations.shp* and *idoutlgd* to Tasks 4&5. First explore the semivariogram cloud. Click the Geostatistical Analyst dropdown arrow, point to Explore Data, and select Semivariogram/Covariance Cloud. Select *stations* for the layer and ANN_PREC for the attribute. To view all possible pairs of control points in the cloud, enter 82,000 for the lag size and 12 for the number of lags. Use the mouse pointer to drag a box around the point to the far right of the cloud. Check *stations* in the ArcMap window. The highlighted pair consists of the control points that are farthest apart in *stations*. The semivariogram shows a typical pattern of spatially correlated data: the semivariance increases rapidly to a distance of about 200,000 meters (2.00×10^5) and then gradually decreases.

2. To zoom in on the distance range of 200,000 meters, change the lag size to 10,000 and the number of lags to 20. The semivariance actually starts to level off at about 125,000 meters. To see if the semivariance has the directional influence, check the box to show search direction. You can change the search direction by either entering the angle direction or using the direction controller in the graphic. Drag the direction controller in the counterclockwise direction from $0°$ to $180°$ but pause at different angles to check the semivariogram. The fluctuation in the semivariance tends to increase from northwest ($315°$) to southwest ($225°$). This indicates that the semivariance has the directional influence. Close the Semivariance/Covariance Cloud window.

3. Select Geostatistical Wizard from the Geostatistical Analyst menu. Select *stations* for the input data and ANN_PREC for the attribute. Click Kriging in the Methods frame. Click Next. The Step 1 panel lets you select the kriging method. Select Ordinary Kriging/Prediction Map. Click Next.

4. The Step 2 panel shows the semivariogram/covariance view, which is similar to the semivariogram/covariance cloud except that the semivariance data have been averaged by distance and direction (i.e., binned). The Models frame lets you choose a mathematical model to fit the empirical semivariogram. First change the lag size to 40,000 and the number of lags to 12. A general guideline in choosing the lag size and number of lags is that their product should be about half the longest distance among all pairs of control points. The longest distance in *stations* is slightly over 960,000 meters. Click on Exponential and then check the box for Anisotropy. (The Exponential model has the best cross-validation statistics overall.) Geostatistical Analyst automatically calculates the optimal angle direction for anisotropy. You can use the Semivariogram/Covariance Surface view and a bandwidth of 6 to see if the calculated angle direction yields the best fit between the model (represented by the yellow line) and the empirical semivariogram. Click Next.

5. The Step 3 panel lets you choose the number of neighbors (control points), and the sampling method. Click Next.

6. The Step 4 panel shows the cross-validation results. The Chart frame offers four types of scatter plots (Predicted versus Measured values, Error versus Measured values, Standardized Error versus Measured values, and Quantile–Quantile plot for Standardized Error against Normal values). The Prediction Errors frame lists cross-validation statistics, including the RMS statistic.

Q6. What is the RMS value?

Q7. Try a different mathematical model (e.g., Spherical, Gaussian, etc.) and see if a better set of cross-validation statistics than the Exponential model is obtainable.

7. Click Finish in the Step 4 panel. Click OK in the Output Layer Information dialog. *Ordinary Kriging Prediction Map* is added to the map. To derive a prediction standard

error map, you will click the Ordinary Kriging/Prediction Standard Error Map in the Step 1 panel and repeat panels 2 to 4.

8. You can follow the same steps as in Task 1 to convert *Ordinary Kriging Prediction Map* and *Ordinary Kriging Prediction Standard Error Map* to rasters, to clip the rasters by using *idoutlgd* as the analysis mask, and to create isolines from the clipped rasters.

Task 5: Use Universal Kriging for Interpolation

What you need: *stations.shp* and *idoutlgd*.

In Task 5 you will run universal kriging on *stations.shp*. The trend to be removed from the kriging process is the first-order trend surface.

1. Click the Geostatistical Analyst dropdown arrow and select Geostatistical Wizard. Select *stations* for the input data and ANN_PREC for the attribute. Click Kriging in the Methods frame. Click Next.

2. In the Step 1 panel, click Universal Kriging/ Prediction Map in the Geostatistical Methods frame. Select First from the Order of Trend dropdown list. Click Next.

3. The Step 2 panel shows the first-order trend that will be removed from the kriging process. Click Next.

4. In the Step 3 panel, enter 10,000 for the lag size and 15 for the number of lags. Click on Spherical and then check the box for Anisotropy. (The Spherical model has the best cross-validation statistics overall.) Click Next.

5. Take the default values for the number of neighbors and the sampling method. Click Next.

6. The Step 5 panel shows the cross-validation results. Although the RMS value is about the same as ordinary kriging in Task 4, the standardized RMS value is lower than ordinary kriging. This means that the estimated standard error from universal kriging is not as reliable as that from ordinary kriging.

Q8. What is the standardized RMS value from the Step 5 panel?

7. Click Finish in the Step 5 panel. Click OK in the Output Layer Information dialog. *Universal Kriging Prediction Map* is an interpolated map from universal kriging. To derive a prediction standard error map, you will click the Universal Kriging/Prediction Standard Error Map in the Step 1 panel and repeat panels 2 to 5.

8. You can follow the same steps as in Task 1 to convert *Universal Kriging Prediction Map* and *Universal Kriging Prediction Standard Error Map* to rasters, to clip the rasters by using *idoutlgd* as the analysis mask, and to create isolines from the clipped rasters.

Challenge Task

What you need: *stations.shp* and *idoutlgd*.

This challenge task asks you to compare the interpolation results from two spline methods in Geostatistical Analyst. Except for the interpolation method, you will use the default values for the challenge task. The task has three parts: one, create an interpolated raster using regularized splines; two, create an interpolated raster using thin-plate splines with tension; and three, use a local operation to compare the two rasters. The result can show the difference between the two interpolation methods.

1. Create a Radial Basis Functions Prediction map by using the kernel function of Completely Regularized Spline. Convert the map to a raster, and save the raster as *regularized*.

2. Create a Radial Basis Functions Prediction map by using the kernel function of Spline with Tension. Convert the map to a raster, and save the raster as *tension*.

3. Select Options from the Spatial Analyst menu. On the General tab, select *idoutlgd* for the analysis mask.

4. Use Raster Calculator from the Spatial Analyst menu to subtract *tension* from *regularized*.

5. The calculation result shows the difference in cell values between the two rasters within *idoutlgd*. Display the difference raster in three classes: lowest value to −0.5, −0.5 to 0.5, and 0.5 to highest value.

Q1. What is the range of cell values in the difference raster?

Q2. What does a positive cell value in the difference raster mean?

Q3. Is there a pattern in terms of where high cell values, either positive or negative, are distributed in the difference raster?

REFERENCES

Bailey, T. C., and A. C. Gatrell. 1995. *Interactive Spatial Data Analysis*. Harlow, England: Longman Scientific & Technical.

Burrough, P. A., and R. A. McDonnell. 1998. *Principles of Geographical Information Systems*. Oxford, England: Oxford University Press.

Carroll, S. S., and N. Cressie. 1996. A Comparison of Geostatistical Methodologies Used to Estimate Snow Water Equivalent. *Water Resources Bulletin* 32: 267–78.

Chang, K., and Z. Li. 2000. Modeling Snow Accumulation with a Geographic Information System. *International Journal of Geographical Information Science* 14: 693–707.

Cressie, N. 1991. *Statistics for Spatial Data*. Chichester, England: Wiley.

Daly, C., R. P. Neilson, and D. L. Phillips. 1994. A Statistical-Topographic Model for Mapping Climatological Precipitation over Mountainous Terrain. *Journal of Applied Meteorology* 33: 140–58.

Davis, J. C. 1986. *Statistics and Data Analysis in Geology,* 2d ed. New York: Wiley.

Declercq, F. A. N. 1996. Interpolation Methods for Scattered Sample Data:

Accuracy, Spatial Patterns, Processing Time. *Cartography and Geographic Information Science* 23: 128–44.

Eriksson, M., and P. P. Siska. 2000. Understanding Anisotropy Computations. *Mathematical Geology* 32: 683–700.

Franke, R. 1985. Thin Plate Splines with Tension. *Computer-Aided Geometrical Design* 2: 87–95.

Franke, R. 1982. Smooth Interpolation of Scattered Data by Local Thin Plate Splines. *Computers and Mathematics with Applications* 8: 273–81.

Garen, D. C., G. L. Johnson, and C. J. Hanson. 1994. Mean Areal Precipitation for Daily Hydrologic Modeling in Mountainous Regions. *Water Resources Bulletin* 30: 481–91.

Goovaerts, P. 1997. *Geostatistics for Natural Resources Evaluation*. New York: Oxford University Press.

Griffith, D. A., and C. G. Amrhein. 1991. *Statistical Analysis for Geographers*. Englewood Cliffs, NJ: Prentice Hall.

Gringarten E., and C. V. Deutsch. 2001. Teacher's Aide: Variogram Interpretation and Modeling. *Mathematical Geology* 33: 507–34.

Hutchinson, M. F. 1995. Interpolating Mean Rainfall Using Thin Plate Smoothing Splines. *International Journal of Geographical Information Systems* 9: 385–403.

Isaaks, E. H., and R. M. Srivastava. 1989. *An Introduction to Applied Geostatistics*. Oxford, England: Oxford University Press.

Jarvis, C. H., N. Stuart, and W. Cooper. 2003. Infometric and Statistical Diagnostics to Provide Artificially-Intelligent Support for Spatial Analysis: The Example of Interpolation. *International Journal of Geographical Information Science* 17: 495–516.

Martinez-Cob, A. 1996. Multivariate Geostatistical Analysis of Evapotranspiration and Precipitation in Mountainous Terrain. *Journal of Hydrology* 174: 19–35.

Mitas, L., and H. Mitasova. 1988. General Variational Approach to the Interpolation Problem. *Computers and Mathematics with Applications* 16: 983–92.

Mitasova, H., and L. Mitas. 1993. Interpolation by Regularized Spline with Tension: I. Theory and Implementation. *Mathematical Geology* 25: 641–55.

Nelson, T., B. Boots, M. Wulder, and R. Feick. 2004. Predicting Forest Age Classes from High Spatial Resolution Remotely Sensed Imagery Using Voronoi Polygon Aggregation. *GeoInformatica* 8: 143–55.

Phillips, D. L., J. Dolph, and D. Marks. 1992. A Comparison of Geostatistical Procedures for Spatial Analysis of Precipitation in Mountainous Terrain. *Agricultural and Forest Meteorology* 58: 119–41.

Robinson, A. H., J. L. Morrison, P. C. Muehrcke, A. J. Kimerling, and S. C. Guptill. 1995. *Elements of Cartography,* 6th ed. New York: Wiley.

Rogerson, P. A. 2001. *Statistical Methods for Geography.* London: Sage Publications.

Scott, D. W. 1992. *Multivariate Density Estimation: Theory, Practice, and Visualization.* New York: Wiley.

Silverman, B. W. 1986. *Density Estimation.* London: Chapman and Hall.

Tabios, G. Q., III, and J. D. Salas. 1985. A Comparative Analysis of Techniques for Spatial Interpolation of Precipitation. *Water Resources Bulletin* 21: 365–80.

Watson, D. F. 1992. *Contouring: A Guide to the Analysis and Display of Spatial Data.* Oxford, England: Pergamon Press.

Webster, R., and M. A. Oliver. 2001. *Geostatistics for Environmental Scientists.* Chichester, England: Wiley.

Webster, R., and M. A. Oliver. 1990. *Statistical Methods in Soil and Land Resource Survey.* Oxford, England: Oxford University Press.

Wickham, J. D., R. V. O'Neill, and K. B. Jones. 2000. A Geography of Ecosystem Vulnerability. *Landscape Ecology* 15: 495–504.

Yang, X., and T. Hodler. 2000. Visual and Statistical Comparisons of Surface Modeling Techniques for Point-Based Environmental Data. *Cartography and Geographic Information Science* 27: 165–75.

Zimmerman, D., C. Pavlik, A. Ruggles, and M. P. Armstrong. 1999. An Experimental Comparison of Ordinary and Universal Kriging and Inverse Distance Weighting. *Mathematical Geology* 31: 375–90.

CHAPTER 17

GEOCODING AND DYNAMIC SEGMENTATION

How can we quickly find the location of a street address? We go to a website such as MapQuest (**http://www.mapquest.com/**), type the address, and get a map with a point symbol representing the address. How can we find a nearby bank in an unfamiliar city? With a mobile phone, we can contact a location-based service company, provide the closest street intersection, and wait for a map showing nearby banks. Although not apparent to us, both activities involve geocoding, the process of plotting a street address or intersection as a point feature on a map. Geocoding has become perhaps the most commercialized GIS-related operation.

Like geocoding, dynamic segmentation can also locate spatial features from a data source that lacks x-, y-coordinates. Dynamic segmentation works with data such as locations of traffic accidents, which are typically reported in linear distances from some known points (e.g., mileposts). Dynamic segmentation is popular with government agencies that produce and distribute geospatial data. Many transportation departments use dynamic segmentation to manage data such as speed limits, rest areas, bridges, and pavement conditions along highway routes. Natural resource agencies also use dynamic segmentation to store and analyze stream reach data and aquatic habitat data.

Chapter 17 covers geocoding and dynamic segmentation in four sections. Section 17.1 discusses geocoding including the reference database, the geocoding process, and the matching options. Section 17.2 reviews geocoding applications. Section 17.3 covers the basic elements of routes and events in dynamic segmentation. Section 17.4 describes the applications of dynamic segmentation in data management, data query, data display, and data analysis.

17.1 GEOCODING

Geocoding refers to the process of assigning spatial locations to data that are in tabular format but have fields that describe their locations. The most common type of geocoding is **address geocoding,** which plots street addresses as point features on a map. Address geocoding is also known as *address matching.*

Address geocoding requires two sets of data. The first data set contains individual street addresses in a table, one record per address (Figure 17.1). The second is a reference database that consists of a street map and attributes for each street segment such as the street name, address ranges, and ZIP codes. Address geocoding interpolates the location of a street address by comparing it with data in the reference database.

```
Name,    Address,    ZIP
Iron Horse, 407 E Sherman Ave, 83814
Franlin's Hoagies, 501 N 4th St, 83814
McDonald's, 208 W Appleway, 83814
Rockin Robin Cafe, 3650 N Government way, 83815
Olive Garden, 525 W Canfield Ave, 83815
Fernan Range Station, 2502 E Sherman Ave, 83814
FBI, 250 Northwest Blvd, 83814
ID Fish & Game, 2750 W Kathleen Ave, 83814
ID Health & Welfare, 1120 W Ironwood Dr, 83814
ID Transportation Dept, 600 W Prairie Ave, 83815
```

Figure 17.1

A sample address table records name, address, and ZIP code.

17.1.1 Geocoding Reference Database

Many GIS users in the United States derive a geocoding reference database from the TIGER/Line files, which are extracts of geographic/cartographic information from the TIGER (Topologically Integrated Geographic Encoding and Referencing) database of the U.S. Census Bureau. The TIGER/Line files contain legal and statistical area boundaries such as counties, census tracts, and block groups, as well as streets, roads, streams, and water bodies. More importantly, the TIGER/Line attributes also include for each street segment the street name, the beginning and end address numbers on each side, and the ZIP code on each side (Figure 17.2).

Geocoding reference databases are also available from commercial companies such as Tele Atlas **(http://www.teleatlas.com/landingpage.jsp)** and Geographic Data Technology. (Tele Atlas North America acquired Geographic Data Technology in April 2004.) These companies advertise their address data products to be current, verified, and updated.

17.1.2 Geocoding Process

A geocoding engine refers to a software program that performs geocoding. GIS packages such as ArcGIS and MapInfo have built-in geocoding engines for address matching. When the process starts, the software first locates the street segment in the reference database that contains the address

```
FEDIRP: A direction that precedes a street name.
FENAME: The name of a street.
FETYPE: The street name type such as St, Rd, and Ln.
FRADDL: The beginning address number on the left side of a street segment.
TOADDL: The ending address number on the left side of a street segment.
FRADDR: The beginning address number on the right side of a street segment.
TOADDR: The ending address number on the right side of a street segment.
ZIPL: The zip code for the left side of a street segment.
ZIPR: The zip code for right side of a street segment.
```

Figure 17.2

The TIGER/Line files include the attributes of FEDIRP, FENAME, FETYPE, FRADDL, TOADDL, FRADDR, TOADDR, ZIPL, and ZIPR, which are important for geocoding.

in the input table. Then it interpolates where the address falls within the address range. For example, if the address is 620 and the address range is from 600 to 700 in the database, the address will be located about one-fifth of the street segment from 600 (Figure 17.3). This process is linear interpolation.

Matching an address with a street segment in the reference database is the challenging part of the geocoding process. A street address consists of a number of components. For example, the address "630 S. Main Street, 83843" has the following components:

- Street number (630)
- Prefix direction (S or South)
- Street name (Main)
- Street type (Street)
- ZIP code (83843)

The above example uses a prefix direction. Some addresses have suffixes such as NE following the street name and type. To plot an address successfully, the components of the address must match those of a street segment in the reference database. An unmatched address can occur if there is a problem with address recording, the reference database, or both.

According to Harries (1999), common errors in address recording (for crime mapping) include: misspelling of the street name, incorrect address number, incorrect direction prefix or suffix, incorrect street type, and abbreviation not recognized by the geocoding engine. We can add PO Box type addresses and incorrect or missing ZIP codes to the list of problems. A reference database can have its

own set of problems. A reference database can be outdated because it does not have information on new streets, street name changes, street closings, and ZIP code changes. In some cases, a reference database may even have missing address ranges, gaps in address ranges, or incorrect ZIP codes.

17.1.3 Address Matching Options

Given the possible errors in the address table, the reference database, or both, a geocoding engine must be able to deal with these errors. Typically, a geocoding engine has provisions for relaxing the matching conditions but uses a scoring system to quantify the matches at the same time. The idea is to let the user decide the rigor or accuracy of address matching. The user can decide the minimum candidate score and the minimum match score. The former determines the likely candidates in the reference database for matching, and the latter determines whether an address is matched or not. Adopting a lower match score (i.e., relaxing the matching conditions to a greater degree) can result in more matches but can also introduce more possible geocoding errors.

A GIS package does not explain in detail how the scoring system works (Box 17.1). Some abbreviations apparently do not lower the match score. For example, Ave can be accepted for Avenue, N for North, and 3rd for Third. But if a street type or a prefix is missing (e.g., Appleway instead of Appleway Ave), one can expect a lower score. A lower score is also expected for the misspelling of a street name (e.g., Masachusett Ave instead of Massachusetts Ave). If a geocoding engine uses the Soundex system (a system that codes together names of similar sounds or of differ-ent spellings) to check the spelling, then names such as Smith and Smythe are treated the same. Few GIS programs, however, allow the user to relax street number or ZIP code.

Because of the various matching options, we can expect to run the geocoding process more than once. The first time through, the geocoding engine reports the number of matched addresses based on the number of matched addresses based on the minimum match score, and the number of

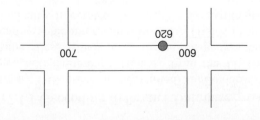

Figure 17.3
Linear interpolation for address geocoding.

<table>
<tr><td>Ⓠ</td><td>*Box* **17.1**</td><td>**Scoring System for Geocoding**</td></tr>
</table>

A GIS package such as ArcGIS does not explain how the scoring system works in geocoding services. Take the example of the spelling sensitivity setting, which has a possible value between 0 and 100. ArcGIS sets the default at 80. But the help document suggests that we use a higher spelling sensitivity if we are sure that the input addresses are spelled correctly and a lower setting if we think that the input addresses may contain spelling errors. Although the suggestion makes sense, it does not say exactly how to set a spelling sensitivity with a numeric value.

There are probably too many scoring rules used by a geocoding engine to be listed in a help document. But we may occasionally discover how the scoring system works in a specific case. For example, if the street name "Appleway Ave" is spelled "Appleway" for a street address, the scoring system in ArcGIS deducts 15 from the match score.

---

unmatched addresses. For each unmatched address, the geocoding engine lists candidates, if any, based on the minimum candidate score. We can accept a candidate for a match or modify the unmatched address before running the geocoding process again.

What is an acceptable hit rate (i.e., percentage of addresses matched)? Although the acceptable level depends on the project, at least one researcher has stated that a 60% hit rate is unacceptable for crime mapping and analysis (Harries 1999). Using the updated TIGER/Line files and a commercial GIS package, we can probably expect to have an initial hit rate of 70% or higher. To bring the rate to 95% or better, as required for competitive location-based services, a current and accurate reference database and additional effort in validating street addresses would be required.

The geocoding process converts street addresses to point features (Figure 17.4). Both ArcGIS and MapInfo also have the side offset and end offset options for the output (Figure 17.5). The side offset places a geocoded point at a specified distance from the side of a street segment. This option is useful for point-in-polygon analysis such as linking addresses to census tracts or land parcels (Ratcliffe 2001). The end offset places a point feature from the end point of a street segment

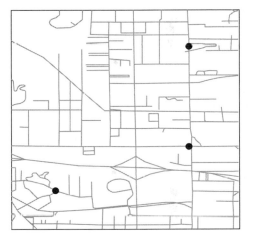

Figure 17.4
Address geocoding plots street addresses as points on a map.

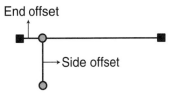

Figure 17.5
The end offset moves a geocoded point away from the end point of a street segment, and the side offset places a geocoded point away from the side of a street segment.

based on a specified percentage of the length of a street segment. This end and offset option can prevent geocoded points from falling on top of a cross street.

17.1.4 Other Types of Geocoding

Intersection matching matches address data with street intersections on a map (Figure 17.6). GIS packages such as ArcGIS and MapInfo offer intersection matching in addition to address matching. Also called *corner matching*, intersection matching is the geocoding method for almost all motor vehicle accidents in cities (Levine and Kim 1998). An address entry for intersection matching includes two streets such as "E Sherman Ave & N 4th St." Using such input data, a geocoding engine locates the street intersection by matching the end point of E. Sherman Ave with the end point of N. 4th St.

Like address matching, intersection matching can run into problems. A reference database may not include all street intersections. For example, interstate highways in nontopological shapefiles can be continuous lines without nodes (i.e., intersections) at road crossings. Using shapefiles for

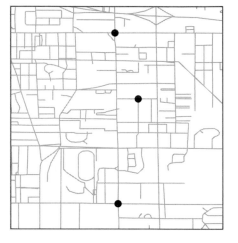

Figure 17.6
An example of intersection matching.

the reference database in this case can fail to geocode many street intersections. A winding street crossing another street more than once can also present a unique problem. It requires additional data such as address numbers or adjacent ZIP codes to determine which street intersection to plot.

Address matching and intersection matching represent street-level geocoding, a process that involves a street network. **ZIP code geocoding** refers to the process of matching a ZIP code to its centroid location. Instead of using a street network, ZIP code geocoding uses a reference database that contains the x-, y-coordinates, either geographic or projected, of ZIP code centroids.

17.2 APPLICATIONS OF GEOCODING

Geocoding is perhaps the most commercialized GIS-related operation. Geocoding plays an important role in Internet mapping and location-based services. MapQuest is only one of many websites that provide address-matching services. Others include Google, Yahoo, Microsoft's MapPoint, and the U.S. Census Bureau. Besides address matching, geocoding can be extended to land parcel mapping (Ratcliffe 2001) and implementation of enhanced 911 services. Some geocoding services are more imaginative such as finding friends and friends-of-friends or locating campaign contributions (Chapter 1).

Geocoding is also a tool for preparing data for spatial analysis. A spatial pattern analysis of real estate prices can be based on point features geocoded from transactions of homes (Basu and Thibodeau 1998). Geographic access to health services can be measured from the travel times and distances between geocoded point locations for the subjects and the medical providers (Fortney et al. 2000). In recent years, geocoding has become a vital tool for crime mapping and analysis because crime records almost always have street addresses or other locational attributes (Harries

---

Box 17.2 | Linear Location Referencing System

Transportation agencies normally use a linear location referencing system to locate events such as accidents and potholes, and facilities such as bridges and culverts, along roads and transit routes. To locate events, a linear referencing system uses distance measures from known points, like the beginning of a route, a milepost, or a road intersection. The address of an accident location, for example, contains a route name and a distance offset from a milepost. The address for a linear event, on the other hand, uses distance offsets from two reference posts. A vector-based GIS uses x-, y-coordinates to locate points, lines, and polygons on a projected coordinate system. The dynamic segmentation model uses routes, measures, and events to bring together a projected coordinate system and a linear referencing system, two fundamentally different measuring systems.

1999; Ratcliffe 2004). Geocoded crime locations are input data to software packages such as CrimeStat (**http://www.icpsr.umich.edu/NACJD/crimestat.htm**) for "hot spot" analysis and space time analysis (Chapter 12).

17.3 DYNAMIC SEGMENTATION

Dynamic segmentation refers to the process of computing the location of events along a route (Nyerges 1990). A **route** is a linear feature with a linear measurement system stored with its geometry. Examples of routes include streets, highways, and streams. **Events** are linearly referenced data that occur along routes. The location of linearly referenced data is based on an offset distance, or a measure from a known point. In the United States, milepost referencing uses distances from a point of origin (e.g., mile 0) to locate mile points along a highway (Box 17.2). Common events include speed limits and traffic accidents along highway routes and fishery habitat conditions along streams. The dynamic segmentation process links events to a route so that these events can be displayed and analyzed with other geographically referenced data. Sections 17.3.1 to 17.3.4 discuss routes and events and how they are created and managed in a GIS.

17.3.1 Routes

To be used with linearly referenced event data, a route must have a built-in measurement system. The way a measurement system is stored with a route differs between the coverage model and the geodatabase data model, two contrasting data models from ESRI, Inc.

The coverage model treats a route as a subclass in a line coverage. A route subclass is based on the hierarchical structure of route, section, and arc, and its measurement system is based on the geometric length of arcs (Miller and Shaw 2001). Figure 17.7 illustrates how linear measures of a route subclass are derived. Figure 17.7a shows three arcs (7, 8, and 9) that make up a linear feature. Figure 17.7b shows a route with three sections, each in a thick line symbol, superimposed on the linear feature. Figure 17.7c tabulates the linear measures of the three sections based on the underlying arcs. Section 1 covers the entire length of arc 7, as indicated by the F-Pos of 0 and the T-Pos of 100 (%). Linear measures of section 1 start at 0 (F-Meas) and end at 40 (T-Meas). Section 2 also covers the entire length of arc 8. Linear measures of section 2 start at the end measure of section 1 (i.e., 40) and end at 170. Section 3 covers 80% of the length of arc 9. Therefore, the end measure of section 3 is 170 + (length of arc 9) × 0.8, or 210.

The geodatabase data model treats a route as a polyline feature class. Instead of working through sections, a route feature class simply stores linear measures (m values) in the geometry field. The geometry of a route feature class therefore includes x, y, and m values: the x and y values locate the linear feature in a coordinate system, and the m values

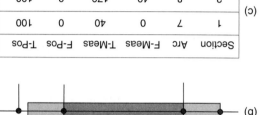

Figure 17.7
An example of a route subclass using the coverage model. See Section 17.3.1 for explanation.

Section	Arc	F-Meas	T-Meas	F-Pos	T-Pos
1	7	0	40	0	100
2	8	40	170	0	100
3	9	170	210	0	80

relate to the route's linear measurement system. The m values can be based on the geometric length of line segments. But they can also be interpolated from mileposts or other reference markers. Figure 17.8 shows the same route as in Figure 17.7 but as a polyline feature class. Each point has a pair of x- and y-coordinates and an m value.

Although route subclasses are still common among data that can be downloaded from a government agency's website, at least one federal program is turning to route feature class to take advantage of the simpler route data structure (Box 17.3).

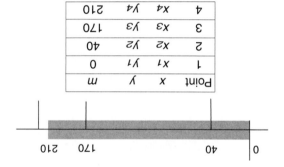

Figure 17.8
An example of a geodatabase route feature class.

Point	x	y	m
1	x1	y1	0
2	x2	y2	40
3	x3	y3	170
4	x4	y4	210

Box 17.3 Route Feature Classes

The National Hydrography Dataset (NHD) is a program that is migrating from the coverage model (NHDinARC) to the geodatabase data model (NHDinGEO) (http://nhd.usgs.gov/geodatabase_review.html). NHDinARC data use a route subclass to store the transport and coastline reaches (route.rch). In contrast, NHDinGEO data has a new NHD-Flowline feature class, which stores m values with its feature geometry. These m values apply to other feature classes that use the reach reference.

Route feature classes have also been adopted by transportation agencies for data delivery. An example is the Washington State Department of Transportation (WSDOT). WSDOT has built state highway routes with linear measures. These highway routes provide the basis for locating features such as rest areas, speed limits, and landscape types along the routes. Tasks 2 and 3 in the applications section use data sets downloaded from the WSDOT website. (http://www.wsdot.wa.gov/mapsdata/geodatacatalog/default.htm).

Box 17.4 | **Create Routes Using ArcGIS**

ArcGIS offers both the conversion method and the interactive method for creating route feature classes from existing lines. The Create Routes tool in ArcToolbox can convert all linear features or selected features from a coverage, a shapefile, or a geodatabase feature class into routes. The output routes can be saved in a shapefile or a geodatabase feature class. ArcToolbox also has other tools for calibrating routes, dissolving route events, locating features along routes, making route event layer, overlaying route events, and transforming route events. ArcMap has the Route Editing toolbar, which offers tools for creating and calibrating routes interactively.

17.3.2 Creating Routes

Routes can be created interactively or through data conversion (Box 17.4). Using the interactive method, we must first digitize a route or select existing lines from a layer that make up a route (Figure 17.9). Then we can apply a measuring command to the route to compute the route measures.

The data conversion method can create routes at once from all linear features or from features selected by a data query. For example, we can create a route system for each numbered highway in a state or for each interstate highway (after the interstate highways are selected from the data set). Figure 17.10 shows five interstate highway routes in Idaho created through the conversion method.

Figure 17.9
The interactive method requires the selection or digitizing of the line segments that make up a route (shown in a thicker line symbol).

Figure 17.10
Interstate highway routes in Idaho.

When creating routes, we must be aware of different types of routes. Routes may be grouped into the following four types:

- **Simple route:** a route follows one direction and does not loop or branch.
- **Combined route:** a route is joined with another route.
- **Split route:** a route subdivides into two routes.
- **Looping route:** a route intersects itself.

Simple routes are straightforward but the other three types require special handling. An example of combined routes is an interstate highway, which has different traffic conditions depending on the traffic direction. In this case, two routes can be built for the same interstate highway, one for each direction. An example of split routes is the split of a business route from an interstate highway (Figure 17.11). At the point of the split, two separate linear measures begin: one continues on the interstate highway, and the other stops at the end of the business route.

A looping route is a route with different parts. Unless the route is dissected into route parts, the route measures will be off. A bus route in Figure 17.12 serves as an example. The bus route has two loops. We can dissect the bus route into three parts: (1) from the origin on the west side of the

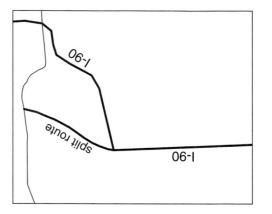

Figure 17.11
An example of a split route.

city to the first crossing of the route, (2) between the first and the second crossing, and (3) from the second crossing back to the origin. Each part is created and measured separately. Following the completion of all three parts, a remeasurement can be applied to the entire route so that it has a consistent and continuous measuring system.

17.3.3 Events

Events are linearly referenced data that are usually stored in an event table. One option to display events is to break up a linear feature (e.g., a highway) into segments so that each segment corresponds to an event value (e.g., a pavement condition). This option, however, requires a large amount of data editing in segmentation. Moreover, it requires one specific data set for each type of event. The dynamic segmentation process overcomes both difficulties by linking events to a route through a linear measurement system, a system that is independent of the x-, y-coordinate system. The process does not have to split a highway into segments, and

Figure 17.12
A looping route divided into three parts for the purpose of route measuring.

it allows different types of events such as pavement conditions, traffic volumes, and traffic accidents to be plotted along the same highway.

Events may be point or line events:

- **Point events,** such as accidents and stop signs, occur at point locations. To relate point events to a route, a point event table must have the route ID, the location measures of the events, and attributes describing the events.
- **Line events,** such as pavement conditions, cover portions of a route. To relate line events to a route, a line event table must have the route ID and the from- and to-measures.

17.3.4 Creating Event Tables

There are two common methods for creating event tables. The first method creates an event table from an existing table that already has data on route ID and linear measures. The second method creates an event table by locating point or polygon features along a route. Data availability is the main consideration in deciding which method to use.

The first method assumes that event data are present in tabular format and the linear measures are available through milepost, river mile, or other data. For example, the Washington State Department of Transportation's website has event tables of rest areas, speed limits, landscape types, and other data (**http://www.wsdot.wa.gov/mapsdata/ geodatacatalog/default.htm**). These tables are in dBASE format. Other formats such as comma delimited text format are also acceptable.

The second method for creating event tables is similar to a vector-based overlay operation (Chapter 12). The input layers to the overlay operation are a route layer and a point or polygon layer. Instead of creating an output layer, the procedure creates an event table.

Figure 17.13 shows rest areas along a highway route. To convert these rest areas into point events, we prepare a point layer containing the rest areas, use a measuring command to measure the locations of the rest areas along the highway route,

FID	Route-ID	Measure	RDLR
1	90	161.33	R
2	90	161.82	L
3	90	198.32	R
4	90	198.32	L

Figure 17.13

An example of converting point features to point events. See Section 17.3.4 for explanation.

and write the information to a point event table. The point event table lists the feature ID (FID), route-ID, measure value, and the side of the road (RDLR). Rest areas 3 and 4 have the same measure value but on different sides of the highway. To determine whether a point is a point event or not, the computer uses a user-defined search radius. If a point is within the search radius from the route, then the point is a point event.

Figure 17.14 shows a stream network and a slope layer with four slope classes. To create a line event table showing slope classes along the stream route, the computer calculates the intersection between the route and the slope layer, assigns each segment of the route the slope class of the polygon it crosses, and writes the information to an event table. The line event table shows the route-ID, the from-measure (F-Meas), the to-measure (T-Meas), and the slope-code for each stream segment. For example, the slope-code is 1 for the first 7638 meters of the stream route 1, changes to the slope-code of 2 for the next 160 meters (7798 − 7638), and so on.

attributes such as highway district and management unit to the table. And, if there are new rest areas, we can add them to the same event table so long as the measure values of these new rest areas are known.

17.4 APPLICATIONS OF DYNAMIC SEGMENTATION

Dynamic segmentation can store attributes along routes. Once attributes are associated with a route, they can be displayed, queried, and analyzed. In other words, dynamic segmentation is useful for data management, data display, data query, and data analysis.

17.4.1 Data Management

From the data management's perspective, dynamic segmentation is useful in two ways. First, different routes can be built on the same linear features. A transit network can be developed from the same street database for modeling transit demand (Choi and Jang 2000). Similarly, multiple trip paths can be built on the same street network, each with a starting and end point, to provide disaggregate travel data for travel demand modeling (Shaw and Wang 2000).

Second, different events can reference the same route or, more precisely, the same linear measurement system stored with the route. Fishery biologists can use dynamic segmentation to store various environmental data with a stream network for habitat studies. And, as shown in Box 17.3, dynamic segmentation is an efficient method for delivering hydrologic data sets and for locating features along highway routes.

17.4.2 Data Display

Once an event table is linked to a route, the table is georeferenced and can be used as a feature layer. This conversion is similar to geocoding an address table or displaying a table with x-, y-coordinates. For data display, an event layer is the same as a

Whether prepared from an existing table or generated from locating features along a route, an event table can be easily edited and updated. For example, after a point event table is created for rest areas along a highway route, we can add other

See Section 17.3.4 for explanation.

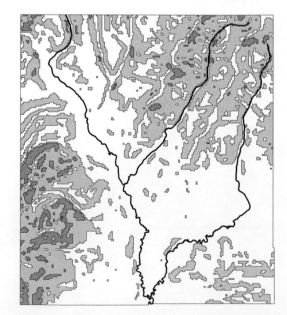

Figure 17.14

An example of creating a line event table by overlaying a route layer and a polygon layer. See Section 17.3.4 for explanation.

Route-ID	F-Meas	T-Meas	Slope-Code
1	0	7638	1
1	7638	7798	2
1	7798	7823	1
1	7823	7832	2
1	7832	8487	1
1	8487	8561	1
1	8561	8561	2
1	8561	8586	1
1	8586	8639	1
1	8639	8643	2

Figure 17.15
The thicker, solid line symbol represents those portions of Washington State's highway network that have the legal speed limit of 70 miles per hour.

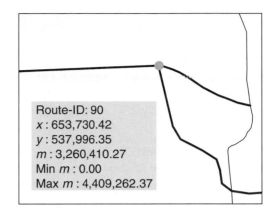

Figure 17.16
Data query at a point, shown here by the small circle, shows the route-ID, the *x*- and *y*-coordinates, and the measure (*m*) value at the point location. Additionally, the beginning and ending measure values of the route are also listed.

stream layer or a city layer. We can use point symbols to display point events and line symbols to display line events (Figure 17.15).

17.4.3 Data Query

We can perform both attribute data query and spatial data query on an event table and its associated event layer. For example, we can query a point event table to select recent traffic accidents along a highway route. The attribute query result can be displayed in the event table as well as in the event layer. To determine if these recent accidents follow a similar pattern as past accidents, we can perform a spatial query and select past accidents within 10 miles of recent accidents for the investigation.

Data query can also be performed on a route to find the measure value of a point. This is similar to clicking a feature to get the attribute data about the feature (Figure 17.16).

17.4.4 Data Analysis

Both routes and events can be used as inputs for data analysis. Therefore, to study the efficiency or equity of public transit routes, we can first buffer the routes with a distance of one-half mile and

overlay the buffer zone with census blocks or block groups. We can then use demographic data from the buffer zone (e.g., population density, income levels, number of vehicles owned per family, and commuting time to work) for the planning and operation of the public transit system.

Events, after they are converted to an event layer, can be analyzed in the same way as any point or linear feature layer. For example, road accidents can be located by dynamic segmentation and then analyzed in terms of their concentration patterns (Steenberghen et al. 2004).

Data analysis can also be performed between two event layers. Thus we can analyze the relationship between traffic accidents and speed limits along a highway route. The analysis is similar to a vector-based overlay operation: the output layer shows the location of each accident and its associated legal speed limit. A query on the output layer can determine whether accidents are more likely to happen on portions of the highway route with higher speed limits. But data analysis can become complex when more than two events are involved and each event has a large number of sections (Huang 2003).

KEY CONCEPTS AND TERMS

Address geocoding: A process of plotting street addresses in a table as point features on a map. Also called *address matching*.

Combined route: A route that is joined with another route.

Dynamic segmentation: The process of computing the location of events along a route.

Events: Attributes occurring along a route.

Geocoding: A process of assigning spatial locations to data that are in tabular format but have fields that describe their locations.

Intersection matching: A process of plotting street intersections as point features on a map. Also called *corner matching*.

Line events: Events that occur along a portion of a route, such as pavement conditions.

Looping route: A route that intersects itself.

Point events: Events that occur at point locations on a route, such as accidents and stop signs.

Route: A linear feature that has a linear measurement system stored with its geometry.

Simple route: A route that follows one direction and does not loop or branch.

Split route: A route that subdivides into two routes.

ZIP code geocoding: A process of matching ZIP codes to their centroid locations.

REVIEW QUESTIONS

1. Describe the two required inputs for address geocoding.

2. List attributes in the TIGER/Line files that are important for geocoding.

3. Go to the MapQuest website (**http://www.mapquest.com/**). Type the street address of your bank for search. Does it work correctly?

4. Try a street intersection at the MapQuest website. Does it work correctly?

5. What address matching options are usually available in a GIS package?

6. What geocoding output options are usually available in a GIS package?

7. Explain the difference between the side offset and end offset options in geocoding.

8. What is ZIP code geocoding?

9. Geocoding is one of the most commercialized GIS-related activities. Can you think of a commercial application example of geocoding besides those that have already been mentioned in Chapter 17?

10. Explain in your own words how events are located along a route using dynamic segmentation.

11. Go to ESRI's geodatabase Web page (**http://support.esri.com/datamodels**). Download the custom template of the transportation data model. Use the search function in Adobe Reader to look for the keyword "route." How many route feature classes are included in the data model?

12. Explain the difference between the coverage model and the geodatabase data model in storing a route's linear measure system.

13. Describe methods for creating routes from existing linear features.

14. How can you tell a point event table from a line event table?

15. Suppose you are asked to prepare an event table that shows portions of Interstate 5 in California crossing earthquake-prone zones (polygon features). How can you complete the task?

16. Check whether the transportation department in your state maintains a website for distributing GIS data. If it does, what kinds of data are available for dynamic segmentation applications?

APPLICATIONS: GEOCODING AND DYNAMIC SEGMENTATION

This applications section covers five tasks. Task 1 asks you to geocode ten street addresses by using a reference database derived from the 2000 TIGER/Line files. Task 2 lets you display and query highway routes and events downloaded from the Washington State Department of Transportation website. Again using data from the Washington State Department of Transportation website, you will analyze the spatial relationship between two event layers in Task 3. In Tasks 4 and 5, you will build routes from existing line features and locate features along routes. Task 4 locates slope classes along a stream route, and Task 5 locates cities along Interstate 5.

Task 1: Geocode Street Addresses

What you need: *cda_add.txt*, a text file containing street addresses of 5 restaurants and 5 government offices in Coeur d'Alene; and *kootenai,* an ArcInfo coverage for Kootenai County, Idaho, converted from the 2000 TIGER/Line files and measured in NAD83 geographic coordinates. Coeur d'Alene is the largest city in Kootenai County.

In Task 1, you will learn how to create point features from street addresses. Address geocoding requires an address table and a reference dataset. The address table contains a list of street addresses to be located. The reference dataset has the address information for locating street addresses. Task 1 includes the following four steps: view the input data; create a geocoding service; run geocoding; and rerun geocoding for the unmatched addresses.

1. Start ArcCatalog, and make connection to the Chapter 17 database. Launch ArcMap. Add *kootenai arc* and *cda_add.txt* to Layers and rename Layers Task 1. Right-click *kootenai arc* and open its attribute table. Because *kootenai arc* is derived from the TIGER/Line files, it has all the attributes from the original data. Figure 17.2 has the descriptions of the attributes. Right-click *cda_add.txt*, and open the table. The table contains the fields of Name, Address, and Zip. Close both tables.

2. Start ArcCatalog if necessary. Expand Address Locators. Double-click on Create New Address Locator. Select US Streets with Zone (File) and click OK in the next dialog. The US Streets with Zone style is closest to the geocoding style of *kootenai arc*.

3. Start with the left side of the New US Streets with Zone (File) Address Locator dialog. Enter Task 1 for the name. Then navigate to *kootenai arc* and select it for the reference data. Select FRADDL for the field of House From Left, TOADDL for House To Left, FRADDR for House From Right, and TOADDR for House To Right. Move to the right side of the dialog. In the Input Address Fields frame, first match Street with Address and delete others. To delete Addr, highlight it and click the Delete button. Delete Street. Next match Zone with Zip. Use the same procedure to delete Zipcode, City, and Zone. Keep the default values for spelling sensitivity, minimum candidate score, and minimum match score in the Matching Options frame. Click OK to dismiss the dialog.

Q1. The default spelling sensitivity value is 80. If you were to change it to 60, how would the change affect the geocoding process?

Q2. What output options are available?

4. The new address locator you have just created shows up in the Catalog Tree with

Task 1 prefixed by the user name on your computer.

5. Go back to ArcMap. Click the Tools menu, point to Geocoding, and select Geocoding Addresses. Click Add in the next dialog, and add the address locator of xxx.Task 1. Click OK.

6. In the next dialog, make sure that cda_add.txt is the address table and the address input fields are correct. Provide the path and save the output shapefile as cda_geocode.shp in the Chapter 17 database. Click OK to geocode the addresses.

7. The Review/Rematch Addresses dialog shows that 9 out of 10 addresses in cda_add.txt are matched with score 80–100 and 1 is unmatched. Click the Match Interactively button. The Interactive Review dialog shows that the unmatched record is that of the ID Fish & Game (IDFG). The IDFG address shows that the ZIP code is 83814, which is incorrect. Change the ZIP code to 83815 in the Zone window, and then click the Modify button. The Candidate window shows a perfect candidate. Highlight the candidate and click Match. Click Close to dismiss the dialog.

8. The Review/Rematch Addresses dialog now shows 100% of addresses matched with score 80–100. Click Done to dismiss the dialog.

9. You need to zoom in on the city of Coeur d'Alene to see the 10 geocoded point features.

Task 2: Display and Query Routes and Events

What you need: *decrease24k.shp*, a shapefile showing Washington state highways; and *SpeedLimitsDecAll.dbf*, a dBASE file showing legal speed limits on the highways.

Both data sets for Task 2 came from the Washington State Department of Transportation website. Originally in geographic coordinates, *decrease24k.shp* has been projected onto the Washington State Plane, South Zone, NAD83, and Units feet. The linear measurement system is in miles. You will learn how to display and query routes and events in Task 2.

1. Insert a new data frame in ArcMap and rename it Task 2. Add *decrease24k.shp* and *SpeedLimitsDecAll.dbf* to Task 2. Open the attribute table of *decrease24k.shp*. The table shows state route identifiers (SR) and route measure attributes (Polyline M). Close the table.

2. This step adds the Identify Route Locations tool. The tool does not appear on any toolbar by default. You need to add it. Select Customize from the Tools menu. On the Commands tab, select the category of Linear Referencing. The Commands frame shows five commands. Drag and drop the Identify Route Locations command to a toolbar (e.g., the standard toolbar). Close the Customize dialog.

3. Open the attribute table of *decrease24k*, and select "SR" = '026'. Highway 26 should now be highlighted in the map. Zoom in on Highway 26. Click the Identify Route Locations tool, and then click a point along Highway 26. This opens the Identify Route Location Results dialog and shows the measure value of the point you clicked as well as other information.

Q3. What is the total length of Highway 26 in miles?

Q4. In which direction is the route mileage accumulated?

4. Clear the selected features. Now you will work with the speed limits event table. Select Add Route Events from the Tools menu. In the next dialog, from top to bottom, select *decrease24k* for the route reference, SR for the route identifier. *SpeedLimitsDecAll* for the event table, SR for the route identifier, line events, B_ARM for the route identifier,

(beginning accumulated route mileage) for the from-measure, and E_ARM (end accumulated route mileage) for the to-measure. Click OK to dismiss the dialog. A new layer, *SpeedLimitsDecAll Events*, is added to Task 2.

5. Right-click *SpeedLimitsDecAll Events* and select Properties. On the Symbology tab, choose Quantities/Graduate colors in the Show frame and LEGSPDDEC (legal speed description) for the value in the Fields frame. Choose a color ramp that can better distinguish the speed limits classes. Click OK to dismiss the dialog. Task 2 now shows the speed limits data on top of the state highway routes.

Q5. How many records of *SpeedLimitsDecAll Events* have speed limits > 60?

Task 3: Analyze Two Event Layers

What you need: *decrease24k.shp*, same as Task 2; *RoadsideAll.dbf*, a dBASE file showing classification of roadside landscape along highways; and *RestAreasAll.dbf*, a dBASE file showing rest areas.

In Task 3, you will use ArcToolbox to overlay the event tables of rest areas and the roadside landscape classes. The output event table can then be added to ArcMap as an event layer.

1. Insert a new data frame in ArcMap and rename it Task 3. Add *decrease24k.shp*, *RoadsideAll.dbf*, and *RestAreasAll.dbf* to Task 3. Open *RoadsideAll*. The CLASSIFICA field stores five landscape classes: forested, open, rural, semi-urban, and urban. (Ignore the coincident features.) Open *RestAreasAll*. The table has a large number of attributes including the name of the rest area (FEATDESCR) and the county (COUNTY). Close the tables.

2. Click Show/Hide ArcToolbox Window to open ArcToolbox. Set the Chapter 17 database for the current workspace. Double-click the Overlay Route Events tool in the Linear Reference Tools toolset.

The Overlay Route Events dialog consists of three parts: the input event table, the overlay event table, and the output event table. Select *RestAreasAll* for the input event table, SR for the route identifier field, POINT for the event table, and ARM for the measure field. Select *RoadsideAll* for the overlay event table, SR for the route identifier field, LINE for the event type, BEGIN_ARM for the from-measure field, and END_ARM for the to-measure field. Select INTERSECT for the type of overlay. Enter *Rest_Roadside.dbf* for the output event table, SR for the route identifier field, and ARM for the measure field. Click OK to perform the overlay operation.

3. Open *Rest_Roadside*. The table combines attributes from *RestAreasAll* and *RoadsideAll*.

Q6. How many rest areas are located in forested areas?

Q7. Are any rest areas located in urban areas?

4. Similar to Task 3, you can use the Add Route Events tool to add *Rest_Roadside* as an event layer. The event layer can display the rest areas and their attributes such as the landscape classification.

Task 4: Create A Stream Route and Analyze Slope Along the Route

What you need: *plne*, an elevation raster; and *streams.shp*, a stream shapefile.

Task 4 lets you analyze slope classes along a stream. Task 4 consists of several subtasks: (1) use Spatial Analyst to create a slope polygon shapefile from *plne*; (2) import a select stream from *streams.shp* as a feature class to a new geodatabase; (3) use ArcToolbox to create a route from the stream feature class; and (4) run an overlay operation to locate slope classes along the stream route.

1. Insert a new data frame in ArcMap and rename the data frame Task 4. Add *plne* to Task 4. Use Spatial Analyst to create a

percent slope raster from *plne*. Use Spatial Analyst to reclassify the percent slope raster into the following five classes: <10%, 10–20%, 20–30%, 30–40%, and >40%. Then convert the reclassified slope raster to a polygon shapefile and name the shapefile *slope*. The field GRIDCODE in *slope* shows the five slope classes.

2. Right-click the Chapter 17 database in ArcCatalog, point to New, and select Personal Geodatabase. Name the geodatabase *stream.mdb*.

3. Next import one of the streams in *streams.shp* as a feature class in *stream.mdb*. Right-click *stream.mdb*, point to Import, and click *streams.shp*. Select Feature Class (single). Select *streams.shp* for the input features, enter *stream165* for the output feature class name, and click on the SQL button for the expression. In the Query Builder dialog, enter the following expression: "USGH_ID" = 165 and click OK. Click OK in the Feature Class to Feature Class dialog to run the import operation. Expand *stream.mdb*, and *stream165* should appear as a feature class in the database.

4. Double-click the Create Routes tool in the Linear Referencing Tools toolset in ArcMap. Select *stream165* for the input line features, select *USGH_ID* for the route identifier field, enter *StreamRoute* for the output route feature class in *stream.mdb*, and click OK. *StreamRoute* is added to the map.

5. Now you will run an operation to locate slope classes along *StreamRoute*. Double-click the Locate Features Along Routes tool in the Linear Referencing Tools toolset. Select *slope* for the input features, select *StreamRoute* for the input route features, enter *Stream_Slope.dbf* for the output event table in the Chapter 17 database, and click OK to run the overlay operation.

6. Select Add Route Events tool from the Tools menu. In the next dialog, make sure that *StreamRoute* is the route reference,

USGH_ID is the route identifier. *Stream_Slope* is the event table, and RID is the route identifier for the event table. Click Line Events. Then click OK to add the event layer.

7. Turn off all layers except *Stream_Slope Events* in the table of contents. Right-click *Stream_Slope Events* and select Properties. On the Symbology tab, select Categories and Unique Values for the Show option, select GRIDCODE for the value field, click on Add All Values, and click OK. Zoom in on *Stream_Slope Events* to view the changes of slope classes along the route.

Q8. How many records are in the *Stream_Slope Events* layer?

Q9. How many records in *Stream_Slope Events* have the GRIDCODE value of 5 (i.e., slope > 40%)?

Task 5: Locate Cities Along U.S. Interstate 5

What you need: *interstates.shp*, a line shapefile containing interstate highways in the conterminous United States; and *uscities.shp*, a point shapefile containing cities in the conterminous United States. Both shapefiles are based on projected coordinates in meters.

In Task 5, you will locate cities along Interstate 5 that runs from Washington State to California. The task consists of three subtasks: extract Interstate 5 from *interstates.shp* to create a new shapefile; create a route from the Interstate 5 shapefile; and locate cities in *uscities.shp* that are within 10 miles of Interstate 5.

1. In ArcCatalog, click Show/Hide ArcToolbox window to open ArcToolbox. Double-click the Select tool in the Analysis Tools/Extract toolset. Select *interstates.shp* for the input features, specify *I5.shp* for the output feature class, and click on the SQL button for the expression. Enter the following expression in the Query Builder: "RTE_NUM1" = ' 5'. (There are two spaces before 5.) Click OK to dismiss the

dialogs. Next add a numeric field to *I5* for the route identifier. Double-click the Add Field tool in the Data Management Tools/ Fields toolset. Select *I5.shp* for the input table, enter RouteNum for the field name, select DOUBLE for the field type, and click OK. Double-click the Calculate Field tool. Select *I5.shp* for the input table, RouteNum for the field name, enter 5 for the expression, and click OK.

2. Insert a new data frame in ArcMap and rename it Task 5. Add *I5.shp* and *uscities. shp* to Task 5. Open ArcToolbox in ArcMap. Double-click the Create Routes tool in the Linear Referencing Tools toolset. Select *I5* for the input line features, select RouteNum for the route identifier field, specify *Route5.shp* for the output route feature class, select LENGTH for the measure source, enter 0.00062137119 for the measure factor, and click OK. The measure factor converts the linear measure units from meters to miles.

3. This step locates cities within 10 miles of *Route5*. Double-click the Locate Features Along Routes tool in the Linear Referencing Tools toolset. Select *uscities* for the input features, select *Route5* for the input route features, select RouteNum for the Route Identifier Field, enter 10 miles for the search radius, specify *Route5_cities.dbf* for the output event table, and click OK.

4. Add *Route5_cities.dbf* to Task 5. Click Add Route Events in the Tools menu. Make sure that *Route5* is the route reference, *Route5_cities* is the event table, and the type of events is point events. Click OK to add the event layer.

Q10. How many cities in Oregon are within 10 miles of *Route5*?

Challenge Task

What you need: access to the Internet and Winzip.

This challenge task asks you to geocode ten business addresses in your city.

1. Write down ten addresses of restaurants and retail stores from the phonebook of your city. Follow the same format as *cda_add.txt* in Task 1 to prepare a comma delimited text file with the addresses.

2. Go to the U.S. Census Bureau website (**http://www.census.gov**) and follow the path TIGER > 108th CD Census 2000 TIGER/Line Files > (your state) > (the file for your county) to download the TIGER/ Line files for your county. (The TIGER/Line files use the 5-digit FIPS code to denote the county; therefore, you should first find out the FIPS code for your county.)

3. Unzip the TIGER/Line file you have downloaded. Use ArcToolbox to derive a coverage from the TIGER/Line files.

4. Use the coverage from Step 3 to geocode the addresses from Step 1.

Q1. How many of the 10 addresses are matched successfully the first time?

Q2. How many of the 10 addresses are matched at the end?

Q3. What kinds of problems, if any, have you encountered in geocoding?

REFERENCES

Basu, S., and T. G. Thibodeau. 1998. Analysis of Spatial Autocorrelation in House Prices. 1998. *The Journal of Real Estate Finance and Economics* 17: 61–85.

Choi, K., and W. Jang. 2000. Development of a Transit Network from a Street Map Database with Spatial Analysis and Dynamic Segmentation. *Transportation Research Part C:* *Emerging Technologies* 8: 129–46.

Fortney, J., K. Rost, and J. Warren. 2000. Comparing Alternative Methods of Measuring

Geographic Access to Health Services. *Health Services & Outcomes Research Methodology* 1: 173–84.

Harries, K. 1999. *Mapping Crime: Principles and Practice.* Washington, DC: U.S. Department of Justice.

Huang, B. 2003. An Object Model with Parametric Polymorphism for Dynamic Segmentation. *International Journal of Geographical Information Science* 17: 343–60.

Levine, N., and K. E. Kim. 1998. The Location of Motor Vehicle Crashes in Honolulu: A Methodology for Geocoding Intersections. *Computers, Environment and Urban Systems* 22: 557–76.

Miller, H. J., and S. Shaw. 2001. *Geographic Information Systems for Transportation: Principles and Applications.* New York: Oxford University Press.

Nyerges, T. L. 1990. Locational Referencing and Highway Segmentation in a Geographic Information System. *ITE Journal* 60: 27–31.

Ratcliffe, J. H. 2001. On the Accuracy of TIGER-Type Geocoded Address Data in Relation to Cadastral and Census Areal Units. *International Journal of Geographical Information Science* 15: 473–85.

Ratcliffe, J. H. 2004. Geocoding Crime and a First Estimate of a Minimum Acceptable Hit Rate. *International Journal of Geographical Information Science* 18: 61–72.

Shaw, S., and D. Wang. 2000. Handling Disaggregate Spatiotemporal Travel Data in GIS. *Geoinformatica* 4: 161–78.

Steenberghen, T., T. Dufays, I. Thomas, and B. Flahaut. 2004. Intra-urban Location and Clustering of Road Accidents Using GIS: A Belgian Example. *International Journal of Geographical Information Science* 18: 169–81.

PATH ANALYSIS AND NETWORK APPLICATIONS

Chapter 18 covers path analysis and network applications, both dealing with movement and linear features. Path analysis is raster-based and has a narrower focus. Using a cost raster that defines the cost of moving through each cell, path analysis finds the least cost path between cells. Path analysis is useful as a planning tool for locating a new road or a new pipeline that is least costly in terms of the construction costs as well as the potential costs of environmental impacts.

Network applications require a network that is vector-based and topologically connected. Perhaps the most common network application is shortest path analysis, which is used, for example,

in in-vehicle navigation systems to help drivers find the shortest route between an origin and a destination. Network applications also include finding closest facility, evaluating allocation, and solving location–allocation problems.

Path analysis and network applications such as shortest path analysis share some common terms and concepts. But they differ in data format and data analysis environment. Path analysis uses raster data to locate a "virtual" least cost path. In contrast, shortest path analysis finds the shortest path between stops on an existing network. By having path analysis and network applications in the same chapter, we can compare raster data and vector data for GIS applications.

Chapter 18 includes the following five sections. Section 18.1 introduces path analysis and its basic elements. Section 18.2 covers applications of path analysis. Section 18.3 examines the basic structure of a road network. Section 18.4 describes how to put together a road network with appropriate attributes for network applications. Section 18.5 provides an overview of network applications.

18.1 PATH ANALYSIS

A path analysis requires a source raster, a cost raster, cost distance measures, and an algorithm for deriving the least accumulative cost path.

18.1.1 Source Raster

A **source raster** defines the source cell(s). Only the source cell has a cell value in the source raster; all other cells are assigned no data. Similar to physical distance measure operations (Chapter 13), cost distance measures spread from the source cell. But in the context of path analysis, one can consider the source cell as an end point of a path, either the origin or the destination. A path analysis derives for a cell the least cost path to the source cell, or to the closest source cell if two or more source cells are present.

18.1.2 Cost Raster

A **cost raster** defines the cost or impedance to move through each cell. A cost raster has two characteristics. First, the cost for each cell is usually the sum of different costs. As an example, Box 18.1 summarizes the cost for constructing a pipeline, which may include the construction and opera-tional costs as well as the potential costs of environmental impacts.

Second, the cost may represent the actual or relative cost. Relative costs are ranked values. For example, costs may be ranked from 1 to 5, with 5 being the highest cost value. A project such as a pipeline project typically involves a wide variety of cost factors. Some factors can be measured in actual costs but others such as aesthetics, wildlife habitats, and cultural resources are difficult to measure in actual costs. Relative costs are therefore a means of standardizing different cost factors.

To put together a cost raster, we start by compiling and evaluating a list of cost variables. We then make a raster for each cost variable and use a local operation to sum the individual cost rasters. The local sum is the total cost necessary to traverse each cell.

18.1.3 Cost Distance Measures

The cost distance measure in a path analysis is based on the node-link cell representation (Figure 18.1). A node represents the center of a cell, and a link—either a lateral link or a diagonal link—connects the node to its adjacent cells. A lateral link connects a cell to one of its four immediate neighbors, and a

---

Box 18.1 Cost Raster for a Site Analysis of Pipelines

A site analysis of a pipeline project must consider the construction and operational costs. Some of the variables that can influence the costs include the following:

- Distance from source to destination
- Topography, such as slope and grading
- Geology, such as rock and soils
- Number of stream, road, and railroad crossings
- Right-of-way costs
- Proximity to population centers

In addition, the site analysis should consider the potential costs of environmental impacts during con-struction and liability costs that may result from accidents after the project has been completed. Environmental impacts of a proposed pipeline project may involve the following:

- Cultural resources
- Land use, recreation, and aesthetics
- Vegetation and wildlife
- Water use and quality
- Wetlands

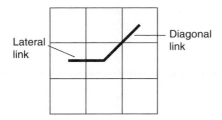

Figure 18.1

Cost distance measures follow the node-link cell representation: a lateral link connects two direct neighbors, and a diagonal link connects two diagonal neighbors.

diagonal link connects the cell to one of the corner neighbors. The distance is 1.0 cell for a lateral link and 1.414 cells for a diagonal link.

The cost distance to travel from one cell to another through a lateral link is 1.0 cell times the average of the two cost values:

$$1 \times [(C_i + C_j)/2]$$

where C_i is the cost value at cell i, and C_j the cost value at neighboring cell j. The cost distance to travel from one cell to another through a diagonal link, on the other hand, is 1.414 cells times the average of the two cost values (Figure 18.2).

The node-link cell representation limits the movement to between neighboring cells and the direction of movement to the eight principal directions. This is why a least cost path often displays a zigzag pattern. It is possible to connect a cell to

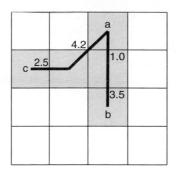

Figure 18.3

The accumulative cost from cell a to cell b is the sum of 1.0 and 3.5, the costs of two lateral links. The accumulative cost from cell a to cell c is the sum of 4.2 and 2.5, the costs of a diagonal link and a lateral link.

more neighbors (i.e., beyond the eight neighboring cells). But this can create nonintuitive intersecting paths, in addition to requiring more computer processing (Miller and Shaw 2001).

18.1.4 Deriving the Least Accumulative Cost Path

Given a cost raster, we can calculate the accumulative cost between two cells by summing the costs associated with each link that connects the two cells (Figure 18.3). But finding the least accumulative cost path is more challenging. This is because many different paths can connect two cells that are

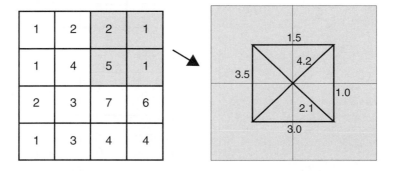

Figure 18.2

The cost distance of a lateral link is the average of the costs in the linked cells, for example, $(1 + 2)/2 = 1.5$. The cost distance of a diagonal link is the average cost times 1.414, for example, $1.414 \times [(1 + 5)/2] = 4.2$.

not immediate neighbors. The least accumulative cost path can only be derived after each possible path has been evaluated.

Finding the least accumulative cost path is an iterative process. The process begins by activating cells adjacent to the source cell and by computing costs to the cells. The cell with the lowest cost distance is chosen from the active cell list, and its value is assigned to the output raster. Next, cells adjacent to the chosen cell are activated and added to the active cell list. Again, the lowest cost cell is chosen from the list and its neighboring cells are activated. Each time a cell is reactivated, meaning that the cell is accessible to the source cell through a different path, its accumulative cost must be re-computed. The lowest accumulative cost is then assigned to the reactivated cell. This process continues until all cells in the output raster are assigned with their least accumulative costs to the source cell.

Figure 18.4 illustrates the cost distance measure operation. Figure 18.4a shows a raster with the source cells at the opposite corners. Figure 18.4b represents a cost raster. To simplify the computation, both rasters are set to have a cell size of 1. Figure 18.4c shows the cost of each lateral link and the cost of each diagonal link. Figure 18.4d shows for each cell the least accumulative cost. Box 18.2 explains how Figure 18.4d is derived.

A cost distance measure operation can result in different types of output. The first is a least

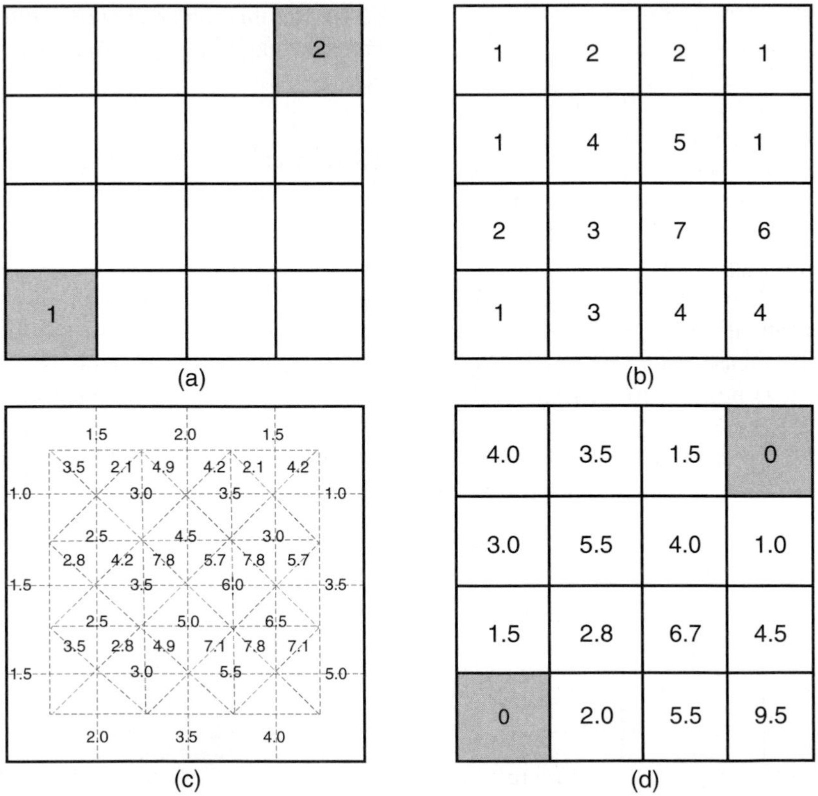

(a)

			2
1			

(b)

1	2	2	1
1	4	5	1
2	3	7	6
1	3	4	4

(c)

(d)

4.0	3.5	1.5	0
3.0	5.5	4.0	1.0
1.5	2.8	6.7	4.5
0	2.0	5.5	9.5

Figure 18.4
The cost distance for each link (c) and the least accumulative cost distance for each cell (d) are derived using the source cells (a) and the cost raster (b). See Box 18.2 for the derivation.

Box 18.2 | Derivation of the Least Accumulative Cost Path

Step 1. Activate cells adjacent to the source cells, place the cells in the active list, and compute cost values for the cells. The cost values for the active cells are as follows: 1.0, 1.5, 1.5, 2.0, 2.8, and 4.2.

		1.5	0
		4.2	1.0
1.5	2.8		
0	2.0		

Step 2. The active cell with the lowest value is assigned to the output raster, and its adjacent cells are activated. The cell at row 2, column 3, which is already on the active list, must be reevaluated, because a new path has become available. As it turns out, the new path with a lateral link from the chosen cell yields a lower accumulative cost of 4.0 than the previous cost of 4.2. The cost values for the active cells are as follows: 1.5, 1.5, 2.0, 2.8, 4.0, 4.5, and 6.7.

		1.5	0
		4.0	1.0
1.5	2.8	6.7	4.5
0	2.0		

Step 3. The two cells with the cost value of 1.5 are chosen, and their adjacent cells are placed in the active list. The cost values for the active cells are as follows: 2.0, 2.8, 3.0, 3.5, 4.0, 4.5, 5.7, and 6.7.

	3.5	1.5	0
3.0	5.7	4.0	1.0
1.5	2.8	6.7	4.5
0	2.0		

Step 4. The cell with the cost value of 2.0 is chosen and its adjacent cells are activated. Of the three adjacent cells activated, two have the accumulative cost values of 2.8 and 6.7. The values remain the same because the alternative paths from the chosen cell yield higher cost values (5 and 9.1, respectively). The cost values for the active cells are as follows: 2.8, 3.0, 3.5, 4.0, 4.5, 5.5, 5.7, and 6.7.

	3.5	1.5	0
3.0	5.7	4.0	1.0
1.5	2.8	6.7	4.5
0	2.0	5.5	

Step 5. The cell with the cost value of 2.8 is chosen. Its adjacent cells all have accumulative cost values assigned from the previous steps. These values remain unchanged because they are all lower than the values computed from the new paths.

	3.5	1.5	0
3.0	5.7	4.0	1.0
1.5	2.8	6.7	4.5
0	2.0	5.5	

Step 6. The cell with the cost value of 3.0 is chosen. The cell to its right has an assigned cost value of 5.7, which is higher than the cost of 5.5 via a lateral link from the chosen cell.

4.0	3.5	1.5	0
3.0	5.5	4.0	1.0
1.5	2.8	6.7	4.5
0	2.0	5.5	

Step 7. All cells have been assigned with the least accumulative cost values, except the cell at row 4, column 4. The least accumulative cost for the cell is 9.5 from either source.

4.0	3.5	1.5	0
3.0	5.5	4.0	1.0
1.5	2.8	6.7	4.5
0	2.0	5.5	9.5

accumulative cost raster as shown in Figure 18.4*d*. The second is a direction raster, showing the direction of the least cost path for each cell. The third is an allocation raster, showing the assignment of each cell to a source cell on the basis of cost distance measures. The fourth type is a shortest path raster, which shows the least cost path from each cell to a source cell. Using the same data as in Figure 18.4, Figure 18.5*a* shows two examples of the least cost path and Figure 18.5*b* shows the

and aspect in different directions. To provide a more realistic analysis of how we traverse the terrain, several new variables have been suggested for cost distance measure operations (Xu and Lathrop 1994, 1995; Collischonn and Pilar 2000; DeMers 2002; Yu et al. 2003).

The first variable is the surface distance. Calculated from an elevation raster, the surface distance measures the ground or actual distance that must be covered from one cell to another. The surface distance increases when the elevation difference (gradient) between two cells increases. Two other variables are the vertical factor and horizontal factor. The vertical factor determines the difficulty of overcoming the vertical elements (e.g., uphill or downhill) in moving from one cell to another. The horizontal factor incorporates the difficulty of overcoming the horizontal elements (e.g., crosswinds) in moving from one cell to another.

ArcGIS uses the term **path distance** to describe the cost distance based on the surface distance, vertical factor, and horizontal factor. Box 18.3 describes path analysis and path distance analysis that are available in ArcGIS.

18.1.5 Improvements for Cost Distance Measures

Cost distance measures presented so far assume an isotropic cost surface, a cost surface that is uniform for all directions. But in reality, the cost surface is not uniform: the terrain changes in elevation, slope,

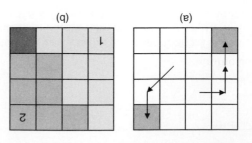

Figure 18.5
The least cost cost paths (a) and the allocation raster (b) are derived using the same input data as in Figure 18.4.

assignment of each cell to a source cell. The darkest cell in Figure 18.5b can be assigned to either one of the two sources.

Box 18.3 Cost Distance Measure Operations in ArcGIS

Both Spatial Analyst and ArcToolbox in ArcGIS offer cost distance measure operations. The Distance menu of Spatial Analyst provides Cost Weighted and Shortest Path. Cost Weighted calculates for each cell the least accumulative cost, over a cost raster, to its closest source cell. The Cost Weighted operation can also produce a direction raster, showing the direction of the least cost path for each cell, and an allocation raster, showing the assignment of each cell to a source. Shortest Path uses the distance and direction rasters created by the cost distance measure operation to generate the least cost path for any cell.

Spatial Analyst Tools in ArcToolbox has a Distance toolset. The toolset includes Cost Allocation, Cost Back Link, Cost Distance, and Cost Path. These tools are similar to those in Spatial Analyst. ArcGIS 9.0 has introduced a set of new tools on Path Distance in ArcToolbox. The cost distance in a path distance analysis is calculated by multiplying the surface distance, vertical factor, and horizontal factor. The surface distance is calculated from an elevation raster. The vertical factor and the horizontal factor are user-defined and optional. A path distance analysis is designed to provide a more realistic analysis of how we traverse the terrain. ArcToolbox offers Path Distance, Path Distance Allocation, and Path Distance Back Link in the Distance toolset.

18.2 APPLICATIONS OF PATH ANALYSIS

Path analysis is useful for planning roads, pipelines, canals, and transmission lines (Collischonn and Pilar 2000; Yu et al. 2003). Path analysis can also be applied to other types of movements. For example, Hepner and Finco (1995) build a model of gas dispersion using an impedance (cost) raster that represents the product of surface distance, slope impedance, and wind impedance. The accumulative impedance in their study represents the difficulty of gas travel between the release point (source cell) and any given cell in the study area. And equal steps of the accumulative impedance values supposedly define the progression over time of a dense gas cloud over the terrain.

Path analysis has also been used for extracting linear features such as roads and rivers from low-resolution digital imagery. Dillabaugh et al. (2002) create a cost raster from edge detection indices and run path analysis over the cost raster to connect rivers that are single pixel in width but are broken in places.

18.3 NETWORK

A **network** is a system of linear features that has the appropriate attributes for the flow of objects. A road system is a familiar network. Other networks include railways, public transit lines, bicycle paths, and streams. A network is typically topology-based: lines (arcs) meet at intersections (nodes), lines cannot have gaps, and lines have directions.

Because many network applications involve road systems, Section 18.3 concentrates on road networks. It starts by describing the geometry and attribute data of a road network including link impedance, turns, one-way streets, and overpasses/underpasses. Then it shows how these data can be put together to form a street network for a real-world example.

18.3.1 Link and Link Impedance

A **link** refers to a road segment defined by two end points. Also called edges, links are the basic geo-

metric features of a network. **Link impedance** is the cost of traversing a link. A simple measure of the cost is the physical length of the link. But the length may not be a reliable measure of cost, especially in cities where speed limits and traffic conditions vary significantly along different streets. A better measure of link impedance is the travel time estimated from the length and the speed limit of a link. For example, if the speed limit is 30 miles per hour and the length is 2 miles, then the link travel time is 4 minutes ($2/30 \times 60$ minutes).

There are variations in measuring the link travel time. The travel time may be directional: the travel time in one direction may be different from the other direction. In that case, the travel time can be entered separately depending on the traffic direction. The travel time may also vary by the time of the day and by the day of the week, thus requiring the setup of different network attributes for different applications.

18.3.2 Node and Turn Impedance

A **node** refers to an end point of a link. If a link has an associated direction, the starting point of the link is the from-node and the end point is the to-node. Like links, nodes are also the basic geometric features of a network. A turn is a transition from one street segment to another in a network. In other words, a turn takes place at a node where two links intersect or meet.

Turn impedance is the time it takes to complete a turn, which is significant in a congested street network (Ziliaskopoulos and Mahmassani 1996). Turn impedance is directional. For example, it may take 5 seconds to go straight, 10 seconds to make a right turn, and 30 seconds to make a left turn at a stoplight. A negative turn impedance value means a prohibited turn, such as turning the wrong way onto a one-way street.

Because a network typically has many turns with different conditions, a **turn table** is used to assign the turn impedance values in the network. A driver approaching a street intersection can go straight, turn right, turn left, and, in some cases, make a U-turn. Assuming the intersection involves four street segments, as most intersections do, this means at least 12 possible turns at the intersection, excluding U-turns. Depending on the level of detail for a study, we may not have to include every intersection and every possible turn in a turn table. A partial turn table listing only turns at intersections with stoplights may be sufficient for a network application.

18.3.3 One-Way or Closed Streets

The value of a designated field in a network's attribute table can show one-way or closed streets. FT, for instance, means that travel is permitted in the direction from the from-node to the to-node of a street segment. TF means that travel is permitted from the to-node to the from-node. And N means that travel is not allowed in either direction.

18.3.4 Overpasses and Underpasses

There are two methods for including overpasses and underpasses in a network. The first method uses nonplanar features: an overpass and the street underneath it are both represented as continuous lines without a node at their intersection (Figure 18.6). The second method treats overpasses and underpasses as planar features: the two arcs representing

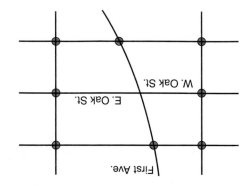

Figure 18.6
First Ave. crosses Oak St. with an overpass. A nonplanar representation with no nodes is used at the intersection of Oak St. and First Ave.

the overpass meet at one node, and the two arcs representing the street underneath the overpass meet at another (Figure 18.7). To differentiate the overpass from the street, it is common to assign a higher "elevation" value (e.g., 1) to the node used by the overpass than the node used by the street (e.g., 0).

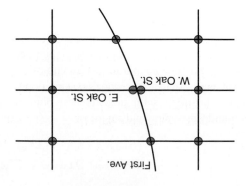

Figure 18.7
First Ave. crosses Oak St. with an overpass. A planar representation with two nodes is used at the intersection: one for First Ave., and the other for Oak St. First Ave. has 1 for the T-elev and F-elev values, indicating that the overpass is on First Ave.

Street name	F-elev	T-elev
First Ave.	0	1
First Ave.	1	0
W. Oak St.	0	0
E. Oak St.	0	0

18.4 PUTTING TOGETHER A NETWORK

Putting together a road network involves three tasks: gather the linear features of the network, build the necessary topology for the network, and attribute the network features. Discussions in Section 18.4 are based on a real-world example from Moscow, Idaho. A university town of 25,000 people, Moscow, Idaho has more or less the same network features as other larger cities except for overpasses or underpasses.

Box 18.4 | Building Network Topology in ArcGIS

ArcGIS has two options for building the topology of a network: the coverage model and the geodatabase data model. The coverage model enforces the topological relationships of connectivity, contiguity, and area definition (Chapter 3). Connectivity is essential for network applications because it makes sure that arcs connect to each other at nodes. We can build the topology of a line coverage by using either Clean or Build.

The geodatabase data model uses a geometric network to store the geometric features of a network and a logical network to store the connectivity between geometric features (Zeiler 1999). The geometric features of the network include edges and junctions.

Edges are like arcs and junctions are like nodes in the coverage model. The logical network stores the information as to how edges and junctions of the geometric network are connected. Although the logical network cannot be viewed in ArcCatalog or ArcMap, we can imagine that it looks like the arc-node list for the coverage model (Chapter 3).

The data structure of a geometric network follows the hierarchical structure of a geodatabase: the network is a feature dataset, and edges and junctions are separate feature classes. The attribute tables of edge and junction features can store attributes such as link impedance, turn impedance, and one-way or closed streets for network analysis.

18.4.1 Gathering Linear Features

The TIGER/Line files from the U.S. Census Bureau are a common data source for making preliminary street networks in the United States. The TIGER/Line files use longitude and latitude values as measurement units and NAD83 (North American Datum of 1983) as the datum. Therefore a preliminary network compiled from the TIGER/Line files must be converted to projected coordinates. Road network data can also be digitized or purchased from commercial companies.

18.4.2 Building Topology

A network coverage converted from the TIGER/Line files has the built-in topology: The streets are connected at nodes and the nodes are designated as either from-nodes or to-nodes. If topology is not available in a data set (e.g., a shapefile or a CAD file), one can use a GIS package to build the topology. ArcGIS, for example, has the coverage model or the geodatabase data model for building network topology (Box 18.4).

Editing and updating the road network is the next step. Mistakes from the TIGER/Line files must be corrected, and new streets must be added. Also, pseudo nodes (i.e., nodes that are not required topologically) must be removed so that a street segment between two intersections is not unnecessarily broken up. But a pseudo node is needed at the location where one street changes into another along a continuous arc (rare but does happen). Without the pseudo node, the continuous arc will be treated as the same street.

18.4.3 Attributing the Network Features

The physical length of a street segment is included in a default field in a line coverage or a geodatabase line feature class. But as described in Section 18.3.1, the link impedance value is usually based on the travel time. Roads derived from the TIGER/Line files have a functional classification field that can be used to assign speed limits. Moscow, Idaho has three speed limits: 35 miles/hour for principal arterials, 30 miles/hour for minor

Moscow, Idaho has two one-way streets serving as the northbound and the southbound lanes of a state highway. These one-way streets are denoted by the value of 1 in a direction field. The direction of all street segments that make up a one-way street must be consistent and pointing in the right direction. Street segments that are incorrectly oriented must be flipped.

Making a turn table is next. ArcInfo Workstation has the Turntable command that can generate a turn table with all intersections and possible turns in a network. The turn table includes the following four default fields: the node (intersection) number, the street segment (arc) numbers involved in a turn, and the turn angle. The turn angle indicates the type of turn: 90 for a left turn, −90 for a right turn, and 0 for going straight. Using the turn angle value, we can add the turn impedance value in minutes or seconds to a turn (Box 18.5).

Figure 18.8 shows a street intersection at node 341 with no restrictions for all possible turns except U-turns. This example uses two turn impedance values: 30 seconds or 0.5 minute for a left turn, and 15 seconds or 0.25 minute for either a right turn or going straight. Some street intersections do not allow certain types of turns. Figure 18.9 shows a street intersection at node 265 with stop signs posted only for the east–west traffic. Therefore, the turn impedance values are assigned only to turns to be made from arc 342 and arc 340.

Figure 18.10 shows a street intersection with stoplights between a southbound one-way street and an east–west two-way street. An impedance value of − 1 means a prohibited turn, such as a right turn from arc 467 to arc 461 or a left turn from arc 462 to arc 461.

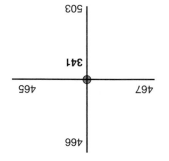

Figure 18.8
Possible turns at node 341.

Node#	Arc1#	Arc2#	Angle	Minutes
341	503	467	90	0.500
341	503	466	0	0.250
341	503	465	−90	0.250
341	467	503	−90	0.250
341	467	466	90	0.500
341	467	465	0	0.250
341	466	503	0	0.250
341	466	467	−90	0.250
341	466	465	90	0.500
341	465	503	90	0.500
341	465	467	0	0.250
341	465	466	−90	0.250

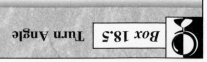

Box 18.5 Turn Angle

The turn angle is the angle between the "from" arc and the "to" arc of a turn. In reality, the angle is often 90° for a right or left turn and 0° for going straight. As measured from the coverage prepared from the TIGER/Line files, the angle is likely to deviate from 90° or 0°. Because the purpose of the angle is simply to determine the type of turn, we can ignore the deviation.

Node#	Arc1#	Arc2#	Angle	Minutes
265	339	342	−87.412	0.000
265	339	340	92.065	0.000
265	339	385	7.899	0.000
265	342	339	87.412	0.500
265	342	340	−0.523	0.250
265	342	385	−84.689	0.250
265	340	339	−92.065	0.250
265	340	342	0.523	0.250
265	340	385	95.834	0.500
265	385	339	−7.899	0.000
265	385	342	84.689	0.000
265	385	340	−95.834	0.000

Figure 18.9
Node 265 has stop signs for the east–west traffic. Turn impedance applies only to turns in the shaded rows.

Node#	Arc1#	Arc2#	Angle	Minutes
339	467	501	90.190	0.500
339	467	462	1.152	0.250
339	467	461	−92.197	−1.000
339	462	501	−90.962	0.250
339	462	467	−1.152	0.250
339	462	461	86.651	−1.000
339	461	501	2.386	0.250
339	461	467	92.197	0.500
339	461	462	−86.651	0.250

Figure 18.10
Node 339 is an intersection between a southbound one-way street and an east–west two-way street. Notice −1 (no turn) in the shaded rows.

18.5 NETWORK APPLICATIONS

A network with the appropriate attributes can be used for a variety of applications. Section 18.5 discusses path finding and accessibility measures. It also links a network to solving location–allocation problems and transportation planning modeling. Some applications are directly accessible through commands in a GIS package. Others require linking a GIS package to a discipline-specific software package, such as a transportation planning software package.

18.5.1 Shortest Path Analysis
Shortest path analysis finds the path with the minimum cumulative impedance between nodes on a network. The path may connect just two nodes—an origin and a destination—or have specific stops between the nodes. Shortest path analysis can help a van driver set up a schedule for dozens of deliveries, or an emergency service connect a dispatch station, accident location, and hospital. With an in-vehicle navigation system, shortest path analysis can also help a driver plan routine trips. Shortest path analysis is studied in operations research, computer science, spatial analysis, and transportation engineering.

Shortest path analysis typically begins with an impedance matrix, in which a value represents the impedance of a direct link between two nodes on a network and a ∞ (infinity) means no direct connection. The analysis then follows Dijkstra's algorithm (1959) to find the shortest distance from node 1 to all other nodes in an iterative process. Although

Dijkstra's algorithm is the common algorithm for solving the shortest path, different computational methods have been proposed (Dreyfus 1969).

Figure 18.11 illustrates a shortest path analysis with six cities on a road network. The impedance matrix of travel time measured in minutes is shown in Table 18.1. The value of ∞ above and below the principal diagonal in the impedance matrix means no direct path between two nodes. Suppose we want to find the shortest path from node 1 to all other nodes in Figure 18.11. We can solve the problem by using an iterative procedure (Lowe and Moryadas 1975). At each step, we choose the shortest path from a list of candidate paths and place the node of the shortest path in the solution list.

The first step is to choose the minimum among the three paths from node 1 to nodes 2, 3, and 4, respectively:

$$\min (p_{12}, p_{13}, p_{14}) = \min (20, 53, 58)$$

We choose p_{12} because it has the minimum impedance value among the three candidate paths. We then place node 2 in the solution list with node 1.

The second step is to prepare a new candidate list of paths that are directly or indirectly connected to nodes in the solution list (nodes 1 and 2):

$$\min (p_{13}, p_{14}, p_{12} + p_{23}) = \min (53, 58, 59)$$

We choose p_{13} and add node 3 to the solution list. To complete the solution list with other nodes on the network we need to go through the following steps:

$$\min (p_{14}, p_{13} + p_{34}, p_{13} + p_{36}) = \min (58, 78, 72)$$

$$\min (p_{13} + p_{36}, p_{14} + p_{45}) = \min (72, 71)$$

$$\min (p_{13} + p_{36}, p_{14} + p_{45} + p_{56}) = \min (72, 84)$$

Table 18.2 summarizes the solution to the shortest path problem from node 1 to all other nodes.

The similarity between shortest path analysis and path analysis in Section 18.1 is obvious. The algorithm for solving the least accumulative cost path is in fact an adaptation of Dijkstra's algorithm.

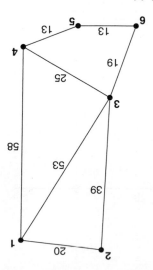

Figure 18.11

Link impedance values between cities on a road network.

TABLE 18.1 | The Impedance Matrix Among Six Nodes in Figure 18.11

	(1)	(2)	(3)	(4)	(5)	(6)
(1)	∞	20	53	58	∞	∞
(2)	20	∞	39	∞	∞	∞
(3)	53	39	∞	25	∞	19
(4)	58	∞	25	∞	13	∞
(5)	∞	∞	∞	13	∞	13
(6)	∞	∞	19	∞	13	∞

TABLE 18.2 | Shortest Paths from Node 1 to All Other Nodes in Figure 18.11

From-node	To-node	Shortest path	Minimum cumulative impedance
1	2	p_{12}	20
1	3	p_{13}	53
1	4	p_{14}	58
1	5	$p_{14} + p_{45}$	71
1	6	$p_{13} + p_{36}$	72

But the two analyses differ in the data model and data analysis environment. Shortest path analysis is vector-based and uses an existing network. Path analysis is raster-based and uses a source raster and a cost raster to search for the least cost path for facilities such as roads and pipelines that are still at the planning stage.

The **traveling salesman problem** is a routing problem, which adds two constraints to shortest path analysis: (1) the salesman must visit each of the select stops only once and (2) the salesman may start from any stop but must return to the original stop. The objective is to determine which route, or tour, the salesman can take to minimize the total impedance value. A common solution to the traveling salesman problem uses a heuristic method (Lin 1965). Beginning with an initial random tour, the method runs a series of locally optimal solutions by swapping stops that yield reductions in the cumulative impedance. The iterative process ends when no improvement can be found by swapping stops. This heuristic approach can usually create a tour with a minimum, or near minimum, cumulative impedance.

18.5.2 Closest Facility

One type of shortest path analysis is to find the closest facility, such as a hospital, fire station, or ATM, to any location on a network. A **closest facility** algorithm first computes the shortest paths from the select location to all candidate facilities, and then chooses the closest facility among the candidates. Figure 18.12 shows, for example, the closest fire station to a street address.

Closest facility analysis is applicable to location-based services (LBS), an application heavily promoted by GIS companies in recent years. An LBS user can use a Web-enabled mobile phone to be located and to access, through mobile location services, information that is relevant to the user's location. For example, if the user wants to find the closest ATM, the LBS provider can run a closest facility analysis and send the information to the user.

Figure 18.12

Shortest path from a street address to its closest fire station, shown by the square symbol.

18.5.3 Allocation

Allocation is the study of the spatial distribution of resources through a network. Resources in allocation studies often refer to public facilities, such as fire stations or schools. Because the distribution of the resources defines the extent of the service area, the main objective of spatial allocation analysis is to measure the efficiency of these resources.

A common measure of efficiency in the case of emergency services is the response time—the time it takes for a fire truck or an ambulance to reach an incident. Figure 18.13, for example, shows areas of a small city (Moscow, Idaho) covered by two existing fire stations within a 2-minute response time. The map shows a large portion of the city is outside the 2-minute response zone. The response time to the city's outer zone is about 5 minutes (Figure 18.14).

If residents of the city demand that the response time to any part of the city be 2 minutes or less, then the options are either to relocate the fire stations or, more likely, to build new fire stations.

A new fire station should be located to cover the largest portion of the city unreachable in 2 minutes by the existing fire stations. The problem then becomes a location and allocation problem, which is covered in Section 18.5.4.

18.5.4 Location–Allocation

An important topic in operations research and spatial analysis, **location–allocation** solves problems of matching the supply and demand by using sets of objectives and constraints. Location–allocation algorithms used to be available in stand-alone computer programs, but GIS packages, such as ArcInfo Workstation and MGE Network, have incorporated some algorithms as application tools.

The private sector offers many location–allocation examples. Suppose a company operates soft-drink distribution facilities to serve supermarkets. The objective in this example is to minimize the total distance traveled, and a constraint, such as a 2-hour drive distance, may be imposed on the problem. A location–allocation analysis is to match the distribution facilities and the supermarkets while meeting both the objective and the constraint.

Location–allocation is also important in the public sector. For instance, a local school board may decide that (1) all school-age children should be within 1 mile of their schools and (2) the total distance traveled by all children should be minimized. In this case, schools represent the supply, and school-age children represent the demand. The objective of this location–allocation analysis is to provide equitable service to a population while maximizing efficiency in the total distance traveled.

The setup of a location–allocation problem requires inputs in supply, demand, and distance measures. The supply consists of facilities or centers at point locations. The demand may consist of points, lines, or polygons, depending on the data source and the level of spatial data aggregation. For example, the locations of school-age children may be represented as individual points, aggregate points along streets, or aggregate points (centroids) in unit areas such as census block groups. Use of aggregate demand points is likely to introduce

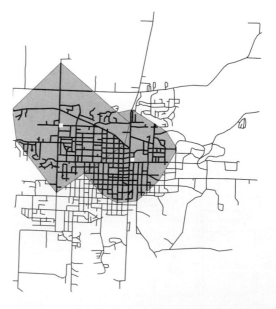

Figure 18.13
Service areas of two fire stations within a 2-minute response time.

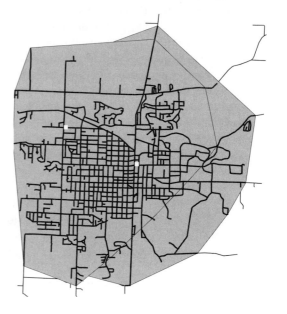

Figure 18.14
Service areas of two fire stations within a 5-minute response time.

errors in location–allocation analysis (Francis et al. 1999). Distance measures between the supply and demand are often included in a distance matrix or a distance list. Distances may be measured along the shortest path between two points on a road network or along the straight line connecting two points. In location–allocation analysis, shortest path distances are likely to yield more accurate results than straight-line distances.

Perhaps the most important input to a location–allocation problem is the model or algorithm for solving the problem. Two common models are minimum distance and maximum covering (Ghosh et al. 1995; Church 1999). The **minimum distance model,** also called the *p*-median location model, minimizes the total distance traveled from all demand points to their nearest supply centers (Hakimi 1964). The *p*-median model has been applied to a variety of facility location problems, including food distribution facilities, public libraries, and health facilities. The **maximum covering model** maximizes the demand covered within a specified time or distance (Church and ReVelle 1974). Public sector location problems, such as the location of emergency medical and fire services, are ideally suited for the maximum covering model. The model is also useful for many convenience-oriented retail facilities, such as movie theaters, banks, and fast-food restaurants.

Both the minimum distance and maximum covering models may have added constraints or options. A maximum distance constraint may be imposed on the minimum distance model so that the solution, while minimizing the total distance traveled, ensures that no demand is beyond the specified maximum distance. A desired distance option may be used with the maximum covering model to cover all demand points within the desired distance.

Here we will use public libraries in a rural county (Latah County, Idaho) as an example to examine three location–allocation scenarios. The demand for public libraries consists of 33 demand points, each representing the total number of households in a census block group (Figure 18.15). Distances are measured along straight lines between the demand points and public library locations.

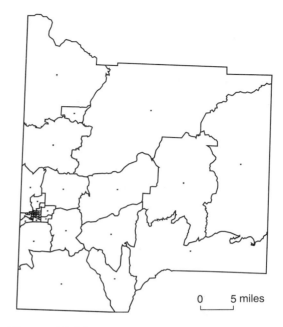

Figure 18.15
Demand points for a public library location problem.

The first scenario assumes that (1) every block group is a possible candidate for a public library and (2) the county government wants to set up three libraries using the maximum covering model with a desired distance of 10 miles. Figure 18.16 shows the three selected candidates. The solution leaves one demand point with 344 households uncovered. To cover every demand point, the county government must either increase the number of public libraries or relax the desired distance.

The second scenario has the same setup as the first except that the county government wants to use the minimum distance model. Figure 18.17 shows the three selected candidates, one of which is not included in Figure 18.16 from the maximum covering model.

The third scenario assumes that one of the three library locations is fixed or has been previously chosen. Figure 18.18 shows the two selected new sites using the minimum distance model with the maximum distance constraint of 10 miles. The solution leaves three demand points with 1124 households uncovered.

Figure 18.16
Three candidates selected by the maximum covering model and the demand points served by the candidates.

Figure 18.18
One fixed and two selected candidates using the minimum distance model with the maximum distance constraint of 10 miles. The fixed candidate is the one to the west.

Figure 18.17
Three candidates selected by the minimum distance model and the demand points served by the candidates.

The location–allocation algorithms and other network analysis functions available in ArcInfo Workstation are summarized in Box 18.6.

18.5.5 Urban Transportation Planning Model

An urban transportation planning (UTP) model typically uses the four-step process of trip generation, trip distribution, modal choice, and trip assignment (Nyerges 1995; Khatib et al. 2001; Miller and Shaw 2001). Trip generation estimates the number of trips to and from each traffic analysis zone (TAZ) in the model. Trip distribution uses the number of trips generated in each TAZ and the distance or travel time between TAZs to produce a trip interchange matrix between TAZs. Modal choice splits trips by travel mode. And trip assignment loads estimated traffic volumes from the model on the network.

UTP modeling is a complex process. Each step in the four-step process involves selecting

Box 18.6 | Network Analysis Using ArcInfo Workstation

The Network Analyst extension to ArcView 3.x can solve shortest path, closest facilities, and allocation problems. But the extension is not available to ArcGIS 9.0. Before ESRI, Inc. adds the Network Analyst extension to ArcGIS, network analysis functions are only available through ArcInfo Workstation commands.

Path and Tour are two commands for solving the shortest path problem. Path uses a user-defined order of stops to solve the problem, whereas Tour determines the order in which stops are visited to minimize the cumulative impedance. ArcInfo Workstation saves the shortest path solution as a route, which can be displayed on top of a network. Both Path and Tour offer the option of providing directions for the shortest path with street names, turns, and distances.

Allocate works with the allocation problem by assigning arcs and/or nodes on a network to centers (e.g., fire stations). Locateallocate works with the location–allocation problem. ArcInfo Workstation offers six location–allocation models including the minimum distance model, the maximum covering model, and their variations. ArcInfo Workstation also offers two location–allocation computing algorithms: GRIA (global regional interchange algorithm) (Densham and Rushton 1992), and TEITZBART (Teitz and Bart 1968). Both algorithms are heuristics. These algorithms can solve location–allocation problems reasonably fast but may not produce optimal solutions, especially with many supply and demand points. The maximum covering model in ArcInfo Workstation has a required parameter of maximum distance and an optional parameter of desired distance. Although the objective is maximum covering, the model can actually leave some demand points uncovered, as seen in Figure 18.16. This apparent discrepancy is a result of using a minimum distance solution technique to solve a maximum covering model (Church 1999).

algorithms or models and requires a good understanding of transportation planning from the software developer and the user (Waters 1999). This is why specialized software packages are normally used for UTP modeling.

One of the necessary inputs to UTP modeling is a transportation network. The network is used to measure the travel impedance between TAZs. Both trip distribution and modal choice in the four-step process need the travel impedance data. The network is also used in trip assignment because at this step the estimated traffic volume is assigned to every link of the network.

To be used for UTP modeling a network must have the arc–node data structure. Nodes are road intersections and critical turns. Arcs or links are road segments between nodes. Link attributes needed for transportation planning usually include distance, speed limit, capacity, and functional classification.

KEY CONCEPTS AND TERMS

Allocation: A study of the spatial distribution of resources on a network.

Closest facility: A network analysis, which computes the shortest paths from the select location to all candidate facilities and then finds the closest facility among the candidates.

Cost raster: A raster that defines the cost or impedance to move through each cell.

Link: A segment separated by two nodes in a road network.

Link impedance: The cost of traversing a link, which may be measured by the physical length or the travel time.

Location–allocation: A spatial analysis, which matches the supply and demand by using sets of objectives and constraints.

Maximum covering model: An algorithm for solving the location–allocation problem by maximizing the demand covered within a specified time or distance.

Minimum distance model: An algorithm for solving the location–allocation problem by minimizing the total distance traveled from all demand points to their nearest supply centers. Also called the p-median location model.

Network: A system of linear features that has the appropriate attributes for the flow of objects such as traffic flow.

Node: An end point of a link.

Path distance: A term used in ArcGIS to describe the cost distance that is calculated from the surface distance, the vertical factor, and the horizontal factor.

Shortest path analysis: A network analysis, which finds the path with the minimum cumulative impedance between nodes on a network.

Source raster: A raster that defines the source, to which the least cost path from each cell is calculated.

Traveling salesman problem: A network analysis, which finds the best route with the conditions of visiting each stop only once, and returning to the original stop where the journey starts.

Turn impedance: The cost of completing a turn on a road network, which is usually measured by the time delay.

Turn table: A table that can be used to assign the turn impedance values on a road network.

REVIEW QUESTIONS

1. What is a source raster for path analysis?

2. Define a cost raster.

3. Box 18.1 summarizes the various costs for a pipeline project. Use Box 18.1 as a reference and list the types of costs that can be associated with a new road project.

4. Cost distance measure operations are based on the node-link cell representation. Explain the representation by using a diagram.

5. Refer to Figure 18.4d. Explain how the cell value of 5.5 (row 2, column 2) is derived. Is it the least accumulative cost possible?

6. Refer to Figure 18.4d. Show the least accumulative cost path for the cell at row 2, column 3.

7. Refer to Figure 18.5b. The cell at row 4, column 4 can be assigned to either one of the two source cells. Show the least cost path from the cell to each source cell.

8. What is an allocation raster?

9. How does the surface distance differ from the regular (planimetric) cost distance?

10. Explain the difference between a network and a line shapefile.

11. What is link impedance?

12. What is turn impedance?

13. What kinds of topological relationships are important for shortest path analysis?

14. What fields are normally included in a turn table?

15. Suppose the impedance value between nodes 1 and 4 is changed from 58 to 40 (because of lane widening) in Figure 18.11. Will it change the result in Table 18.2?

16. Define location–allocation analysis.

17. Explain the difference between the minimum distance model and the maximum covering model in location–allocation analysis.

APPLICATIONS: PATH ANALYSIS AND NETWORK APPLICATIONS

This applications section has five tasks. Tasks 1 and 2 cover path analysis. You will work with the least accumulative cost distance in Task 1 and the path distance in Task 2. Task 3 shows you how to convert a line shapefile to a geometric network. Tasks 4 and 5 use ArcInfo Workstation to run a shortest path analysis and an allocation analysis. ArcGIS 9.0 does not include the Network Analyst extension. Therefore, Tasks 4 and 5 are set up using ArcInfo Workstation.

Task 1: Compute the Least Accumulative Cost Distance

What you need: *sourcegrid* and *costgrid*, the same rasters as in Figure 18.4; and *pathgrid*, a raster to be used with the shortest path function. All three rasters are sample rasters and do not have the projection file.

In Task 1, you will use the same inputs as in Figures 18.4*a* and 18.4*b* to create the same outputs as in Figures 18.4*d*, 18.5*a*, and 18.5*b*.

1. Make connection to the Chapter 18 database in ArcCatalog. Launch ArcMap. Rename the data frame Task 1, and add *sourcegrid*, *costgrid*, and *pathgrid* to Task 1.

2. Make sure that the Spatial Analyst toolbar is available. Click the Spatial Analyst downward arrow, point to Distance, and select Cost Weighted. In the Cost Weighted dialog, do the following: select *sourcegrid* for distance to, select *costgrid* for the cost raster, check to create direction, check to create allocation, and opt for temporary rasters including the output raster. Click OK to run the commands.

3. *CostDistance to sourcegrid* shows the least accumulative cost distance from each cell to a source cell. You can use the Identify tool to click a cell and find its accumulative cost.

Q1. Are the cell values in *CostDistance to sourcegrid* the same as those in Figure 18.4*d*?

4. *CostDirection to sourcegrid* shows the least cost path from each cell to a source cell. The cell value in the raster indicates which neighboring cell to traverse to reach a source cell. The directions are coded 1 to 8, with 0 representing the cell itself (Figure 18.19).

6	7	8
5	0	1
4	3	2

Figure 18.19
Direction measures in a direction raster are numerically coded. The focal cell has a code of 0. The numeric codes 1 to 8 represent the direction measures of 90°, 135°, 180°, 225°, 270°, 315°, 360°, and 45° in a clockwise direction.

5. *CostAllocation to sourcegrid* shows the allocation of cells to each source cell. The output raster is the same as Figure 18.5*b*.

6. Click the Spatial Analyst downward arrow, point to Distance, and select Shortest Path. In the Shortest Path dialog, select *pathgrid* for path to, *costgrid* for the cost distance raster, *CostDirection to sourcegrid* for the cost direction raster, and specify the path type for each cell. Specify *Path.shp* for the output features. Click OK to run the command. *Path* shows the path from each cell in *pathgrid* to its closest source. *Path* is the same as Figure 18.5*a*.

Task 2: Compute the Path Distance

What you need: *emidasub*, an elevation raster; *peakgrid*, a source raster with one source cell; and *emidapathgd*, a path raster that contains two cell values. All three rasters are projected onto UTM coordinates in meters.

In Task 2, you will find the least cost path from each of the two cells in *emidapathgd* to the source cell in *peakgrid*. The least cost path is based on the path distance. Calculated from an elevation raster, the path distance measures the ground or actual distance that must be covered between cells. The source cell is at a higher elevation than the two cells in *emidapathgd*. Therefore, you can imagine the objective of Task 2 is to find the least cost hiking path from each of the two cells in *emidapathgd* to the source cell in *peakgrid*.

1. Insert a new data frame in ArcMap, and rename it Task 2. Add *emidasub*, *peakgrid*, and *emidapathgd* to Task 2. Select Properties from the context menu of *emidasub*. On the Symbology tab, right-click the Color Ramp box to uncheck Graphic View. Then select Elevation #1. As shown in the map, the source cell in *peakgrid* is located near the summit of the elevation surface and the two cells in *emidapathgd* are located in low elevation areas.

2. Open ArcToolbox in ArcMap. Select the Chapter 18 database for the current workspace. Double-click the Path Distance tool in the Spatial Analyst Tools/Distance toolset. Select *peakgrid* for the input raster, specify *pathdist1* for the output distance raster, select *emidasub* for the input surface raster, specify *backlink1* for the output backlink raster, and click OK to run the command.

Q2. What is the range of cell values in *pathdist1*?

Q3. If a cell value in *pathdist1* is 900, what does the value mean?

3. Double-click the Cost Path in the Spatial Analyst Tools/Distance toolset. Select *emidapathgd* for the input raster, select *pathdist1* for the input cost distance raster, select *backlink1* for the input cost backlink raster, specify *path1* for the output raster, and click OK.

4. Open the attribute table of *path1*. Click the left cell of the first record. As shown in the map, the first record is the cell in *peakgrid*.

Click the second record, which is the least cost path from the first cell in *emidapathgd* (in the upper-right corner) to the cell in *peakgrid*.

Q4. What does the third record in the *path1* attribute table represent?

Task 3: Convert a Line Shapefile to a Geodatabase Geometric Network

What you need: *moscowst.shp*, a line shapefile containing a street network of Moscow, Idaho.

Task 3 shows you how to convert a line shapefile to a geodatabase geometric network in ArcCatalog. Without the Network Analyst extension, it is difficult to use the generated geometric network for network applications such as shortest path analysis.

1. Right-click the Chapter 18 database in ArcCatalog, point to New, and select Personal Geodatabase. Rename the geodatabase *Moscow.mdb*. Right-click *Moscow.mdb*, point to New, and select Feature Dataset. Enter *MoscowNet* for the Feature Dataset name, and click the Edit button for the spatial reference. Import the spatial reference of *moscowst.shp* to be the feature dataset's spatial reference. Dismiss the dialogs.

2. Right-click *MoscowNet*, point to Import, and select Feature Class (single). Select *moscowst.shp* for the input features, specify *Street* for the output feature class name, and click OK to import.

3. Right-click *MoscowNet*, point to New, and select the Build Geometric Network. The Build Geometric Network Wizard opens. Read the instructions in the first panel and click Next. In the second panel, make sure that you want to build a geometric network from existing features. In the third panel, make sure that *Street* is the feature class to build the network from and then enter *MoscowSt* for the name of the network. Click Next in the fourth panel. Check Yes to snap junctions in the fifth panel. Skip the next panel. Click Finish. A report suggests that

4 features are invalid. (All 4 features are loops attached to the end of dead-end streets.)

4. Expand *MoscowNet*. The feature dataset has three feature classes: *MoscowSt*, *MoscowSt_Junctions*, and *Street*. *MoscowSt* is the geometric network. Right-click *MoscowSt* and select Properties. The General tab shows that *MoscowSt_Junctions*, a junction feature class, and *Street*, an edge feature class, are the two participating feature classes in the network.

5. You can preview the geography and the table of *MoscowSt_Junctions* and *Street*. As explained in Chapter 18 the feature attribute table can be used to store attributes for network applications.

Task 4: Run Shortest Path Analysis in ArcInfo Workstation

What you need: *uscities*, a point coverage, and *interstates*, a line coverage.

uscities contains cities in the conterminous United States, and *interstates* is a network of the interstate highways. The objective of Task 4 is to find the best (shortest) route between any two nodes or cities. The shortest path analysis in ArcInfo Workstation cannot use *uscities* as input but must first find nodes on *interstates* that are closest to the points of interest in *uscities*. Identification of nodes actually occupies a major portion of the task. This task uses travel time to derive the shortest path. The speed limit for calculating the travel time is 65 miles/hour, and the travel time considers only the link impedance. The cities to be queried are Helena, Montana, and Raleigh, North Carolina. In the following instructions, explanation of a command comes after /*.

1. First build a node attribute table (NAT) for *interstates*.

   Arc: build interstates nodes

2. Go to ArcPlot and set up the draw environment.

   Arc: arcplot
   Arcplot: display 9999

Arcplot: mapex interstates /* set the map extent as that of *interstates*
Arcplot: arcs interstates /* plot *interstates*
Arcplot: nodes interstates /* plot nodes of *interstates*

3. Next plot Helena and Raleigh.

   Arcplot: reselect uscities point city_name = 'Helena' or city_name = 'Raleigh' /* select Helena and Raleigh from *uscities*
   Arcplot: pointmarkers uscities 3 /* plot Helena and Raleigh in green

4. Find the closest node on *interstates* to Helena.

   Arcplot: mapex * /* zoom in the area around Helena
   Arcplot: clear /* clear the display window
   Arcplot: arcs interstates /* plot *interstates* around Helena
   Arcplot: nodes interstates /* plot nodes of *interstates* around Helena
   Arcplot: pointmarkers uscities 3 /* plot Helena in green
   Arcplot: identify interstates node * /* click node closest to Helena and record the INTERSTATES-ID of the node (33)

5. Clear the display window and repeat the same procedure to find the INTERSTATES-ID of the node closest to Raleigh (328). After the nodes representing Helena and Raleigh are identified, use the following commands to run a shortest path analysis and to plot the result.

   Arcplot: mapex interstates /* reset the map extent as that of *interstates*
   Arcplot: arcs interstates /* plot *interstates*
   Arcplot: nodes interstates /* plot nodes of *interstates*
   Arcplot: pointmarkers uscities 3 /* plot Helena and Raleigh in green
   Arcplot: netcover interstates trip /* specify *interstates* for the network coverage and *trip* for the route system
   Arcplot: impedance minutes /* specify minutes as the link impedance

Arcplot: path 33 328 end 1 /* specify
the stops of 33 and 328 and the route-ID of 1
Arcplot: routelines interstates trip 4 /* plot
the shortest path in blue
Arcplot: directions 1 route minute /* get the
travel directions and the total travel time

Task 5: Run Allocation in ArcInfo Workstation

What you need: a network coverage called *moscowst* and a point coverage called *firestat*.

moscowst contains a street network and *firestat* has two fire stations in Moscow, Idaho. Task 5 deals with an allocation problem. The objective is to measure the efficiency of the two fire stations. Similar to Task 4, you cannot use *firestat* directly as input. Instead, the nodes of *moscowst* closest to the two fire stations are used as inputs to Allocate. These two nodes have been identified as having the moscow-ID values of 603 and 848. The efficiency or the service area of the two fire stations can be measured in physical distance or response time. For this task, you will use a response time of 3 minutes.

1. Go to ArcPlot and set up the draw environment.

   Arc: arcplot
   Arcplot: display 9999
   Arcplot: mapex moscowst /* set the extent as that of *moscowst*
   Arcplot: arcs moscowst /* plot *moscowst*
   Arcplot: nodes moscowst /* plot nodes of *moscowst*

Arcplot: pointmarkers firestat 3 /* plot *firestat* in green

2. Specify the network and impedance before running Allocate, and plot the results.

   Arcplot: netcover moscowst fire /* specify *moscowst* for the network coverage and *fire* for the route system
   Arcplot: impedance minutes /* specify minutes as the link impedance
   Arcplot: allocate 603 848 end 3 out /* specify 603 and 848 for the stations and 3 minutes for the response time
   Arcplot: routelines moscowst fire 2 /* plot the streets within the service area in red

Challenge Task

What you need: *menan-b1*, an elevation raster showing the south butte of Menan Buttes; *MenanPeak.shp*, a shapefile containing a point near the top of *menan-b1*; and *MenanPath.shp*, a shapefile containing two points near the base of *menan-b1*.

This challenge task asks you to find the least cost path from each point in *MenanPath.shp* to the point in *MenanPeak.shp*. The cost distance should be based on the actual distance (i.e., path distance).

Q1. What are the least accumulative cost distances from your analysis?

Q2. Plot the least accumulative cost distance paths connecting each point in *MenanPath.shp* to the point in *MenanPeak.shp*.

REFERENCES

Church, R. L. 1999. Location Modelling and GIS. In P. A. Longley, M. F. Goodchild, D. J. MaGuire, and D. W. Rhind, eds., *Geographical Information Systems,* 2d ed., pp. 293–303. New York: Wiley.

Church, R. L., and C. S. ReVelle. 1974. The Maximal Covering Location Problem. *Papers of the Regional Science Association* 32: 101–18.

Collischonn, W., and J. V. Pilar. 2000. A Direction Dependent

Least Costs Path Algorithm for Roads and Canals. *International Journal of Geographical Information Science* 14: 397–406.

DeMers, M. N. 2002. *GIS Modeling in Raster*. New York: Wiley.

Densham, P. J., and G. Rushton. 1992. A More Efficient Heuristic for Solving Large *p*-Median Problems. *Papers in Regional Science* 71: 307–29.

Dijkstra, E. W. 1959. A Note on Two Problems in Connexion with Graphs. *Numerische Mathematik* 1: 269–71.

Dillabaugh, C. R., K. O. Niemann, and D. E. Richardson. 2002. Semi-Automated Extraction of Rivers from Digital Imagery. *GeoInformatica* 6: 263–84.

Dreyfus, S. 1969. An Appraisal of Some Shortest Path Algorithms. *Operations Research* 17: 395–412.

Francis, R. L., T. J. Lowe, G. Rushton, and M. B. Rayco. 1999. A Synthesis of Aggregation Methods for Multifacility Location Problems: Strategies for Containing Error. *Geographical Analysis* 31: 67–87.

Ghosh, A., S. McLafferty, and C. S. Craig. 1995. Multifacility Retail Networks. In Z. Drezner, ed., *Facility Location: A Survey of Applications and Methods,* pp. 301–30. New York: Springer.

Hakimi, S. L. 1964. Optimum Locations of Switching Centers and the Absolute Centers and Medians of a Graph. *Operations Research* 12: 450–59.

Hepner, G. H., and M. V. Finco. 1995. Modeling Dense Gas Contaminant Pathways over Complex Terrain Using a Geographic Information System. *Journal of Hazardous Materials* 42: 187–99.

Khatib, Z., K. Chang, and Y. Ou. 2001. Impacts of Analysis Zone Structures on Modeled Statewide Traffic. *Journal of Transportation Engineering* 127: 31–38.

Lin, S. 1965. Computer Solutions of the Travelling Salesman Problem. *Bell System Technical Journal* 44: 2245–69.

Lowe, J. C., and S. Moryadas. 1975. *The Geography of Movement.* Boston: Houghton Mifflin.

Miller, H. J., and S. Shaw. 2001. *Geographic Information Systems for Transportation: Principles and Applications.* New York: Oxford University Press.

Nyerges, T. L. 1995. Geographical Information System Support for Urban/Regional Transportation Analysis. In S. Hanson, ed., *The Geography of Urban Transportation,* 2d ed., pp. 240–65. New York: Guilford Press.

Teitz, M. B., and P. Bart. 1968. Heuristic Methods for Estimating the Generalized Vertex Median of a Weighted Graph. *Operations Research* 16: 953–61.

Waters, N. 1999. Transportation GIS: GIS-T. In P. A. Longley, M. F. Goodchild, D. J. MaGuire, and D. W. Rhind, eds., *Geographical Information Systems,* 2d ed., pp. 827–44. New York: Wiley.

Xu, J., and R. G. Lathrop. 1994. Improving Cost-Path in a Raster Data Format. *Computers & Geoscience* 20: 1455–65.

Xu, J., and R. G. Lathrop. 1995. Improving Simulation Accuracy of Spread Phenomena in a Raster-Based Geographic Information System. *International Journal of Geographical Information Science* 9: 153–68.

Yu, C., J. Lee, and M. J. Munro-Stasiuk. 2003. Extensions to Least Cost Path Algorithms for Roadway Planning. *International Journal of Geographical Information Science* 17: 361–76.

Zeiler, M. 1999. *Modeling Our World: The ESRI Guide to Geodatabase Design.* Redlands, CA: ESRI Press.

Ziliaskopoulos, A. K., and H. S. Mahmassani. 1996. A Note on Least Time Path Computation Considering Delays and Prohibitions for Intersection Movements. *Transportation Research B* 30: 359–67.

GIS MODELS AND MODELING

CHAPTER OUTLINE

Previous chapters have presented tools for exploring, manipulating, and analyzing vector data and raster data. One of many uses of these tools is to build models. What is a model? A **model** is a simplified representation of a phenomenon or a system. Several types of models have already been covered in this book. A map is a model. So are the vector and raster data models for representing spatial features and the relational model for representing a database system. A model helps us better understand a phenomenon or a system by retaining the significant features and relationships of reality.

Chapter 19 discusses use of GIS for building models. Two points must be clarified at the start.

First, Chapter 19 deals with models using geographically referenced data or geospatial data. Some researchers have used the term "spatially explicit models" to describe these models. Second, the emphasis is the use of GIS in modeling rather than the models. Although Chapter 19 covers a number of models, the intent is simply to use them as examples. A basic requirement in modeling is the modeler's interest and knowledge of the system to be modeled (Hardisty et al. 1993). This is why many models are discipline specific. For example, models of the environment usually consist of atmospheric, hydrologic, land surface/subsurface, and ecological models. It would be impossible for an introductory GIS book to discuss each of these environmental models, not to mention models from other disciplines.

Chapter 19 is divided into the following five sections. Section 19.1 discusses the basic elements of GIS modeling. Sections 19.2 and 19.3 cover binary models and index models, respectively. Section 19.4 deals with regression models, both linear and logistic. Section 19.5 introduces process models including soil erosion and other environmental models. Although these four types of models—binary, index, regression, and process—differ in degree of complexity, they share two common elements: a set of selected spatial variables, and the functional or mathematical relationship between the variables.

19.1 BASIC ELEMENTS OF GIS MODELING

Before we build a GIS model, we must have a basic understanding of type of model, the modeling process, and the role of GIS in the modeling process.

19.1.1 Classification of GIS Models

It is difficult to classify many models used by GIS users. DeMers (2002), for example, classifies models by purpose, methodology, and logic. But the boundary between the classification criteria is not always clear. Rather than proposing an exhaustive classification, Section 19.1.1 covers some broad categories of models by purpose. It serves as an introduction to the models to be discussed later.

A model may be **descriptive** or **prescriptive.** A descriptive model describes the existing conditions of spatial data, and a prescriptive model offers a prediction of what the conditions could be or should be. If we use maps as analogies, a vegetation map would represent a descriptive model and a potential natural vegetation map would represent a prescriptive model. The vegetation map shows existing vegetation, whereas the potential natural vegetation map predicts the vegetation that could occupy a site without disturbance or climate change.

A model may be **deterministic** or **stochastic.** Both deterministic and stochastic models are mathematical models represented by equations with parameters and variables. A stochastic model considers the presence of some randomness in one or more of its parameters or variables, but a deterministic model does not. As a result of random processes, the predictions of a stochastic model

can have measures of error or uncertainty, typically expressed in probabilistic terms. This is why a stochastic model is also called a probabilistic or statistical model. Among the local interpolation methods covered in Chapter 16, for instance, only kriging represents a stochastic model. Besides producing a prediction map, a kriging interpolator can also generate a standard error for each predicted value.

A model may be **static** or **dynamic.** A dynamic model emphasizes the changes of spatial data and the interactions between variables, whereas a static model deals with the state of spatial data at a given time. Time is important to show the process of changes in a dynamic model (Peuquet 1994). Simulation is a technique that can generate different states of spatial data over time. Many environmental models such as groundwater pollution and soil water distribution are best studied as dynamic models (Rogowski and Goyne 2002).

A model may be **deductive** or **inductive.** A deductive model represents the conclusion derived from a set of premises. These premises are often based on scientific theories or physical laws. An inductive model represents the conclusion derived from empirical data and observations. To assess the potential for a landslide, for example, one can use a deductive model based on laws in physics or use an inductive model based on recorded data from past landslides.

19.1.2 The Modeling Process

The development of a model follows a series of steps. The first step is to define the goals of the model. This is analogous to defining a research problem. What is the phenomenon to be modeled? Why is the model necessary? What spatial and time scales are appropriate for the model? The modeler can organize the essential structure of a model by using a conceptual diagram.

The second step is to break down the model into elements and to define the properties of each element and the interactions between the elements. A flowchart is a useful tool for linking the elements. Also at this step, the modeler gathers mathematical equations of the model and commands in a GIS (or the computer code) to carry out the computation.

The third step is the implementation and calibration of the model. The modeler needs data to run and calibrate the model. Model calibration is an iterative process, a process that repeatedly compares the output from the model to the observed data, adjusts the numeric values of the parameters, and reruns the model. Uncertainties in model prediction are a major problem in calibrating a deterministic model. Sensitivity analysis is a technique that can quantify these uncertainties by measuring the effects of input changes on the output.

A calibrated model is a tool ready for prediction. But the model must be validated before it can be generally accepted. Model validation assesses the model's ability to predict under conditions that are different from those used in the calibration phase. Model validation therefore requires a different set of data from those used for developing the model. The modeler can split observed data into two subsets: one subset for developing the model and the other subset for model validation (e.g., Chang and Li 2000). But in many cases the required additional data set presents a problem and forces the modeler to forgo the validation step. A model that has not been validated is likely to be ignored by other researchers (Brooks 1997).

19.1.3 The Role of GIS in Modeling

GIS can assist the modeling process in several ways. First, a GIS is a tool that can process, display, and integrate different data sources including maps, digital elevation models (DEMs), GPS (global positioning system) data, images, and tables. These data are needed for the implementation, calibration, and validation of a model. A GIS can function as a database management tool and, at the same time, is useful for modeling-related tasks such as exploratory data analysis and data visualization.

Second, models built with a GIS can be vector-based or raster-based. The choice depends on the nature of the model, data sources, and the computing algorithm. A raster-based model is preferred if the spatial phenomenon to be modeled varies continuously over the space such as soil erosion and snow accumulation. A raster-based model is also preferred if satellite images and DEMs constitute a major portion of the input data, or if the modeling involves intense and complex computations. But raster-based models are not recommended for studies of travel demand, for example, because travel demand modeling requires the use of a topology-based road network (Chang et al. 2002). Vector-based models are generally recommended for spatial phenomena that involve well-defined locations and shapes.

Third, the distinction between raster-based and vector-based models does not preclude modelers from integrating both types of data in the modeling process. Algorithms for conversion between vector and raster data are easily available in GIS packages. The decision about which data format to use in analysis should be based on the efficiency and the expected result, rather than the format of the original data. For instance, if a vector-based model requires a precipitation layer (e.g., an isohyet layer) as the input, it would be easier to interpolate a precipitation raster from known points and then derive the precipitation layer from the raster.

Fourth, the process of modeling may take place in a GIS or require the linking of a GIS to other computer programs. Many GIS packages including ArcGIS, GRASS, IDRISI, ILWIS, and MFworks have extensive analytical functions for modeling. But a GIS package cannot accommodate statistical analysis as well as a statistical analysis package can, or perform dynamic simulation efficiently. In those cases, the modeler would want to link a GIS to a statistical analysis package or a simulation program.

19.1.4 Integration of GIS and Other Modeling Programs

There are three scenarios for linking a GIS to other computer programs (Corwin et al. 1997). Modelers may encounter all three scenarios in the modeling process, depending on the tasks to be accomplished.

A **loose coupling** involves transfer of data files between the GIS and other programs. For example, one can export data to be run in a statistical analysis package from the GIS and import results from statistical analysis back to the GIS for data visualization or display. Under this scenario, the modeler must create and manipulate data files to be exported or imported unless the interface has already been established between the GIS and the target program. A **tight coupling** gives the GIS and other programs a common user interface. For instance, the GIS can have a menu selection to run a simulation program on soil erosion. An **embedded system** bundles the GIS and other programs with shared memory and a common interface. The Geostatistical Analyst extension to ArcGIS is an example of having geostatistical functions embedded into a GIS environment. The other option is to embed selected GIS functions into a statistical analysis environment (Zhang and Griffith 2000).

Among the four types of GIS models included in Chapter 19, regression and process models usually require the coupling of a GIS with other programs. Binary and index models, on the other hand, can be built entirely in a GIS.

19.2 BINARY MODELS

A **binary model** uses logical expressions to select spatial features from a composite feature layer or multiple rasters. The output of a binary model is in binary format: 1 (true) for spatial features that meet the selection criteria and 0 (false) for features that do not. We may consider a binary model an extension of data query.

A vector-based binary model requires that overlay operations be performed to combine geometries and attributes to be used in data query into a composite feature layer (Figure 19.1). In contrast, a raster-based binary model can be derived directly from querying multiple rasters, with each raster representing a criterion (Figure 19.2).

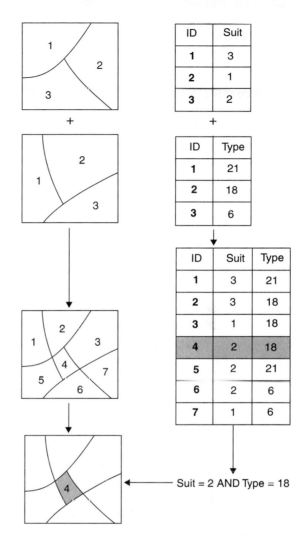

Figure 19.1
To build a vector-based binary model, first overlay the layers so that their spatial features and attributes (Suit and Type) are combined. Then, use the query statement, Suit = 2 AND Type = 18, to select Polygon 4 and save it to the output layer.

The binary model has many uses. A simple application is change detection. We can develop a binary model on land-use change by overlaying two land-use layers and querying the attribute data of the composite layer to find where, for example, forested land has been converted to housing development.

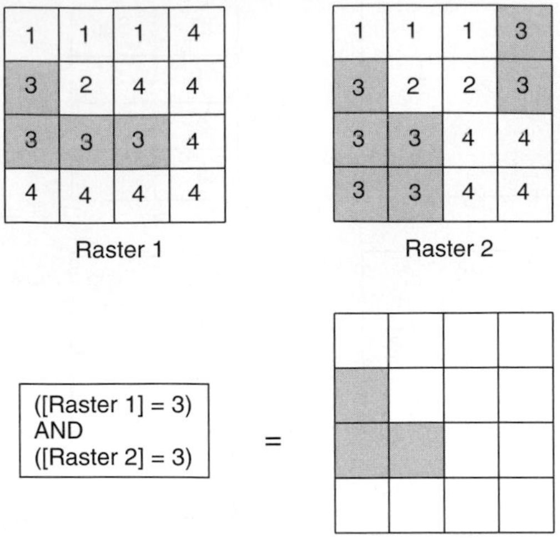

1	1	1	4
3	2	4	4
3	3	3	4
4	4	4	4

Raster 1

1	1	1	3
3	2	2	3
3	3	4	4
3	3	4	4

Raster 2

([Raster 1] = 3) AND ([Raster 2] = 3) =

Figure 19.2
To build a raster-based binary model, use the query statement, [Raster 1] = 3 AND [Raster 2] = 3, to select three cells (shaded) and save them to the output raster.

This type of operation can also be performed with raster data by querying two rasters for a specific change or by creating unique combinations of land uses from two rasters.

Siting analysis is probably the most common application of the binary model. A siting analysis determines if a unit area (i.e., a polygon or a cell) meets a set of selection criteria for locating a landfill, a ski resort, or a university campus. There are at least two approaches to running a siting analysis. One is to evaluate a set of nominated or preselected sites, and the other is to evaluate all potential sites. Although the two approaches may use different sets of selection criteria (e.g., more stringent criteria for evaluating preselected sites), they follow the same procedure for evaluation.

Suppose a county government wants to select potential industrial sites that meet the following criteria:

- At least 5 acres in size
- Commercial zones
- Vacant or for sale
- Not subject to flooding

- Not more than 1 mile from a heavy-duty road
- Less than 10 percent slope

Operationally, the task involves the following four steps:

1. Gather all digital data sets relevant to the selection criteria. Some of these data sets may require preprocessing. For instance, we may derive a commercial zone layer from a land-use layer. (The alternative is to use the land-use layer directly and to query the commercial zone later.)
2. Create a 1-mile buffer zone layer by buffering heavy-duty roads.
3. Perform Intersect operations to combine the road buffer zone layer and other layers.
4. Query the composite feature layer to find which parcels are potential industrial sites.

The above procedure uses Intersect, instead of other overlay operations, because Intersect can limit the output to only those parcels that meet the criteria. The procedure also uses well-defined or "crisp" threshold values to remove land from consideration. The road buffer is exactly 1 mile and the minimum parcel size is exactly 5 acres. These threshold values automatically exclude parcels that are more than 1 mile from heavy-duty roads or are smaller than 5 acres.

Government programs such as the Conservation Reserve Program (CRP) (Box 19.1) are often characterized by their detailed and explicit guidelines. Crisp threshold values simplify the process of siting analysis. But they can also become too restrictive or arbitrary in real-world applications. In an example cited by Steiner (1983), local residents questioned whether any land could meet the criteria of a county comprehensive plan on rural housing. To mitigate the suspicion, a study was made to show that there were indeed sites available. An alternative to crisp threshold values is to use fuzzy sets (Hall et al. 1992; Hall and Arnberg 2002; Stoms et al. 2002). Fuzzy sets do not use sharp boundaries. Rather than being placed in a class (e.g., outside the road buffer), a unit area is associated with a group of membership grades,

Box 19.1 | **The Conservation Reserve Program**

The Conservation Reserve Program (CRP) is a voluntary program administered by the Farm Service Agency (FSA) of the U.S. Department of Agriculture (**http://www.fsa.usda.gov/dafp/cepd/crp.htm**). Its main goal is to reduce soil erosion on marginal croplands (Osborne 1993). Land eligible to be placed in the CRP includes cropland that is planted to an agricultural commodity during two of the five most recent crop years. Additionally, the cropland must meet the following criteria:

- Have an erosion index of 8 or higher or be considered highly erodible land
- Be considered a cropped wetland

- Be devoted to any of a number of highly beneficial environmental practices, such as filter strips, riparian buffers, grass waterways, shelter belts, wellhead protection areas, and other similar practices
- Be subject to scour erosion
- Be located in a national or state CRP conservation priority area, or be cropland associated with or surrounding noncropped wetlands

The difficult part of implementing the CRP in a GIS is putting together the necessary map layers, unless they are already available in a statewide database (Wu et al. 2002).

which suggest the extents to which the unit area belongs to different classes. Therefore, fuzziness is a way to handle uncertainty and complexity in multicriteria evaluation.

19.3 INDEX MODELS

An **index model** calculates the index value for each unit area and produces a ranked map based on the index values. An index model is similar to a binary model in that both involve multicriteria evaluation and both depend on overlay operations for data processing. But an index model produces for each unit area an index value rather than a simple yes or no.

19.3.1 The Weighted Linear Combination Method

The primary consideration in developing an index model, either vector- or raster-based, is the method for computing the index value. The **weighted linear combination** method is probably the most common method for computing the index value (Saaty 1980; Banai-Kashani 1989; Malczewski 2000). Following the analytic hierarchy process

proposed by Saaty (1980), the method of weighted linear combination involves evaluation at three levels (Figure 19.3).

First, the relative importance of each criterion, or factor, is evaluated against other criteria. Many studies have used expert-derived paired comparison for evaluating criteria (Saaty 1980; Banai-Kashani 1989; Pereira and Duckstein 1993; Jiang and Eastman 2000; Basnet et al. 2001). This method involves performing ratio estimates for each pair of criteria. For instance, if criterion A is considered to be three times more important than criterion B, then 3 is recorded for A/B and 1/3 is recorded for B/A. Using a criterion matrix of ratio estimates and their reciprocals as the input, the paired comparison method derives a weight for each criterion. The criterion weights are expressed in percentages, with the total equaling 100 percent or 1.0. Paired comparison is available in various software packages (e.g., Expert Choice).

Second, data for each criterion are standardized. A common method for data standardization is linear transformation. For example, the following formula can convert interval data into a standardized scale of 0.0 to 1.0:

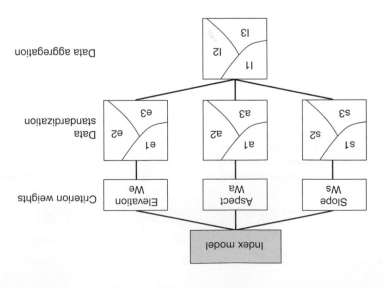

Figure 19.3

To build an index model with the selection criteria of slope, aspect, and elevation, the weighted linear combination method involves evaluation at three levels. The first level of evaluation determines the criterion weights (e.g., Ws for slope). The second level of evaluation determines standardized values for each criterion (e.g., s1, s2, and s3 for slope). The third level of evaluation determines the index (aggregate) value for each unit area.

$$S_i = \frac{X_i - X_{min}}{X_{max} - X_{min}}$$

(19.1)

where S_i is the standardized value for the original value X_i, X_{min} is the lowest original value, and X_{max} is the highest original value. We cannot use Eq. (19.1) if the original data are nominal or ordinal data. In those cases, a ranking procedure based on expertise and knowledge can convert the data into a standardized range such as 0 to 1, 1 to 5, or 0 to 100.

Third, the index value is calculated for each unit area by summing the weighted criterion values and dividing the sum by the total of the weights:

$$I = \frac{\sum_{i=1}^{n} w_i x_i}{\sum_{i=1}^{n} w_i}$$

(19.2)

where I is the index value, n is the number of crite-ria, w_i is the weight for criterion i, and x_i is the standardized value for criterion i.

Figure 19.4 shows the procedure for develop-ing a vector-based index model, and Figure 19.5 shows a raster-based index model. As long as cri-terion weighting and data standardization are well defined, it is not difficult to use the weighted lin-ear combination method to build an index model in a GIS. But we must document standardized val-ues and criterion weights in detail. User interface to simplify the process of building an index model is therefore welcome (Box 19.2).

19.3.2 Other Methods

There are many alternatives to the weighted linear combination method. These alternatives mainly deal with the issues of independence of factors, criterion weights, data aggregation, and data standardization. Weighted linear combination cannot deal with interdependence between factors (Hopkins 1977). A land suitability model may include soils and slope in a linear equation and treat them as independent factors. But in reality soils and slope are dependent of one another. One solution to the interdependence problem is to use a nonlinear function and express

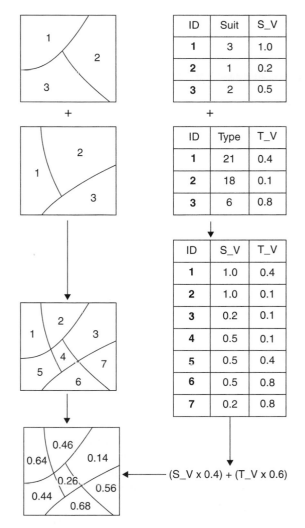

Figure 19.4

Building a vector-based index model requires several steps. First, standardize the Suit and Type values of the input layers into a scale of 0.0 to 1.0. Second, overlay the layers. Third, assign a weight of 0.4 to the layer with Suit and a weight of 0.6 to the layer with Type. Finally, calculate the index value for each polygon in the output by summing the weighted criterion values. For example, Polygon 4 has an index value of 0.26 ($0.5 \times 0.4 + 0.1 \times 0.6$).

the relationship between factors mathematically. But a nonlinear function is usually limited to two factors rather than multiple factors as required in an index model. Another solution is the rule of combi-

nation method proposed by Hopkins (1977). Using rules of combination, we would assign suitability values to sets of combinations of environmental factors and express them through verbal logic instead of numeric terms. The rule of combination method has been widely adopted in land suitability studies (e.g., Steiner 1983), but the method can become unwieldy given a large variety of criteria and data types (Pereira and Duckstein 1993).

Paired comparison for determining criterion weights is sometimes called direct assessment. An alternative to direct assessment is trade-off weighting (Hobbs and Meier 1994; Xiang 2001). Trade-off weighting determines the criterion weights by asking participants to state how much of one criterion they are willing to give up to obtain a given improvement in another criterion. In other words, trade-off weighting is based on the degree of compromise one is willing to make between two criteria when an ideal combination of the two criteria is not attainable. Although realistic in some real-world applications, trade-off weighting has shown to be more difficult to understand and use than direct assessment (Hobbs and Meier 1994).

Data aggregation refers to the derivation of the index value. Weighted linear combination calculates the index value by summing the weighted criterion values. One alternative is to skip the computation entirely and assign the lowest value, the highest value, or the most frequent value among the criteria to the index value (Chrisman 2001). Another alternative is the ordered weighted averaging (OWA) operator, which uses ordered weights instead of criterion weights in computing the index value (Yager 1988; Jiang and Eastman 2000). Suppose that a set of weights is {0.6, 0.4}. If the ordered position of the criteria at location 1 is {A, B}, then criterion A has the weight of 0.6 and criterion B has the weight of 0.4. If the ordered position of the criteria at location 2 is {B, A}, then B has the weight of 0.6 and A has the weight of 0.4. Ordered weighted averaging is therefore more flexible than weighted linear combination. Flexibility in data aggregation can also be achieved by use of fuzzy sets, which allow a unit area to belong to different classes with membership grades (Hall et al. 1992; Stoms et al. 2002).

Data standardization converts the values of each criterion into a standardized scale. A common method is linear transformation as shown in Eq. (19.1), but there are other quantitative methods. In their habitat evaluation study, Pereira and Duckstein (1993) use expert-derived value functions for data standardization, with a specific, often nonlinear, function for each criterion. Jiang and Eastman (2000) use a fuzzy set membership function to transform data into fuzzy measures for their analysis of industrial allocation.

Criterion weights, data aggregation, and data standardization are the same issues that a spatial decision support system (SDSS) must deal with. Designed to work with spatial data, an SDSS can assist the decision maker in making a choice from a set of alternatives according to given evaluation criteria (Densham 1991; Jankowski 1995; Malczewski 1999). Although its emphasis is on decision making, an SDSS shares the same methodology as an index model. An index model developer can therefore benefit from becoming familiar with the SDSS literature.

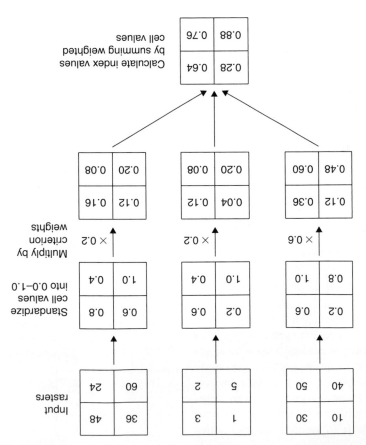

Figure 19.5
Building a raster-based index model requires the following steps. First, standardize the cell values of each input raster into a scale of 0.0 to 1.0. Second, multiply each input raster by its criterion weight. Finally, calculate the index values in the output raster by summing the weighted criterion values. For example, the index value of 0.28 is calculated by: 0.12 + 0.04 + 0.12.

Box 19.2 | **Geoprocessing Tools in ArcGIS 9.0**

ArcGIS 9.0 has combined Buffer Wizard and Geoprocessing Wizard in ArcGIS 8.x with additional tools into ArcToolbox. No longer a separate application, ArcToolbox is now available in both ArcCatalog and ArcMap. Tools in ArcToolbox can be used in several different ways. The simplest way is to use dialogs. Other methods include command lines, models, and scripts. The applications section of Chapter 19 covers all different methods of using tools for building binary and index models.

ModelBuilder, which was first introduced in ArcView 3.2, is back in ArcGIS 9.0. ModelBuilder is no longer just a tool for building a raster-based index model but can work with raster-based as well as vector-based tools. The user interface, however, remains the same. We can build a model diagram by stringing together a series of inputs, tools, and outputs. Once a model is built, it can be saved and reused with different model parameters or inputs.

19.3.3 Applications of the Index Model

Index models are commonly used for suitability analysis and vulnerability analysis. Here we will look at six examples from different disciplines.

Example 1 In 1981 the Natural Resources Conservation Service (NRCS), known then as the Soil Conservation Service, proposed the Land Evaluation and Site Assessment (LESA) system, a tool intended to be used by state and local planners in determining the conditions that justify conversion of agricultural land to other uses (Wright et al. 1983). LESA consists of two sets of factors: LE measures the inherent soil-based qualities of land for agricultural use, and SA measures demands for nonagricultural uses. Many state and local governments have developed, or are in the process of developing, their own LESA models. As an example, the California LESA model completed in 1997 **(http://www.consrv.ca.gov/DLRP/qh_lesa.htm)** uses the following factors and factor weights (in parentheses):

1. Land evaluation factors
   - Land capability classification (25 percent)
   - Storie index rating (25 percent)
2. Site assessment factors
   - Project size (15 percent)
   - Water resource availability (15 percent)

- Surrounding agricultural lands (15 percent)
- Surrounding protected resource lands (5 percent)

For a given location, one would first rate (standardize) each of the factors on a 100-point scale and then sum the weighted factor scores to derive a single index value.

Example 2 The U.S. Environmental Protection Agency developed the DRASTIC model for evaluating groundwater pollution potential (Aller et al. 1987). The acronym DRASTIC stands for the seven factors used in weighted linear combination: Depth to water, net Recharge, Aquifer media, Soil media, Topography, Impact of the vadose zone, and hydraulic Conductivity. The use of DRASTIC involves rating each parameter, multiplying the rating by a weight, and summing the total score by:

(19.3)

$$\text{Total Score} = \sum_{i=1}^{7} W_i P_i$$

where P_i is factor (input parameter) i and W_i is the weight applied to P_i. A critical review of the DRASTIC model in terms of the selection of factors and the interpretation of numeric scores and weights is available in Merchant (1994).

Example 3 Lathrop and Bognar (1998) prioritize areas for conservation protection in Sterling Forest on the New York–New Jersey border by using the following five factors:

- Development limitations due to soil conditions/steep slopes/flooding
- Nonpoint source pollution potential due to proximity to water/wetlands
- Habitat fragmentation potential due to distance from existing roads and development
- Sensitive wildlife habitat areas
- Aesthetic impact (visibility) from the Appalachian and Sterling Ridge Trails)

After making a raster for each factor, they rank the environmental cost/development constraints for each raster from 1 to 5, with 1 for very slight and 5 for very severe. They then assign the maximum value of any of the five factors to each cell in the output raster. Cells with low values are suitable for conservation protection.

Example 4 Habitat Suitability Index (HSI) models typically evaluate habitat quality by using weighted linear combination and factors considered to be important to the wildlife species (Brooks 1997). The following is an HSI model for pine marten developed by Kliskey et al. (1999):

$$
\text{HSI} = \text{sqrt} \left([(3SR_{BSZ} + SR_{SC} + SR_{DS})/6] \right.
$$
$$
\left. [(SR_{CC} + SR_{SS}) / 2] \right)
$$
(19.4)

where SR_{BSZ}, SR_{SC}, SR_{DS}, SR_{CC}, and SR_{SS} are the ratings for biogeoclimatic zone, site class, dominant species, canopy closure, and seral stage, respectively. The model is scaled so that the HSI values range from 0 for unsuitable habitat to 1 for optimal habitat.

Example 5 Chuvieco and Congalton (1989) construct a forest fire hazard index model for a study area in the Mediterranean coast of Spain by using the following five factors:

- Vegetation species, classified according to fuel class, stand conditions, and site (v)
- Elevation (e)
- Slope (s)
- Aspect (a)
- Proximity to roads and trails, campsites, or housing (r)

They standardize the values of each factor by using 0, 1, and 2 for high, medium, and low fire hazard, respectively. They then use the following weighted linear equation to compute the hazard index H:

$$
H = 1 + 100v + 30s + 10a + 5r + 2e
$$
(19.5)

To make the forest fire hazard index model more operational, a follow-up study includes new variables such as weather data, more objective criteria for weighting variables, and a new scheme for integrating the variables (Chuvieco and Salas 1996).

Example 6 Finco and Hepner (1999) build a model of human vulnerability to chemical accident by including the following demographic factors:

- Total population
- Number of people younger than 18 years of age and older than 65 years
- Economic status as measured by household income
- Proximity to sensitive institutions such as schools, hospitals, and health clinics

They rate (standardize) each factor from 0 to 10, with 10 being most vulnerable. The total population and sensitive population factors are positively related to vulnerability, whereas the economic status factor is inversely related to vulnerability. They assume 100 meters as the zone of influence for sensitive institutions. They then combine the four factors into a single measure of vulnerability.

19.4 REGRESSION MODELS

A **regression model** relates a dependent variable to a number of independent (explanatory) variables in an equation, which can then be used for prediction or estimation (Rogerson 2001). Like an index model, a regression model can use overlay operations in a GIS to combine variables needed for the analysis. There are two types of regression model: linear regression and logistic regression. Some GIS packages are capable of performing linear or logistic regression analysis. Both ArcGIS and IDRISI have commands to build raster-based linear or logistic models. GRASS has a command to build linear regression models. Unlike statistical analysis packages, these GIS commands do not offer choices of methods for running linear or logistic regression analysis.

19.4.1 Linear Regression Models

A multiple linear regression model is defined by:

(19.6)

$$y = a + b_1x_1 + b_2x_2 + \cdots + b_nx_n$$

where y is the dependent variable, x_i is the independent variable i, and $b_1, \ldots, b_n$ are the regression coefficients. All variables in the equation are numeric variables. They can also be the transformation of some variables. Common transformations include square, square root, and logarithmic.

The primary purpose of linear regression is to predict values of y from values of x_i. But linear regression requires several assumptions about the error, or residual, between the predicted value and the actual value (Miles and Shevlin 2001):

- The errors have a normal distribution for each set of values of the independent variables.
- The errors have the expected (mean) value of zero.
- The variance of the errors is constant for all values of the independent variables.
- The errors are independent of one another.

An additional assumption in the case of multiple linear regression is that the correlation among the independent variables should not be high.

Numerous linear regression models are available in the literature. As an example, Chang and Li (2000) use linear regression to model snow accu-

mulation: Snow water equivalent (SWE) is the dependent variable and location and topographic variables derived from a DEM are the independent variables. One of their watershed-level models is expressed as:

(19.7)

$$\text{SWE} = a + b_1 \text{ EASTING} + b_2 \text{ SOUTHING} + b_3 \text{ ELEV}$$

where a, b_1, b_2, and b_3 are the regression coefficients, EASTING is the column number of a cell, SOUTHING is the row number of a cell, and ELEV is the elevation value of a cell.

19.4.2 Logistic Regression Models

Logistic regression is used when the dependent variable is categorical (e.g., presence or absence) and the independent variables are categorical, numeric, or both (Menard 2002). Although having the same form as linear regression, logistic regression uses the logit of y as the dependent variable:

(19.8)

$$\text{logit}(y) = a + b_1x_1 + b_2x_2 + b_3x_3 + \cdots$$

The logit of y is the natural logarithm of the odds (also called odds ratio):

(19.9)

$$\text{logit}(y) = \ln(p/(1 - p))$$

where ln is the natural logarithm, $p/(1 - p)$ is the odds, and p is the probability of the occurrence of y. To convert logit (y) back to the odds or the probability, Eq. (19.9) can be rewritten as:

(19.10)

$$p/(1 - p) = e^{(a + b_1x_1 + b_2x_2 + b_3x_3 + \cdots)}$$

(19.11)

$$p = e^{(a + b_1x_1 + b_2x_2 + b_3x_3 + \cdots)} / [1 + e^{(a + b_1x_1 + b_2x_2 + b_3x_3 + \cdots)}]$$

or,

19.5 PROCESS MODELS

A **process model** integrates existing knowledge about the environmental processes in the real world into a set of relationships and equations for quantifying the processes (Beck et al. 1993). Modules or submodels are often needed to cover different components of a process model. Some of these modules may use mathematical equations derived from empirical data, whereas others may use equations derived from physics. A process model offers both a predictive capability and an explanation that is inherent in the proposed processes (Hardisty et al. 1993). Therefore, process models are by definition predictive and dynamic models.

Examples of process models covered in Section 19.5 all fall under the category of environmental models. Environmental models are typically process models because they must deal with the interaction of many variables including physical variables such as climate, topography, vegetation, and soils as well as cultural variables such as land management. As to be expected, environmental models are complex and data-intensive and usually face issues of uncertainty to a greater extent than traditional natural science or social science models (Couclelis 2002).

19.5.1 Soil Erosion Models

Soil erosion is an environmental process that involves climate, soil properties, topography, soil surface conditions, and human activities. A well-known model of soil erosion is the Revised Universal Soil Loss Equation (RUSLE), the updated version of the Universal Soil Loss Equation (USLE) (Wischmeier and Smith 1965, 1978; Renard et al. 1997). RUSLE predicts the average soil loss carried by runoff from specific field slopes in specified cropping and management systems and from rangeland.

RUSLE is a multiplicative model with six factors:

$$A = R\,K\,L\,S\,C\,P \tag{19.15}$$

$$p = 1/[1 + e^{-(a + b_1x_1 + b_2x_2 + b_3x_3 + \ldots)}] \tag{19.12}$$

where e is the exponent.

The main advantage of using logistic regression is that it does not require the assumptions needed for linear regression. Many logistic regression models are also available in the literature. A habitat suitability model for red squirrel developed by Pereira and Itami (1991) is based on the following model:

$$\text{logit }(y) = 0.002 \text{ elevation} - 0.228 \text{ slope} + 0.685 \text{ canopy1} + 0.443 \text{ canopy2} + 0.481 \text{ canopy3} + 0.009 \text{ aspectE–W} \tag{19.13}$$

where canopy1, canopy2, and canopy3 represent three categories of canopy.

Another example is a logistic regression model built by Mladenoff et al. (1995) to estimate the amount and spatial distribution of favorable gray wolf habitat. They first create wolf pack areas (presence) and nonpack areas (absence) in polygon coverages. Wolf pack areas are home ranges derived from telemetry data of wolf location points. Nonpack areas are randomly located in the study area at least 10 kilometers from known pack territories. Both pack and nonpack areas are then overlaid with the landscape coverages of human population density, prey density, road density, land cover, and land ownership. Using the composite coverage, stepwise logistic regression analysis converge on the following model:

$$\text{logit }(y) = 6.5988 - 14.6189\,R \tag{19.14}$$

where R is road density. This logistic regression model, which is based on data from Wisconsin, is later applied to a three-state region (Wisconsin, Minnesota, and Michigan) to map the amount and distribution of favorable wolf habitat. The same model is also used in a subsequent study to predict habitat suitable for wolves in the Northeast (Mladenoff and Sickley 1998).

Box 19.3 Explanation of the Six Factors in RUSLE

The rainfall–runoff erosivity factor (R) quantifies the effect of raindrop impact and also measures the amount and rate of runoff likely to be associated with the rain. In the United States, the R factor can be interpolated from the isoerodent map compiled by RUSLE developers from analysis of rainfall data. Separate equations are available for calculating R for cropland in the Northwestern wheat and range region and other special areas.

The soil erodibility factor (K) measures change in the soil per unit of applied external force or energy. Specifically, it is the rate of soil loss per rainfall erosion index unit as measured on an experimental plot, which has a length of 72.6 feet and has a slope of 9 percent. RUSLE developers recommend that the K factor be obtained from direct measurements on natural runoff plots. But studies have shown that K factors can be estimated from soil and soil profile parameters. In the United States, the K factor is available in the Soil Survey Geographic (SSURGO) database.

The topographic factor (LS) represents the ratio of soil loss on a given slope length (L) and steepness (S) to soil loss from a slope that has a length of 72.6 feet and a steepness of 9 percent. RUSLE developers recommend that L be measured from samples taken in the field. The procedure for converting L and S into the LS factor varies, depending on whether the slope is uniform, irregular, or segmented.

The crop management factor (C) represents the effect of cropping and management practices on soil erosion. It uses an area under clean-tilled continuous-fallow conditions as a standard to measure soil loss under actual conditions. The calculation of C considers the change of crop and soil parameters over time as well as the changing climate variables.

The support practice factor (P) is the ratio of soil loss with a specific support practice to the corresponding loss with upslope and downslope tillage. The support practices for cultivated land include contouring, strip cropping, terracing, and subsurface drainage.

---

where A is the average soil loss, R is the rainfall–runoff erosivity factor, K is the soil erodibility factor, L is the slope length factor, S is the slope steepness factor, C is the crop management factor, and P is the support practice factor. Box 19.3 includes a more detailed description of the six factors in RUSLE.

To use RUSLE, we must first understand the basic structure of the model:

- The dimension of A is expressed in ton per acre per year ($ton.acre^{-1}.yr^{-1}$) in the United States, which is determined by the dimensions of R and K.
- L and S are combined into a single topographic factor LS.
- LS, C, and P are dimensionless, and their values represent deviations from a standard derived by RUSLE developers from experimental data.

- RUSLE considers the interactions between the factors. For example, both C and P are affected by the rainfall condition.
- RUSLE can compute a soil loss value for each time period over which the crop and soil parameters as well as the climate variables are assumed to remain constant.
- USLE was originally developed from analyses of field data and precipitation records for cropland east of the Rocky Mountains. RUSLE has incorporated modifications of the original equations for areas in the West. RUSLE has also included estimation of the K factor using regression analysis with soil and soil profile data.

Among the six factors in RUSLE, the slope length factor L poses more questions than other factors (Renard et al. 1997). Slope length is defined as the horizontal distance from the point of origin of overland flow to the point where either the slope gradient decreases enough that deposition begins or the flow is concentrated in a defined channel (Wischmeier and Smith 1978). RUSLE developers recommend that slope length be measured from samples taken in the field. But it is a problem to measure slope length in an area of complex and irregular topography (Desmet and Govers 1996b), or to select "representative" transects in a farm field (Busacca et al. 1993).

Moore and Burch (1986) have proposed a method for estimating LS. Based on the unit stream power theory, the method uses the equation:

(19.16)

$$LS = (A_s/22.13)^m (\sin \beta/0.0896)^n$$

where A_s is the upslope contributing area, β is the slope angle, m is the slope length exponent, and n is the slope steepness exponent. The exponents m and n are estimated to be 0.6 and 1.3, respectively. When implemented in a raster-based GIS, the LS factor for each cell can be calculated from the slope and the catchment area of the cell (Moore and Wilson 1992; Moore et al. 1993; Moore and Wilson 1994; Gertner et al. 2002).

RUSLE developers, however, recommend that the L and S components be separated in the computational procedure for the LS factor (Foster 1994; Renard et al. 1997). The equation for L is:

(19.17)

$$L = (\lambda/72.6)^m$$

where λ is the measured slope length, and m is the slope length exponent. The exponent m is calculated by:

$$m = \beta/(1 + \beta)$$

$$\beta = (\sin \theta/0.0896) / [3.0 (\sin \theta)^{0.8} + 0.56]$$

where β is the ratio of rill erosion (caused by flow) to interrill erosion (principally caused by raindrop impact), and θ is the slope angle. The equation for S is:

(19.18)

$$S = 10.8 \sin \theta + 0.03, \text{ for slopes of less than 9\%}$$

$$S = 16.8 \sin \theta - 0.50, \text{ for slopes of 9\% or steeper}$$

Both L and S also need to be adjusted for special conditions, such as the adjustment of the slope length exponent m for the erosion of thawing, cultivated soils by surface flow, and the use of a different equation than Eq. (19.17) for slopes shorter than 15 feet.

There are at least two methods that follow the same procedure and equations as proposed by RUSLE developers but still use a GIS to automate the estimation of L and S. One method proposed by Hickey et al. (1994) uses DEMs as the input and estimates for each cell the slope steepness by computing the maximum downhill slope and the slope length by iteratively calculating the cumulative downhill slope length. Another method proposed by Desmet and Govers (1996a) also uses DEMs as the input and estimates for each cell the slope steepness by using a quadratic surface-fitting method and the slope length by calculating the unit-contributing area.

USLE and RUSLE have evolved over the past 50 years. This soil erosion model has gone through numerous cycles of model development, calibration, and validation. The process continues. A new, updated model called WEPP (Water Erosion Prediction Project) is expected to replace RUSLE (Laflen et al. 1991; Laflen et al. 1997). In addition to modeling soil erosion on hillslopes, WEPP can model the deposition in the channel system. Integration of WEPP and GIS is desirable because a GIS can be used to extract hillslopes and channels as well as identify the watershed (Cochrane and Flanagan 1999, 2003). A description and download site of WEPP is available at http://topsoil.nserl.purdue.edu/nserlweb/weppmain/wepp.html.

19.5.2 Other Process Models

The AGNPS (Agricultural Nonpoint Source) model analyzes nonpoint source pollution and estimates runoff water quality from agricultural watersheds

(Young et al. 1987). AGNPS is event-based and operates on a cell basis. Using various types of input data, the model simulates runoff, sediment, and nutrient transport. For example, AGNPS can incorporate a modified form of USLE for estimating upland erosion for single storms (Young et al. 1989):

(19.19)

$$SL = (EI) \, K \, LS \, C \, P \, (SSF)$$

where SL is the soil loss, EI is the product of the storm total kinetic energy and maximum 30-minute intensity, K is the soil erodibility, LS is the topographic factor, C is the cultivation factor, P is the supporting practice factor, and SSF is a factor to adjust for slope shape within the cell. Detached sediment calculated from the equation is routed through the cells according to yet another equation based on the characteristics of the watershed.

The SWAT (Soil and Water Assessment Tool) model predicts the impact of land management practices on water quality and quantity, sediment, and agricultural chemical yields in large complex watersheds (Srinivasan and Arnold 1994). SWAT is a process-based continuous simulation model. Inputs to SWAT include land management practices such as crop rotation, irrigation, fertilizer use, and pesticide application rates, as well as the physical characteristics of the basin and subbasins such as precipitation, temperature, soils, vegetation, and topography. The creation of the input data files requires substantial knowledge at the subbasin level. Model outputs include simulated values of surface water flow, groundwater flow, crop growth, sediment, and chemical yields.

Also available in the literature are process models on nonpoint source pollutants in the vadose zone (Corwin et al. 1997), landslides (Montgomery et al. 1998), and groundwater contamination (Loague and Corwin 1998).

19.5.3 GIS and Process Models

Process models are typically raster-based. The role of a GIS in building a process model depends on the complexity of the model. A simple process model may be prepared and run entirely within a GIS. But more often a GIS is delegated to the role of performing modeling-related tasks such as data visualization, database management, and exploratory data analysis. The GIS is then linked to other computer programs for complex and dynamic analysis.

Commercial GIS packages do not offer commands for building process models. GRASS has commands for preparing input variables to AGNPS, running the model, and viewing the model. GRASS also has a command that can produce the LS and S factors in RUSLE. The NRCS has a demonstration project using the Soil Survey Geographic (SSURGO) database to develop SWAT models of five small watersheds in Iowa (**http://waterhome.tamu.edu/NRCSdata/ SWAT_SSURGO**). SWAT is included in the Better Assessment Science Integrating point and Nonpoint Sources (BASINS) system developed by the U.S. Environmental Protection Agency, which can be downloaded at **http://www.epa. gov/waterscience/BASINS/**.

KEY CONCEPTS AND TERMS

Binary model: A GIS model that uses logical expressions to select features from a composite feature layer or multiple rasters.

Deductive model: A model that represents the conclusion derived from a set of premises.

Descriptive model: A model that describes the existing conditions of spatial data.

Deterministic model: A mathematical model that does not involve randomness.

REVIEW QUESTIONS

1. Describe the difference between a descriptive model and a prescriptive model.

2. How does a static model differ from a dynamic model?

3. Describe the steps involved in the modeling process.

4. Suppose you use kriging to build an interpolation model. How do you calibrate the model?

5. In many instances, you can build a GIS model that is either vector-based or raster-based. What general guidelines should you use in deciding which model to build?

6. What does loose coupling mean in the context of linking a GIS to another software package?

7. Why is a binary model often considered an extension of data query?

8. Provide an example of a binary model from your discipline.

9. How does an index model differ from a binary model?

10. Many index models use the weighted linear combination method to calculate the index value. Explain the steps one follows in using the weighted linear combination method.

11. What are the general shortcomings of the weighted linear combination method?

12. Provide an example of an index model from your discipline.

13. What kinds of variables can be used in a logistic regression model?

14. What is an environmental model?

15. Provide an example of a process model from your discipline. Can the model be built entirely in a GIS?

16. Go to the SWAT input web page (http://www.brc.tamus.edu/swat/swatinp.html) and examine the inputs to SWAT. Which input data can be prepared or created in a GIS?

Dynamic model: A model that emphasizes the changes of spatial data over time and the interactions between variables.

Embedded system: A GIS is bundled with other computer programs in a system with shared memory and a common interface.

Index model: A GIS model that uses the index value calculated from a composite feature layer or multiple rasters to produce a layer with ranked data.

Inductive model: A model that represents the conclusion derived from empirical data and observations.

Loose coupling: The process for linking a GIS and other computer programs through the transfer of data files.

Model: A simplified representation of a phenomenon or a system.

Prescriptive model: A model that offers a prediction of what the conditions of spatial data could be or should be.

Process model: A GIS model that integrates existing knowledge into a set of relationships and equations for quantifying the physical processes.

Regression model: A GIS model that uses a dependent variable and a number of independent variables in a regression equation for prediction or estimation.

Static model: A model that deals with the state of spatial data at a given time.

Stochastic model: A mathematical model that considers the presence of some randomness in one or more of its parameters or variables.

Tight coupling: The process for linking a GIS and other computer programs through a common user interface.

Weighted linear combination: A method that computes the index value for each unit area by summing the products of the standardized value and the weight for each criterion.

This applications section covers four tasks. Tasks 1 and 2 let you build binary models using vector data and raster data, respectively. Tasks 3 and 4 let you build index models using vector data and raster data, respectively. Different options for running the geoprocessing operations are covered in this section. Tasks 1 and 4 use the ModelBuilder. Task 2 uses the Raster Calculator in the Spatial Analyst extension. Task 3 uses command lines. In Task 4, you also have a chance to examine a Python script, a text file created by exporting the model.

Task 1: Build a Vector-Based Binary Model

What you need: *elevzone.shp*, an elevation zone shapefile; and *stream.shp*, a stream shapefile.

Task 1 asks you to locate the potential habitats of a plant species. Both *elevzone.shp* and *stream.shp* are measured in meters and spatially registered. The field ZONE in *elevzone.shp* shows three elevation zones. The potential habitats must meet the following criteria: (1) in elevation zone 2 and (2) within 200 meters of streams. You will use the Model-Builder to complete the task.

1. Start ArcCatalog, and make connection to the Chapter 19 database. Launch ArcMap. Rename the data frame Tasks 1&2. Add *stream.shp* and *elevzone.shp* to Tasks 1&2. Open the ArcToolbox window. Set the Chapter 19 database for the current workspace. Because the ModelBuilder is run from a toolbox, you will first create a new toolbox. Right-click ArcToolbox and select New Toolbox. Rename the new toolbox Chap19.

Q1. The context menu of the Chap19 toolbox has the selections of New and Add. Besides model, what other types of tools can the toolbox handle?

2. Right-click Chap19, point to New, and select Model. Rename Model in the Chap19 toolbox Task1. Select Model Properties from the Model menu in the Model window. On

the General tab, change both the name and label to Task1 and click OK.

3. The first step is to buffer *streams* with a buffer distance of 200 meters. Drag the Buffer tool from the Analysis Tools/Proximity toolset and drop it in the Model window. Right-click Buffer and select Open. In the Buffer dialog, select *stream* from the dropdown list for the input features, name the output feature class *strmbuf.shp*, enter 200 (meters) for the distance, and select ALL for the dissolve type. Click OK.

4. The visual objects in the Model window are color-coded. The input is coded blue, the tool gold, and the output green. The model can be executed one tool (function) at a time or as an entire model. Run the Buffer tool first. Right-click Buffer and select Run. The tool turns red during processing. After processing, both the tool and the output have the added drop shadow. Right-click *strmbuf.shp*, and select Add to Display.

5. Next overlay *elevzone* and *strmbuf*. Drag the Intersect tool from the Analysis Tools/Overlay toolset and drop it in the Model window. Right-click Intersect and select Open. In the Intersect dialog, select *strmbuf.shp* and *elevzone* from the dropdown list for the input features, name the output feature class *pothab.shp*, and click OK.

6. Right-click Intersect and select Run. After the overlay operation is done, right-click *pothab.shp* and add it to display. Turn off all layers except *pothab* in ArcMap's table of contents.

7. The final step is to select areas from *pothab* that are in elevation zone 2. Drag the Select tool from the Analysis Tools/Extract toolset and drop it in the Model window. Right-click Select and select Open. In the Select dialog, select *pothab.shp* for the input features, name the output feature class *final.shp*, and

click the SQL button for the expression. Enter the following SQL statement in the expression box: "ZONE" = 2. Click OK to dismiss the dialogs. Right-click Select and select Run. Add *final.shp* to display.

8. Select Auto Layout from the Model window's View menu and let the ModelBuilder rearrange the model diagram. Finally, select Save from the Model menu before closing the window. To run the Task1 model next time, right-click Task1 in the Chap19 toolbox and select Edit.

Task 2: Build a Raster-Based Binary Model

What you need: *elevzone_gd*, an elevation zone grid; and *stream_gd*, a stream grid.

Task 2 tackles the same problem as Task 1 but uses raster data. Both *elevzone_gd* and *stream_gd* have the cell resolution of 30 meters. The cell value in *elevzone_gd* corresponds to the elevation zone. The cell value in *stream_gd* corresponds to the stream ID. You will use the Raster Calculator in Spatial Analyst for Task 2, because it is more efficient than other geoprocessing options.

1. Add *stream_gd* and *elevzone_gd* to Tasks 1&2. Make sure that the Spatial Analyst extension is selected and its toolbar is checked. The first step is to create continuous distance measures from *stream_gd*. Click the Spatial Analyst dropdown arrow, point to Distance, and select Straight Line. Make sure that *stream_gd* is the raster to calculate the distance to, enter 30 (meters) for the output cell size, and opt for a temporary output raster. Click OK. *Distance to stream_gd* is the temporary output raster.

2. Now you can query *elevzone_gd* and *Distance to stream_gd* to locate the potential habitats. Select Raster Calculator from the Spatial Analyst dropdown list. Enter the following expression in the expression box: [Distance to stream_gd] <= 200 AND [elevzone_gd] = 2. Click Evaluate.

3. *Calculation* shows the potential habitats with the value of 1. Compare *Calculation* with *final* from Task 1. They should cover the same areas.

Q2. What tool in ArcToolbox can you use to complete Step 1 of Task 2?

Task 3: Build a Vector-Based Index Model

What you need: *soil.shp*, a soil shapefile; *landuse.shp*, a land-use shapefile; and *depwater.shp*, a depth to water shapefile.

Task 3 simulates a project on mapping groundwater vulnerability. The project assumes that groundwater vulnerability is related to three factors: soil characteristics, depth to water, and land use. Each factor has been rated on a standard-ized scale from 0 to 5. These standardized values are stored in SOILRATE in *soil.shp*, DWRATE in *depwater.shp*, and LURATE in *landuse.shp*. The score of 9.9 is assigned to areas such as urban and built-up areas in *landuse.shp*, which should not be included in the model. The project also assumes that the soil factor is more important than the other two factors and is therefore assigned a weight of 0.6 (60%), compared to 0.2 (20%) for the other two factors. The index model can be expressed as Index value = 0.6 × SOILRATE + 0.2 × LURATE + 0.2 × DWRATE. In Task 3, you will use a geodatabase, command lines, and attribute data analysis to create the index model.

1. First create a new personal geodatabase and import the three input shapefiles as feature classes to the geodatabase. In ArcCatalog, right-click the Chapter 19 database, point to New, and select Personal Geodatabase. Rename the geodatabase *Task3.mdb*. Right-click *Task3.mdb*, point to Import, and select Feature Class (multiple). Use the browser in the next dialog to select *soil.shp*, *landuse.shp*, and *depwater.shp* for the input features. Make sure that *Task3.mdb* is the output geodatabase. Click OK to run the import operation.

2. The main part of Task 3 is to overlay all three feature classes. You will use command

lines to perform the overlay operations. Click Show/Hide Command Line Window on ArcCatalog's standard toolbar to open the Command Line window. Several features about use of command lines should be mentioned. One, you can type the command (i.e., tool) or drag and drop the command from ArcToolbox to the Command Line window. Two, a command uses a number of arguments. The Command Line window shows the command syntax; it also highlights the argument to be entered following a space. But the window does not explain the function of each argument. To get the explanation, you can use the ArcGIS Desktop Help document and look for the command line syntax. Three, you can take the default for an argument by entering #. Four, the Command Line window allows you to run one command at a time or several commands together. You can enter two or more command lines by using Ctrl-Enter, and press Enter to execute the commands.

3. Enter the following two command lines in the Command Line window:
workspace c:\Chapter19\Task3.mdb
Intersect_analysis depwater;landuse;soil vulner ALL # INPUT
The workspace command assumes that the current workspace is c:\Chapter19\Task3. mdb. If you omit the line, you will get an error message stating that the operation fails to create the overlay output (e.g., *vulner*). The first required argument for the Intersect_ analysis command is the names of the input features, separated by semicolons. The second required argument is the name of the output feature class. The three optional arguments relate to the joining of attributes, the cluster tolerance, and the output type. Press Enter.

Q3. How many feature classes have you specified for the Intersect operation?

4. The remainder of Task 3 consists of attribute data operations. Preview the attribute table of *vulner*. The table has all three rates needed for computing the index value. But you must go through a couple of steps before computation: add a new field for the index value, and exclude areas with the LURATE value of 9.9 from computation.

5. Open the ArcToolbox window in ArcCatalog. Double-click the Add Field Tool in the Data Management Tools/Fields toolset. In the next dialog, select *vulner* for the input table, enter TOTAL for the field name, select DOUBLE for the field type, enter 11 for the field precision, and 3 for the field scale. Click OK. Make sure that TOTAL has been added.

6. Insert a new data frame in ArcMap, and rename it Task 3. Add *soil*, *landuse*, *depwater*, and *vulner* from *Task3.mdb* to Task 3. (You will only be working with *vulner*; the other three feature classes are for reference.) Open the attribute table of *vulner*. TOTAL appears in the table with Nulls. Click the Options dropdown arrow and choose Select By Attributes. Enter the following SQL statement: "LURATE" < > 9.9. Click Apply. Right-click TOTAL and select Calculate Values. Click Yes in the Field Calculator message box. Enter the following expression in the Field Calculator dialog: 0.6 × [SOILRATE] + 0.2 × [LURATE] + 0.2 × [DWRATE]. Click OK to dismiss the dialog. The field TOTAL is populated with the calculated index values. You can assign a TOTAL value of −99 to urban areas. Click the Options dropdown arrow and select Switch Selection. Right-click TOTAL and select Calculate Values. Enter −99 in the expression box, and click OK. At this point, you have completed calculating the index value. Select Clear Selection from the Options dropdown arrow. Close the table.

Q4. Excluding –99 for urban areas, what is the value range of TOTAL?

7. This step is to display the index values of *vulner*. Select Properties from the context

menu of *vulner* in ArcMap. On the Symbology tab, choose Quantities and Graduated colors in the Show box. Click the Value dropdown arrow and select TOTAL. Click Classify. In the Classification dialog, select 6 classes and enter 0, 3.0, 3.5, 4.0, 4.5, and 5.0 as Break Values. Double-click the default symbol for urban areas (range –99–0) in the Layer Properties dialog and change it to a Hollow symbol for areas not analyzed.

8. Once the index value map is made, you can modify the classification so that the grouping of index values may represent a rank order such as very severe (5), severe (4), moderate (3), slight (2), very slight (1), and not applicable (–99). You can then convert the index value map into a ranked map by doing the following: save the rank of each class under a new field called RANK, and then use the Dissolve tool from the Data Management Tools/Generalization toolset to remove boundaries of polygons that fall within the same rank. The ranked map should look much simpler than the index value map.

Task 4: Build a Raster-Based Index Model

What you need: *soil*, a soils raster; *landuse*, a land-use raster; and *depwater*, a depth to water raster.

Task 4 performs the same analysis as Task 3 but uses raster data. All three rasters have the cell resolution of 90 meters. The cell value in *soil* corresponds to SOILRATE, the cell value in *landuse* corresponds to LURATE, and the cell value in *depwater* corresponds to DWRATE. The only difference is that urban areas in *landuse* are already classified as no data. In Task 4, you will use the ModelBuilder to create a model and then export the model to a Python script.

1. Insert a new data frame and rename it Task 4. Add *soil*, *landuse*, and *depwater* to Task 4. If necessary, use Add Toolbox in the context menu of ArcToolbox to add the Chap19 toolbox from My Toolboxes to the menu of ArcToolbox. Right-click Chap19, point to New, and select Model. Rename the new model Task4. Open the Model Properties dialog and, on the General tab, change both the name and label of the model to Task4.

2. Click the Add Data or Tools button, and add *depwater*, *landuse*, and *soil* to the Task4 window. Drag the Single Output Map Algebra tool from the Spatial Analyst Tools/Map Algebra toolset and drop it in the model window. Right-click the Single Output Map Algebra tool and select Properties. On the Preconditions tab, click the Select All button and then OK. The three input rasters are now connected to the tool. (An alternative is to use the Add Connection tool to connect each input raster to the Single Output Map Algebra tool.) Right-click the Single Output Map Algebra tool and select Open. Enter the following Map Algebra expression in the next dialog: $0.6 \times [soil] + 0.2 \times [landuse] + 0.2 \times [depwater]$. Use the browser to specify *vulner* for the output raster. Click OK to dismiss the dialog.

3. Right-click the Single Output Map Algebra tool and select Run. Then right-click *vulner*, and add it to display. In the table of contents of ArcMap, *vulner* shows a value range from 2.904 to 5.0.

4. This step is to reclassify *vulner* so that the output shows 1 (≤ 3.00) for "very slight," 2 (3.01–3.50) for "slight," 3 (3.51–4.00) for "moderate," 4 (4.01–4.50) for "severe," and 5 (4.51–5.00) for "very severe." Drag the Reclassify tool from the Spatial Analyst Tools/Reclass toolset and drop it in the model window. Right-click the Reclassify tool and select Properties. On the Preconditions tab, check *vulner* and click OK. *vulner* is now connected to the Reclassify tool. Right-click the Reclassify tool and select Open. In the Reclassify dialog, select *vulner* for the input raster and then click the Classify button. In the Classification dialog, select 5 classes, enter 3, 3.5, 4, 4.5, and 5 for the break values, and click OK. In

the Reclassify dialog, enter *reclass_vuln* for the output raster and click OK.

5. Right-click the Reclassify tool and select Run. Then right-click *reclass_vuln*, and add it to display. Right-click *reclass_vuln* in the table of contents, and select Properties. On the Symbology tab, change the label of 1 to Very slight, 2 to Slight, 3 to Moderate, 4 to Severe, and 5 to Very severe. Click OK. Now the raster layer is shown with the proper labels.

Q5. What percentage of the study area is labeled "Very severe"?

6. Select Save from the Model menu to save the Task4 model. Then click the Model menu again, point to Export, point to To Script, and select Python. Save the script as *Task4.py* in the Chapter 19 database.

7. Use Notepad or any other word processing software to open *Task4.py*. The script should look as follows:

```
# - - - - - - - - - - - - - - - - - - - - - - - - - -

# Task4.py
# Created on: Thu Jul 22 2004 05:44:58 PM
# (generated by ArcGIS/ModelBuilder)

# - - - - - - - - - - - - - - - - - - - - - - - - - -

# Import system modules
import sys, string, os, win32com.client

# Create the Geoprocessor object
gp = win32com.client.Dispatch
("esriGeoprocessing.GpDispatch.1")
# Check out any necessary licenses
gp.CheckOutExtension("spatial")

# Load required toolboxes...
gp.AddToolbox("C:/Program Files/ArcGIS/
ArcToolbox/Toolboxes/Spatial Analyst
Tools.tbx")

# Local variables...
landuse = "C:/chap19/landuse"
depwater = "C:/chap19/depwater"
soil = "C:/chap19/soil"
vulner = "C:/chap19/vulner"
reclass_vuln = "C:/chap19/reclass_vuln"
```

```
# Process: Single Output Map Algebra...
gp.SingleOutputMapAlgebra_sa("0.6*[soil]
+ 0.2*[landuse] + 0.2*[depwater]", vulner,"")

# Process: Reclassify...
gp.Reclassify_sa(vulner, "Value",
"2.9040000438690186 3 1;3 3.5 2;3.5 4 3;4
4.5 4;4.5 5 5", reclass_vuln, "DATA")
```

Task4.py is a text file that documents every command you have used in Task 4. The line that starts with # is a comment line. The gp object is the geoprocessing object. And sa at the end of each tool name stands for Spatial Analyst. Otherwise, the lines in *Task4.py* are self-explanatory.

8. To run *Task4.py* with changes of the input raster or the computational formula for the index model, you can do the following:
   a. Right-click Chap19 in the ArcToolbox, point to Add, and select Script. The dialogs that follow allow you to specify *Task4.py* for the script.
   b. After Script is added to the Chap19 toolbox, right-click Script and select Edit. The PythonWin opens with *Task4.py* in the window. You can make changes in the command lines and then click the Run tool to run the script.

Challenge Task

What you need: *soil*, *landuse*, and *depwater*, same as Task 4.

The Spatial Analyst Tools/Overlay toolset has a tool called Weighted Overlay, which can overlay integer rasters to produce an index model. This challenge task asks you to use this tool and the ModelBuilder to create a raster-based index model similar to that in Task 4. Here are some tips for completing the challenge task.

1. Because the Weighted Overlay tool only works with integer rasters, you must use the Int tool in the Spatial Analyst Tools/Math toolset to convert *depwater*, *landuse*, and *soil* to integer rasters. The Int tool simply

truncates the decimal digits. For example, both 3.2 and 3.6 are converted to 3.

2. The weighted overlay table has four columns. The Raster column lists the input integer rasters. The % Influence column lists weights associated with each input raster, which must add up to 100 (%). The Field column lists the cell values in each input raster. And the Scale Value column lists the scale values corresponding to the cell values.

Because the cell values for the input rasters have already been scaled from 1 to 5, you just have to make sure that the scale value is the same as the cell value.

Q1. What cell values are in the output raster?

Q2. Describe the difference between the challenge task and Task 4 in terms of the output raster.

REFERENCES

Aller, L., T. Bennett, J. H. Lehr, R. J. Petry, and G. Hackett 1987. DRASTIC: A Standardized System for Evaluating Groundwater Pollution Potential Using Hydrogeologic Settings. U.S. Environmental Protection Agency, EPA/600/2-87/035, pp. 622.

Banai-Kashani, R. 1989. A New Method for Site Suitability Analysis: The Analytic Hierarchy Process. Environmental Management 13: 685-93.

Basnet, B. B., A. A. Apan, and S. R. Raine. 2001. Selecting Suitable Sites for Animal Waste Application Using a Raster GIS. Environmental Management 28: 519-31.

Beck, M. B., A. J. Jakeman, and M. J. McAleer. 1993. Construction and Evaluation of Models of Environmental Systems. In A. J. Jakeman, M. B. Beck, and M. J. McAleer, eds., Modelling Change in Environmental Systems, pp. 3-35. Chichester, England: Wiley.

Brooks, R. P. 1997. Improving Habitat Suitability Index Models. Wildlife Society Bulletin 25: 163-67.

Busacca, A. J., C. A. Cook, and D. J. Mulla. 1993. Comparing Landscape-Scale Estimation of Soil Erosion in the Palouse Using Cs-137 and RUSLE. Journal of Soil and Water Conservation 48: 361-67.

Chang, K., and Z. Li. 2000. Modeling Snow Accumulation with a Geographic Information System. International Journal of Geographical Information Science 14: 693-707.

Chang, K., Z. Khatib, and Y. Ou. 2002. Effects of Zoning Structure and Network Detail on Traffic Demand Modeling. Environment and Planning B 29: 37-52.

Chrisman, N. 2001. Exploring Geographic Information Systems, 2d ed. New York: Wiley.

Chuvieco, E., and R. G. Congalton. 1989. Application of Remote Sensing and Geographic Information Systems to Forest Fire Hazard Mapping. Remote Sensing of the Environment 29: 147-59.

Chuvieco, E., and J. Salas. 1996. Mapping the Spatial Distribution of Forest Fire Danger Using GIS. International Journal of Geographical Information Systems 10: 333-45.

Cochrane, T. A., and D. C. Flanagan. 2003. Representative Hillslope Methods for Applying the WEPP Model with DEMs and GIS. Transactions of the ASAE 46: 1041-49.

Cochrane, T. A., and D. C. Flanagan. 1999. Assessing Water Erosion in Small Watersheds Using WEPP with GIS and Digital Elevation Models. Journal of Soil and Water Conservation 54: 678-85.

Corwin, D. L., P. J. Vaughan, and K. Loague. 1997. Modeling Nonpoint Source Pollutants in the Vadose Zone with GIS. Environmental Science & Technology 31: 2157-75.

Coucelis, H. 2002. Modeling Frameworks, Paradigms and Approaches. In K. C. Clarke, B. O. Parks, and M. P. Crane, eds., Geographic Information Systems and Environmental Modeling, pp. 36-50. Upper Saddle River, NJ: Prentice Hall.

DeMers, M. N. 2002. GIS Modeling in Raster. New York: Wiley.

Densham, P. J. 1991. Spatial Decision Support Systems. In D. J. Maguire, M. F. Goodchild, and D. W. Rhind, eds., *Geographical Information Systems: Principles and Applications*, Vol 1., pp. 403–12. London: Longman.

Desmet, P. J. J., and G. Govers. 1996a. Comparison of Routing Systems for DEMs and Their Implications for Predicting Ephemeral Gullies. *International Journal of Geographical Information Systems* 10: 311–31.

Desmet, P. J. J., and G. Govers. 1996b. A GIS Procedure for Automatically Calculating the USLE LS Factor on Topographically Complex Landscape Units. *Journal of Soil and Water Conservation* 51: 427–33.

Finco, M. V., and G. F. Hepner. 1999. Investigating U.S.–Mexico Border Community Vulnerability to Industrial Hazards: A Simulation Study in Ambos Nogales. *Cartography and Geographic Information Science* 26: 243–52.

Foster, G. R. 1994. Comments on "Length-Slope Factor for the Revised Universal Soil Loss Equation: Simplified Method of Estimation." *Journal of Soil and Water Conservation* 49: 171–73.

Gertner, G., G. Wang, S. Fang, and A. B. Anderson. 2002. Effect and Uncertainty of Digital Elevation Model Spatial Resolutions on Predicting the Topographic Factor for Soil Loss Estimation. *Journal of Soil and Water Conservation* 57: 164–74.

Hall, O., and W. Arnberg. 2002. A Method for Landscape Regionalization Based On Fuzzy Membership Signatures.

Landscape and Urban Planning 59: 227–40.

Hall, G. B., F. Wang, and Subaryono. 1992. Comparison of Boolean and Fuzzy Classification Methods in Land Suitability Analysis by Using Geographical Information Systems. *Environment and Planning A* 24: 497–516.

Hardisty, J., D. M. Taylor, and S. E. Metcalfe. 1993. *Computerized Environmental Modelling*. Chichester, England: Wiley.

Hickey, R., A. Smith, and P. Jankowski. 1994. Slope Length Calculations from a DEM Within ARC/INFO GRID. *Computers, Environment and Urban Systems* 18: 365–80.

Hobbs, B. F., and P. M. Meier. 1994. Multicriteria Methods for Resource Planning: An Experimental Comparison. *IEEE Transactions on Power Systems* 9 (4): 1811–17.

Hopkins, L. D. 1977. Methods for Generating Land Suitability Maps: A Comparative Evaluation. *Journal of the American Institute of Planners* 43: 386–400.

Jankowski, J. 1995. Integrating Geographic Information Systems and Multiple Criteria Decision-Making Methods. *International Journal of Geographical Information Systems* 9: 251–73.

Jiang, H., and J. R. Eastman. 2000. Application of Fuzzy Measures in Multi-Criteria Evaluation in GIS. *International Journal of Geographical Information Science* 14: 173–84.

Kliskey, A. D., E. C. Lofroth, W. A. Thompson, S. Brown, and H. Schreier. 1999. Simulating and Evaluating Alternative Resource-Use Strategies Using GIS-Based

Habitat Suitability Indices. *Landscape and Urban Planning* 45: 163–75.

Laflen, J. M., W. J. Elliot, D. C. Flanagan, C. R. Meyer, and M. A. Nearing. 1997. WEPP— Predicting Water Erosion Using a Process-Based Model. *Journal of Soil and Water Conservation* 52: 96–102.

Laflen, J. M., L. J. Lane, and G. R. Foster. 1991. WEPP: A New Generation of Erosion Prediction Technology. *Journal of Soil and Water Conservation* 46: 34–38.

Lathrop, R. G., Jr., and J. A. Bognar. 1998. Applying GIS and Landscape Ecologic Principles to Evaluate Land Conservation Alternatives. *Landscape and Urban Planning* 41: 27–41.

Loague, K., and D. L. Corwin. 1998. Regional-Scale Assessment of Non-Point Source Groundwater Contamination. *Hydrologic Processes* 12: 957–65.

Malczewski, J. 2000. On the Use of Weighted Linear Combination Method in GIS: Common and Best Practice Approaches. *Transactions in GIS* 4: 5–22.

Malczewski, J. 1999. *GIS and Multicriteria Decision Analysis*. New York: Wiley.

Menard, S. 2002. *Applied Logistic Regression Analysis*, 2d ed. Thousand Oaks, CA: Sage.

Merchant, J. W. 1994. GIS-Based Groundwater Pollution Hazard Assessment: A Critical Review of the DRASTIC Model. *Photogrammetric Engineering & Remote Sensing* 60: 1117–27.

Miles, J., and Shevlin, M. 2001. *Applying Regression & Correlation: A Guide for Students and Researchers*. London: Sage.

Mladenoff, D. J., and T. A. Sickley. 1998. Assessing Potential Gray Wolf Restoration in the Northeastern United States: A Spatial Prediction of Favorable Habitat and Potential Population Levels. *Journal of Wildlife Management* 62: 1–10.

Mladenoff, D. J., T. A. Sickley, R. G. Haight, and A. P. Wydeven. 1995. A Regional Landscape Analysis and Prediction of Favorable Gray Wolf Habitat in the Northern Great Lakes Regions. *Conservation Biology* 9: 279–94.

Montgomery, D. R., K. Sullivan, and H. M. Greenberg. 1998. Regional Test of a Model for Shallow Landsliding. *Hydrologic Processes* 12: 943–55.

Moore, I. D., and G. J. Burch. 1986. Physical Basis of the Length-Slope Factor in the Universal Soil Loss Equation. *Soil Science Society of America Journal* 50: 1294–98.

Moore, I. D., A. K. Turner, J. P. Wilson, S. K. Jenson, and L. E. Band. 1993. GIS and Land-Surface-Subsurface Process Modelling. In M. F. Goodchild, B. O. Park, and L. T. Styaert, eds., *Environmental Modelling with GIS*, pp. 213–30. Oxford, England: Oxford University Press.

Moore, I. D., and J. P. Wilson. 1994. Reply to Comments by Foster on ''Length-Slope Factor for the Revised Universal Soil Loss Equation: Simplified Method of Estimation.'' *Journal of Soil and Water Conservation* 49: 174–80.

Moore, I. D., and J. P. Wilson. 1992. Length-Slope Factor for the Revised Universal Soil Loss Equation: Simplified Method of Estimation. *Journal of Soil and Water Conservation* 47: 423–28.

Osborne, C. T. 1993. CRP Status, Future and Policy Options. *Journal of Soil and Water Conservation* 48: 272–78.

Pereira, J. M. C., and L. Duckstein. 1993. A Multiple Criteria Decision-Making Approach to GIS-Based Land Suitability Evaluation. *International Journal of Geographical Information Systems* 7: 407–24.

Pereira, J. M. C., and R. M. Itami. 1991. GIS-Based Habitat Modeling Using Logistic Multiple Regression: A Study of the Mt. Graham Red Squirrel. *Photogrammetric Engineering & Remote Sensing* 57: 1475–86.

Pequet, D. J. 1994. It's About Time: A Conceptual Framework for the Representation of Temporal Dynamics in Geographic Information Systems. *Annals of the Association of American Geographers* 84: 441–61.

Renard, K. G., G. R. Foster, G. A. Weesies, D. K. McCool, and D. C. Yoder, coordinators. 1997. Predicting Soil Erosion by Water: A Guide to Conservation Planning with the Revised Universal Soil Loss Equation (RUSLE). *Agricultural Handbook 703*. Washington, DC: U.S. Department of Agriculture.

Rogerson, P. A. 2001. *Statistical Methods for Geography.* London: Sage.

Rogowski, A., and J. Goyne. 2002. Dynamic Systems Modeling and Four Dimensional Geographic Information Systems. In K. C. Clarke, B. O. Parks, and M. P. Crane, eds., *Geographic Information Systems and Environmental Modeling,* pp. 122–59. Upper Saddle River, NJ: Prentice Hall.

Saaty, T. L. 1980. *The Analytic Hierarchy Process.* New York: McGraw-Hill.

Srinivasan, R., and J. G. Arnold. 1994. Integration of a Basin-Scale Water Quality Model with GIS. *Water Resources Bulletin* 30: 453–62.

Steiner, F. 1983. Resource Suitability: Methods for Analyses. *Environmental Management* 7: 401–20.

Stoms, D. M., J. M. McDonald, and F. W. Davis. 2002. Fuzzy Assessment of Land Suitability for Scientific Research Reserves. *Environmental Management* 29: 545–58.

Wischmeier, W. H., and D. D. Smith. 1978. Predicting Rainfall Erosion Losses: A Guide to Conservation Planning. *Agricultural Handbook 537.* Washington, DC: U.S. Department of Agriculture.

Wischmeier, W. H., and D. D. Smith. 1965. Predicting Rainfall-Erosion Losses from Cropland East of the Rocky Mountains: Guide for Selection of Practices for Soil and Water Conservation. *Agricultural Handbook 282.* Washington, DC: U.S. Department of Agriculture.

Wright, L. E., W. Zitzmann, K. Young, and R. Googins. 1983. LESA—Agricultural Land Evaluation and Site Assessment. *Journal of Soil and Water Conservation* 38: 82–89.

Wu, J., M. D. Randolm, M. D. Nellis, G. J. Kluitenberg, H. L. Seyler, and B. C. Rundquist. 2002. Using GIS to Assess and Manage the Conservation Reserve Program in Finney County, Kansas.

Photogrammetric Survey and Remote Sensing 68: 735–44.

Xiang, W. 2001. Weighting-by-Choosing: A Weight Elicitation Method for Map Overlays. *Landscape and Urban Planning* 56: 61–73.

Yager, R. 1988. On Ordered Weighted Averaging Aggregation Operators in Multicriteria Decision Making. *IEEE Transactions on Systems, Man, and Cybernetics* 18: 183–90.

Young, R. A., C. A. Onstad, D. D. Bosch, and W. P. Anderson. 1989. AGNPS: A Nonpoint-Source Pollution Model for Evaluating Agricultural Watersheds. *Journal of Soil and Water Conservation* 44: 168–73.

Young, R. A., C. A. Onstad, D. D. Bosch, and W. P. Anderson. 1987. AGNPS, Agricultural Nonpoint-Source Pollution Model: A Large Watershed Analysis Tool. *Conservation Research Report 35*. Washington, DC: Agricultural Research Service, U.S. Department of Agriculture.

Zhang, Z., and D. A. Griffith. 2000. Integrating GIS Components and Spatial Statistical Analysis in DBMSs. *International Journal of Geographical Information Science* 14: 543–66.

INDEX